Mechanical Properties of Engineered Materials

MECHANICAL ENGINEERING
A Series of Textbooks and Reference Books

Founding Editor

L. L. Faulkner

Columbus Division, Battelle Memorial Institute
and Department of Mechanical Engineering
The Ohio State University
Columbus, Ohio

1. *Spring Designer's Handbook*, Harold Carlson
2. *Computer-Aided Graphics and Design*, Daniel L. Ryan
3. *Lubrication Fundamentals*, J. George Wills
4. *Solar Engineering for Domestic Buildings*, William A. Himmelman
5. *Applied Engineering Mechanics: Statics and Dynamics*, G. Boothroyd and C. Poli
6. *Centrifugal Pump Clinic*, Igor J. Karassik
7. *Computer-Aided Kinetics for Machine Design*, Daniel L. Ryan
8. *Plastics Products Design Handbook, Part A: Materials and Components; Part B: Processes and Design for Processes*, edited by Edward Miller
9. *Turbomachinery: Basic Theory and Applications*, Earl Logan, Jr.
10. *Vibrations of Shells and Plates*, Werner Soedel
11. *Flat and Corrugated Diaphragm Design Handbook*, Mario Di Giovanni
12. *Practical Stress Analysis in Engineering Design*, Alexander Blake
13. *An Introduction to the Design and Behavior of Bolted Joints*, John H. Bickford
14. *Optimal Engineering Design: Principles and Applications*, James N. Siddall
15. *Spring Manufacturing Handbook*, Harold Carlson
16. *Industrial Noise Control: Fundamentals and Applications*, edited by Lewis H. Bell
17. *Gears and Their Vibration: A Basic Approach to Understanding Gear Noise*, J. Derek Smith
18. *Chains for Power Transmission and Material Handling: Design and Applications Handbook*, American Chain Association
19. *Corrosion and Corrosion Protection Handbook*, edited by Philip A. Schweitzer
20. *Gear Drive Systems: Design and Application*, Peter Lynwander
21. *Controlling In-Plant Airborne Contaminants: Systems Design and Calculations*, John D. Constance
22. *CAD/CAM Systems Planning and Implementation*, Charles S. Knox
23. *Probabilistic Engineering Design: Principles and Applications*, James N. Siddall
24. *Traction Drives: Selection and Application*, Frederick W. Heilich III and Eugene E. Shube
25. *Finite Element Methods: An Introduction*, Ronald L. Huston and Chris E. Passerello

26. *Mechanical Fastening of Plastics: An Engineering Handbook*, Brayton Lincoln, Kenneth J. Gomes, and James F. Braden
27. *Lubrication in Practice: Second Edition*, edited by W. S. Robertson
28. *Principles of Automated Drafting*, Daniel L. Ryan
29. *Practical Seal Design*, edited by Leonard J. Martini
30. *Engineering Documentation for CAD/CAM Applications*, Charles S. Knox
31. *Design Dimensioning with Computer Graphics Applications*, Jerome C. Lange
32. *Mechanism Analysis: Simplified Graphical and Analytical Techniques*, Lyndon O. Barton
33. *CAD/CAM Systems: Justification, Implementation, Productivity Measurement*, Edward J. Preston, George W. Crawford, and Mark E. Coticchia
34. *Steam Plant Calculations Manual*, V. Ganapathy
35. *Design Assurance for Engineers and Managers*, John A. Burgess
36. *Heat Transfer Fluids and Systems for Process and Energy Applications*, Jasbir Singh
37. *Potential Flows: Computer Graphic Solutions*, Robert H. Kirchhoff
38. *Computer-Aided Graphics and Design: Second Edition*, Daniel L. Ryan
39. *Electronically Controlled Proportional Valves: Selection and Application*, Michael J. Tonyan, edited by Tobi Goldoftas
40. *Pressure Gauge Handbook*, AMETEK, U.S. Gauge Division, edited by Philip W. Harland
41. *Fabric Filtration for Combustion Sources: Fundamentals and Basic Technology*, R. P. Donovan
42. *Design of Mechanical Joints*, Alexander Blake
43. *CAD/CAM Dictionary*, Edward J. Preston, George W. Crawford, and Mark E. Coticchia
44. *Machinery Adhesives for Locking, Retaining, and Sealing*, Girard S. Haviland
45. *Couplings and Joints: Design, Selection, and Application*, Jon R. Mancuso
46. *Shaft Alignment Handbook*, John Piotrowski
47. *BASIC Programs for Steam Plant Engineers: Boilers, Combustion, Fluid Flow, and Heat Transfer*, V. Ganapathy
48. *Solving Mechanical Design Problems with Computer Graphics*, Jerome C. Lange
49. *Plastics Gearing: Selection and Application*, Clifford E. Adams
50. *Clutches and Brakes: Design and Selection*, William C. Orthwein
51. *Transducers in Mechanical and Electronic Design*, Harry L. Trietley
52. *Metallurgical Applications of Shock-Wave and High-Strain-Rate Phenomena*, edited by Lawrence E. Murr, Karl P. Staudhammer, and Marc A. Meyers
53. *Magnesium Products Design*, Robert S. Busk
54. *How to Integrate CAD/CAM Systems: Management and Technology*, William D. Engelke
55. *Cam Design and Manufacture: Second Edition*; with cam design software for the IBM PC and compatibles, disk included, Preben W. Jensen
56. *Solid-State AC Motor Controls: Selection and Application*, Sylvester Campbell
57. *Fundamentals of Robotics*, David D. Ardayfio
58. *Belt Selection and Application for Engineers*, edited by Wallace D. Erickson
59. *Developing Three-Dimensional CAD Software with the IBM PC*, C. Stan Wei
60. *Organizing Data for CIM Applications*, Charles S. Knox, with contributions by Thomas C. Boos, Ross S. Culverhouse, and Paul F. Muchnicki

61. *Computer-Aided Simulation in Railway Dynamics*, by Rao V. Dukkipati and Joseph R. Amyot
62. *Fiber-Reinforced Composites: Materials, Manufacturing, and Design*, P. K. Mallick
63. *Photoelectric Sensors and Controls: Selection and Application*, Scott M. Juds
64. *Finite Element Analysis with Personal Computers*, Edward R. Champion, Jr., and J. Michael Ensminger
65. *Ultrasonics: Fundamentals, Technology, Applications: Second Edition, Revised and Expanded*, Dale Ensminger
66. *Applied Finite Element Modeling: Practical Problem Solving for Engineers*, Jeffrey M. Steele
67. *Measurement and Instrumentation in Engineering: Principles and Basic Laboratory Experiments*, Francis S. Tse and Ivan E. Morse
68. *Centrifugal Pump Clinic: Second Edition, Revised and Expanded*, Igor J. Karassik
69. *Practical Stress Analysis in Engineering Design: Second Edition, Revised and Expanded*, Alexander Blake
70. *An Introduction to the Design and Behavior of Bolted Joints: Second Edition, Revised and Expanded*, John H. Bickford
71. *High Vacuum Technology: A Practical Guide*, Marsbed H. Hablanian
72. *Pressure Sensors: Selection and Application*, Duane Tandeske
73. *Zinc Handbook: Properties, Processing, and Use in Design*, Frank Porter
74. *Thermal Fatigue of Metals*, Andrzej Weronski and Tadeusz Hejwowski
75. *Classical and Modern Mechanisms for Engineers and Inventors*, Preben W. Jensen
76. *Handbook of Electronic Package Design*, edited by Michael Pecht
77. *Shock-Wave and High-Strain-Rate Phenomena in Materials*, edited by Marc A. Meyers, Lawrence E. Murr, and Karl P. Staudhammer
78. *Industrial Refrigeration: Principles, Design and Applications*, P. C. Koelet
79. *Applied Combustion*, Eugene L. Keating
80. *Engine Oils and Automotive Lubrication*, edited by Wilfried J. Bartz
81. *Mechanism Analysis: Simplified and Graphical Techniques, Second Edition, Revised and Expanded*, Lyndon O. Barton
82. *Fundamental Fluid Mechanics for the Practicing Engineer*, James W. Murdock
83. *Fiber-Reinforced Composites: Materials, Manufacturing, and Design, Second Edition, Revised and Expanded*, P. K. Mallick
84. *Numerical Methods for Engineering Applications*, Edward R. Champion, Jr.
85. *Turbomachinery: Basic Theory and Applications, Second Edition, Revised and Expanded*, Earl Logan, Jr.
86. *Vibrations of Shells and Plates: Second Edition, Revised and Expanded*, Werner Soedel
87. *Steam Plant Calculations Manual: Second Edition, Revised and Ex panded*, V. Ganapathy
88. *Industrial Noise Control: Fundamentals and Applications, Second Edition, Revised and Expanded*, Lewis H. Bell and Douglas H. Bell
89. *Finite Elements: Their Design and Performance*, Richard H. MacNeal
90. *Mechanical Properties of Polymers and Composites: Second Edition, Revised and Expanded*, Lawrence E. Nielsen and Robert F. Landel
91. *Mechanical Wear Prediction and Prevention*, Raymond G. Bayer

92. *Mechanical Power Transmission Components*, edited by David W. South and Jon R. Mancuso
93. *Handbook of Turbomachinery*, edited by Earl Logan, Jr.
94. *Engineering Documentation Control Practices and Procedures*, Ray E. Monahan
95. *Refractory Linings Thermomechanical Design and Applications*, Charles A. Schacht
96. *Geometric Dimensioning and Tolerancing: Applications and Techniques for Use in Design, Manufacturing, and Inspection*, James D. Meadows
97. *An Introduction to the Design and Behavior of Bolted Joints: Third Edition, Revised and Expanded*, John H. Bickford
98. *Shaft Alignment Handbook: Second Edition, Revised and Expanded*, John Piotrowski
99. *Computer-Aided Design of Polymer-Matrix Composite Structures*, edited by Suong Van Hoa
100. *Friction Science and Technology*, Peter J. Blau
101. *Introduction to Plastics and Composites: Mechanical Properties and Engineering Applications*, Edward Miller
102. *Practical Fracture Mechanics in Design*, Alexander Blake
103. *Pump Characteristics and Applications*, Michael W. Volk
104. *Optical Principles and Technology for Engineers*, James E. Stewart
105. *Optimizing the Shape of Mechanical Elements and Structures*, A. A. Seireg and Jorge Rodriguez
106. *Kinematics and Dynamics of Machinery*, Vladimír Stejskal and Michael Valášek
107. *Shaft Seals for Dynamic Applications*, Les Horve
108. *Reliability-Based Mechanical Design*, edited by Thomas A. Cruse
109. *Mechanical Fastening, Joining, and Assembly*, James A. Speck
110. *Turbomachinery Fluid Dynamics and Heat Transfer*, edited by Chunill Hah
111. *High-Vacuum Technology: A Practical Guide, Second Edition, Revised and Expanded*, Marsbed H. Hablanian
112. *Geometric Dimensioning and Tolerancing: Workbook and Answerbook*, James D. Meadows
113. *Handbook of Materials Selection for Engineering Applications*, edited by G. T. Murray
114. *Handbook of Thermoplastic Piping System Design*, Thomas Sixsmith and Reinhard Hanselka
115. *Practical Guide to Finite Elements: A Solid Mechanics Approach*, Steven M. Lepi
116. *Applied Computational Fluid Dynamics*, edited by Vijay K. Garg
117. *Fluid Sealing Technology*, Heinz K. Muller and Bernard S. Nau
118. *Friction and Lubrication in Mechanical Design*, A. A. Seireg
119. *Influence Functions and Matrices*, Yuri A. Melnikov
120. *Mechanical Analysis of Electronic Packaging Systems*, Stephen A. McKeown
121. *Couplings and Joints: Design, Selection, and Application, Second Edition, Revised and Expanded*, Jon R. Mancuso
122. *Thermodynamics: Processes and Applications*, Earl Logan, Jr.
123. *Gear Noise and Vibration*, J. Derek Smith
124. *Practical Fluid Mechanics for Engineering Applications*, John J. Bloomer
125. *Handbook of Hydraulic Fluid Technology*, edited by George E. Totten
126. *Heat Exchanger Design Handbook*, T. Kuppan

127. *Designing for Product Sound Quality*, Richard H. Lyon
128. *Probability Applications in Mechanical Design*, Franklin E. Fisher and Joy R. Fisher
129. *Nickel Alloys*, edited by Ulrich Heubner
130. *Rotating Machinery Vibration: Problem Analysis and Troubleshooting*, Maurice L. Adams, Jr.
131. *Formulas for Dynamic Analysis*, Ronald L. Huston and C. Q. Liu
132. *Handbook of Machinery Dynamics*, Lynn L. Faulkner and Earl Logan, Jr.
133. *Rapid Prototyping Technology: Selection and Application*, Kenneth G. Cooper
134. *Reciprocating Machinery Dynamics: Design and Analysis*, Abdulla S. Rangwala
135. *Maintenance Excellence: Optimizing Equipment Life-Cycle Decisions*, edited by John D. Campbell and Andrew K. S. Jardine
136. *Practical Guide to Industrial Boiler Systems*, Ralph L. Vandagriff
137. *Lubrication Fundamentals: Second Edition, Revised and Expanded*, D. M. Pirro and A. A. Wessol
138. *Mechanical Life Cycle Handbook: Good Environmental Design and Manufacturing*, edited by Mahendra S. Hundal
139. *Micromachining of Engineering Materials*, edited by Joseph McGeough
140. *Control Strategies for Dynamic Systems: Design and Implementation*, John H. Lumkes, Jr.
141. *Practical Guide to Pressure Vessel Manufacturing*, Sunil Pullarcot
142. *Nondestructive Evaluation: Theory, Techniques, and Applications*, edited by Peter J. Shull
143. *Diesel Engine Engineering: Thermodynamics, Dynamics, Design, and Control*, Andrei Makartchouk
144. *Handbook of Machine Tool Analysis*, Ioan D. Marinescu, Constantin Ispas, and Dan Boboc
145. *Implementing Concurrent Engineering in Small Companies*, Susan Carlson Skalak
146. *Practical Guide to the Packaging of Electronics: Thermal and Mechanical Design and Analysis*, Ali Jamnia
147. *Bearing Design in Machinery: Engineering Tribology and Lubrication*, Avraham Harnoy
148. *Mechanical Reliability Improvement: Probability and Statistics for Experimental Testing*, R. E. Little
149. *Industrial Boilers and Heat Recovery Steam Generators: Design, Applications, and Calculations*, V. Ganapathy
150. *The CAD Guidebook: A Basic Manual for Understanding and Improving Computer-Aided Design*, Stephen J. Schoonmaker
151. *Industrial Noise Control and Acoustics*, Randall F. Barron
152. *Mechanical Properties of Engineered Materials*, Wolé Soboyejo
153. *Reliability Verification, Testing, and Analysis in Engineering Design*, Gary S. Wasserman
154. *Fundamental Mechanics of Fluids: Third Edition*, I. G. Currie

Additional Volumes in Preparation

HVAC Water Chillers and Cooling Towers: Fundamentals, Application, and Operations, Herbert W. Stanford III

Handbook of Turbomachinery: Second Edition, Revised and Expanded, Earl Logan, Jr., and Ramendra Roy

Progressing Cavity Pumps, Downhole Pumps, and Mudmotors, Lev Nelik

Gear Noise and Vibration: Second Edition, Revised and Expanded, J. Derek Smith

Intermediate Heat Transfer, Kau-Fui Vincent Wong

Mechanical Engineering Software

Spring Design with an IBM PC, Al Dietrich

Mechanical Design Failure Analysis: With Failure Analysis System Software for the IBM PC, David G. Ullman

Mechanical Properties of Engineered Materials

Wolé Soboyejo
Princeton University
Princeton, New Jersey

CRC Press is an imprint of the
Taylor & Francis Group, an **informa** business

First published 2003 by Marcel Dekker. Inc.

Published 2019 by CRC Press
Taylor & Francis Group
6000 Broken Sound Parkway NW, Suite 300
Boca Raton, FL 33487-2742

First issued in paperback 2019

ISBN 13: 978-0-367-44693-2 (pbk)
ISBN 13: 978-0-8247-8900-8 (hbk)

Visit the Taylor & Francis Web site at
http://www.taylorandfrancis.com

and the CRC Press Web site at
http://www.crcpress.com

Preface

My primary objective in this book is to provide a simple introduction to the subject of mechanical properties of engineered materials for undergraduate and graduate students. I have been encouraged in this task by my students and many practicing engineers with a strong interest in the mechanical properties of materials and I hope that this book will satisfy their needs. I have endeavored to cover only the topics that I consider central to the development of a basic understanding of the mechanical properties of materials. It is not intended to be a comprehensive review of all the different aspects of mechanical properties; such a task would be beyond the capabilities of any single author. Instead, this book emphasizes the fundamental concepts that must be mastered by any undergraduate or graduate engineer before he or she can effectively tackle basic industrial tasks that require an understanding of mechanical properties. This book is intended to bridge the gap between rigorous theory and engineering practice.

The book covers essential principles required to understand and interpret the mechanical properties of different types of materials (i.e., metals, ceramics, intermetallics, polymers, and their composites). Basic concepts are discussed generically, except in cases where they apply only to specific types/classes of materials. Following a brief introduction to materials science and basic strength of materials, the fundamentals of elasticity and plasticity are presented, prior to a discussion of strengthening mechanisms (including composite strengthening concepts). A simple introduction to the subject of fracture mechanics is then presented along with fracture and toughening mechanisms and a description of the effects of fatigue and the environment.

The book concludes with an overview of time-dependent viscoelastic/visco-plastic behavior, creep, and creep crack growth phenomena. Wherever possible, the text is illustrated with worked examples and case studies that show how to apply basic principles to the solution of engineering problems.

This book has been written primarily as a text for a senior undergraduate course or first-level graduate course on mechanical properties of materials. However, I hope that it will also be useful to practicing engineers, researchers, and others who want to develop a working understanding of the basic concepts that govern the mechanical properties of materials. To ensure a wide audience, I have assumed only a basic knowledge of algebra and calculus in the presentation of mathematical derivations. The reader is also assumed to have a sophomore-level understanding of physics and chemistry. Prior knowledge of basic materials science and strength of materials concepts is not assumed, however. The better-prepared reader may, therefore, skim through some of the elementary sections in which these concepts are introduced.

Finally, I would like to acknowledge a number of people that have supported me over the years. I am grateful to my parents, Alfred and Anthonia, for the numerous sacrifices that they made to provide me with a good education. I am indebted to my teachers, especially John Knott, Anthony Smith, David Fenner, and Stan Earles, for stimulating my early interest in materials and mechanics. I am also thankful to my colleagues in the field of mechanical behavior who have shared their thoughts and ideas with me over the years. In particular, I am grateful to Frank McClintock for his critical review of the first five chapters, and his suggestions for the book outline.

I also thank my colleagues in the mechanical behavior community for helping me to develop my basic understanding of the subject over the past 15 years. I am particularly grateful to Anthony Evans, John Hutchinson, Paul Paris, Robert Ritchie, Richard Hertzberg, Gerry Smith, Ali Argon, Keith Miller, Rod Smith, David Parks, Lallit Anand, Shankar Sastry, Alan Needleman, Charlie Whitsett, Richard Lederich, T. S. Srivatsan, Pranesh Aswath, Zhigang Suo, David Srolovitz, Barrie Royce, Noriko Katsube, Bob Wei, Campbell Laird, Bob Hayes, Rajiv Mishra, and many others who have shared their understanding with me in numerous discussions over the years.

I am indebted to my past and present staff scientists and postdoctoral research associates (Chris Mercer, Seyed Allameh, Fan Ye, Pranav Shrotriya, and Youlin Li) and personal assistants (Betty Adam, Alissa Horstman, Jason Schymanski, Hedi Allameh, and Yingfang Ni) for their assistance with the preparation of the text and figures. Betty Adam deserves special mention since she helped put the book together. I simply cannot imagine how this project could have been completed without her help.

I am grateful to my students and colleagues at Princeton University, MIT, and The Ohio State University who have provided me with a stimulating working environment over the past few years. In particular, I thank Lex Smits, my current department chair, and all my colleagues. My interactions with colleagues and students have certainly been vital to the development of my current understanding of the mechanical behavior of materials.

Partial financial support for the preparation of this book was provided by the National Science Foundation (DMR 0075135 and DMR 9458018). I would like to thank the Program Managers, Dr. Bruce McDonald and Dr. K. L. Murty, for providing the financial support and encouragement that made this book possible. Appreciation is also extended to Prof. Tom Eager and Prof. Nam Suh of MIT for inviting me to spend a sabbatical year as Visiting Martin Luther King Professor in the departments of Materials Science and Engineering and Mechanical Engineering at MIT. The sabbatical year (1997–1998) at MIT provided me with a stimulating environment for the development of the first few chapters of this book.

I also thank Dawn Wechsler, Janet Sachs, Elizabeth Curione, and Rita Lazzazzaro of Marcel Dekker, Inc., for their patience and understanding. This project would certainly not have been completed (by me) without their vision, patience, and encouragement.

Finally, I thank my wife, Morenike, for giving me the freedom and the time to write this book. This was time that I should have spent with her and our young family. However, as always, she was supportive of my work, and I know that this book could have never been completed without her forebearance and support.

Wolé Soboyejo

Contents

1

Overview of Crystal/Defect Structure and Mechanical Properties and Behavior

1.1 INTRODUCTION

The mechanical behavior of materials describes the response of materials to mechanical loads or deformation. The response can be understood in terms of the basic effects of mechanical loads on defects or atomic motion. A simple understanding of atomic and defect structure is, therefore, an essential prerequisite to the development of a fundamental understanding of the mechanical behavior of materials. A brief introduction to the structure of materials will be presented in this chapter. The treatment is intended to serve as an introduction to those with a limited prior background in the principles of materials science. The better prepared reader may, therefore, choose to skim this chapter.

1.2 ATOMIC STRUCTURE

In ancient Greece, Democritus postulated that atoms are the building blocks from which all materials are made. This was generally accepted by philosophers and scientists (without proof) for centuries. However, although the small size of the atoms was such that they could not be viewed directly with the available instruments, Avogadro in the 16th century was able to determine that one mole of an element consists of 6.02×10^{23} atoms. The periodic table of elements was also developed in the 19th century before the imaging of crystal structure was made possible after the development of x-ray techniques later that century. For the first time, scientists were able to view the effects of atoms that had been postulated by the ancients.

A clear picture of atomic structure soon emerged as a number of dedicated scientists studied the atomic structure of different types of materials. First, it became apparent that, in many materials, the atoms can be grouped into unit cells or building blocks that are somewhat akin to the pieces in a Lego set. These building blocks are often called crystals. However, there are many materials in which no clear grouping of atoms into unit cells or crystals can be identified. Atoms in such amorphous materials are apparently randomly distributed, and it is difficult to discern clear groups of atoms in such materials. Nevertheless, in amorphous and crystalline materials, mechanical behavior can only be understood if we appreciate the fact that the atoms within a solid are held together by forces that are often referred to as chemical bonds. These will be described in the next section.

1.3 CHEMICAL BONDS

Two distinct types of chemical bonds are known to exist. Strong bonds are often described as primary bonds, and weaker bonds are generally described as secondary bonds. However, both types of bonds are important, and they often occur together in solids. It is particularly important to note that the weaker secondary bonds may control the mechanical behavior of some materials, even when much stronger primary bonds are present. A good example is the case of graphite (carbon) which consists of strong primary bonds and weaker secondary bonds (Fig. 1.1). The relatively low strength of graphite can be attributed to the low shear stress required to induce the sliding of strongly (primary) bonded carbon layers over each other. Such sliding is easy because the bonds between the sliding (primary bonded) carbon planes are weak secondary bonds.

1.3.1 Primary Bonds

Primary bonds may be ionic, covalent, or metallic in character. Since these are relatively strong bonds, primary bonds generally give rise to stiff solids. The different types of primary bonds are described in detail below.

1.3.1.1 Ionic Bonding

Ionic bonds occur as a result of strong electrostatic Coulomb attractive forces between positively and negatively charged ions. The ions may be

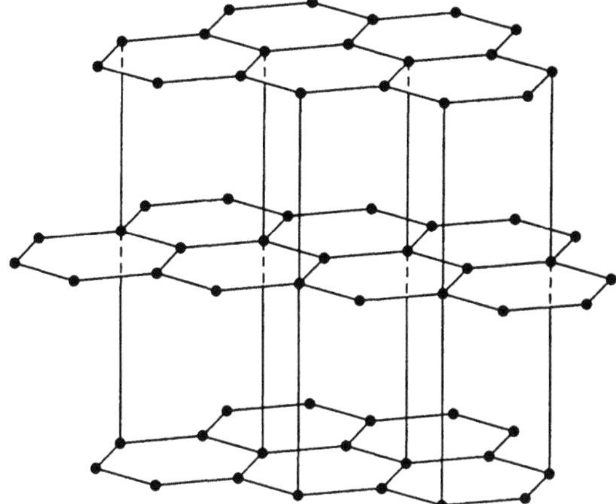

FIGURE 1.1 Schematic of the layered structure of graphite. (Adapted from Kingery et al., 1976. Reprinted with permission from John Wiley and Sons.)

formed by the donation of electrons by a cation to an anion (Fig. 1.2). Note that both ions achieve more stable electronic structures (complete outer shells) by the donation or acceptance of electrons. The resulting attractive force between the ions is given by:

$$F = -\alpha \frac{Q_1 Q_2}{r^2} \tag{1.1}$$

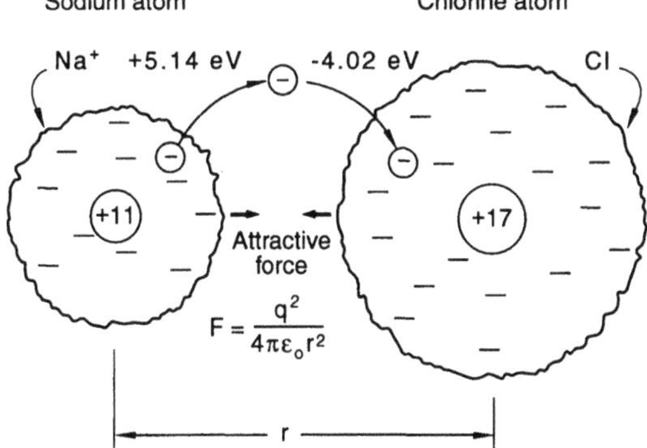

FIGURE 1.2 Schematic of an ionic bond—in this case between a sodium atom and a chlorine atom to form sodium chloride. (Adapted from Ashby and Jones, 1994. Reprinted with permission from Pergamon Press.)

where a is a proportionality constant, which is equal to $1/(4\pi\varepsilon_0)$, ε_0 is the permitivity of the vacuum (8.5×10^{-12} F/m), Q_1 and Q_2 are the respective charges of ions 1 and 2, and r is the ionic separation, as shown in Fig. 1.2. Typical ionic bond strengths are between 40 and 200 kcal/mol. Also, due to their relatively high bond strengths, ionically bonded materials have high melting points since a greater level of thermal agitation is needed to shear the ions from the ionically bonded structures. The ionic bonds are also nonsaturating and nondirectional. Such bonds are relatively difficult to break during slip processes that after control plastic behavior (irreversible deformation). Ionically bonded solids are, therefore, relatively brittle since they can only undergo limited plasticity. Examples of ionically bonded solids include sodium chloride and other alkali halides, metal oxides, and hydrated carbonates.

1.3.1.2 Covalent Bonds

Another type of primary bond is the covalent bond. Covalent bonds are often found between atoms with nearly complete outer shells. The atoms typically achieve a more stable electronic structure (lower energy state) by sharing electrons in outer shells to form structures with completely filled outer shells [(Fig. 1.3(a)]. The resulting bond strengths are between 30 and 300 kcal/mol. A wider range of bond strengths is, therefore, associated with covalent bonding which may result in molecular, linear or three-dimensional structures.

One-dimensional linear covalent bonds are formed by the sharing of two outer electrons (one from each atom). These result in the formation of molecular structures such as Cl_2, which is shown schematically in Figs 1.3b and 1.3c. Long, linear, covalently bonded chains, may form between quadrivalent carbon atoms, as in polyethylene [Figs 1.4(a)]. Branches may also form by the attachment of other chains to the linear chain structures, as shown in Fig. 1.4(b). Furthermore, three-dimensional covalent bonded

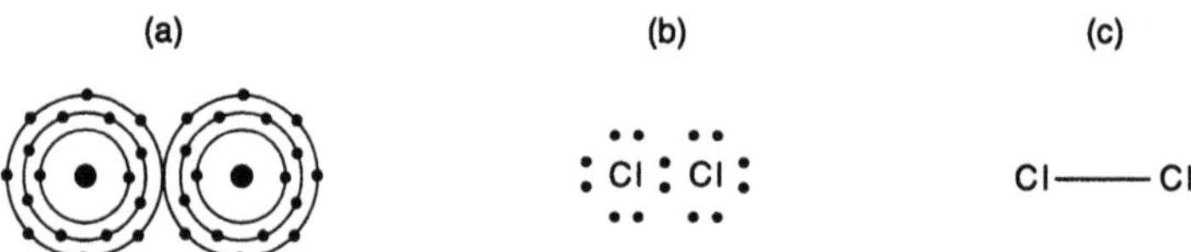

FIGURE 1.3 The covalent bond in a molecule of chlorine (Cl_2) gas: (a) planetary model; (b) electron dot schematic; (c) "bond-line" schematic. (Adapted from Shackleford, 1996. Reprinted with permission from Prentice-Hall.)

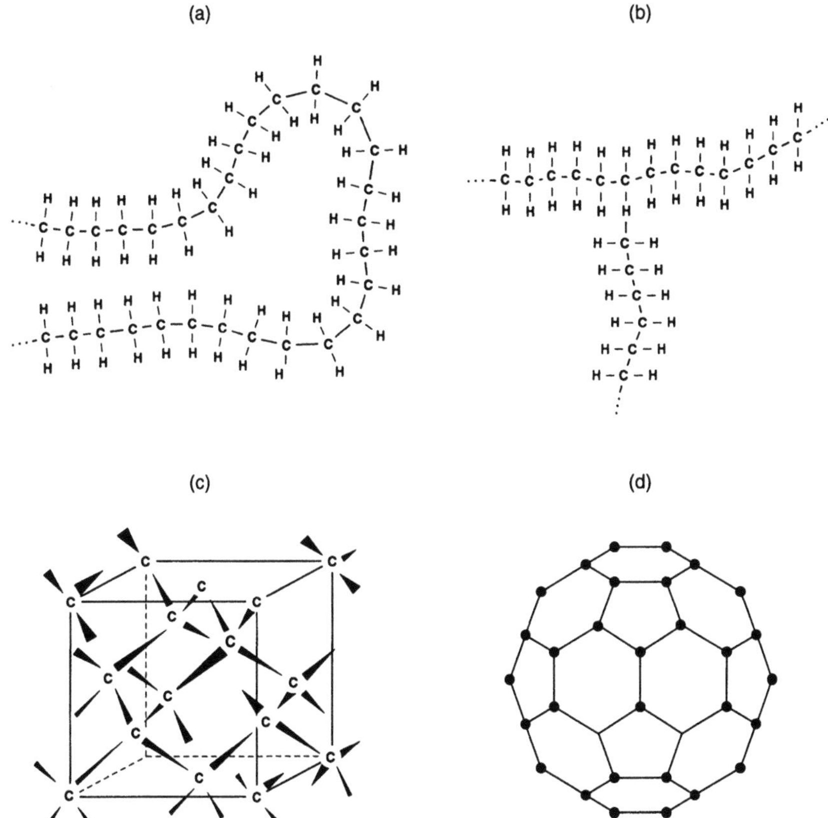

Figure 1.4 Typical covalently bonded structures: (a) three-dimensional structure of diamond; (b) chain structure of polyethylene; (c) three-dimensional structure of diamond; (d) buckeyball structure of C_{60}. (Adapted from Shackleford, 1996. Reprinted with permission from Prentice-Hall.)

structures may form, as in the case of diamond [Fig. 1.4(c)] and the recently discovered buckeyball structure [Fig. 1.4(d)].

Due to electron sharing, covalent bonds are directional in character. Elasticity in polymers is associated with the stretching and rotation of bonds. The chain structures may also uncurl during loading, which generally gives rise to elastic deformation. In the case of elastomers and rubber-like materials, the nonlinear elastic strains may be in excess of 100%. The elastic moduli also increase with increasing temperature due to changes in entropy that occur on bond stretching.

Plasticity in covalently bonded materials is associated with the sliding of chains consisting of covalently bonded atoms (such as those in polymers) or covalently bonded layers (such as those in graphite) over each other [Figs 1.1 and 1.4(a)]. Plastic deformation of three-dimensional covalently bonded structures [Figs 1.4(c) and 1.4(d)] is also difficult because of the inherent resistance of such structures to deformation. Furthermore, chain sliding is restricted in branched structures [Fig. 1.4(b)] since the branches tend to restrict chain motion.

1.3.1.3 Metallic Bonds

Metallic bonds are the third type of primary bond. The theory behind metallic bonding is often described as the Drüde–Lorenz theory. Metallic bonds can be understood as the overall effect of multiple electrostatic attractions between positively charged metallic ions and a "sea" or "gas" of delocalized electrons (electron cloud) that surround the positively charged ions (Fig. 1.5). This is illustrated schematically in Fig. 1.5. Note that the outer electrons in a metal are delocalized, i.e., they are free to move within the metallic lattice. Such electron movement can be accelerated by the application of an electric field or a temperature field. The electrostatic forces between the positively charged ions and the sea of electrons are very strong. These strong electrostatic forces give rise to the high strengths of metallically bonded materials.

Metallic bonds are nonsaturating and nondirectional in character. Hence, line defects within metallically bonded lattices can move at relatively low stresses (below those required to cause atomic separation) by slip processes at relatively low stress levels. The mechanisms of slip will be discussed later. These give rise to the ductility of metals, which is an important property for machining and fabrication processes.

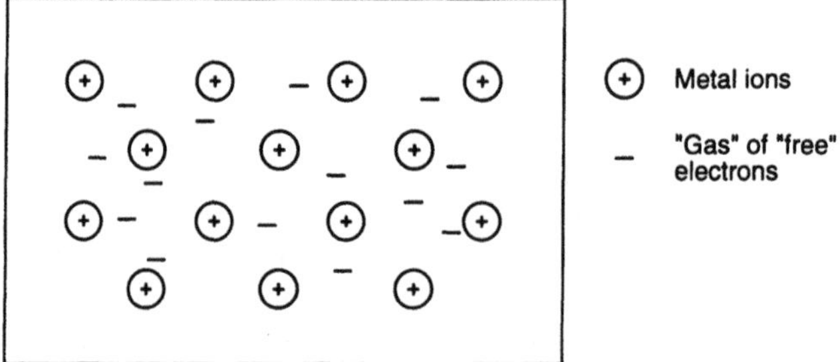

Figure 1.5 Schematic of metallic bonding. (Adapted from Ashby and Jones, 1994. Reprinted with permission from Pergamon Press.)

1.3.2 Secondary Bonds

Unlike primary bonds, secondary bonds (temporary dipoles and Van der Waals' forces) are relatively weak bonds that are found in several materials. Secondary bonds occur due to so-called dipole attractions that may be temporary or permanent in nature.

1.3.2.1 Temporary Dipoles

As the electrons between two initially uncharged bonded atoms orbit their nuclei, it is unlikely that the shared electrons will be exactly equidistant from the two nuclei at any given moment. Hence, small electrostatic attractions may develop between the atoms with slightly higher electron densities and the atoms with slightly lower electron densities [Fig. 1.6(a)]. The slight perturbations in the electrostatic charges on the atoms are often referred to as temporary dipole attractions or Van der Waals' forces [Fig. 1.6(a)]. However, spherical charge symmetry must be maintained over a period of time, although asymmetric charge distributions may occur at particular moments in time. It is also clear that a certain statistical number of these attractions must occur over a given period.

Temporary dipole attractions result in typical bond strengths of ~ 0.24 kcal/mol. They are, therefore, much weaker than primary bonds.

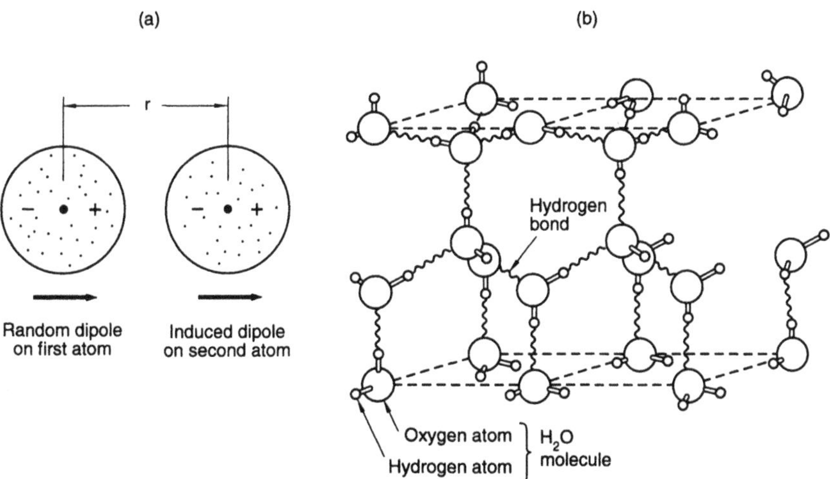

FIGURE 1.6 Schematics of secondary bonds: (a) temporary dipoles/Van der Waals' forces; (b) hydrogen bonds in between water molecules. (Adapted from Ashby and Jones, 1994. Reprinted with permission from Pergamon Press.)

Nevertheless, they may be important in determining the actual physical states of materials. Van der Waals' forces are found between covalently bonded nitrogen (N_2) molecules. They were first proposed by Van der Waals to explain the deviations of real gases from the ideal gas law. They are also partly responsible for the condensation and solidification of molecular materials.

1.3.2.2 Hydrogen Bonds

Hydrogen bonds are induced as a result of permanent dipole forces. Due to the high electronegativity (power to attract electrons) of the oxygen atom, the shared electrons in the water (H_2O) molecule are more strongly attracted to the oxygen atom than to the hydrogen atoms. The hydrogen atom therefore becomes slightly positively charged (positive dipole), while the oxygen atom acquires a slight negative charge (negative dipole). Permanent dipole attractions, therefore, develop between the oxygen and hydrogen atoms, giving rise to bridging bonds, as shown in Fig. 1.6(b). Such hydrogen bonds are relatively weak (0.04–0.40 kcal/mol). Nevertheless, they are required to keep water in the liquid state at room-temperature. They also provide the additional binding that is needed to keep several polymers in the crystalline state at room temperature.

1.4 STRUCTURE OF SOLIDS

The bonded atoms in a solid typically remain in their lowest energy configurations. In several solids, however, no short- or long-range order is observed. Such materials are often described as amorphous solids. Amorphous materials may be metals, ceramics, or polymers. Many are metastable, i.e., they might evolve into more ordered structures on subsequent thermal exposure. However, the rate of structural evolution may be very slow due to slow kinetics.

1.4.1 Polymers

The building blocks of polymers are called mers [Figs 1.7(a) and 1.7(b)]. These are organic molecules, each with hydrogen atoms and other elements clustered around one or two carbon atoms. Polymers are covalently bonded chain structures that consist of hundreds of mers that are linked together via addition or condensation chemical reactions (usually at high temperatures and pressures). Most polymeric structures are based on mers with covalently bonded carbon–carbon (C–C) bonds. Single (C–C), double (C=C), and triple (C≡C) bonds are found in polymeric structures. Typical chains contain between 100 and 1000 mers per chain. Also, most of the basic properties

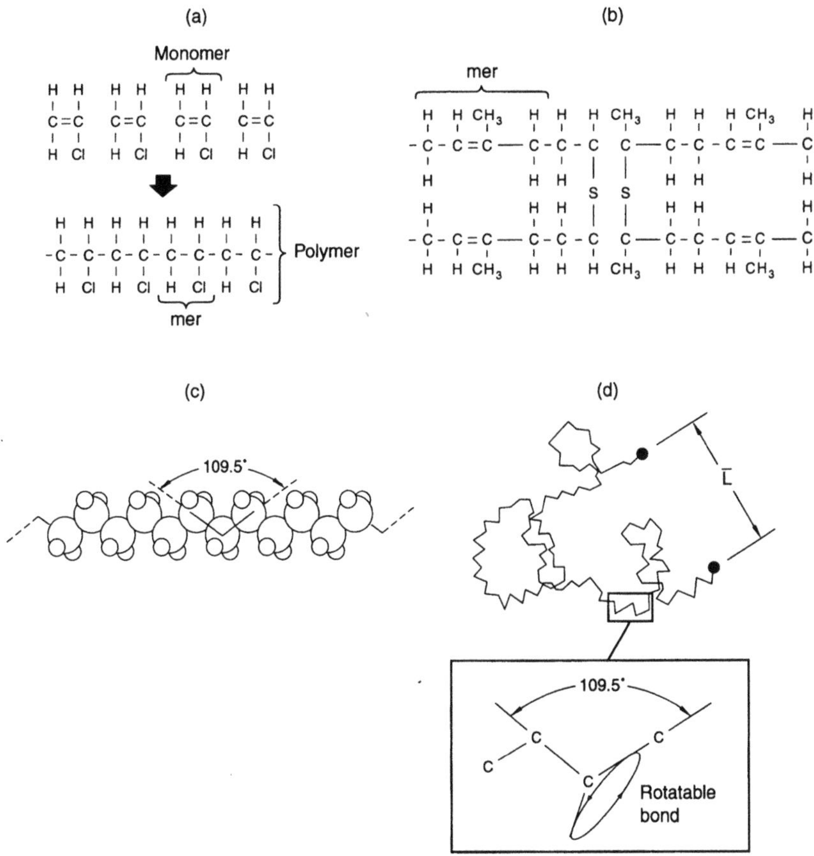

FIGURE 1.7 Examples of polymeric structures: (a) polymerization to form poly(vinyl chloride) $(C_2H_3Cl)_n$; (b) cross-linked structure of polyisoprene; (c) bond angle of 109.5°; (d) bond stretching and rotation within kinked and coiled structure. (Adapted from Shackleford, 1996. Reprinted with permission from Prentice-Hall).

of polymers improve with increasing average number of mers per chain. Polymer chains may also be cross-linked by sulfur atoms (Fig. 1.7(b)]. Such cross-linking by sulfur atoms occurs by a process known as vulcanization, which is carried out at high temperatures and pressures. Commercial rubber (isoprene) is made from such a process.

The spatial configurations of the polymer chains are strongly influenced by the tetrahedral structure of the carbon atom [Fig. 1.7(c)]. In the case of single C–C bonds, an angle of 109.5° is subtended between the carbon atom and each of the four bonds in the tetrahedral structure. The resulting chain structures will, therefore, tend to have kinked and coiled structures, as shown in Figs 1.7(d). The bonds in tetrahedral structure may also rotate, as shown in Fig. 1.7(d).

Most polymeric structures are amorphous, i.e., there is no apparent long-or short-range order to the spatial arrangement of the polymer chains. However, evidence of short- and long-range order has been observed in some polymers. Such crystallinity in polymers is due primarily to the formation of chain folds, as shown in Fig. 1.8. Chain folds are observed typically in linear polymers (thermoplastics) since such linear structures are amenable to folding of chains. More rigid three-dimensional thermoset structures are very difficult to fold into crystallites. Hence, polymer crystallinity is typically not observed in thermoset structures. Also, polymer chains with large side groups are difficult to bend into folded crystalline chains.

In general, the deformation of polymers is elastic (fully reversible) when it is associated with unkinking, uncoiling or rotation of bonds [Fig. 1.7(d)]. However, polymer chains may slide over each other when the applied stress or temperature are sufficiently large. Such sliding may be restricted by large side groups [Fig. 1.4(b)] or cross-links [Fig. 1.7(b)]. Permanent, plastic, or viscous deformation of polymers is, thus, associated with chain sliding, especially in linear (thermoplastic) polymers. As discussed

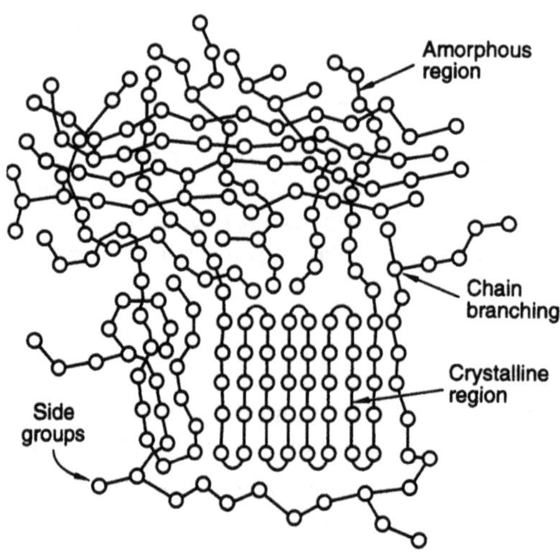

FIGURE 1.8 Schematic of amorphous and crystalline regions within long-chain polymeric structure. (Adapted from Ashby and Jones, 1994. Reprinted with permission from Pergamon Press.)

earlier, chain sliding is relatively difficult in three-dimensional (thermoset) polymers. Hence, thermosets are relatively rigid and brittle compared to thermoplastics.

Long-chain polymeric materials exhibit a transition from rigid glass-like behavior to a viscous flow behavior above a temperature that is generally referred to as the glass transition temperature, T_g. This transition temperature is usually associated with change in coefficient of thermal expansion which may be determined from a plot of specific volume versus temperature (Fig. 1.9). It is also important to note that the three-dimensional structures of thermosets (rigid network polymers) generally disintegrate at elevated temperatures. For this reason, thermosets cannot be reused after temperature excursions above the critical temperature levels required for structural disintegration. However, linear polymers (thermoplastics) do not disintegrate so readily at elevated temperatures, although they may soften considerably above T_g. They can thus be re-used after several elevated-temperature exposures.

1.4.2 Metals and Ceramics

Metals are usually solid elements in the first three groups of the periodic table. They contain de-localized outer electrons that are free to "swim about" when an electric field is applied, as discussed in Sect. 1.3.1.3 on metallic bonding. Ceramics are compounds formed between metals and

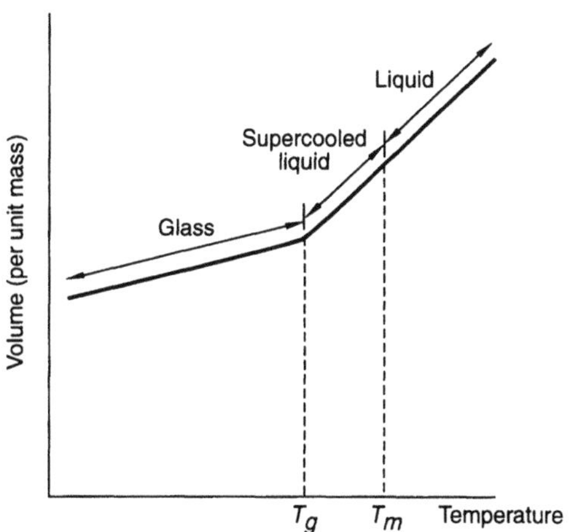

Figure 1.9 Schematic illustration of ductile-to-brittle transition in plot of specific volume versus temperature (Adapted from Shackleford, 1996. Reprinted with permission from Prentice-Hall.)

nonmetals. Ceramics may have ionic, covalent, or mixed ionic and covalent bonds. The relatively high compressive strengths of most ceramics may also be attributed largely to strong ionic and/or covalent bonds. Unfortunately, however, most ceramics are brittle due to their inability to accommodate strains in the presence of crack tips (especially under tensile loading where cracks tend to open up).

Metals and ceramics usually have long-range, ordered, crystalline structures. However, amorphous structures may also form under certain processing conditions. In the case of crystalline metallic and ceramic materials, the atoms within each crystal all have the same orientation. A crystalline lattice, consisting of regular repeated units (somewhat akin to Lego building blocks in a child's play kit) in a regular lattice, is observed. Each repeated unit is usually referred to as a unit cell, and the unit cell is generally chosen to highlight the symmetry of the crystal.

An example of a two-dimensional unit cell is shown in Fig. 1.10(a). This illustrates the two-dimensional layered structure of graphite which is one of the allotropes of carbon. Note that each carbon atom in the graphite structure is surrounded by three near neighbors. However, the orientations of the near neighbors to atoms A and B are different. Atoms similar to A are found at N and Q, and atoms similar to B are found at M and P. In any case, we may arbitrarily choose a unit cell, e.g., OXAY, that can be moved to various positions until we fill the space with identical units. If the repetition of the unit is understood to occur automatically, only one unit must be described to describe fully the crystal.

The unit chosen must also be a parallelogram in two dimensions, or a parallelepiped in three dimensions. It is referred to as a mesh or a net in two dimensions, or a unit cell in three dimensions. Since atoms at the corner of the mesh or unit cell may be shared by adjacent nets/unit cells, the total number of atoms in a unit cell may depend on the sum of the fractions of atoms that are present with an arbitrarily selected unit cell. For example, the mesh shown in Fig. 1.10(a) contains $(4 \times 1/4)$ 1 atom. The three-dimensional parallelepiped also contains $(8 \times 1/8)$ 1 atom per unit cell [Fig. 1.10(b)].

It is also important to note here that the origin of the unit cell is at the corner of the unit cell [Fig. 1.10(c)]. The sides of the unit parallelepiped also correspond to the cartesian x, y, z axes, and the angles α, β, γ are the axial angles. The arrangement of atoms may therefore be described by a three-dimensional grid of straight lines that divide the space into parallelepipeds that are equal in size. The intersection of lines is called a space lattice, and the crystal is constructed by stacking up the unit cells in a manner somewhat analogous the stacking of Lego pieces [Fig. 1.10(c)].

The most common crystalline lattices in metallic materials are the body-centered cubic, face-centered cubic, hexagonal closed-packed and

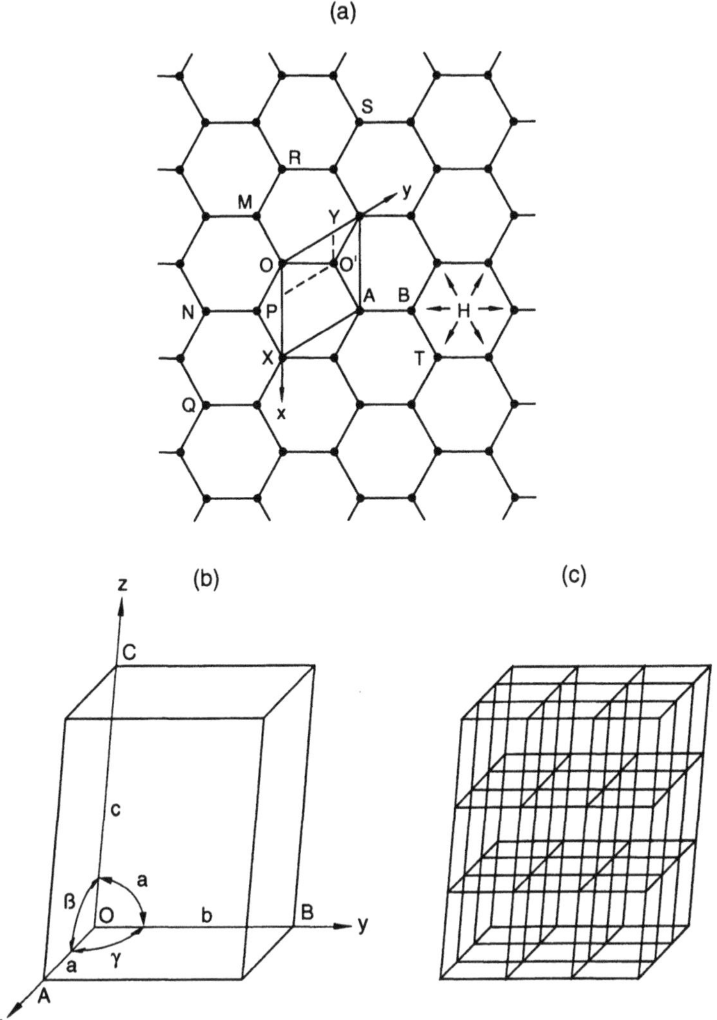

FIGURE 1.10 Schematics of possible unit cells. (a) Two-dimensional structure of graphite. (From Kelly and Groves, 1970.) (b) Parrellepiped/unit cell showing axes and angles. (c) A space lattice. (Adapted from Hull and Bacon, 1984. Reprinted with permission from Pergamon Press.)

the simple cubic structures. These are shown schematically in Fig. 1.11. The hexagonal and cubic structures can be constructed by the stacking of individual crystals, as shown in [Figs 1.12 (a and b)]. Note that the hexagonal closed packed (h.c.p.) have an ABABAB stacking sequence, while the

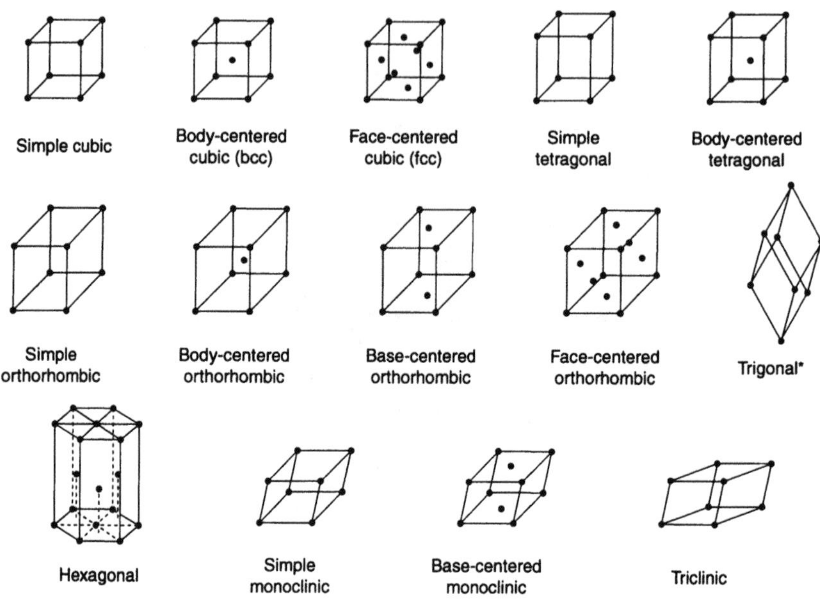

FIGURE 1.11 Schematics of 14 Bravais lattices. (Adapted from Shackleford, 1996. Reprinted with permission from Prentice-Hall.)

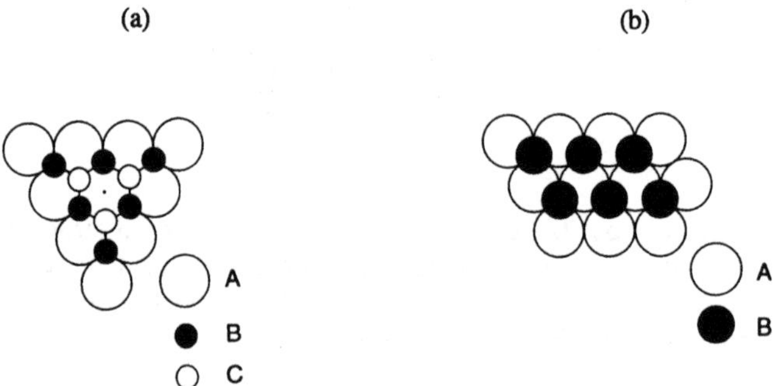

FIGURE 1.12 Schematic of stacking sequence in closed packed lattices: (a) hexagonal closed packed structure; (b) face-centered cubic structure. (Adapted from Hull and Bacon, 1984. Reprinted with permission from Pergamon Press.)

(f.c.c.) structure has an ABCABC stacking sequence. Also, both f.c.c. and h.c.p. structures are closed packed structures with several closed-packed planes on which plastic flow can occur.

Crystalline ceramic materials generally have more complex structures with lower symmetry. In general, however, 14 Bravais lattices are possible in crystalline materials, as shown in Fig. 1.11. Note the increasing complexity and the reduction in symmetry of the Bravais lattices that do not have cubic or hexagonal crystal structures. Crystalline cermics have less symmetric structures. The loss of symmetry is partly responsible for the relatively brittle behavior that is typically observed in ceramic systems. This will be discussed later. Further discussion on Bravais lattices can be found in any standard text on crystallography.

1.4.3 Intermetallics

Intermetallics are compounds formed between different metals. The bonds are often mixed metallic and covalent bonds. However, most intermetallics are often metallic-like in character. Intermetallics are, therefore, generally strong, but brittle as a result of their mixed bonding character. They also have predominantly noncubic (nonsymmetric) structures. Nevertheless, (relatively) light weight, high-temperature intermetallics such as gamma-based titanium aluminides (TiAl) and niobium aluminides (Nb_3Al) are of commercial interest, especially in the aerospace industry where they are being considered as possible replacements for heavier nickel-, iron-, or cobalt-base superalloys, themselves often containing intermetallics (Fig. 1.13). Some recent improvements in the balance of properties of these materials suggests that they may be used in the next generation of aircraft turbines.

1.4.4 Semiconductors

These are typically group IV elements (or their compounds) that have four outer electrons. The outer electrons can be excited into the conduction bands by application of electric fields. Electrical conductivity in semiconductors occurs either by electron or hole movement [Figs 1.14(a) and 1.14(b)]. In recent years, semiconductor packages have been applied in several electronic devices consisting of layered configurations of semiconductors deposited on metallic substrates within polymeric or ceramic encapsulants (Fig. 1.15). Such layered structures behave very much like structural materials. The mechanical properties of semiconductor devices has thus emerged as one of the fastest growing areas in the field of mechanical behavior.

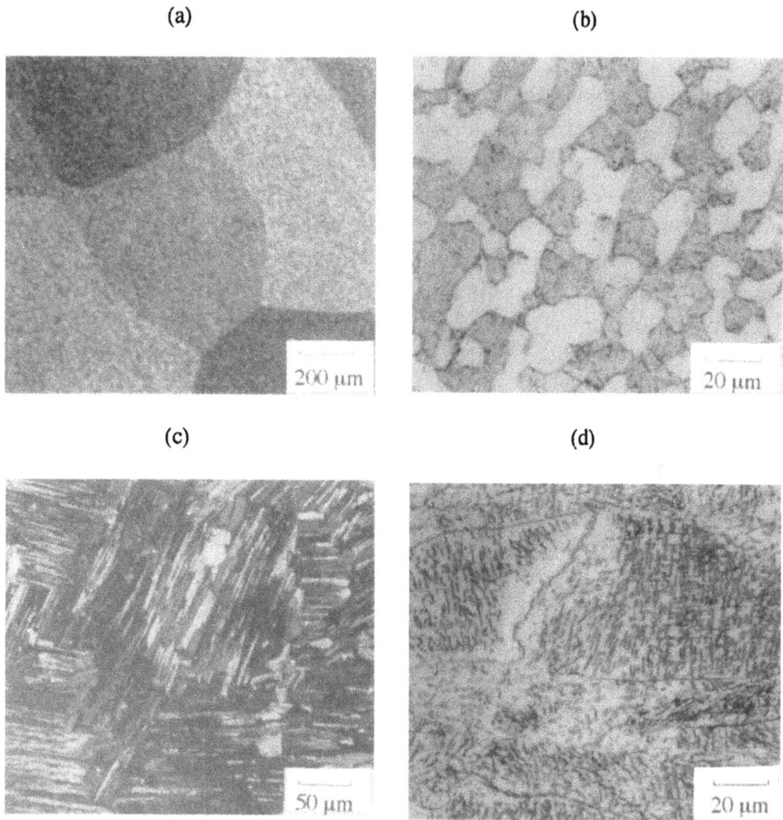

Figure 1.13 Microstructures of some metallic and intermetallic materials: (a) grains of single phase α niobium metal; (b) duplex $\alpha+\beta$ microstructure of Ti–6Al–4V alloy; (c) eutectoid $\alpha+\gamma$ microstructure of gamma-based titanium aluminide intermetallic (Ti–48Al–2Cr); (d) intermetallic δ' (Ni_3Nb) precipitates in a γ-nickel solid solution matrix within IN718 superalloy. (Courtesy of Dr Christopher Mercer.)

1.4.5 Composites

Composites are mixtures of two or more phases [Figs 1.15(a)–(d)]. Usually, the continuous phase is described as the matrix phase, while the discontinuous phase is described as the reinforcement phase. Composites constitute the great majority of materials that are encountered in nature. However, they may also be synthetic mixtures. In any case, they will tend to have mechanical properties that are intermediate between those of the matrix and

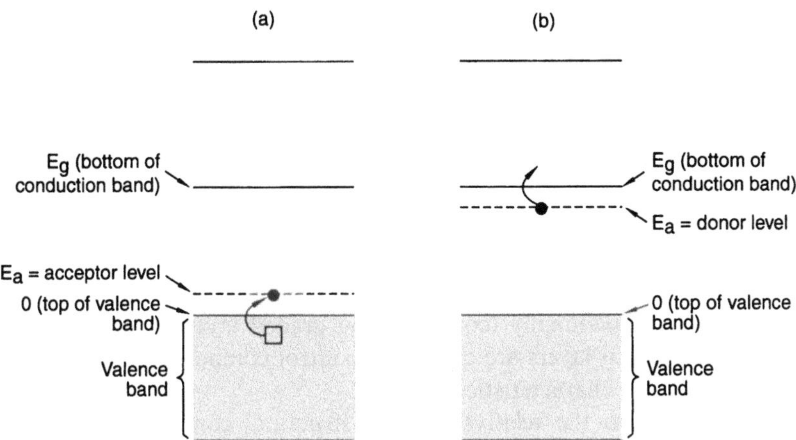

FIGURE 1.14 Schematic illustration of semiconduction via: (a) electron movement; (b) hole movement. (Adapted from Shackleford, 1996. Reprinted with permission from Prentice-Hall.)

reinforcement phases. Mixture rules are sometimes used to predict the mechanical properties of composites. The fracture behavior but not the stiffness of most composites are also strongly affected by the interfacial properties between the matrix and reinforcement phases. The interfaces, along with the matrix, must be engineered carefully to obtain the desired balance of mechanical properties.

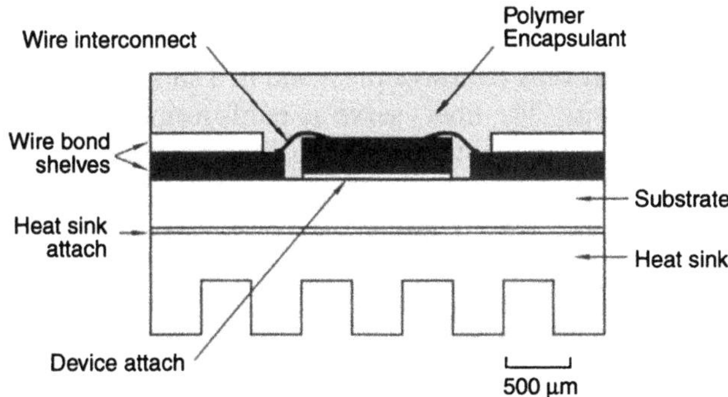

FIGURE 1.15 Schematic of typical semiconductor package. (Courtesy of Dr Rheiner Dauskardt.)

An almost infinite spectrum of composite materials should be readily apparent to the reader. However, with the exception of natural composites, only a limited range of composite materials have been produced for commercial purposes. These include: composites reinforced with brittle or ductile particles, whiskers, fibers, and layers [Figs 1.16(a)–(d)]. Composites reinforced with co-continuous, interpenetrating networks of reinforcement matrix and reinforcement materials have also been produced, along with woven fiber composites from which most textiles are made. Most recently, there has also been considerable interest in the development of functionally (continuously) graded layered composites in which the interfacial layers are graded to control composite residual stress levels and thermal characteristics.

In addition to the relatively exotic structural composites described above, more conventional materials have also been fabricated from composites. These include construction materials such as concrete, which is a mixture of sand, gravel, and cement. Reinforced concrete is another example of a composite material that consists of steel rods buried in concrete structures. Such composites may also be prestressed or post-tensioned to increase the inherent resistance to fracture. Concrete composites reinforced with steel or carbon fibers have also been developed in recent years. Since these fibers have very high strengths, concrete composites with extremely high strengths have been developed for a range of civil and structural engineering applications. Concrete composites have been used recently in various bridge deck designs and in other structural engineering applications. Polymer matrix composites (polymer matrices reinforced with stiff fibers) have also been considered for possible use in bridge applications due to their attractive combinations of mechanical properties. However, there are some concerns about their ability to withstand impact loading.

Wood is an example of a commonly used composite material. It is composed of tube-like cells that are aligned vertically or horizontally along the height of a tree. The tubular cells reinforce the wood in a similar manner to fibers in a synthetic composite. The fibers serve as reinforcements within the matrix, which consists of lignin and hemicellulose (both polymeric materials). Other natural composite materials include natural fibers such as silk, cotton and wool. These are all polymeric composites with very complex layered structures.

Layered composite structures also exist in all of the electronic packages that are used in modern electronic devices. Since the reliability of these packages is often determined by the thermal and mechanical properties of the individual layers and their interfaces, a good understanding of composite concepts is required for the design of such packages. Electronic packages typically consist of silica (semiconductor) layers deposited on metallic substrates within polymeric composites (usually silica filled epoxies). The other materials that have been used in electronic packaging include: alumina, aluminum, silicon nitride, and a wide range of other materials. The layered materials have been selected due to their combinations of heat conductivity (required for Joule/I^2R heat dissipation) and electrical properties.

1.5 STRUCTURAL LENGTH SCALES: NANOSTRUCTURE, MICROSTRUCTURE, AND MACROSTRUCTURE

It should be clear from the above discussion that the structure of solids can be considered at different length scales (from atomic to microscopic and macroscopic scales). Physicists often work at the atomic level, while most materials scientists work on the microscopic level. Unfortunately, however, most engineers tend to have only a macroscopic level of understanding of structure [Figs 16(a)–(d)]. They are often unaware of the atomic and microstructural constituents that can affect the mechanical behavior of materials and of the role mechanics plays even on the atomic scale. Failure to recognize the potential importance of these issues can lead to bad design. In the worst cases, failure to understand the effects of microscale constituents on the mechanical properties of materials has led to plane crashes, bridge failures, and shipwrecks. An understanding of mechanical behavior on different length scales is, therefore, essential to the safe design of structures.

The major challenge in this area is how to link existing theoretical models on different length scales, i.e., it is generally difficult to link atomistic models to microscopic models, or microscopic models to macroscopic models. At crack tips, all length scales may enter into the problem. The engineer must appreciate the relevant aspects of mechanical behavior at the different length scales. Furthermore, the size scale of the structure can affect the mechanical behavior of the materials, and the length scales may range from nanometers (close to atomic dimensions) to millimeters (easily viewed with the naked eye).

Unfortunately, however, there are no unifying concepts that bridge the gap between the different length scales. Quantitative models must, therefore, be developed at the appropriate length scales. This book presents the basic concepts required for a fundamental understanding of mechanical behavior at the different scales. However, since the mechanical behavior of materials is strongly affected by structure and defects, a brief review of defect structures and microstructures is provided in Chap. 2 along with the indicial notation required for the description of atomic structure.

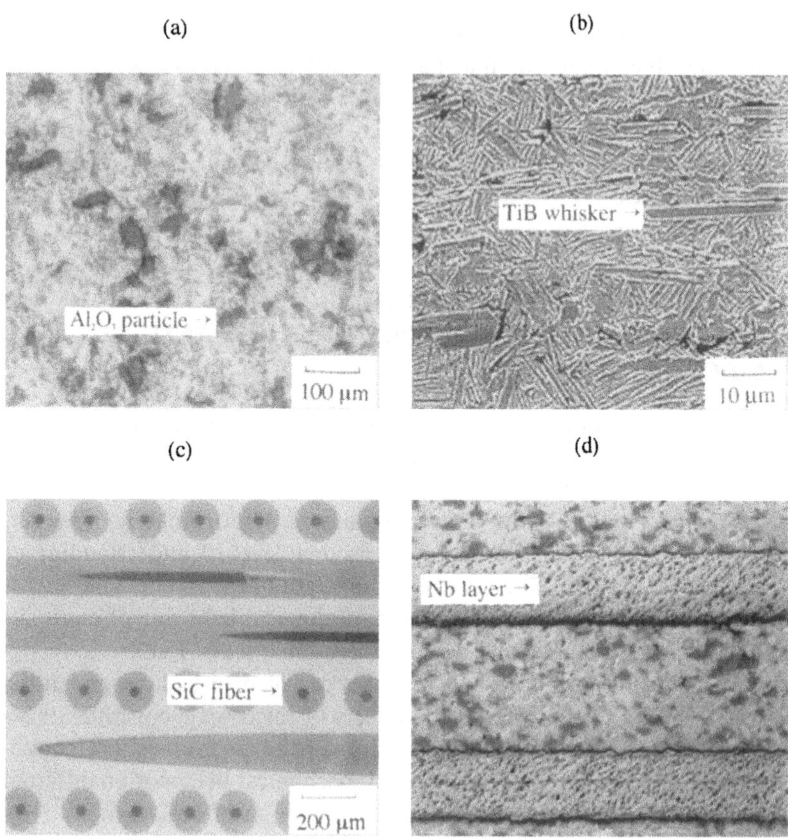

Figure 1.16 Examples of composite microstructures. (a) Al_2O_3 particulate-reinforced Al composite. (Courtesy of Prof. T. S. Srivatsan.) (b) TiB whisker-reinforced Ti–6Al–4V composite. (c) SiC fiber-reinforced Ti–15V–3Cr–3Al–3Sn composite. (d) Layered $MoSi_2$/Nb composite.

1.6 SUMMARY

A brief introduction to the structure of materials has been presented in this chapter. Following a review of the structure of crystalline and amorphous materials, the different classes of materials (metals, polymers, ceramics, intermetallics, and semiconductors and composites) were introduced. The chapter then concluded with an introduction to structural length scales related to nanostructure, microstructure, and macrostructure.

BIBLIOGRAPHY

Ashby, M. F. and Jones, D. R. H. (1994) Engineering Materials: An Introduction to Their Properties and Applications. Pergamon Press, New York.

This provides a simple introduction to the structure and properties of materials. The text is well illustrated with diagrams and case studies that make it easier to understand the basic concepts.

Askeland, D. (1996) The Science and Engineering of Materials. 3rd ed., PWS-Kent Publishing, Boston, MA.

This is a very good introductory textbook. In particular, the treatment of ferrous and nonferrous metallurgy is excellent.

Callister, W. D. (1994) Materials Science and Engineering: An Introduction. 5th ed. John Wiley, New York.

This is an excellent introductory textbook to materials science and engineering. The text is easy to read, and it covers most of the introductory concepts with some rigor.

Hull, D., and Bacon, D. J. (1984) Introduction to Dislocations. 3rd ed. Pergamon Press, New York.

Kelly, A. and G. W. Groves, G. W. (1970) Crystallography and Crystal Defects. Addison-Wesley, Boston, MA.

This classic book provides a rigorous introduction to the structure of crystalline solids and their defects.

Kingery, W. D., Bowen, H. K., and Uhlmann, D. R. (1976) Introduction to Ceramics. 2nd ed. John Wiley, New York.

This is perhaps the best introductory text to the structure and properties of ceramic materials.

Matthews, F. L. and Rawlings, R. D. (1993) Composite Materials: Engineering and Science, Chapman and Hall, New York.

This book provides a very clear introduction to composite materials. The mechanics concepts are presented in a very logical sequence that is easy to understand. The book is also well illustrated with worked examples that help the reader to develop a strong grasp of the basic concepts.

McClintock, F. A. and Argon, A.S. (1993) Mechanical Behavior of Material. Tech Books, Fairfax, VA.

This classical text provides a more advanced treatment of the fundamentals of the mechanical behavior of materials. The text provides a rigorous review of the fundamental mechanics aspects of the mechanical behavior of materials.

Shackleford, J. F. (1996) Introduction to Materials Science for Engineers. 4th ed. Prentice-Hall.

This is one of the best introductory texts on materials science and engineering. The book is generally well written and well illustrated with figures and tables.

Taya, M. and Arsenault, R.J. (1989) Metal Matrix Composites, Pergamon Press, New York.

This concise book provides a very good introduction to the structure and properties of metal matrix composites. The book covers both the mechanics and materials aspects of metal matrix composites.

2

Defect Structure and Mechanical Properties

2.1 INTRODUCTION

Since the mechanical behavior of solids is often controlled by defects, a brief review of the different types of defects is presented in this chapter along with the indicial notation that is often used in the characterization of atomic planes and dimensions. The possible defect length scales are also discussed before presenting a brief introduction to diffusion-controlled phase transformations. Finally, an overview of the mechanical behavior of materials is presented in an effort to prepare the reader for more detailed discussion in subsequent chapters. The material described in this chapter is intended for those with limited prior background in the principles of materials science. The better prepared reader may, therefore, choose to skim this chapter and move on to Chap. 3 in which the fundamentals of stress and strain are presented.

2.2 INDICIAL NOTATION FOR ATOMIC PLANES AND DIRECTIONS

Abbreviated notation for the description of atomic planes and directions are presented in this section. The so called Miller indicial notation is presented first for cubic lattices. This is followed by a brief introduction to Miller–Bravais notation, which is generally used to describe atomic planes and directions in hexagonal closed packed structures.

2.2.1 Miller Indicial Notation

Miller indicial notation is often used to describe the planes and directions in a cubic lattice. The Miller indices of a plane can be obtained simply from the reciprocal values of the intercepts of the plane with the x, y, and z axes. This is illustrated schematically in Figs 2.1 and 2.2. The reciprocals of the intercepts are then multiplied by appropriate scaling factors to ensure that all the resulting numbers are integer values corresponding to the least common factors. The least common factors are used to represent the Miller indices of a plane. Any negative numbers are represented by bars over them. A single plane is denoted by $(x\ y\ z)$ and a family of planes is usually represented as $\{x\ y\ z\}$.

Similarly, atomic directions may be specified using Miller indices. These are vectors with integer values that represent the particular atomic direction $[u\ v\ w]$, as illustrated in Fig. 2.3(a). The square brackets are generally used to denote single directions, while angular brackets are used to represent families of directions. An example of the $\langle 111 \rangle$ family of directions is given in Fig. 2.3(b).

The Miller indices of planes and directions in cubic crystals may be used to determine the unit vectors of the direction and the plane normal, respectively. Unit vectors are given simply by the direction cosines $[l\ m\ n]$ to be

$$n = l\hat{\mathbf{i}} + m\hat{\mathbf{j}} + n\hat{\mathbf{k}} \tag{2.1}$$

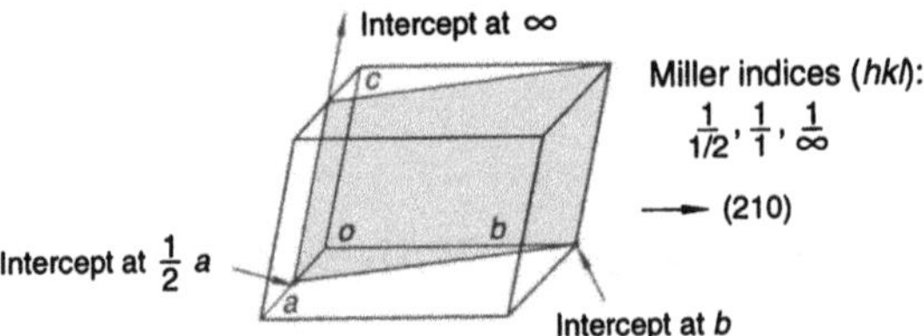

Figure 2.1 Determination of Miller indices for crystal planes. (Adapted from Shackleford, 1996. Reprinted with permission from Prentice-Hall.)

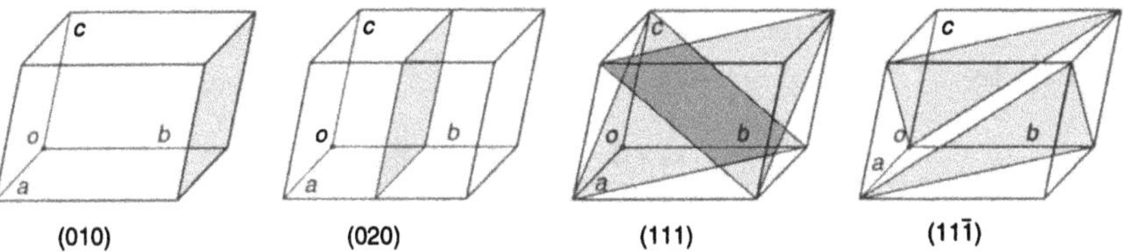

FIGURE 2.2 Examples of crystal planes. (Adapted from Shackleford, 1996. Reprinted with permission from Prentice-Hall.)

In the case of a direction, d_1, described by unit vector $[x_1\ y_1\ z_1]$, the direction cosines are given by

$$\hat{d}_1 = \frac{x_1\hat{\mathbf{i}} + y_1\hat{\mathbf{j}} + z_1\hat{\mathbf{k}}}{\sqrt{x_1^2 + y_1^2 + z_1^2}} \tag{2.2}$$

In the case of a plane with a plane normal with a unit vector, $\hat{\mathrm{n}}_2$, that has components $(u_1\ v_1\ w_1)$, the unit vector, $\hat{n}_1$, is given by

$$\hat{n}_1 = \frac{u_1\hat{\mathbf{i}} + v_1\hat{\mathbf{j}} + w_1\hat{\mathbf{k}}}{\sqrt{u_1^2 + v_1^2 + w_1^2}} \tag{2.3}$$

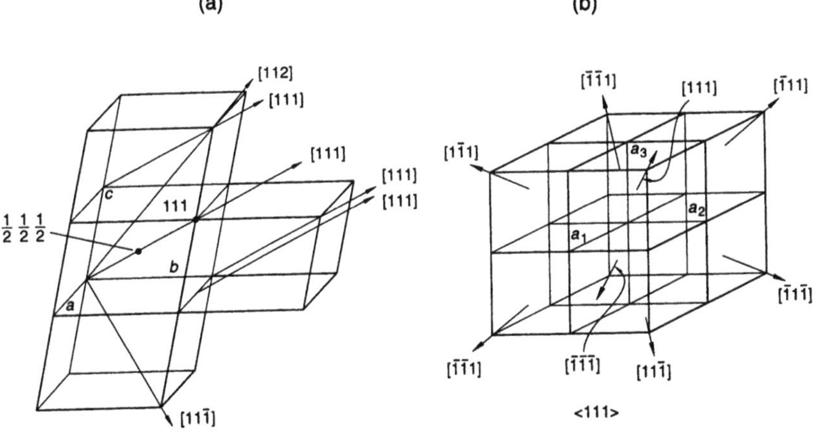

FIGURE 2.3 Determination of crystal directions: (a) single [111] directions; (b) family of ⟨111⟩ directions. (Adapted from Shackleford, 1996. Reprinted with permission from Prentice-Hall.)

The angle, ϕ, between two directions $d_1 = [x_1\ y_1\ z_1]$ and $d_2 = [x_2\ y_2\ z_2]$ is given by

$$\cos\phi = \hat{d}_1 \cdot \hat{d}_2 = \frac{x_1x_2\hat{\mathbf{i}} + y_1y_2\hat{\mathbf{j}} + z_1z_2\hat{\mathbf{k}}}{\sqrt{(x_1^2 + y_1^2 + z_1^2)(x_2^2 + y_2^2 + z_2^2)}} \tag{2.4}$$

Similarly, the angle λ between two planes with plane normals given by $n_1 = (u_1\ v_1\ w_1)$ and $n_2 = (u_2\ v_2\ w_2)$, is given by

$$\cos\lambda = \frac{u_1u_2\hat{\mathbf{i}} + v_1v_2\hat{\mathbf{j}} + w_1w_2\hat{\mathbf{k}}}{\sqrt{(u_1^2 + v_1^2 + w_1^2 + w_1^2)(u_2^2 + v_2^2 + w_2^2)}} \tag{2.5}$$

The direction of the line of intersection of two planes n_1 and n_2 is given by the vector cross product, $n_3 = n_1 \times n_2$, which is given by

$$\cos\lambda = \frac{1}{\sqrt{(u_1^2 + v_1^2 + w_1^2)(u_2^2 + v_2^2 + w_2^2)}} \begin{vmatrix} \hat{\mathbf{i}} & \hat{\mathbf{j}} & \hat{\mathbf{k}} \\ u_1 & v_1 & w_1 \\ u_2 & v_2 & w_2 \end{vmatrix} \tag{2.6}$$

2.2.2 Miller–Bravais Indicial Notation

In the case of hexagonal closed packed lattices, Miller–Bravais indicial notation is used to describe the directions and the plane normals. This type of notation is illustrated schematically in Fig. 2.4. Once again, the direction is described by a vector with the smallest possible integer components. However, three $(a_1\ a_2\ a_3)$ axes are used to specify the directions in the horizontal $(a_1\ a_2\ a_3)$ plane shown in Fig. 2.4(a).

The fourth co-ordinate in Miller–Bravais notation corresponds to the vertical direction, which is often denoted by the letter c. Miller–Bravais indicial notation for direction is thus given by $n = [a_1\ a_2\ a_3\ c]$. Similarly, Miller–Bravais indicial notation for a plane is given by the reciprocals of the intercepts on the a_1, a_2, a_3 and c axes. As before, the intercepts are multiplied by appropriate scaling factors to obtain the smallest possible integer values of the Miller–Bravais indices.

The Miller–Bravais notation for planes is similar to the Miller indicial notation described earlier. However, four indices $(a_1\ a_2\ a_3\ c)$ are needed to describe a plane in Miller–Bravais indicial notation, as shown in Fig. 2.4(b). Note that a_1, a_2, a_3, c correspond to the reciprocals of the intercepts on the a_1, a_2, a_3, and c axes. The indices are also scaled appropriately to represent the planes with the smallest integer indices. However, only two of the three basal plane co-ordinates are independent. The indices a_1, a_2, and a_3 must,

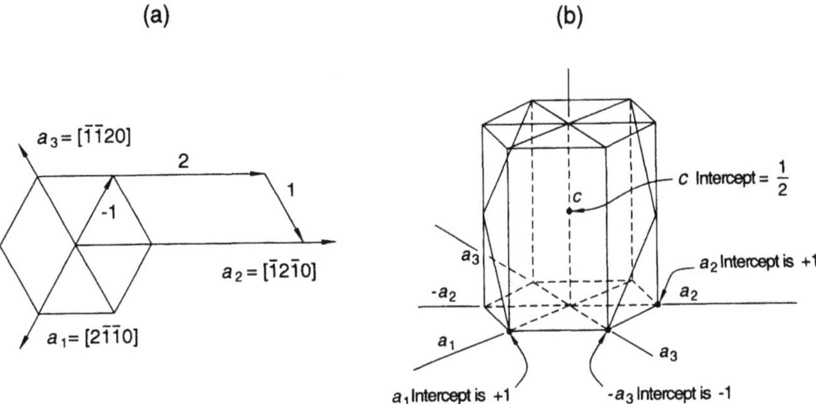

FIGURE 2.4 Miller–Bravais indicial notation for hexagonal closed packed structures: (a) example showing determination of direction indices; (b) example showing determination of plane indices. (Adapted from Shackleford, 1996. Reprinted with permission from Prentice-Hall.)

therefore, be selected such that $a_i + a_j = -a_k$, for sequential values of i, j, and k between 1 and 3.

To assist the reader in identifying the Miller–Bravais indicial notation for directions, two examples of Miller–Bravais direction indices are presented in Fig. 2.4. These show two simple methods for the determination of Miller–Bravais indices of diagonal axes. Fig. 2.5(a) shows diagonal axes of Type I which correspond to directions along any of the axes on the basal plane, i.e., a_1, a_2 or a_3. Note that the Miller–Bravais indices are not [1000] since these violate the requirement that $a_i + a_j = -a_k$. The diagonal axes of Type I, therefore, help us to identify the correct Miller–Bravais indices for the a_1 direction as $[2\bar{1}\bar{1}0]$. Note that the unit vector along the a_1 direction is $1/3[2\bar{1}\bar{1}0]$. Similarly, we may show that the unit vectors along the a_2 and a_3 directions are given by $1/3[\bar{1}2\bar{1}0]$ and $1/3[\bar{1}\bar{1}20]$. The diagonal axis of Type I, therefore, enables us to find the Miller–Bravais indices for any of the directions along the axes on the basal plane.

The other common type of diagonal axis is shown in Fig. 2.5(b). This corresponds to a direction that is intermediate between a_i and $-a_k$. In the example shown in Fig. 2.5(b), the vector s is given simply by the sum of the unit vectors along the a_1 and $-a_3$ directions. The Miller–Bravais indices for this direction are, therefore, given by $1/2[10\bar{1}0]$. It is important to note here that the c component in the Miller–Bravais notation should always be included even when it is equal to zero. The vectors corresponding to different directions may also be treated using standard vector algebra.

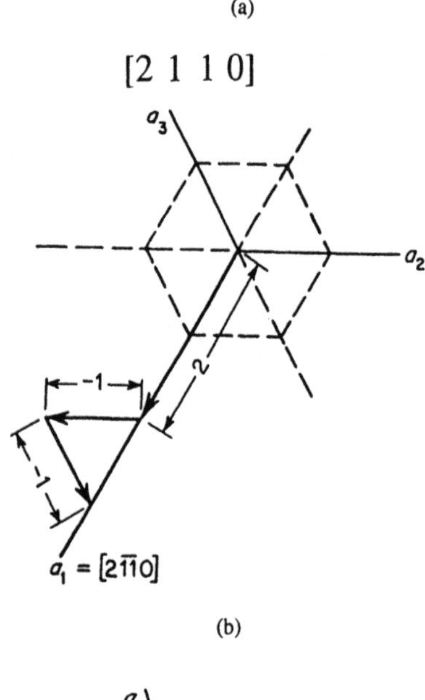

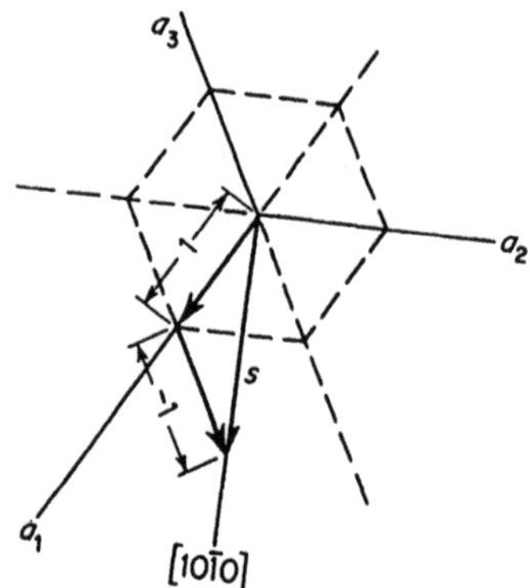

Figure 2.5 Schematic illustration of diagonal axes of (a) Type I, and (b) Type II. (Adapted from Read-Hill and Abbaschian, 1992. Reprinted with permission from PWS Kent.)

2.3 DEFECTS

All solids contain defects. Furthermore, structural evolution and plastic deformation of solids are often controlled by the movement of defects. It is, therefore, important for the student of mechanical behavior to be familiar with the different types of defects that can occur in solids.

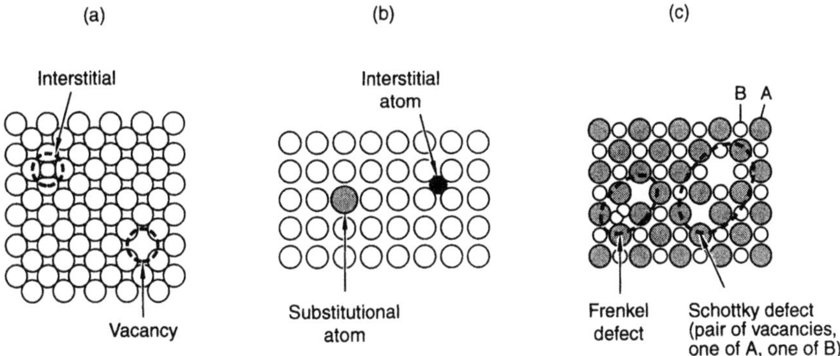

FIGURE 2.6 Examples of point defects: (a)] vacancy and interstitial elements; (b) substitutional element and interstitial impurity element; (c) pairs of ions and vacancies. [(a) and (c) are adapted from Shackleford, 1996—reprinted with permission from Prentice-Hall; (b) is adapted from Hull and Bacon, 1984. Reprinted with permission from Pergamon Press.]

Defects are imperfections in the structure. They may be one-dimensional point defects (Fig. 2.6), line defects (Fig. 2.7), two-dimensional plane defects (Fig. 2.8), or three-dimensional volume defects such as inclusions or porosity, Fig. 1.16(d). The different types of defects are described briefly in this section.

2.3.1 One-Dimensional Point Defects

One-dimensional point defects [Fig. 2.6) may include vacancies [Fig. 2.6(a)], interstitials [Figs 2.6(a) and 2.6(b)], solid solution elements [Fig. 2.6(b)], and pairs or clusters of the foregoing, Fig. 2.6(c). Pairs of ions (Frenkel defects) or vacancies (Schottky defects) are often required to maintain charge neutrality, Fig. 2.6(c). Point defects can diffuse through a lattice, especially at temperatures above approximately 0.3–0.5 of the absolute melting temperature. If the movement of point defects produces a net state change, it causes thermally activated stress-induced deformation, such as creep. The diffusion of point defects such as vacancies may also lead to the growth of grains in a polycrystalline material.

2.3.2 Line Defects

Line defects consist primarily of dislocations, typically at the edges of patches where part of a crystallographic plane has slipped by one lattice

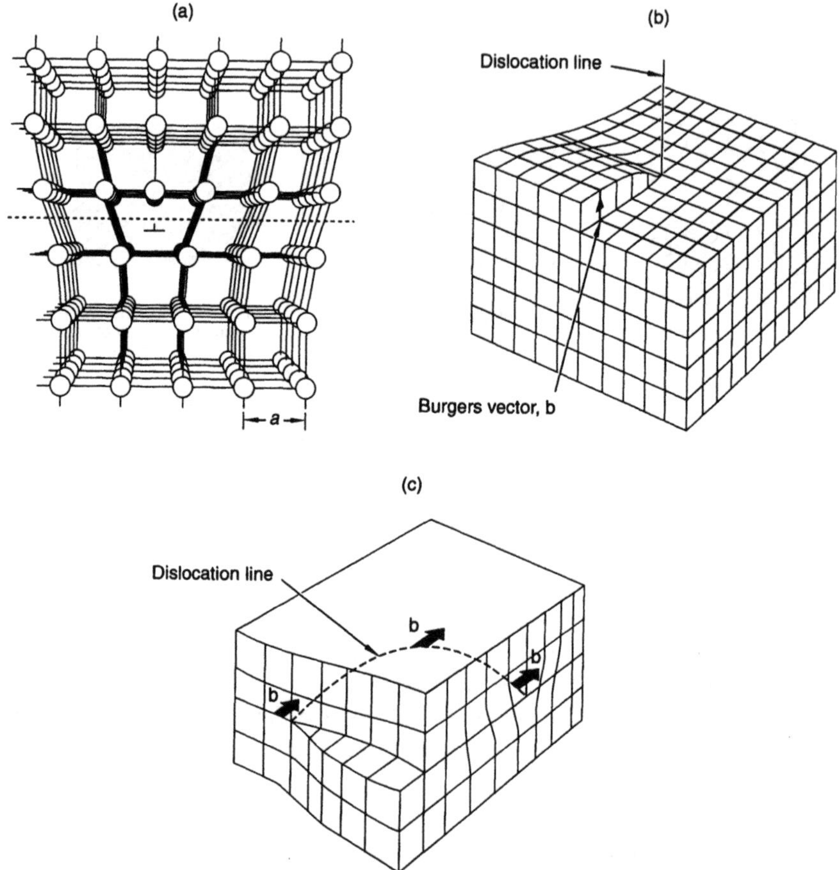

FIGURE 2.7 Examples of line defects: (a) edge dislocations; (b) screw dislocations; (c) mixed dislocations. (Adapted from Hull and Bacon, 1980. Reprinted with permission from Pergamon Press.)

spacing (Fig. 2.7). The two pure types of dislocations are edge and screw, Figs 2.7(a) and 2.7(b). Edge dislocations have slip (Burgers) vectors perpendicular to the dislocation line [Fig. 2.7a)], while screw dislocations have translation vectors parallel to the dislocation line, Fig. 2.7(b). In general, however, most dislocations are mixed dislocations that consist of both edge and screw dislocation components, Fig. 2.7(c). Note that the line segments along the curved dislocation in Fig. 2.7(c) have both edge and screw components. However, the deflection segments are either pure edge or pure screw at either end of the curved dislocation, Fig. 2.7(c).

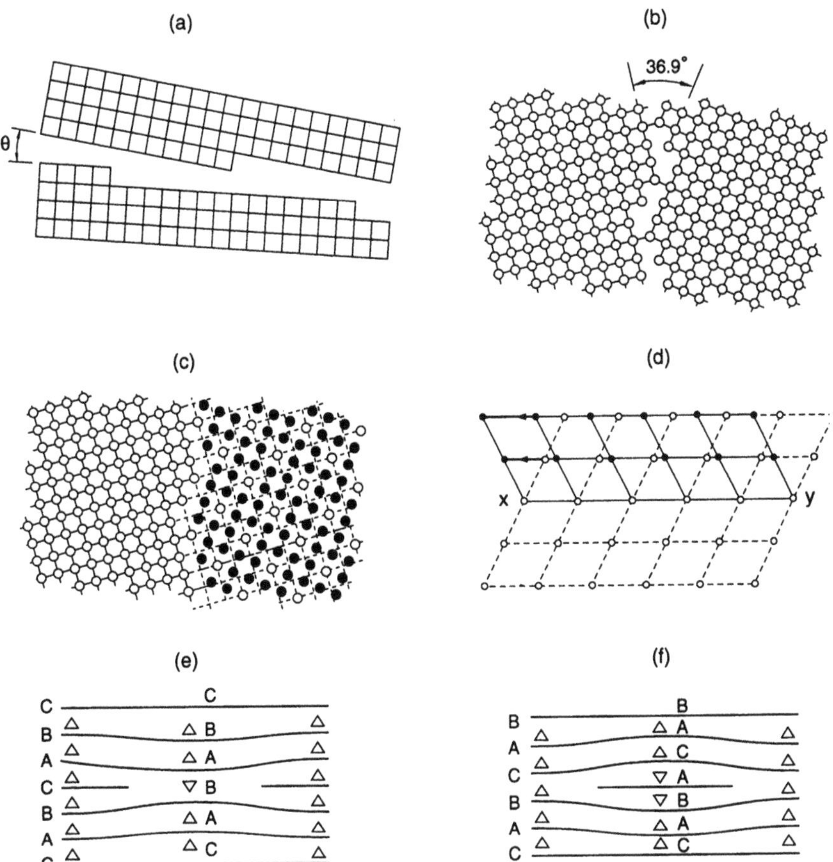

FIGURE 2.8 Examples of surface defects: (a) low-angle tilt boundary; (b) high-angle tilt boundary; (c) $S = 5$ boundary; (d) twin boundary; (e) intrinsic stacking fault; (f) extrinsic stacking fault. (Adapted from Shackleford, 1996. Reprinted with permission from Prentice-Hall.)

2.3.3 Surface Defects

Surface defects are two-dimensional planar defects (Fig. 2.8). They may be grain boundaries, stacking faults, or twin boundaries. These are surface boundaries across which the perfect stacking of atoms within a crystalline lattice changes. High- or low-angle tilt or twist boundaries may involve changes in the crystallographic orientations of adjacent grains, Figs 2.8(a) and 2.8(b). The orientation change across the boundary may be described using the concept of coincident site lattices. For example, a $\Sigma = 5$ or

$\Sigma^{-1} = 1/5$ boundary is one in which 1 in 5 of the grain boundary atoms match, as shown in Fig. 2.8(c).

Twin boundaries may form within crystals. Such boundaries lie across deformation twin planes, as shown in Fig. 2.8(d). Note that the atoms on either side of the twin planes are mirror images. Stacking faults may also be formed when the perfect stacking in the crystalline stacking sequence is disturbed, Figs 2.8(e) and 2.8(f). These may be thought of as the absence of a plane of atoms (intrinsic stacking faults) or the insertion of rows of atoms that disturb the arrangement of atoms (extrinsic stacking faults). Intrinsic and extrinsic stacking faults are illustrated schematically in Figs 2.8(e) and 2.8(f), respectively. Note how the perfect ABCABC stacking of atoms is disturbed by the insertion or absence of rows of atoms.

2.3.4 Volume Defects

Volume defects are imperfections such as voids, bubble/gas entrapments, porosity, inclusions, precipitates, and cracks. They may be introduced into a solid during processing or fabrication processes. An example of volume defects is presented in Fig. 2.9. This shows MnS inclusions in an A707 steel. Another example of a volume defect is presented in Fig. 1.16(d). This shows evidence of ~1–2 vol % of porosity in a molybdenum disilicide composite. Such pores may concentrate stress during mechanical loading. Volume defects can grow or coalesce due to applied stresses or temperature fields. The growth of three-dimensional defects may lead ultimately to catastrophic failure in engineering components and structures.

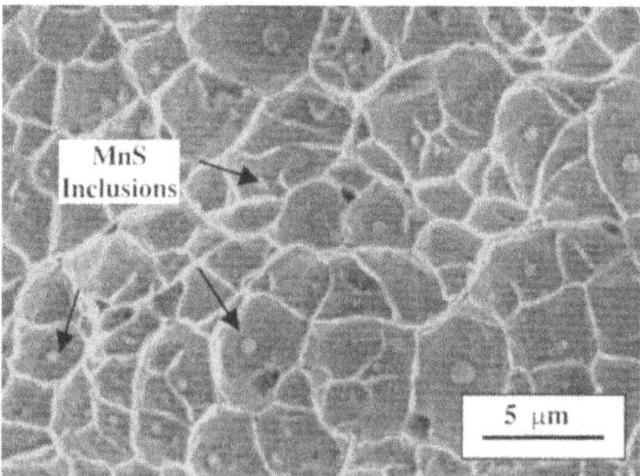

FIGURE 2.9 MnS inclusions in an A707 steel. (Courtesy of Jikou Zhou.)

2.4 THERMAL VIBRATIONS AND MICROSTRUCTURAL EVOLUTION

As discussed earlier, atoms in a crystalline solid are arranged into units that are commonly referred to as grains. The grain size may be affected by the control of processing and heat treatment conditions. Grains may vary in size from nanoscale (~10–100 nm) to microscale (~1–100 μm), or macroscale (~1–10 mm). Some examples of microstructures are presented in Figs 1.13(a–d). Note that the microstructure may consist of single phases [Fig. 1.13(a)] or multiple phases [Figs 1.13(b–d)]. Microstructures may also change due to diffusion processes that occur at temperatures above the so-called recrystallization temperature, i.e., above approximately 0.3–0.5 of the melting temperature in degrees Kelvin.

Since the evolution of microstructure is often controlled by diffusion processes, a brief introduction to elementary aspects of diffusion theory is presented in this section. This will be followed by a simple description of phase nucleation and grain growth. The kinetics of phase nucleation and growth and growth in selected systems of engineering significance will be illustrated using transformation diagrams. Phase diagrams that show the equilibrium proportions of constituent phases will also be introduced along with some common transformation reactions.

2.4.1 Statistical Mechanics Background

At temperatures above absolute zero (0 K), the atoms in a lattice vibrate about the equilibrium positions at the so-called Debye frequency, ν, of $\sim 10^{13}$ s^{-1}. Since the energy required for the lattice vibrations is supplied thermally, the amplitudes of the vibration increase with increasing temperature. For each individual atom, the probability that the vibration energy is greater than q is given by statistical mechanics to be

$$P = e^{-q/kT} \tag{2.7}$$

where k is the Boltzmann constant (1.38×10^{-23} J·atom^{-1}K^{-1}) and T is the absolute temperature in degrees Kelvin. The vibrating lattice atoms can only be excited into particular quantum states, and the energy, q, is given simply by Planck's law ($q = h\nu$). Also, at any given time, the vibrational energy varies statistically from atom to atom, and the atoms continuously exchange energy as they collide with each other due to atomic vibrations. Nevertheless, the average energy of the vibrating atoms in a solid is given by statistical mechanics to be $3kT$ at any given time. This may be sufficient to promote the diffusion of atoms within a lattice.

2.4.2 Diffusion

Diffusion is the thermally- or stress-activated movement of atoms or vacancies from regions of high concentration to regions of low concentration (Shewmon, 1989). It may occur in solids, liquids, or gases. However, we will restrict our attention to considerations of diffusion in solids in the current text. Consider the interdiffusion of two atomic species A and B shown schematically in Fig. 2.10; the probability that n_A atoms of A will have energy greater than or equal to the activation barrier, q, is given by $n_A e^{-q/kT}$. Similarly, the probability that n_B atoms of B will have energy greater than or equal to the activation barrier is given by $n_B e^{-q/kT}$. Since the atoms may move in any of six possible directions, the actual frequency in any given direction is $\nu/6$. The net number of diffusing atoms, n, that move from A to B is thus given by

$$n_d = \frac{\upsilon}{6}(n_A - n_B)e^{-q/KT} \tag{2.8}$$

If the diffusion flux, J, is defined as the net number of diffusing atoms, n_d, per unit area, i.e., $J = n_d/(l_1 l_2)$, and the concentration gradient, dC/dx,

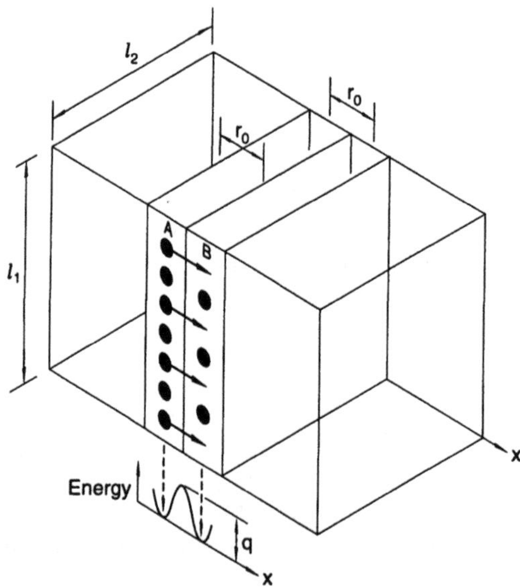

FIGURE 2.10 Schematic illustration of diffusion: activation energy required to cross a barrier. (Adapted from Ashby and Jones, 1994. Reprinted with permission from Pergamon Press).

which is given simply by $-(C_A - C_B)/r_0$, the diffusion flux, J, may then be expressed as

$$J = D_0 \exp\left(\frac{-q}{kT}\right)\left(\frac{dC}{dx}\right) \quad (2.9)$$

If we scale the quantity q by the Avogadro number, then the energy term becomes $Q = N_A q$ and $R = kN_A$. Equation (2.9) may thus be expressed as

$$J = -D_0 \exp\left(\frac{-Q}{RT}\right)\left(\frac{dC}{dx}\right) \quad (2.10)$$

If we now substitute $D = -D_0 \exp\left(\frac{-Q}{RT}\right)$ into Eq. (2.10), we obtain the usual expression for J, i.e., J is given by

$$J = -D\frac{dC}{dx} \quad (2.11)$$

The above expression is Fick's first law of diffusion. It was first proposed by Adolf Hicks in 1855. It is important to note here that the diffusion coefficient for self-diffusion, D, can have a strong effect on the creep properties, i.e., the time-dependent flow of materials at temperatures greater than ~0.3–0.5 of the melting temperature in degrees Kelvin. Also, the activation energy, Q, in Eq. (2.10) is indicative of the actual mechanism of diffusion, which may involve the movement of interstitial atoms [Fig. 2.11(a)] and vacancies [Fig. 2.11(b)].

Diffusion may also occur along fast diffusion paths such as dislocation pipes along dislocation cores [Fig. 2.12(a)] or grain boundaries [Fig. 2.12(b)]. This is facilitated in materials with small grain sizes, d_g, i.e., a large number of grain boundaries per unit volume. However, diffusion in most crystalline materials occurs typically by vacancy movement since the activation energies required for vacancy diffusion (~1 eV) are generally lower than the activation energies required for interstitial diffusion (~2–4 eV). The activation energies for self-diffusion will be shown later to be consistent with activation energies from creep experiments.

2.4.3 Phase Nucleation and Growth

The random motion of atoms and vacancies in solids, liquids, and gases are associated with atomic collisions that may give rise to the formation of small embryos or clusters of atoms, as shown in Figs 2.13(a) and 2.13(b). Since the initial free-energy change associated with the initial formation and growth of such clusters is positive (Read-Hill and Abbaschian, 1992), the initial clusters of atoms are metastable. The clusters may, therefore, disintegrate due to the effects of atomic vibrations and atomic collisions. However, a

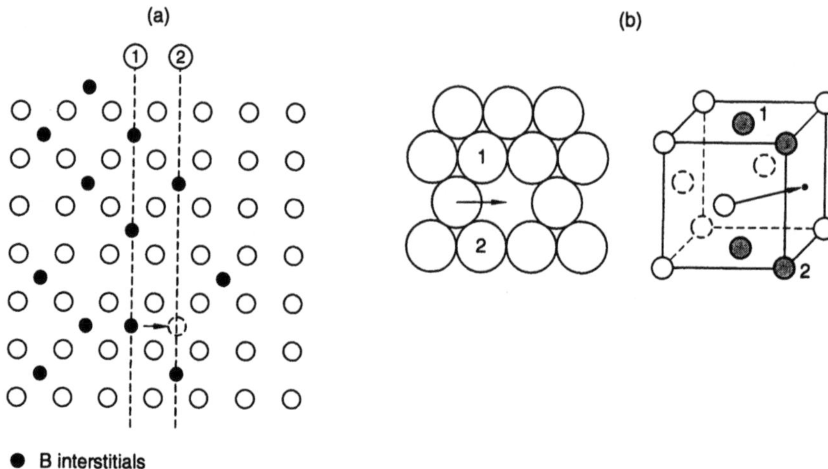

FIGURE 2.11 Schematic illustration of diffusion mechanisms: (a) movement of interstitial atoms; (b) vacancy/solute diffusion. (Adapted from Shewmon, 1989. Reproduced with permission from the Minerals, Metals, and Materials Society.)

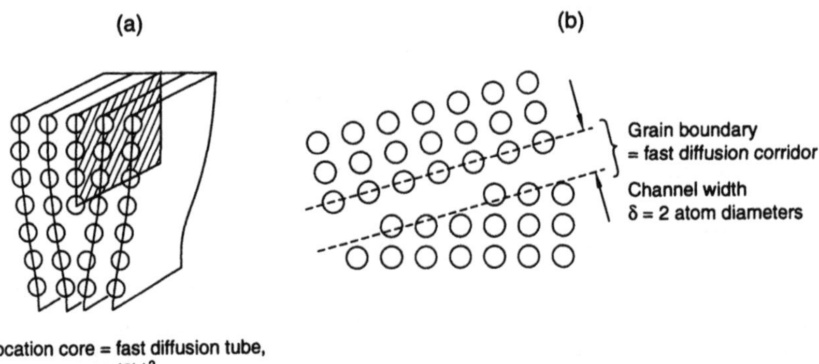

FIGURE 2.12 Fast diffusion mechanisms: (a) dislocation pipe diffusion along dislocation core; (b) grain boundary diffusion. (Adapted from Ashby and Jones, 1980. Reprinted with permission from Pergamon Press.)

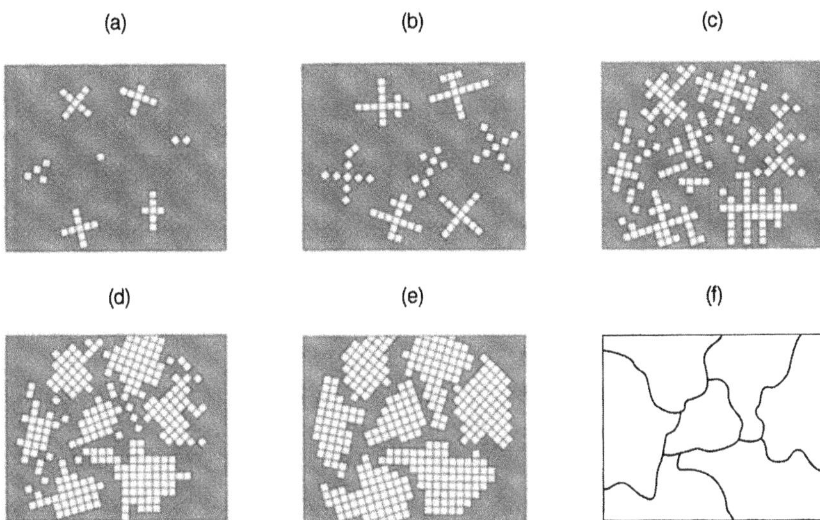

FIGURE 2.13 Schematic illustration of nucleation and growth: (a, b) formation of embryos; (c,d) nuclei growth beyond critical cluster size; (e) impingement of growing grains; (f) polycrystalline structure. (Adapted from Altenpohl, 1998.)

statistical number of clusters or embryos may grow to a critical size, beyond which further growth results in a lowering of the free energy. Such clusters may be considered stable, although random atomic jumps may result in local transitions in cluster size to dimensions below the critical cluster dimension.

Beyond the critical cluster size, the clusters of atoms may be considered as nuclei from which new grains can grow primarily as a result of atomic diffusion processes, Figs 2.13(c) and 2.13(d). The nuclei grow until the emerging grains begin to impinge on each other, Fig. 2.13(e). The growth results ultimately in the formation of a polycrystalline structure, Fig. 2.13(f).

Subsequent grain growth occurs by interdiffusion of atoms and vacancies across grain boundaries. However, grain growth is mitigated by interstitial and solute "atmospheres" that tend to exert a drag on moving grain boundaries. Grain growth is also associated with the disappearance of smaller grains and the enhanced growth of larger grains. Due to the combined effects of these factors, a limiting grain size is soon reached. The rate at which this limiting grain size is reached depends on the annealing duration and the amount of prior cold work introduced during deformation processing via forging, rolling, swaging, and/or extrusion.

The simple picture of nucleation and growth presented above is generally observed in most crystalline metallic materials. However, the rate of nucleation is generally enhanced by the presence of pre-existing nuclei such as impurities on the mold walls, grain boundaries, or other defects. Such defects make it much easier to nucleate new grains heterogeneously, in contrast to the difficult homogeneous nucleation schemes described earlier. In any case, the nuclei may grow by diffusion across grain boundaries to form single-phase or multi-phase microstructures, such as those shown in Fig. 1.13.

A simple model of grain growth may be developed by using an analogy of growing soap bubbles. We assume that the growth of the soap bubbles (analogous to grains) is driven primarily by the surface energy of the bubble/grain boundaries. We also assume that the rate of grain growth is inversely proportional to the curvature of the grain boundaries, and that the curvature itself is inversely proportional to the grain diameter. We may then write:

$$d(D)/dt = k/d \tag{2.12}$$

where D is the average grain size, t is time elapsed, and k is a proportionality constant. Separating the variables and integrating Eq. (2.12) gives the following expression:

$$D^2 = kt + c \tag{2.13}$$

where c is a constant of integration. For an initial grain size of D_0 at time $t = 0$, we may deduce that $c = D_0^2$. Hence, substituting the value of c into Eq. (2.13) gives

$$D^2 - D_0^2 = kt \tag{2.14}$$

Equation (2.14) has been shown to fit experimental data obtained for the growth of soap bubbles under surface tension forces. Equation (2.14) has also been shown to fit the growth behavior of metallic materials when grain growth is controlled by surface energy and the diffusion of atoms across the grain boundaries. In such cases, the constant k in Eq. (2.14) exhibits an exponential dependence which is given by

$$k = k_0 \exp(-Q/RT) \tag{2.15}$$

where k_0 is an empirical constant, Q is the activation energy for the grain growth process, T is the absolute temperature, and R is the universal gas constant. By substituting Eq. (2.15) into Eq. (2.14), the grain growth law may be expressed as

$$D^2 - D_0^2 = tk_0 \exp(-Q/RT) \tag{2.16}$$

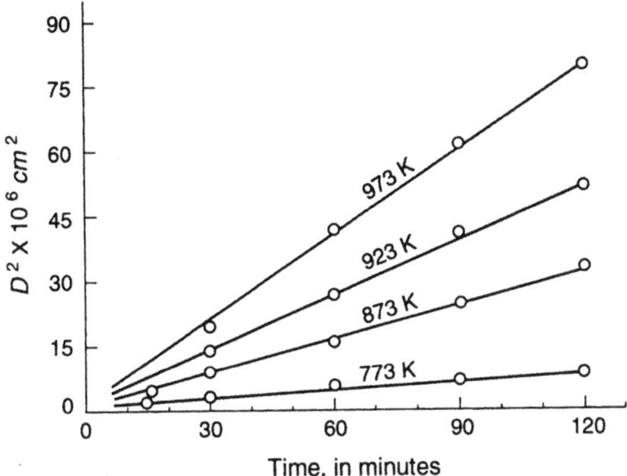

FIGURE 2.14 Grain growth isotherms for α-brass (10% Zn–90% Cu). Note that D^2 varies directly with time. (Adapted from Feltham and Copley, 1958.)

If we assume that $D_0 = 0$, then the grain growth law [Eq. (2.14)] becomes

$$d = (kt)^{1/2} \tag{2.17}$$

Equations (2.16) and (2.17) are consistent with data for grain growth in alpha brass (10% Zn–90% Cu) presented in Fig. 2.14. However, the exponent in the grain growth law is often somewhat different from the value of 1/2 in Eq. (2.17). It is, therefore, common to report the grain growth law in the following form:

$$d = k'(t)^n \tag{2.18}$$

where n is a number that is generally less than the value of 1/2 predicted for diffusion across grain boundaries, and k' is an empirical constant.

2.4.4 Introduction to Phase Diagrams

Let us start by considering a simple two-component system, e.g., a system consisting of Cu and Ni atoms. Since Cu and Ni have similar atomic radii, crystal structures, valence, and electronegativities, they are completely miscible across the complete composition range. The equilibrium structures in the Cu and Ni system can be represented on a phase diagram, Fig. 2.15(a). A phase diagram may be considered as a map that shows the phases (a phase is a physically distinct, homogeneous aspect of a system) that exist in equilibrium as a function of temperature and composition, Fig. 2.15(a).

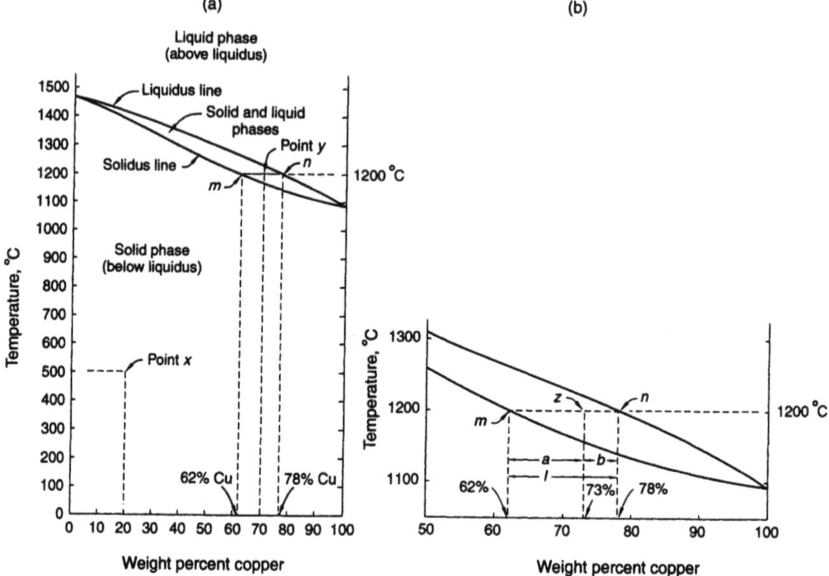

FIGURE 2.15 Cu/Ni phase diagram: (a)] complete phase diagram; (b) illustration of the Lever rule in partial phase diagram. (Adapted from Read-Hill and Abbaschian, 1992. Reprinted with permission from PWS Kent.)

In the case of the Cu–Ni phase diagram shown in Fig. 2.15(a), the mixtures exist either as solid phases (below the solidus line) or liquid phases (above the liquidus line). A mixed two-phase (solid + liquid) region also exists between the solidus and liquidus in Fig. 2.15(a). Since Ni and Cu are completely soluble across the complete composition regime, the phase diagram in Fig. 2.15(a) is often referred to as a binary isomorphous phase diagram.

The compositions at the extreme left and extreme right of the phase diagram [Fig. 2.15(a)] correspond to 100% Ni and 100% Cu, respectively. However, intermediate compositions consist of both Cu and Ni atoms. The phase diagram may be read simply by identifying the phases present at the specified co-ordinates. For example, point x on Fig. 2.15(a) represents a solid phase that contains 80% Ni and 20% Cu at 500°C. It should be clear that the mixtures to the left of the diagram have more Ni than Cu. Vice versa, the mixtures to the right of the phase diagram have more Cu than Ni.

Now, consider the composition of the point y in the two-phase (solid + liquid) regime. This point has a composition that is in between those at points *m* and *n*. The compositions of *m* and *n* may be read directly from the abscissa in Fig. 2.15(a). The points *m* and *n* correspond to mixtures containing 62 and 78% copper, respectively. The liquid + solid mixture at *y* actually contains 70% Cu and 30% Ni, as shown in Fig. 2.15(a).

Now refer to the enlarged region of the solid–liquid region shown in Fig. 2.15(b). The composition at point *z* in Fig. 2.15(b) may be determined using the so-called *Lever rule*. The Lever rule states that the fraction of mixture *m* at point *z* is given simply by $(b/a)] \times 1$, Fig. 2.15(b). Similarly, the fraction of mixture *n* at point *z* is given by $(a/b) \times 1$, Fig. 2.15(b). Note also that the fraction of mixture *n* is equal to one minus the fraction of mixture *m*. Conversely, the fraction of mixture *m* is equal to one minus the fraction of mixture *n*. Hence, as we move across the isothermal (tie) line from *m* to *n*, the composition changes from 100% solid to 100% liquid in Fig. 2.15(b). We may, therefore, use the Lever rule to find the fractions of solid and liquid at point *m*. The reader should be able show that the fraction of solid phase at *z* is b/l. Similarly, the fraction of liquid phase is a/l at point *z*, Fig. 2.15(b).

Intermediate phases may also form due to supersaturation with one of the alloying elements. This may result in the formation of two-phase mixtures in more complex phase diagrams. Fig. 2.16 shows the Al–Cu phase diagram. Note that the α-Al phase becomes quickly supersaturated with Cu.

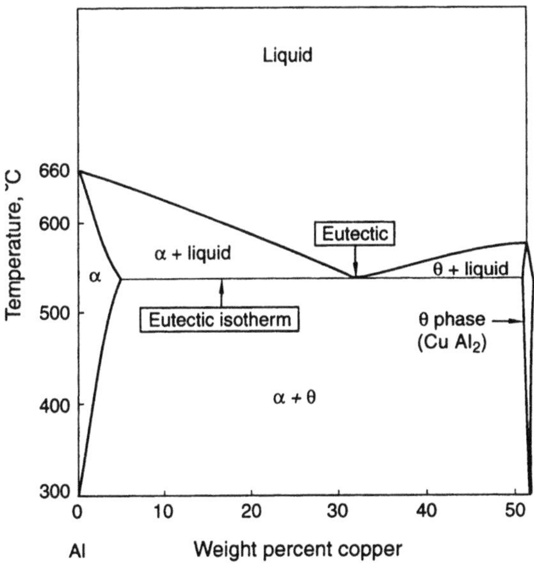

Figure 2.16 Left-hand section of the aluminum–copper phase diagram.

A significant fraction of the phase diagram, therefore, consists of a two-phase regime in which the α phase and θ phase ($CuAl_2$) are in equilibrium. The Al–Cu system also exhibits a minimum melting point at which liquid aluminum undergoes a reaction to form an α phase and a θ phase. Such reactions involving the formation of two solid phases from a single liquid phase are known as eutectic reactions. Eutectic reactions are generally associated with zero freezing range and lowest melting points. They therefore form the basis for the design of low melting point solder and braze alloys.

The left-hand section of the Fe–C phase diagram is shown in Fig. 2.17 (Chipman, 1972). This is a very important phase diagram since it provides the basis for the design of steels and cast irons that are generally the materials of choice in structural applications. Steels are Fe–C mixtures that contain less than 1.4 wt % carbon, whereas cast irons typically contain between 1.8 and 4 wt % carbon. Note also that we may further subdivide steels into low-carbon steels (< 0.3 wt % C), medium-carbon steels (0.3–0.8 wt % C) and high-carbon (0.8–1.4 wt % C) steels.

The five phases observed on the left-hand side of the Fe–C diagram are α-Fe (also called ferrite), γ-Fe (also called austenite), δ-Fe, Fe_3C (also called cementite), and a liquid solution of C in Fe (Brooks, 1996). The α-Fe has a body-centered cubic (b.c.c.) structure. It contains randomly distributed car-

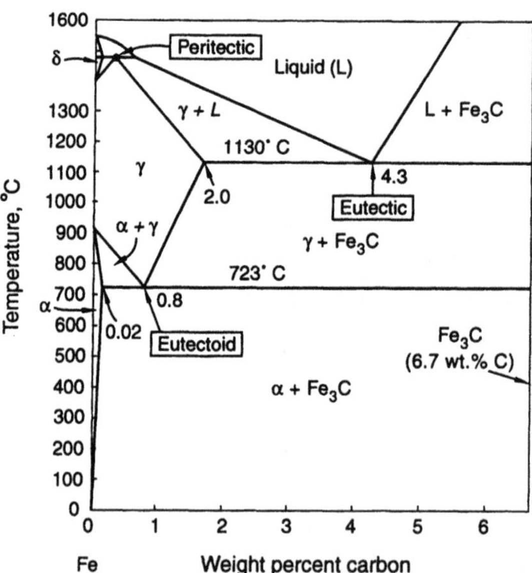

FIGURE 2.17 Left-hand section of the iron–carbon phase diagram.

bon atoms in solution in b.c.c. iron. The maximum solubility of C (in α-Fe) of only 0.035 wt % occurs at 723°C. Pure α-Fe iron is also stable below 914°C. However, above this temperature, γ-Fe is the stable phase. This is a random solid solution of carbon in face-centered cubic (f.c.c) iron. The maximum solid solubility of C in γ-Fe of 1.7 wt % occurs at 1130°C. Body-centered cubic δ ferrite is stable between 1391° and 1536°C. It contains a random interstitial solid solution of C in b.c.c. iron.

A number of important transformations are illustrated on the Fe–C diagram (Fig. 2.17). Note the occurrence of a eutectic reaction (Liquid 1 = Solid 2 + Solid 3) at a carbon content of 4.3 wt %. A similar reaction also occurs at a carbon content 0.8 wt %. This "eutectic-like" reaction involves the formation of two new solids from another solid phase (Solid 1 = Solid 2 + Solid 3). Such a reaction is referred to as a eutectoid reaction. Eutectoid reactions generally result in the formation of lamellar microstructures. In the case of eutectoid steels, a lamellar structure called pearlite is formed as a result of the eutectoid reaction, Figs 2.18(a) and 2.18(b). It is formed by the

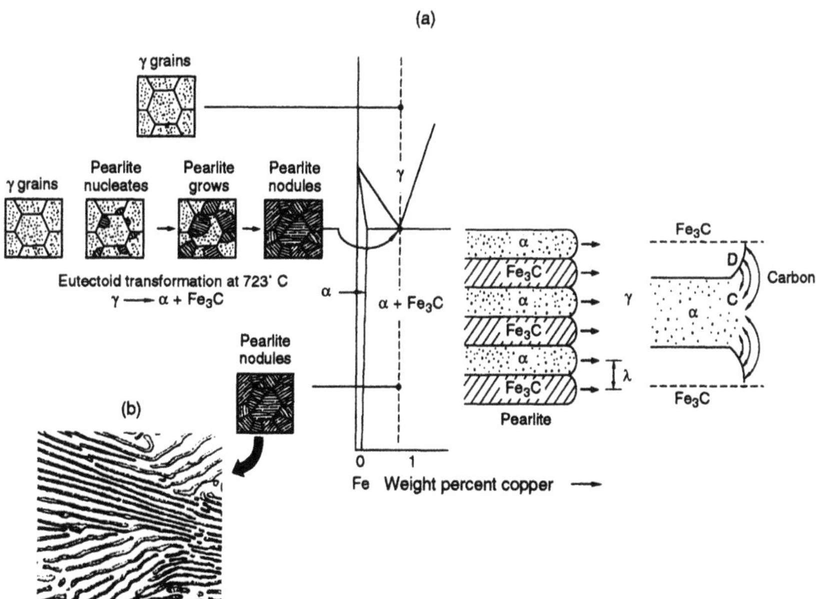

Figure 2.18 Eutectoid reaction in steels: (a)] phase transformations that occur during the cooling of a eutectoid steel; (b) pearlitic microstructure showing alternating layers of ferrite and cementite. (Adapted from Van Vlack, 1980. Reprinted with permission from Addison Wesley.)

decomposition of austenite (γ-Fe) into ferrite (α-Fe) and cementite (Fe_3C). The resulting structure consists of alternating platelets of ferrite and cementite with proportions and spacing given by the Lever rule.

Hypoeutectoid structures are formed for compositions to the left of the eutectoid compositions (<0.8 wt % C), as shown in Fig. 2.19(a). These consist of proeutectoid ferrite (formed before the eutectoid transformation at 723°C) and pearlite (formed as a result of the eutectoid reaction at 723°C). Similarly, hypereutectoid structures are produced for compositions to the right of the eutectoid (>0.8 wt % C), as shown in Fig. 2.19(b). These consist of proeutectoid carbide (formed before the eutectoid transformation at 723°C) and pearlite (formed as a result of the eutectoid reaction at 723°C).

2.4.5 Introduction to Transformation Diagrams

As discussed earlier, phase diagrams show the phases that are present under equilibrium conditions. However, they do not show the phase changes that occur during microstructural evolution towards equilibrium. Since the formation of new phases is strongly influenced by temperature and time, it is helpful to represent the formation of such phases on plots of temperature versus time. Such diagrams are referred to as transformation diagrams since they show the phase transformations that can occur as a function of temperature and time.

2.4.5.1 Time–Temperature–Transformation Diagrams

Phase transformations that occur under isothermal conditions (constant temperature conditions) are represented on temperature–time-transformation (TTT) diagrams or C-curves (Fig. 2.20). Such diagrams are produced by the cooling of preheated material to a given temperature, and subsequent isothermal exposure at that temperature for a specified duration, before quenching (fast cooling) the material to room temperature. The phases formed during the isothermal exposure are then identified during subsequent microstructural analysis. A TTT diagram for pure iron is shown in Fig. 2.20(a). This was produced by fast cooling from the austenitic field (stable f.c.c. iron) to different temperatures in the α field. Note that metastable austenite (f.c.c. iron) is retained initially, Fig. 2.20(a). However, austenite is metastable in the α field. Stable body b.c.c. iron therefore forms by a process of nucleation and growth during the isothermal exposure.

The number 1 shown at the "nose" of the first TTT curve [Fig. 2.20(a)] corresponds to the start of the transformation (1% transformation). Similarly, the curves labeled 25, 50, 75, and 99 correspond, respectively, to different stages of the transformation, i.e., the transformations are 25,

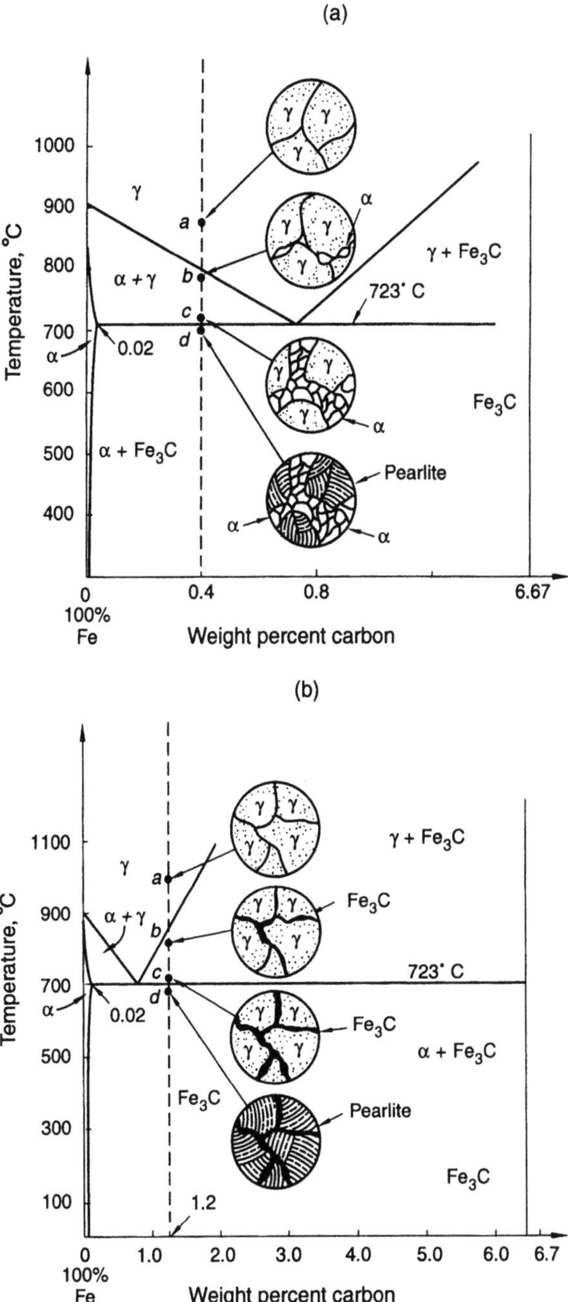

Figure 2.19 Phase transformations that occur during the cooling of (a) hypoeutectoid steel and (b) hypereutectoid steel. (Adapted from Van Vlack, 1980. Reprinted with permission from Addison Wesley.)

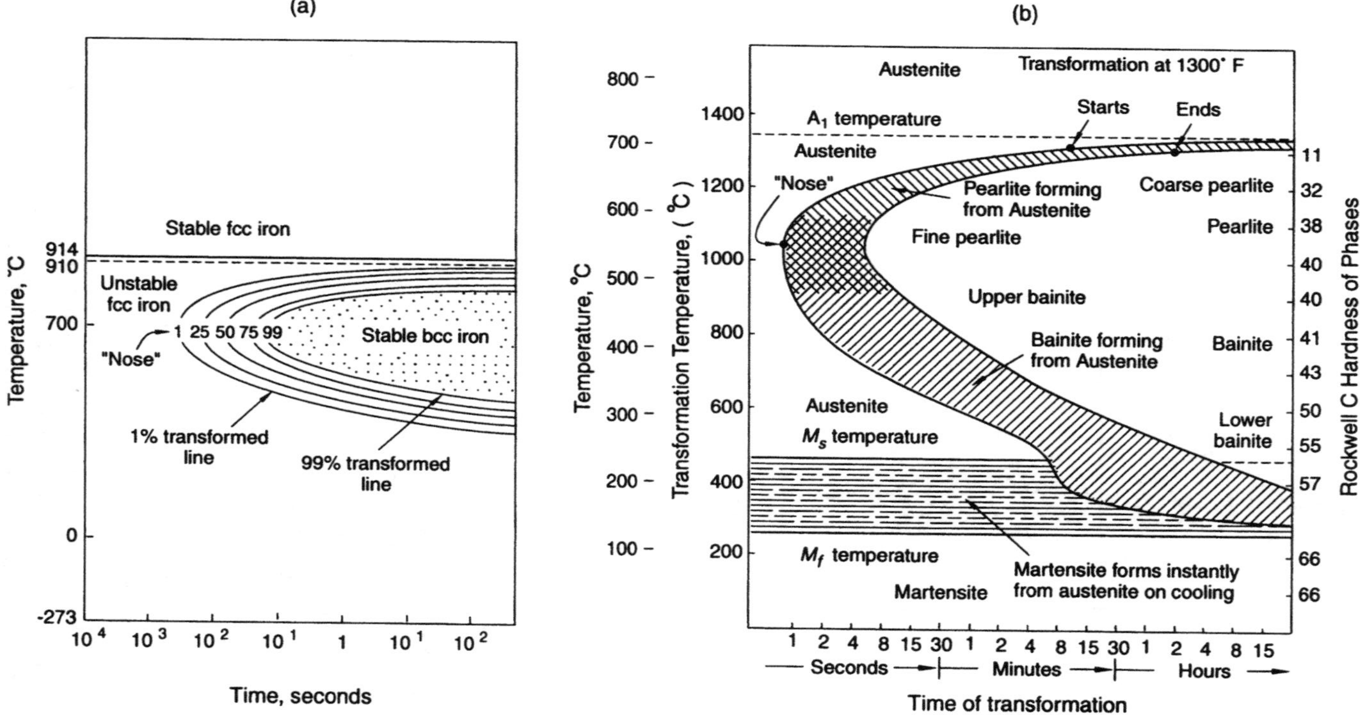

Figure 2.20 Time–temperature–transformation (TTT) curves for (a) pure iron and (b) eutectoid steel. (Adapted from Ashby and Jones, 1994. Reprinted with permission from Pergamon Press.)

50, 75, and 99% complete along these curves. These C-curves, therefore, span the entire range between the start and the finish of the transformations in the semischematic shown in Fig. 2.20(a). They are, therefore, useful in the study of the kinetics (paths) of the phase transformations.

Another example of a TTT diagrams is presented in Fig. 2.20(b). This shows a TTT diagram for a eutectoid steel with the designation 1080 steel, i.e., a 10XX series plain carbon steel that contains 0.80 wt % C. As before [Fig. 2.20(a)], metastable austenite is retained initially after the initial quench from the austenitic field [Fig. 2.20(b)]. The metastable austenite then transforms by nucleation and growth to form α-ferrite and carbide (Fe_3C) after isothermal exposure in the $\alpha + Fe_3C$ field, Fig. 2.20(b).

It is important to note that the morphology of the $\alpha + Fe_3C$ phases in eutectoid steel (Fe–0.08C) depends on whether the isothermal exposure is carried out above or below the nose of the TTT curve, Fig. 2.20(b). For annealing above the nose of the curve, pearlite is formed. The pearlite has a coarse morphology (coarse pearlite) after exposure at much higher temperatures above the nose, and a fine morphology (fine pearlite) after exposure at lower temperatures above the nose of the TTT curve. Bainite is formed after annealing below the nose of the TTT curve. Upper bainite is formed at higher temperatures below the nose, and lower bainite is formed at lower temperatures below the nose, Fig. 2.20(b).

It is also interesting to study the bottom section of TTT diagram for the eutectoid 1080 steel, Fig. 2.20(b). This shows that a phase called martensite is formed after fast cooling (quenching) from the austenitic field. This plate-like phase starts to form at the so-called M_S (martensite start) temperature. Martensite formation is completed by fast cooling to temperatures below the M_F (martensite finish) temperature, Fig. 2.20(b). However, unlike the other phases discussed so far, martensite does not form by a process of nucleation and growth since there is insufficient time for long-range diffusion to occur during fast cooling from the austenitic field.

Instead, martensite forms by a diffusionless or shear transformation that involves only the local shuffling of atoms. Martensite formation is illustrated in Fig. 2.21. First, local shuffling of atoms results in two adjacent f.c.c. cells coming together. This results in the face-centered atom in the f.c.c. unit cell becoming the body-centered atom in the distorted body-centered cubic cell. A body-centered tetragonal unit cell is formed at the center of the two adjacent f.c.c. cells, as shown in Fig. 2.21(a). Furthermore, there is no one-to-one matching (coherence) between the corner atoms in the new cell and the old cells. However, coherency may be obtained by rotating the b.c.c. lattice.

Martensite growth occurs at high speeds (close to the speed of sound), and the parent phase is replaced by the product phase as the martensite

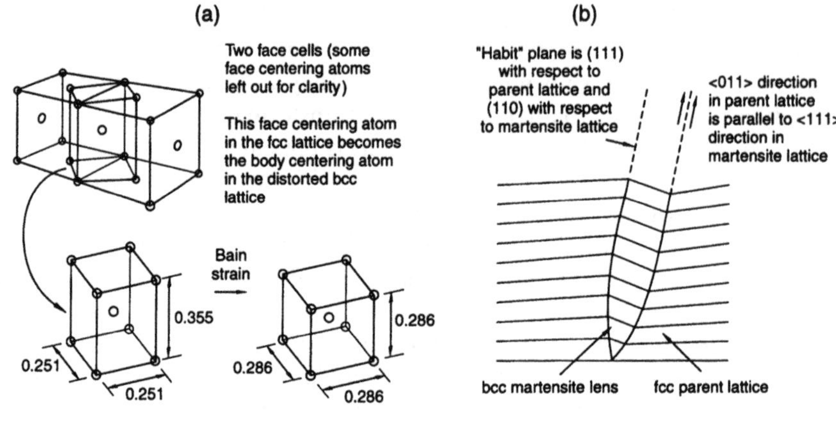

FIGURE 2.21 Formation of martensite: (a) local shuffling brings two f.c.c. lattices together (note that the Bain strain is needed to restore undistorted cubic cell; (b) coherent thin martensite plate. (Adapted from Ashby and Jones, 1994. Reprinted with permission from Pergamon Press.)

advances. Since the interface advances rapidly, no composition change is associated with martensitic phase transformations. Furthermore, no diffusion is required for martensitic phase transformations to occur. It is also important to note that martensitic phase transformations have been shown to occur in ceramic materials and other metallic materials such as titanium. Also, martensites are always coherent with the parent lattice. They grow as thin lenses on preferred planes along directions that cause the least distortion of the lattice. The crystallographic directions for martensites in pure iron are shown in Fig. 2.21(b).

Steel martensites contain a significant amount of interstitial carbon atoms that are locked up in the distorted b.c.c. structure after quenching from the austenitic field. Such interstitial carbon atoms promote significant strengthening by restricting the movement of dislocations. However, they also contribute to the brittle behavior of steel martensites, i.e., martensitic steels are *strong* but *brittle*. Furthermore, since martensite is metastable, subsequent heating (tempering) in the $\alpha + Fe_3C$ phase field will result in the transformation of martensite into a more stable structure consisting of α-ferrite and carbide (Fig. 2.22).

Finally in this section, it is important to note that the tempering of martensitic steels is often used to produce so-called *tempered martensitic steels.* Some martensite may also be retained in such structures, depending

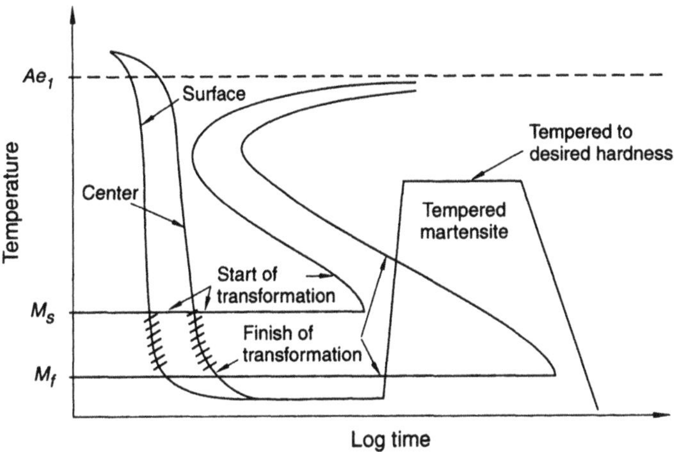

FIGURE 2.22 Schematic illustration of quenching and tempering process on a CCT curve. (Adapted from Van Vlack, 1980. Reprinted with permission from Addison Wesley.)

on the degree of tempering, i.e., the heating duration and temperature. Tempered martensitic steels are usually moderately strong and reasonably ductile. However, they are not as strong as untempered steel martensites. Nevertheless, their attractive combinations of moderate strength and fracture toughness make them the materials of choice in numerous engineering applications of steels.

2.4.5.2 Continuous Cooling–Transformation Diagrams

The TTT diagrams are useful for studying the evolution of phases under isothermal conditions. However, in engineering practice, materials are often heat treated and cooled to room temperature at different rates. For example, a hot piece of steel may be removed from a furnace, and air cooled or water quenched to produce a desired microstructure. The transformations produced after such controlled cooling are generally not predicted by TTT diagrams.

The microstructures produced at controlled cooling rates are generally represented on continuous cooling–transformation (CCT) diagrams. A CCT diagram for a eutectoid steel is shown in Fig. 2.23. For comparison, isothermal transformation curves and times for the same eutectoid steel are also shown in dashed lines in Fig. 2.23. Note that the CCT curves are shifted downwards and to the right, since part of the time was spent at elevated temperature where the nucleation initiated more slowly (in comparison with

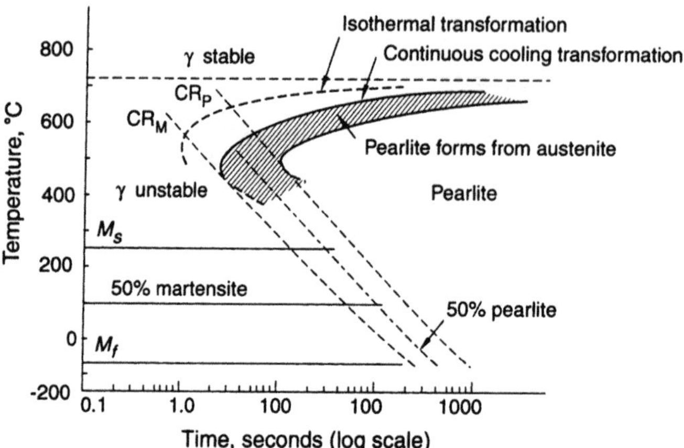

Figure 2.23 Continuous cooling–transformation (CCT) curve for a eutectoid steel. (Adapted from Van Vlack, 1980. Reprinted with permission from Addison Wesley.)

that in the isothermally exposed material). However, as expected, the austenite decomposes into pearlite by a process of nucleation and growth.

Figure 2.23 also shows that martensite is formed after fast cooling at rates that just miss the nose of the transformation curve, i.e., rates that are faster than the slope of the dashed line labeled CR_M in the figure. The formation of martensite in this regime occurs by the diffusionless transformation that was described in the previous section. As before, martensite start (M_s) and martensite finish (M_F), temperatures may also be represented on the CCT diagram. Furthermore, martensite is not produced for cooling rates below the dashed line labeled as CR_P in Fig. 2.23. Cooling rates below this critical level result only in the formation of pearlite.

In *plain carbon eutectoid steels* (Fe–0.08C), the critical cooling rates corresponding to CR_M and CR_P are 140°C/s and 35°C/s, respectively (Fig. 2.23). However, the critical cooling rates are generally lower in more complex *alloyed steels*. Such steels are alloyed to engineer certain combinations of microstructure/mechanical properties, and resistance to environmental attack. The alloying elements in alloyed steels may include Ni, Cr, Mn, W, and Mo. These elements may stabilize either the ferrite (α) or austenite (γ) phases. Nickel is an austenite stabilizer, i.e., it increases the temperature range across which the austenite is stable. Chromium is generally added to promote improved corrosion resistance, and Mn is added to increase fracture toughness, e.g., in the well-known Hadfield steels. Molybdenum and W are often added to hot-work tool steels to improve their strength.

Further details on steels can be found in specialized texts on steels or physical metallurgy (Honeycombe and Bhadesia, 1995).

2.5 OVERVIEW OF MECHANICAL BEHAVIOR

The mechanical behavior of materials depends on their response to loads, temperature, and the environment. In several practical problems, the combined effects of these controlling parameters must be assessed. However, the individual effects of loads (elastic and plastic deformation) must be studied in detail before attempting to develop an understanding of the combined effects of load and temperature, or the effects of load and environment.

The material response may also depend on the nature of the loading. When the applied deformation increases continuously with time (as in a tensile test), then reversible (elastic) deformation may occur at small loads prior to the onset of irreversible/plastic deformation at higher loads. Under reversed loading, the material may also undergo a phenomenon known as "fatigue." This occurs even at stresses below those required for bulk plastic deformation. Fatigue may lead to catastrophic fracture if its effects are not foreseen in the design of most engineering structures and components.

Engineers must be aware of a wide range of possible material responses to load, temperature, and environment (Ashby and Jones, 1994; McClintock and Argon, 1993). These will be discussed briefly in this section prior to more in-depth presentations in the subsequent chapters.

2.5.1 Tensile and Compressive Properties

The tensile and compressive properties of a material describe its response to axial loads along the orthogonal (x, y, z) axes. Loads that stretch the boundaries of a solid are usually described as tensile loads, while those that compress the system boundaries are described as compressive loads. For relatively small displacements, the induced deformation is fully recovered upon removal of the applied loads, and the deformation is called "elastic" deformation. Elastic deformation may be linear or nonlinear, depending on the atomic structure (applied loads are directly proportional to the displacements of the system boundaries in the case of linear elastic deformation).

Also, elastic deformation may be time dependent when time is required for the atoms within a solid to flow to the prescribed displacements. Such time-dependent, fully reversible, elastic deformation is generally described as viscoelastic deformation if the system boundaries flow back to their original positions after some period of time subsequent to the removal of the applied loads. Viscoelastic deformation may occur in polymers, metals, and ceramics. Viscoelastic behavior is associated with underlying molecular flow processes. A fundamental understanding of molecular flow processes is, therefore, helpful to the understanding of viscoelastic behavior.

When the imposed loads per unit area are sufficiently high, the distortions of the systems' boundaries may not be fully recovered upon removal of the imposed loads. This occurs when the atoms cannot flow back to their original positions upon removal of the applied loads. This results in permanent or plastic deformation, since the shape of the material is changed permanently as a result of the applied loads. Plastic deformation may occur under tensile or compressive loading conditions, and it is generally associated with zero volume change. Furthermore, plastic deformation is nearly time independent at temperatures below the recrystallization temperature.

2.5.2 Shear Properties

When a twisting moment is applied to a solid, relative movement is induced across a surface due to the imposed loads. The shear properties of a material describe its response to the imposed shear loads. Both positive and negative (counterclockwise or clockwise) shear may be imposed on a solid, and the resulting shear (angular) displacements are fully reversible for small levels of angular displacement. Instantaneous elastic or viscoelastic shear processes may also occur, as discussed in the previous section on axial properties. Furthermore, yielding may occur under shear loading at stresses that are lower in magnitude than those required for plastic flow under axial loading. In general, however, the conditions required for plastic flow are strongly dependent on the combinations of shear and axial loads that are applied. These will be discussed in detail in subsequent chapters.

2.5.3 Strength

With the exception of metallic materials, most materials derive their strengths from their primary and secondary bond strengths. In general, however, the measured strength levels are less than the theoretical strengths due to chemical bonding. This is due to the effects of defects which generally give rise to premature failure before the theoretical strength levels are reached. The effects of stress concentrators are particularly severe in brittle materials such as ceramics and brittle intermetallics. Strength levels in such brittle materials are often associated with statistical distributions of defects.

The situation is somewhat different in the case of metallic materials. In addition to the inherent strength of metallic bonds, most metallic materials derive their strengths from the interactions of dislocations with defects such as solid solution alloying elements, interstitial elements, other dislocations, grain boundaries, and submicroscopic precipitates. These defects (including the dislocations themselves) strengthen the metallic lattice by impeding the movement of dislocations. Most metals are, therefore, modified by alloying (adding controlled amounts of other elements) and processing to improve their resistance to dislocation motion. However, excessive restriction of dislocation motion may result in brittle behavior. Brittle behavior may also occur due to the effects of cracks or notches which may form during manufacturing.

2.5.4 Hardness

The hardness of a material is a measure of its resistance to penetration by an indenter. Hardness is also a measure of strength and often has the units of stress. The indenter is often fabricated from a hard material such as diamond or hardened steel. The tips of the indenters may be conical, pyramidal, or spherical in shape. The indenter tips may also be relatively small (nano- or micro-indenters) or very large (macroindenters). Since indentation tests are relatively easy to perform (macroindentations require only limited specimen preparation), they are often used to obtain quick estimates of strength.

Micro- and nano-indenters have also been developed. These enable us to obtain estimates of moduli and relative estimates of the strengths of individual phases within a multiphase alloy. However, due to the nature of the constrained deformation around any indenter tip, great care is needed to relate hardness data to strength.

Nevertheless, some empirical and approximate theoretical "rules-of thumb" have been developed to estimate the yield strength from the hardness. One "rule-of-thumb" states that the yield strength (or tensile strength in materials that strain harden) is approximately equal to one-third of the measured hardness level.

Since hardness tests are relatively easy to perform (compared to tensile tests), estimates of the yield strengths (or ultimate strength) are often obtained from hardness measurements. However, the users of such rules must always remember the approximate nature of such relationships between hardness and strength, i.e., they only provide estimates of strength within 20%, adequate for many practical purposes. Furthermore, the measured hardnesses may vary with indenter size when the indents are less than 1 μm. The size dependence of such small indents has been attributed to plasticity length-scale effects which may give rise to strain gradient effects.

2.5.5 Fracture

If the applied monotonic loads (loads that increase continuously with time) are sufficiently large, fracture will occur due to the separation of atomic bonds or due to the growth of holes from defects by plastic deformation. Fracture by pure bond rupture occurs only in the case of brittle cleavage failure in which atomic separation occurs (without plasticity) along low index planes. Slight deviations from planarity lead to fracture surfaces that have "river lines" that are akin to what one might expect to see on a geographical map, Fig. 2.24(a).

(a)

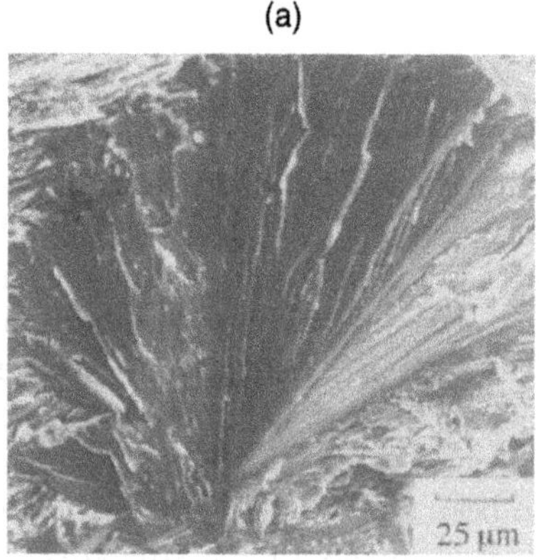

(b)

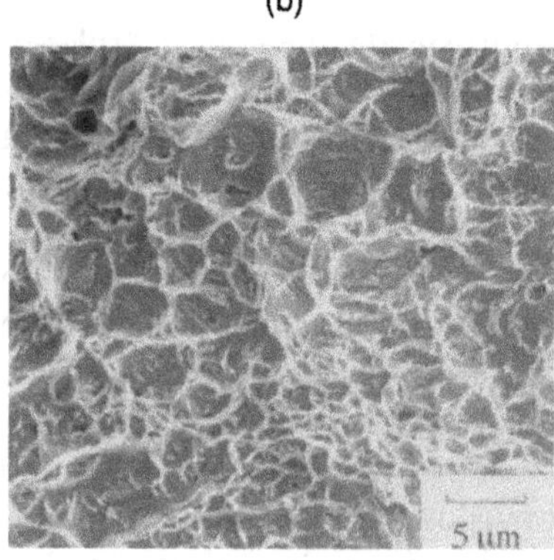

Figure 2.24 Typical fracture modes: (a)] brittle cleavage fracture mode in Ti–48Al; (b) ductile dimpled fracture mode in an IN 718 nickel-based superalloy. (Courtesy of Dr. Chris Mercer).

In most cases, however, fracture is preceded by plastic deformation (even in the case of so-called cleavage fracture), crack nucleation, and crack growth phenomena. Final fracture occurs when cracks grow to a critical size. Since brittle fracture is controlled mainly by the separation of bonds, fracture processes are typically controlled by maximum axial (hydrostatic) stresses.

In contrast, microductile fracture processes are often preceded and accompanied by plastic flow that is controlled mainly by shear stresses which tend to promote the movement of line defects such as dislocations (Fig. 2.7). At the same time, the hole growth is promoted by triaxial tensile stresses. The conditions for the onset of ductile fracture are also strongly dependent on the nucleation and linkage of holes in ductile metals. This often results in a ductile dimpled fracture mode in metallic materials, Fig. 2.24(b). Fracture mechanisms will be discussed in greater detail in subsequent chapters.

2.5.6 Creep

As discussed earlier, at temperatures above the recrystallization temperature (~0.3–0.5 of the absolute melting temperature) thermally and stress activated flow processes may contribute strongly to deformation. These flow processes occur due to the movement of point defects (vacancies) or line defects (dislocations) under static loading. Creep deformation may also be associated with microvoid formation or microvoid coalescence, especially during the final stages of deformation prior to fracture.

Creep deformation occurs in both crystalline and noncrystalline materials, and the time to failure may range from minutes/hours (in materials deformed at high stresses and temperature) to geological time scales (millions of years) in materials within the earth's crust. A study of the micromechanisms of creep and creep fracture is often a guide to the development of phenomenological relations between applied stress and temperature.

Before fracture, the tolerance of precision-made components in high-temperature gas turbines may be lost due to creep deformation. For this reason, creep-resistant high-temperature materials such as nickel- and cobalt-base alloys are used in turbines. The microstructures of the creep-resistant alloys are tailored to improve their inherent resistance to creep deformation. Nevertheless, creep remains one of the major life-limiting factors in the design of turbines because, in the interest of high efficiency, engineers will increase the operating temperature until creep and turbine blade replacement is a problem.

There is, therefore, a strong interest in the development of alternative high-temperature materials such as intermetallic titanium aluminide alloys,

Fig. 1.13(c). These alloys have the potential to replace existing aerospace materials for intermediate temperature applications, due to their attractive combinations of light weight and elevated-temperature strength retention at temperatures up to 800°C. However, they are relatively brittle below ~ 650–700°C.

2.5.7 Fatigue

Fatigue is the response of materials to reversing, or cyclic loads. Fatigue occurs as a result of the formation and growth of cracks at stress levels that may be half the tensile strength or less. The initiation of such macrocracks is associated either with reversing slip on crystallographic planes or with crack growth from pre-existing defects such as notches, gas bubble entrapments, precipitates, or inclusions. The cycle-by-cycle accumulation of localized plasticity often, but not always, leads to microcrack nucleation. In most cases, the most intense damage occurs at the surface, and fatigue crack initiation is usually attributed to surface roughening due to surface plasticity phenomena. Fatigue crack initiation is also associated with environmentally induced chemical reactions or chemisorption processes that limit the extent of reversibility of plastic flow during load reversal.

Upon the initiation of fatigue cracks, the remaining life of a structure is controlled by the number of reversals that are required to grow dominant cracks to failure. Such cracks may grow at stresses that are much lower than the bulk yield stress. Fatigue failure may, therefore, occur at stress levels that are much lower than those required for failure under different loading conditions. Hence, it is essential to assess both the initiation and propagation components of fatigue life in the design of engineering structures and components for service under reversed loading conditions.

2.5.8 Environmental Effects

In some environments, chemical species can initiate or accelerate the crack initiation and growth processes. This results in stress-assisted processes that are commonly referred to as stress corrosion cracking or corrosion fatigue. Stress corrosion cracking may occur at stresses that are much lower than those required for fracture in less aggressive environments. It may occur as a result of hydrogen embrittlement (attacking or redirecting bonds by the diffusion of hydrogen atoms into the regions of high stress concentration ahead of a crack tip) or anodic dissolution processes. In both cases, the useful life and limit loads that can be applied to a structure can be drastically reduced. Stress corrosion cracking may also occur in conjunction with fatigue or creep loading. Furthermore, the irradiation damage induced by particles in nuclear reactors must be assessed for safe use of such reactors.

2.5.9 Creep Crack Growth

As discussed in the section on creep deformation, microvoid nucleation may lead to cracking or accelerated creep that terminates creep life. In structures containing pre-existing cracks, creep crack growth may occur by vacancy coalescence ahead of a dominant crack. This occurs primarily as a result of creep processes immediately ahead of the crack tip. Creep crack growth is, therefore, a potential failure mechanism in elevated-temperature structures and components. This is particularly true when the structures or components operate at temperatures above the recrystallization temperature (~0.3–0.5 of the absolute melting temperature). Hence, creep crack growth may occur by void growth and linkage in gas turbines, nuclear reactors, and components of chemical plants that undergo significant exposures to elevated temperature at intermediate or high stresses.

2.6 SUMMARY

A brief introduction to the Miller and Miller–Bravais indicial notation was presented at the start of this chapter. A statistical mechanics framework was then described before introducing the basic concepts of diffusion-controlled and diffusionless phase transformations. Finally, an overview of the mechanical behavior of materials was presented. Mechanical behavior was introduced as the simple response of materials to mechanical loads. Material response to applied loads was also shown to be dependent on temperature and/or environment. Further details on the mechanical behavior of materials will be presented in subsequent chapters along with the mechanics concepts that are needed to acquire a quantitative understanding.

BIBLIOGRAPHY

Altenpohl, D.G., Das, S.K., Kaufman, J.G. and Bridges, R. (1998) Aluminum: Technology, Applications, and Environment: A Profile of a Modern Metal. Minerals, Metals, and Materials Society.

Ashby, M. F. and Jones, D. R. H. (1994) Engineering Materials: An Introduction to Their Properties and Applications. vols 1 and 2. Pergamon Press, New York.

Brooks, C.R. (1996) Principles of Heat Treatment of Plain Carbon and Low Alloy Steels. ASM International, Materials Park, OH.

Chipman, J. (1972) Metall Trans. vol. 3, p. 55.

Feltham, P. and Copley, G. J. (1958) Acta Metall, vol. 6, p. 539.

Honeycombe, R. F. W. and Bhadesia, H. K. D. H. (1995) Steels: Microstructure and Properties. 2nd ed., Edward Arnold, London.

Hull, D. and Bacon, D. J. (1984) Introduction to Dislocations. 3rd ed. Pergamon Press, New York

McClintock, F. A. and Argon, A. S. (1993) Mechanical Behavior of Materials. Tech Books, Fairfax, VA.

Read-Hill, R. E. and Abbaschian, R. (1992) Physical Metallurgy Principles. PWS Kent, Boston, MA.

Shackleford, J.F. (1996) Introduction to Materials Science for Engineers, 5th ed. Prentice-Hall, Upper Saddle River, NJ.

Shewmon, P. (1989) Diffusion in Solids. 2nd ed. ASM International, Cleveland, OH.

Van Vlack, L. H. (1980) The Elements of Materials Science and Engineering. Addison Wesley, Reading, MA.

3

Basic Definitions of Stress and Strain

3.1 INTRODUCTION

The mechanical properties of materials describe their characteristic responses to applied loads and displacements. However, most texts relate the mechanical properties of materials to stresses and strains. It is, therefore, important for the reader to become familiar with the basic definitions of stress and strain before proceeding on to the remaining chapters of this book. However, the well-prepared reader may choose to skip/skim this chapter, and then move on to Chap. 4 in which the fundamentals of elasticity are introduced.

The basic definitions of stress and strain are presented in this chapter along with experimental methods for the measurement and application of strain and stress. The chapter starts with the relationships between applied loads/displacements and geometry that give rise to the basic definitions of strain and stress. Simple experimental methods for the measurement of strain and stress are then presented before describing the test machines that are often used for the application of strain and stress in the laboratory.

3.2 BASIC DEFINITIONS OF STRESS

The forces applied to the surface of a body may be resolved into components that are perpendicular or parallel to the surface, Figs 3.1(a)–3.1(c). In

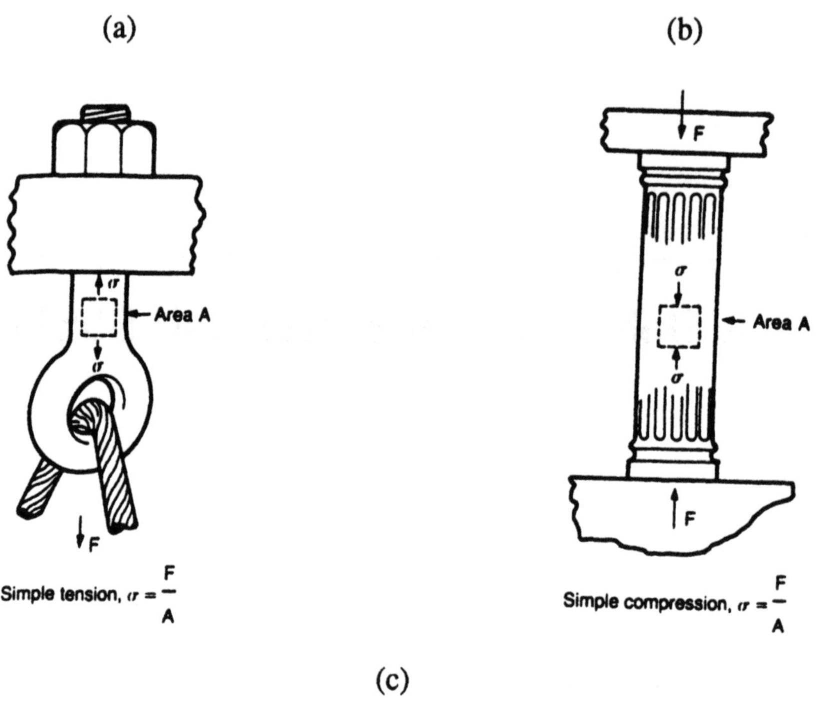

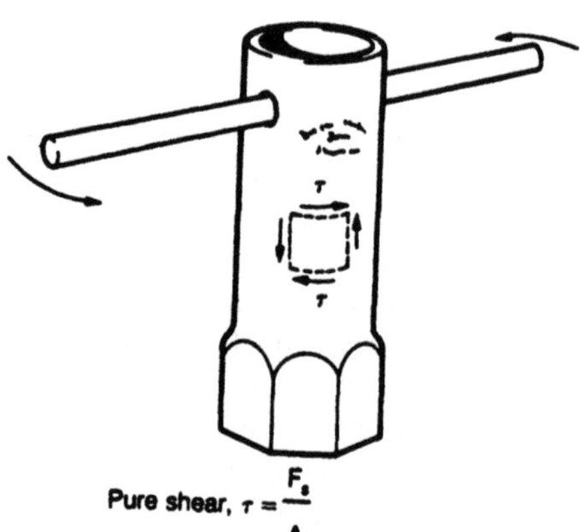

Figure 3.1 Different types of stress: (a)] uniaxial tension; (b) uniaxial compression; (c) twisting moment. (After Ashby and Jones, 1996. Courtesy of Butterworth Heinemann.)

cases where uniform forces are applied in a direction that is perpendicular to the surface, i.e., along the direction normal to the surface, Figs 3.1(a) and 3.1(b), we can define a uniaxial stress, σ, in terms of the normal axial load, P_n, divided by the cross-sectional area, A. This gives

$$\sigma = \frac{\text{Applied load (normal to surface)}}{\text{cross-sectional area}} = \frac{P_n}{A} \tag{3.1}$$

It is also apparent from the above expression that stress has SI units of newtons per square meter (N/m^2) or pascals (Pa)]. Some older texts and most engineering reports in the U.S.A. may also use the old English units of pounds per square inch (psi) to represent stress. In any case, uniaxial stress may be positive or negative, depending on the direction of applied load Figs 3.1(a) and 3.1(b). When the applied load is such that it tends to stretch the atoms within a solid element, the sign convention dictates that the stress is positive or tensile, Fig. 3.1(a)]. Conversely, when the applied load is such that it tends to compress the atoms within a solid element, the uniaxial stress is negative or compressive, Fig. 3.1(b). Hence, the uniaxial stress may be positive (tensile) or negative (compressive), depending on the direction of the applied load with respect to the solid element that is being deformed.

Similarly, the effects of twisting [Fig. 3.1(c)] on a given area can be characterized by shear stress, which is often denoted by the Greek letter, τ, and is given by:

$$\tau = \frac{\text{Applied load (parallel to surface)}}{\text{cross-sectional area}} = \frac{P_s}{A} \tag{3.2}$$

Shear stress also has units of newtons per square meter square ($N/m^{2)}$ or pounds per square inch (psi). It is induced by torque or twisting moments that result in applied loads that are parallel to a deformed area of solid, Fig. 3.1(c). The above definitions of tensile and shear stress apply only to cases where the cross-sectional areas are uniform.

More rigorous definitions are needed to describe the stress and strain when the cross-sectional areas are not uniform. Under such circumstances, it is usual to define uniaxial and shear stresses with respect to infinitesimally small elements, as shown in Fig 3.1. The uniaxial stresses can then be defined as the limits of the following expressions, as the sizes, dA, of the elements tend towards zero:

$$\sigma = \lim_{dA \to 0} \left(\frac{P_n}{dA} \right) \tag{3.3}$$

and

$$\tau = \lim_{dA \to 0} \left(\frac{P_s}{dA} \right) \tag{3.4}$$

where P_n and P_s are the respective normal and shear loads, and dA is the area of the infinitesimally small element. The above terms are illustrated schematically in Figs 3.1(a) and 3.1(c), with F being equivalent to P.

Unlike force, stress is not a vector quantity that can be described simply by its magnitude and direction. Instead, the general definition of stress requires the specification of a direction normal to an area element, and a direction parallel to the applied force. *Stress* is, therefore, a *second rank tensor* quantity, which generally requires the specification of *two direction normals*. An introduction to tensor notation will be provided in Chap. 4. However, for now, the reader may think of the stress tensor as a matrix that contains all the possible components of stress on an element. This concept will become clearer as we proceed in this chapter.

The state of stress on a small element may be represented by orthogonal stress components within a Cartesian co-ordinate framework (Fig. 3.2). Note that there are nine stress components on the orthogonal faces of

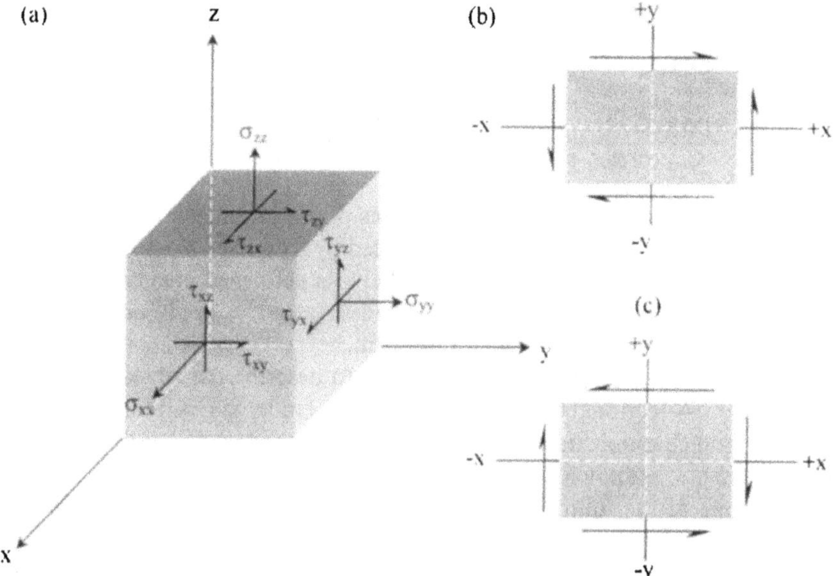

Figure 3.2 (a) States of Stress on an Element, (b) positive shear stress and (c) Negative shear stress. Courtesy of Dr. Seyed M. Allameh.

the cube shown in Fig. 3.2. Hence, a 3×3 matrix may be used to describe all the possible uniaxial and shear stresses that can act on an element. The reader should note that a special sign convention is used to determine the suffixes in Fig. 3.2. The first suffix, i, in the σ_{ij} or τ_{ij} terms corresponds to the direction of normal to the plane, while the second suffix, j, corresponds to the direction of the force. Furthermore, when both directions are positive or negative, the stress term is positive. Similarly, when the direction of the load is opposite to the direction of the plane normal, the stress term is negative.

We may now describe the complete stress tensor for a generalized three-dimensional stress state as

$$[\sigma] = \begin{bmatrix} \sigma_{xx} & \tau_{xy} & \tau_{xz} \\ \tau_{yx} & \sigma_{yy} & \tau_{yz} \\ \tau_{zx} & \tau_{zy} & \sigma_{zz} \end{bmatrix} \tag{3.5}$$

Note that the above matrix, Eq. (3.5), contains only six independent terms since $\tau_{ij} = \tau_{ji}$ for moment equilibrium. The generalized state of stress at a point can, therefore, be described by three uniaxial stress terms (σ_{xx}, σ_{yy}, σ_{zz}) and three shear stress terms (τ_{xy}, τ_{yz}, τ_{zx}). The uniaxial and shear stresses may also be defined for any three orthogonal axes in the Cartesian co-ordinate system. Similarly, cylindrical (r, θ, L) and spherical (r, θ, L) co-ordinates may be used to describe the generalized state of stress on an element.

In the case of a cylindrical co-ordinate system, the stress tensor is given by

$$[\sigma] = \begin{bmatrix} \sigma_{rr} & \tau_{r\theta} & \tau_{rL} \\ \tau_{\theta r} & \sigma_{\theta\theta} & \tau_{\theta L} \\ \tau_{Lr} & \tau_{L\theta} & \sigma_{LL} \end{bmatrix} \tag{3.6}$$

Similarly, for a spherical co-ordinate system, the stress tensor is given by

$$[\sigma] = \begin{bmatrix} \sigma_{rr} & \tau_{r\theta} & \tau_{r\phi} \\ \tau_{\theta r} & \sigma_{\theta\theta} & \tau_{\theta\phi} \\ \tau_{\phi r} & \tau_{\phi\theta} & \sigma_{\phi\phi} \end{bmatrix} \tag{3.7}$$

It should be apparent from the above discussion that the generalized three-dimensional states of stress on an element may be described by any of the above co-ordinate systems. In general, however, the choice of co-ordinate system depends on the geometry of the solid that is being analyzed. Hence, the analysis of a cylindrical solid will often utilize a cylindrical co-ordinate system, while the analysis of a spherical solid will generally be done within a spherical co-ordinate framework.

In all the above co-ordinate systems, six independent stress components are required to fully describe the state of stress on an element. Luckily, most problems in engineering involve simple uniaxial or shear states [Fig. 3.1). Hence, many of the components of the above stress tensors are often equal to zero. This simplifies the computational effort that is needed for the calculation of stresses and strains in many practical problems. Nevertheless, the reader should retain a picture of the generalized state of stress on an element, as we develop the basic concepts of mechanical properties in the subsequent chapters of this book. We will now turn our attention to the basic definitions of strain.

3.3 BASIC DEFINITIONS OF STRAIN

Applied loads or displacements result in changes in the dimensions or shape of a solid. For the simple case of a uniaxial displacement of a solid with a uniform cross-sectional area [Fig. 3.3(a)], the axial strain, ε, is can be defined simply as the ratio of the change in length, u, to the original length, l. This is given by [Fig. 3.3(a)]:

$$\varepsilon = u/l \tag{3.8}$$

Note that uniaxial strain is a dimensionless quantity since it represents the ratio of two length terms. Furthermore, strain as described by Eq. (3.8), is often referred to as the *engineering strain*. It assumes that a uniform displacement occurs across the gauge length [Fig. 3.3(a)]. However, it does not account for the incremental nature of displacement during the deformation process. Nevertheless, the engineering strain is generally satisfactory for most engineering purposes.

Similarly, for small displacements, a shear strain, γ, can be defined as the angular displacement induced by an applied shear stress. The shear strain, γ, is given by

$$\gamma = w/l = \tan\theta \tag{3.9}$$

where γ, w, l and θ are shown schematically in Fig. 3.3(b). The angle γ has units of radians. However, the shear strain is generally presented as a dimensionless quantity.

It is important to note here that the above equations for the engineering strain assume that the stresses are uniform across the area elements or uniform cross-sections that are being deformed. The engineering shear and axial strains must be distinguished from the so-called "true strains" which will be described in Chap. 5.

Similar to stress, the engineering strain may have three uniaxial (ε_{xx}, ε_{yy}, ε_{zz}) and shear (γ_{xy}, γ_{yz}, γ_{zx}) components, Fig. 3.4(a). The three-dimen-

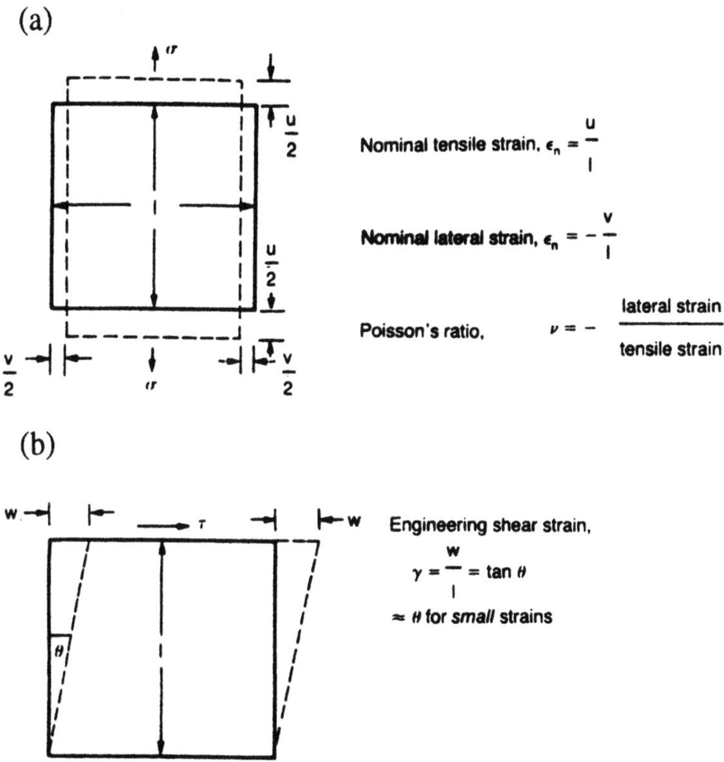

FIGURE 3.3 Definitions of strain: (a)] uniaxial strain; (b) shear strain. (After Ashby and Jones, 1996. Courtesy of Butterworth-Heinemann.)

sional strain components may also be perceived in terms of the simple definitions or axial and shear strains presented earlier (Fig. 3.3). However, the same shape change may also be resolved as an axial or shear strain, depending on the choice of co-ordinate system. Also, note that the displacement vectors along the (x, y, z) axes are usually described by displacement co-ordinates (u, v, w). The uniaxial strain, ε_{xx}, due to displacement gradient in the x direction is given by

$$\varepsilon_{xx} = \frac{\left[u + \left(\frac{\partial u}{\partial x}\right)\mathrm{d}x\right] - u}{\mathrm{d}x} = \frac{\partial u}{\partial x} \tag{3.10}$$

Similarly, the shear strain due to relative displacement gradient in the y-direction is given by Fig. 3.4 to be:

$$\varepsilon_{xy} = \frac{\left[v + \left(\dfrac{\partial v}{\partial x}\right)dx\right] - v}{dx} = \frac{\partial v}{\partial x} \tag{3.11}$$

It should be clear from the above equations that nine strain components can be defined for a generalized state of deformation at a point. These can be presented in the following strain matrix:

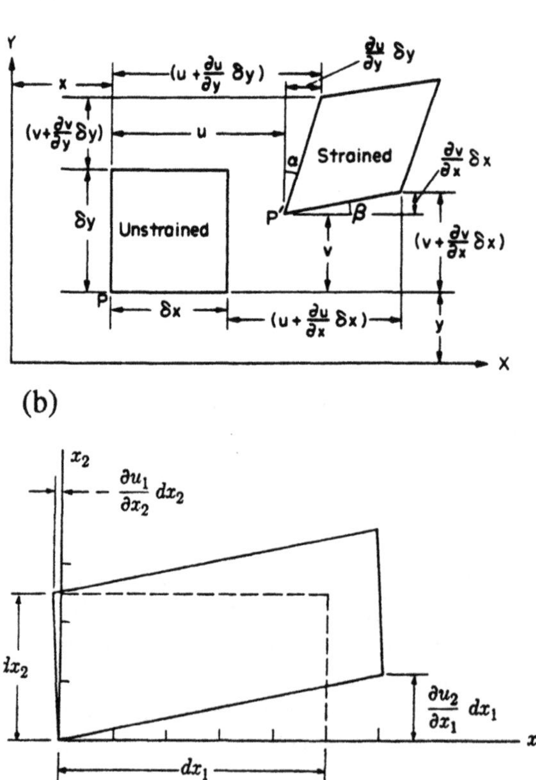

FIGURE 3.4 Definitions of strain and rotation: (a) components of strain; (b) rotation about the *x–y* plane. (After Hearn, 1985—courtesy of Elsevier Science.)

$$[\varepsilon] = \begin{bmatrix} \varepsilon_{xx} & \varepsilon_{xy} & \varepsilon_{xz} \\ \varepsilon_{yx} & \varepsilon_{yy} & \varepsilon_{yz} \\ \varepsilon_{zx} & \varepsilon_{zy} & \varepsilon_{zz} \end{bmatrix} = \begin{bmatrix} \frac{\partial u}{\partial x} & \frac{\partial u}{\partial y} & \frac{\partial u}{\partial z} \\ \frac{\partial v}{\partial x} & \frac{\partial v}{\partial y} & \frac{\partial v}{\partial z} \\ \frac{\partial w}{\partial x} & \frac{\partial w}{\partial y} & \frac{\partial w}{\partial z} \end{bmatrix} \tag{3.12}$$

Note that some texts may use the transposed version of the displacement gradient matrix given above. If this is done, the transposed versions should be maintained to obtain the results of the strain matrix that will be presented subsequently. Also, the above form of the displacement gradient strain matrix is often avoided in problems where strain can be induced as a result of rotation without a stress. This is because we are often concerned with strains induced as a result of applied stresses. Hence, for several problems involving stress-induced deformation, we subtract out the rotation terms to obtain relative displacements that describe the local changes in the shape of the body, Fig. 3.4(b).

The rotation strains may be obtained by considering the possible rotations about any of the three orthogonal axes in a Cartesian co-ordinate system. For simplicity, let us start by considering the special case of deformation by rotation about the z axis, i.e., deformation in the x–y plane. This is illustrated schematically in Fig. 3.4(b). The average rotation in the x–y plane is given by

$$\omega_{xy} = \frac{1}{2}\left(\frac{\partial v}{\partial x} - \frac{\partial u}{\partial y}\right) \tag{3.13}$$

Similarly, we may obtain expressions for ω_{yz} and ω_{zx} by cyclic permutations of the x, y and z position terms and subscripts, and the corresponding (u, v, w) displacement terms. This yields:

$$\omega_{yz} = \frac{1}{2}\left(\frac{\partial w}{\partial y} - \frac{\partial v}{\partial z}\right) \tag{3.14}$$

and

$$\omega_{zx} = \frac{1}{2}\left(\frac{\partial u}{\partial z} - \frac{\partial w}{\partial x}\right) \tag{3.15}$$

The components of the rotation matrix can thus be expressed in the following matrix form:

$$[\omega_{ij}]=\begin{bmatrix} 0 & \omega_{xy} & \omega_{xz} \\ \omega_{yx} & 0 & \omega_{yz} \\ \omega_{zx} & \omega_{zy} & 0 \end{bmatrix}=\begin{bmatrix} 0 & \frac{1}{2}\left(\frac{\partial u}{\partial y}-\frac{\partial v}{\partial x}\right) & \frac{1}{2}\left(\frac{\partial u}{\partial z}-\frac{\partial w}{\partial x}\right) \\ \frac{1}{2}\left(\frac{\partial v}{\partial x}-\frac{\partial u}{\partial y}\right) & 0 & \frac{1}{2}\left(\frac{\partial v}{\partial z}-\frac{\partial w}{\partial y}\right) \\ \frac{1}{2}\left(\frac{\partial w}{\partial x}-\frac{\partial u}{\partial z}\right) & \frac{1}{2}\left(\frac{\partial w}{\partial y}-\frac{\partial v}{\partial z}\right) & 0 \end{bmatrix} \tag{3.16}$$

Subtracting Eq. (3.16) from Eq. (3.12) yields the following matrix for the shape changes:

$$[\varepsilon_{ij}]=\begin{bmatrix} \varepsilon_{xx} & \varepsilon_{xy} & \varepsilon_{xz} \\ \varepsilon_{yx} & \varepsilon_{yy} & \varepsilon_{yz} \\ \varepsilon_{zx} & \varepsilon_{zy} & \varepsilon_{zz} \end{bmatrix}=\begin{bmatrix} \frac{\partial u}{\partial x} & \frac{1}{2}\left(\frac{\partial u}{\partial y}+\frac{\partial v}{\partial x}\right) & \frac{1}{2}\left(\frac{\partial u}{\partial z}+\frac{\partial w}{\partial x}\right) \\ \frac{1}{2}\left(\frac{\partial v}{\partial x}+\frac{\partial u}{\partial y}\right) & \frac{\partial v}{\partial y} & \frac{1}{2}\left(\frac{\partial v}{\partial z}+\frac{\partial w}{\partial y}\right) \\ \frac{1}{2}\left(\frac{\partial w}{\partial x}+\frac{\partial u}{\partial z}\right) & \frac{1}{2}\left(\frac{\partial w}{\partial y}+\frac{\partial v}{\partial z}\right) & \frac{\partial w}{\partial z} \end{bmatrix} \tag{3.17}$$

Note that the sign convention is similar to that described earlier for stress. The first suffix in the ε_{ij} term corresponds to the direction of the normal to the plane, while the second suffix corresponds to the direction of the displacement induced by the applied strain. Similarly, three shear strains are the strain components with the mixed suffixes, i.e., ε_{xy}, ε_{yz}, and ε_{zx}. It is also important to recognize the patterns in subscripts (x, y, z) and the displacements (u, v, w) in Eqs (3.16) and (3.17). This makes it easier to remember the expressions for the possible components of strain on a three-dimensional element. Note also that the factor of 1/2 in Eq. (3.17) is often not included in several engineering problems where only a few strain components are applied. The tensorial strains (ε_{ij} terms) are then replaced by corresponding tangential shear strain terms, γ_{ij} which are given by

$$\gamma_{ij}=2\varepsilon_{ij} \tag{3.18}$$

The strain matrix for stress-induced displacements is thus given by

$$[\gamma_{ij}]=\begin{bmatrix} 0 & \gamma_{xy} & \gamma_{xz} \\ \gamma_{yx} & 0 & \gamma_{yz} \\ \gamma_{zx} & \gamma_{zy} & 0 \end{bmatrix}=\begin{bmatrix} 0 & \frac{\partial u}{\partial y}+\frac{\partial v}{\partial x} & \frac{\partial u}{\partial z}+\frac{\partial w}{\partial x} \\ \frac{\partial v}{\partial x}+\frac{\partial u}{\partial y} & 0 & \frac{\partial v}{\partial z}+\frac{\partial w}{\partial y} \\ \frac{\partial w}{\partial x}+\frac{\partial u}{\partial z} & \frac{\partial w}{\partial y}+\frac{\partial v}{\partial z} & 0 \end{bmatrix} \tag{3.19}$$

The above shear strain components are often important in problems involving plastic flow. Similarly, the volumetric strains are important in brittle fracture problems where atomic separation can occur by bond rupture due to the effects of axial strains. For small strains, the volumetric strain (Fig. 3.5) is given by the sum of the axial strains:

$$\varepsilon_V = \frac{\Delta V}{V} = \varepsilon_{xx} + \varepsilon_{yy} + \varepsilon_{zz} \tag{3.20}$$

The above definitions of strain apply to cases where the displacements are relatively small. More accurate strain formulations, e.g., large-strain Lagrangian formulations, may be needed when the strains are larger. The reader is referred to standard texts on plasticity and experimental mechanics for further details on these formulations.

Finally in this section, it is important to note that stress-free thermal strains may also be induced as a result of the thermal expansion or thermal contraction that can occur, respectively, on heating or cooling to or from a reference temperature. Under such conditions, the thermal strains are given by

$$\Delta\varepsilon_i = \alpha_i \Delta T = \alpha_i (T - T_0) \tag{3.21}$$

where $\Delta\varepsilon_i$ is the thermal strain along an axis i, T is the actual temperature of the solid, and T_0 is a reference stress-free temperature. Thermal strains are particularly important in problems involving surface contact between two materials with different thermal expansion coefficients. When the thermal expansion coefficient mismatch between the two materials in contact is large, large strains/stresses can be induced at the interfaces between the two faces. The mismatch thermal strains can also result in internal residual strains/stresses that are retained in the material on cooling from elevated

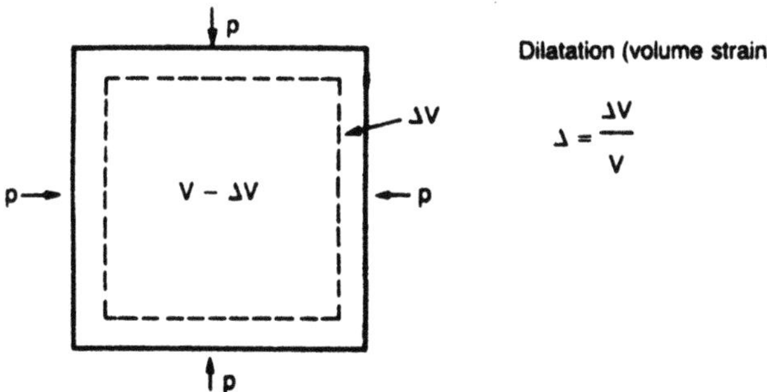

Figure 3.5 Schematic illustration of volumetric strain. (After Ashby and Jones, 1996. Courtesy of Butterworth-Heinemann.)

temperature. Under such conditions, the residual stresses, σ_i, can be estimated from expressions of the form:

$$\sigma_i \sim E\Delta\alpha\Delta T = E_i(\alpha_1 - \alpha_2)(T - T_0) \tag{3.22}$$

where E_i is the Young's modulus in the i direction, α is the thermal expansion coefficient along the direction i, subscripts 1 and 2 denote the two materials in contact, T is the actual/current temperature, and T_0 is the reference stress-free temperature below which residual stresses can build up. Above this temperature, residual stresses are relaxed by flow processes.

Interfacial residual stress considerations are particularly important in the design of composite materials. This is because of the large differences that are typically observed between the thermal expansion coefficients of different materials. Composites must, therefore, be engineered to minimize the thermal residual strains/stresses. Failure to do so may result in cracking if the residual stress levels are sufficiently large. Interfacial residual stress levels may be controlled in composites by the careful selection of composite constituents that have similar thermal expansion coefficients. However, this is often impossible in the real world. It is, therefore, more common for scientists and engineers to control the interfacial properties of composites by the careful engineering of interfacial dimensions and interfacial phases to minimize the levels of interfacial residual stress in different directions.

3.4 MOHR'S CIRCLE OF STRESS AND STRAIN

Let us now consider the simple case of a two-dimensional stress state on an element in a bar of uniform rectangular cross sectional area subjected to uniaxial tension, [Fig. 3.6]. If we now take a slice across the element at an angle, θ, the normal and shear forces on the inclined plane can be resolved using standard force balance and basic trigonometry. The dependence of the stress components on the plane angle, θ, was first recognized by Oligo Mohr. He showed that the stresses, $\sigma_{x'x'}$, $\sigma_{y'y'}$, $\tau_{x'y'}$ along the inclined plane are given by the following expressions:

$$\sigma_{x'x'} = \frac{\sigma_{xx} + \sigma_{yy}}{2} + \left(\frac{\sigma_{xx} - \sigma_{yy}}{2}\right)\cos 2\theta \tag{3.23a}$$

$$\sigma_{y'y'} = \frac{\sigma_{xx} + \sigma_{yy}}{2} + \left(\frac{\sigma_{xx} - \sigma_{yy}}{2}\right)\cos 2\theta \tag{3.23b}$$

$$\tau_{x'y'} = -\left(\frac{\sigma_{xx} - \sigma_{yy}}{2}\right)\sin 2\theta \tag{3.23c}$$

where θ, σ_{xx}, σ_{yy}, τ_{xy}, $\sigma_{x'x'}$, $\sigma_{y'y'}$, and $\tau_{x'y'}$ are stresses shown in Fig. 3.6. The above equations can be represented graphically in the so-called Mohr's circle (Fig. 3.7) which has radius, R, and center, C, given by

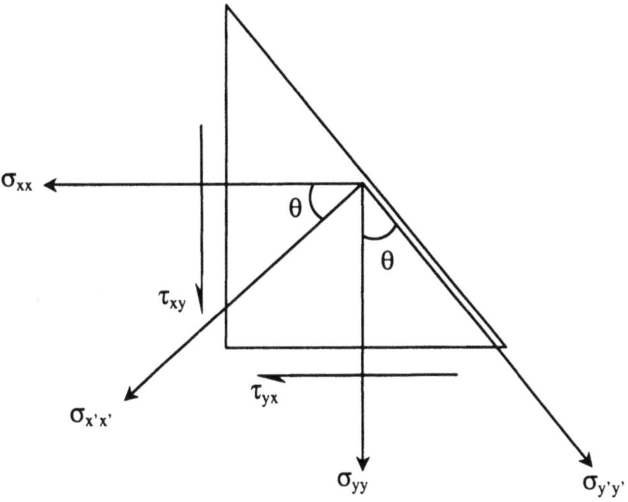

Figure 3.6 Schematic of stresses on a plane inclined across an element. Courtesy of Dr. Seyed M. Allameh

$$R = \sqrt{\left(\frac{\sigma_{xx} - \sigma_{yy}}{2}\right)^2 + \tau_{xy}^2} \tag{3.24}$$

and

$$C = \frac{\sigma_{xx} + \sigma_{yy}}{2} \tag{3.25}$$

Note the sign convention that is used to describe the plane angle, 2θ, and the tensile and shear stress components in Fig. 3.7. It is important to remember this sign convention when solving problems involving the use of Mohr's circle. Failure to do so may result in the wrong signs or magnitudes of stresses. The actual construction of the Mohr's circle is a relatively simple process once the magnitudes of the radius, R, and center position, C, have been computed using Eqs (3.24) and (3.25), respectively. Note that the locus of the circle describes all the possible states of stress on the element at the point, P, for various values of θ between 0° and 180°. It is also important to note that several combinations of the stress components (σ_{xx}, σ_{yy}, τ_{xy}) may result in yielding, as the plane angle, θ, is varied. These combinations will be discussed in Chap. 5.

When a generalized state of triaxial stress occurs, three Mohr's circles [Fig. 3.8(a)] may be drawn to describe all the possible states of stress. These circles can be constructed easily once the principal stresses, σ_1, σ_2, and σ_3,

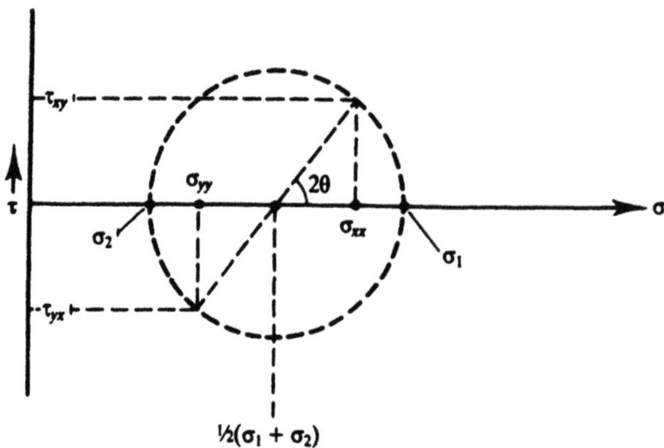

FIGURE 3.7 Mohr's circle of stress—note the sign convention. (After Courtney, 1990. Courtesy of McGraw-Hill.)

are known. The principal stresses σ_1, σ_2, and σ_3 are usually arranged in increasing algebraic order, with $\sigma_1 > \sigma_2 > \sigma_3$.

Similarly, principal strains may be determined from graphical plots of shear strain versus axial strain. However, the tensorial strain components, $\gamma_{xy}/2$, must be plotted on the ordinates of such plots, Fig. 3.8(b). Otherwise, the procedures for the determination of principal strains are the same as those described above for principal stresses.

3.5 COMPUTATION OF PRINCIPAL STRESSES AND PRINCIPAL STRAINS

Although principal stresses and strains may be determined using Mohr's circle, it is more common to compute them using some standard polynomial expressions. It is important to remember that the same form of equations may be used to calculate principal stresses and strains. However, the shear strains must be represented as $\gamma_{xy}/2$ when the polynomial equations are used in the determination of principal strains. Nevertheless, to avoid repetition, the current discussion will focus on the equations for the computation of principal stresses, with the implicit understanding that the same form of equations can be used for the calculation of principal strains. Principal stresses may be determined by solving polynomial equations of the form:

$$\sigma_i^3 - I_1\sigma_i^2 - I_2\sigma_i - I_3 = 0 \tag{3.26}$$

(a)

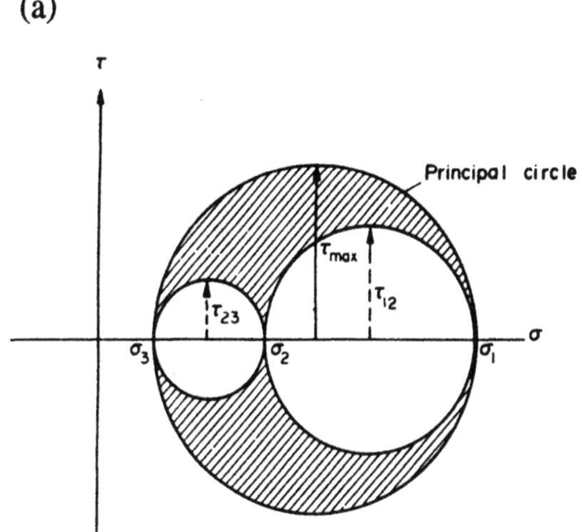

(b)

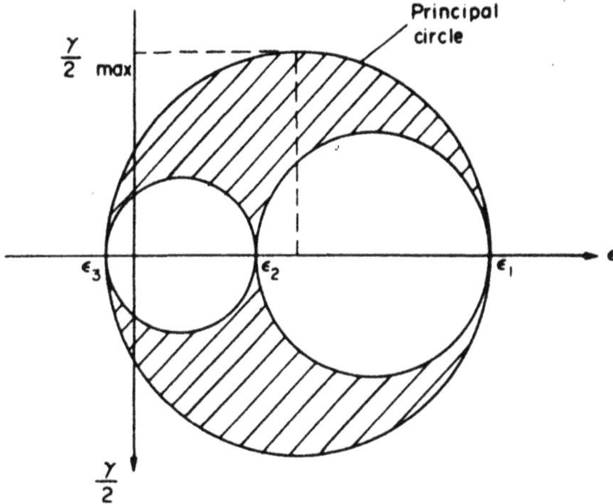

Figure 3.8 Mohr's circle representation of generalized three-dimensional states of (a)] stress and (b) strain. (After Hearn, 1985. Courtesy of Elsevier Science.)

where σ_i represents the total stress on an element, and I_1, I_2, and I_3 correspond to the first, second, and third invariants of the stress tensor. These stress invariants do not vary with the choice of orthogonal axes (x, y, z). Also, the values of the stress invariants (I_1, I_2 and I_3) can be computed from the second rank stress tensor, σ_{ij}, presented earlier. This is given by

$$\sigma_{ij} = \begin{bmatrix} \sigma_{xx} & \tau_{xy} & \tau_{xz} \\ \tau_{yx} & \sigma_{yy} & \tau_{yz} \\ \tau_{zx} & \tau_{zy} & \sigma_{zz} \end{bmatrix} \tag{3.27}$$

The first invariant of the stress tensor, I_1, is given by the sum of the leading diagonal terms in the stress tensor. Hence, I_1 can be determined from:

$$I_1 = \sigma_{xx} + \sigma_{yy} + \sigma_{zz} \tag{3.28}$$

Similarly, I_2, the second invariant of the stress tensor can be obtained from the algebraic sum of the cofactors of the three terms in any of the three rows or columns of the stress tensor. This gives the same value of I_2, which may also be computed from

$$I_2 = -\sigma_{xx}\sigma_{yy} - \sigma_{yy}\sigma_{zz} - \sigma_{zz}\sigma_{xx} + \tau_{xy}^2 + \tau_{yz}^2 + \tau_{zx}^2 \tag{3.29}$$

Note the rotational symmetry in the above equation, i.e., xy is followed by yz, which is followed by zx. It should be easy to remember the equation for I_2 once you recognize the pattern.

When the above equation for I_3 is expanded and simplified, it can be shown that I_3 is given simply by

$$I_3 = \sigma_{xx}\sigma_{yy}\sigma_{zz} + 2\tau_{xy}\tau_{yz}\tau_{zx} - \sigma_{xx}\tau_{yz}^2 - \sigma_{yy}\tau_{zx}^2 - \sigma_{zz}\tau_{xy}^2 \tag{3.30}$$

It is important to note here that the above coefficients I_1, I_2 and I_3 can be obtained by solving the following eigenvalue problem:

$$\sigma_{ij}x = \sigma x \tag{3.31}$$

where σ_{ij} is the stress tensor, x is the eigenvector of σ_{ij} and σ is the corresponding eigenvalue, i.e. the principal stress. Rearranging Equation 3.31 now gives

$$(\sigma_{ij} - \sigma\, \mathrm{I})x = 0 \tag{3.32}$$

where I is the identity matrix and the other terms have their usual meaning. The non-trivial solution to Equation 3.32 is given by Equation

$$\det(\sigma_{ij} - \lambda\sigma) = 0 \tag{3.33a}$$

or

$$\begin{bmatrix} \sigma_{xx} - \sigma & \tau_{xy} & \tau_{xz} \\ \tau_{yx} & \sigma_{yy} - \sigma & \tau_{yz} \\ \tau_{zx} & \tau_{zy} & \sigma_{zz} - \sigma \end{bmatrix} = 0 \tag{3.33b}$$

Writing out the terms of the determinant given by Equation 3.33b gives the characteristic equation, which corresponds to the polynomial expression presented in Equation 3.26.

Once again, it is important to note the rotational symmetry of the x, y, z subscripts in the above equation. The equation for I_3 is relatively easy to remember once the rotational symmetry in the (x, y, z) terms is recognized.

Once the values of I_1, I_2, and I_3 are known, the three principal stresses can be determined by solving Eq. (3.26) to find the values of σ_i for which $\sigma_1 = 0$. The three solutions are the three principal stresses. They can then be ranked algebraically to determine the solutions that correspond to σ_1, σ_2, and σ_3, respectively. Once these stresses are determined, it is relatively easy to construct the Mohr's circle for a three-dimensional state of stress, as shown in Fig. 3.8(a). It is important to note that all the possible states of stress on an element are represented by the shaded area in this figure. The three principal shear stress values can also be deduced from Fig. 3.8(a). Finally, in this section, it is important to note that the Mohr's circle for pure hydrostatic state of stress ($\sigma = \sigma_{xx} = \sigma_{yy} = \sigma_{xx}$) reduces to a point. The reader should verify that this is indeed the case before proceeding on to the next section.

3.6 HYDROSTATIC AND DEVIATORIC STRESS COMPONENTS

The components of stress at a point, σ_i may be separated into hydrostatic, σ_h, and deviatoric, σ_d, stress components, i.e., $\sigma_i = \sigma_h + \sigma_d$. The hydrostatic stress represents the average of the uniaxial stresses along three orthogonal axes. It is very important in brittle fracture processes where failure may occur without shear. This is because the axial stresses are most likely to cause separation of bonds, in the absence of shear stress components that may induce plastic flow. In any case, the hydrostatic stress can be calculated from the first invariant of the stress tensor. This gives

$$\sigma_h = \frac{I_1}{3} = \frac{\sigma_{xx} + \sigma_{yy} + \sigma_{zz}}{3} \tag{3.34}$$

Hence, the hydrostatic stress is equal to the average of the leading diagonal terms in the stress tensor, Eqs (3.5) and (3.34). A state of pure hydrostatic stress is experienced by a fish, at rest in water. This is illustrated in Fig. 3.9. Note that the hydrostatic stress occurs as a result of the average water pressure exerted on the fish. This pressure is the same for any given choice of orthogonal stress axes. The hydrostatic stress tensor, σ''_{ij}, is given by

$$[\sigma''_{ij}] = \begin{bmatrix} \sigma & 0 & 0 \\ 0 & \sigma & 0 \\ 0 & 0 & \sigma \end{bmatrix} \tag{3.35}$$

where $\sigma = \sigma_{xx} = \sigma_{yy} = \sigma_{zz}$. It is important to realize that there are no shear stress components in the hydrostatic stress tensor. However, since most stress states consist of both axial and shear stress components, the generalized three-dimensional state of stress will, therefore, consist of both hydrostatic and deviatoric stress components.

The deviatoric stresses are particularly important because they tend to cause plasticity to occur in ductile solids. Deviatoric stress components, σ'_{ij}, may be represented by the difference between the complete stress tensor, σ_{ij}, and the hydrostatic stress tensor, σ''_{ij}. Hence, $\sigma'_{ij} = \sigma_{ij} - \sigma''_{ij}$. The deviatoric stress tensor, σ'_{ij}, is therefore given by

$$\sigma'_{ij} = \sigma_{ij} - \sigma''_{ij} = \begin{bmatrix} \sigma_{xx} & \tau_{xy} & \tau_{xz} \\ \tau_{yz} & \sigma_{yy} & \tau_{yz} \\ \tau_{zx} & \tau_{zy} & \sigma_{zz} \end{bmatrix} - \begin{bmatrix} \frac{I_1}{3} & 0 & 0 \\ 0 & \frac{I_1}{3} & 0 \\ 0 & 0 & \frac{I_1}{3} \end{bmatrix} \tag{3.36a}$$

or

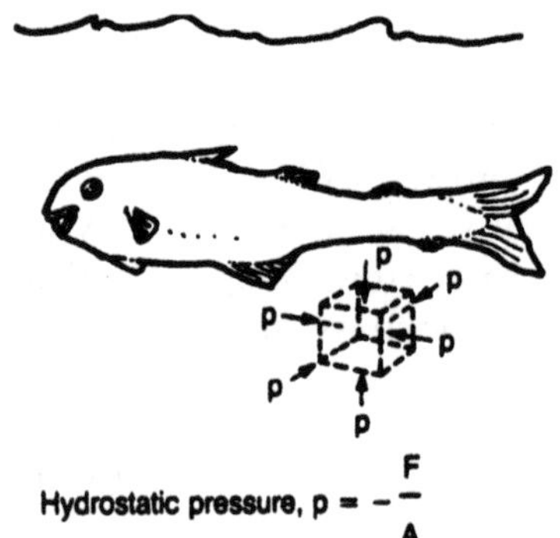

FIGURE 3.9 Hydrostatic stress on fish at rest in water (After Ashby and Jones, 1996. Courtesy of Butterworth-Heinemann.)

$$\sigma'_{ij} = \begin{bmatrix} \sigma_{xx} - \frac{I_1}{3} & \tau_{xy} & \tau_{xz} \\ \tau_{yx} & \sigma_{yy} - \frac{I_1}{3} & \tau_{yz} \\ \tau_{zx} & \tau_{zy} & \sigma_{zz} - \frac{I_1}{3} \end{bmatrix} \tag{3.36b}$$

The maximum values of the deviatoric stresses (principal deviatoric stresses) may also be computed using a polynomial expression similar to the cubic equation presented earlier (Eq. (3.26) for the determination of the principal stresses. Hence, the three principal deviatoric stresses may be computed from:

$$\sigma'^{3}_{ij} - J_1(\sigma'_{ij})^2 - J_2(\sigma'_{ij}) - J_3 = 0 \tag{3.37}$$

where J_1, J_2, and J_3 are the first, second, and third invariants of the deviatoric stress tensor. As before, J_1 may be determined from the sum of the leading diagonal terms, J_2 from the sum of the cofactors, and J_3 from the determinant of the σ'_{ij} tensor. Upon substitution of the appropriate parameters, it is easy to show that $J_1 = 0$.

It is particularly important to discuss the parameter J_2 since it is often encountered in several problems in plasticity. In fact, the conventional theory of plasticity is often to referred to as the J_2 deformation theory, and plasticity is often observed to occur when J_2 reaches a critical value. As discussed, J_2 can be computed from the sum of the cofactors of any of the rows or columns in the deviatoric stress tensor. If the stress components in the first row of the deviatoric stress tensor are used for this purpose, it can be shown that J_2 is given by

$$J_2 = \begin{vmatrix} \sigma_{yy} - \frac{I_1}{3} & \tau_{yz} \\ \tau_{zy} & \sigma_{zz} - \frac{I_1}{3} \end{vmatrix} - \begin{vmatrix} \tau_{yx} & \tau_{yz} \\ \tau_{zx} & \sigma_{zz} - \frac{I_1}{3} \end{vmatrix} + \begin{vmatrix} \tau_{yx} & \sigma_{yy} - \frac{I_1}{3} \\ \tau_{zx} & \tau_{zy} \end{vmatrix} \tag{3.38}$$

Expanding out the determinants in the above equations, and substituting appropriate expressions for I_1, I_2, and I_3 into the resulting equation, it can be shown that J_2 is given by

$$J_2 = \tfrac{1}{6}\left[(\sigma_{xx} - \sigma_{yy})^2 + (\sigma_{yy} - \sigma_{zz})^2 + (\sigma_{zz} - \sigma_{xx})^2 + 6(\tau^2_{xy} + \tau^2_{yz} + \tau^2_{zx})\right] \tag{3.39a}$$

or

$$J_2 = \tfrac{1}{6}[(\sigma_1 - \sigma_2)^2 + (\sigma_2 - \sigma_3)^2 + (\sigma_3 - \sigma_1)^2] \tag{3.39b}$$

A number of empirical plastic flow rules are based on J_2 deformation theory. In particular, the Von Mises yield criterion suggests that yielding occurs under uniaxial or multiaxial loading conditions when the maximum distortional energy reaches a critical value of J_2. These will be discussed in Chap. 5 along with the fundamentals of plasticity theory.

Finally, in this section, it is important to remember that equations with the same form as the above equations (for stress) may be used to calculate the corresponding hydrostatic and deviatoric strain components. However, as before, the shear strain components must be represented by tensorial strain components, $\gamma_{ij}/2$, in such equations.

3.7 STRAIN MEASUREMENT

It is generally difficult to measure stress directly. However, it is relatively easy to measure strain with electric resistance strain gauges connected to appropriate bridge circuits. It is also possible to obtain measurements of strain from extensometers, grid displacement techniques, Moiré interferometry, and a wide range of other techniques that are beyond the scope of this book. The interested reader is referred to standard texts on experimental strain measurement. However, since most experimental strain measurements are obtained from strain gauges, a basic description of strain gauge measurement techniques is presented in this section. Photoelasticity is also described as one example of a stress measurement technique.

3.7.1 Strain Gauge Measurements

The strain gauge is essentially a length of wire of foil that is attached to a nonconducting substrate. The gauge is bonded to the surface that is being strained. The resistance of the wire, R, is given by

$$R = \frac{\rho l}{A} \tag{3.40}$$

where R is the resistivity, l is the length of the wire, and A is the cross-sectional area. Hence, the resistance of the wire will change when the length of wire changes due to applied strain or stress. The changes in gauge resistance may be expressed as

$$\frac{\Delta R}{R} = k\frac{\Delta l}{l} \tag{3.41}$$

where ΔR is the change in resistance, Δl is the change in length, and k is the gauge factor (usually specified by the strain gauge manufacturer).

Alternatively, since $\Delta l/l$ is equivalent to strain, ε, strain may be estimated from the following expression:

$$\varepsilon = \frac{(\Delta R/R)}{k} \tag{3.42}$$

Since the strain levels in most engineering components are relatively low, sensitive Wheatstone bridge circuits are needed to determine the resistance levels (Fig. 3.10). The conditions required for a galvanometer reading of zero (a balanced bridge circuit) are given by

$$R_1 \times R_3 = R_2 \times R_4 \tag{3.43}$$

A number of bridge configurations may be used to measure strain. A half-bridge wiring consists of one active gauge (the gauge that is being strained) and one dummy gauge (attached to an unstrained material that is similar to the unstrained material). The dummy gauge cancels out the effects of temperature changes that may occur during the strain measurements. Such temperature compensation may significantly improve the accuracy of strain measurement in half-bridge circuits. The other two resistances (R_3 and R_4) in the half-bridge circuit are standard resistors. It is important to note here that quarter-bridge and full-bridge circuits may also be used in practice. A quarter-bridge circuit contains only one active resistance with no dummy gauge for temperature compensation, while four (full) bridge configurations contain four active gauges. Furthermore, the Wheatstone bridge

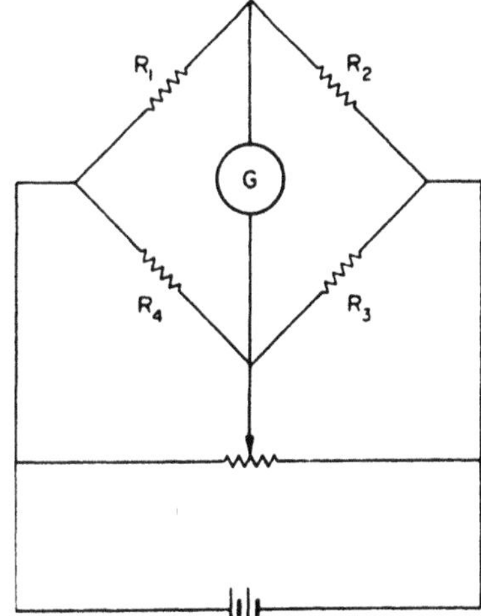

FIGURE 3.10 Wheatstone bridge circuit. (After Hearn, 1985. Courtesy of Elsevier Science.)

may be powered by a direct current or alternating current source. The latter eliminates unwanted noise signals. It also provides a more stable output signal.

3.7.2. Introduction to Photoelasticity

As discussed earlier, it is generally difficult to measure stress. However, the stresses in some transparent materials may be measured using photoelastic techniques. These rely on illumination with plane-polarized light obtained by passing light rays through vertical slots that produce polarized light beams with rays that oscillate only along one vertical plane (Fig. 3.11). When the model is stressed in a direction parallel to the polarizing axis, a fringe pattern is formed against a light (bright field) background. Conversely, when the stress axes are perpendicular to the polarizing axis, a "dark field" or black image is obtained.

In some materials, the application of stress may cause an incident plane-polarized ray to split into two coincident rays with directions that coincide with the directions of the principal axes. Since this phenomenon is only observed during the temporary application of stress, it is known as "temporary birefringence." Furthermore, the speeds of the rays are proportional to the magnitudes of the stresses along the principal directions. Hence, the emerging rays are out of phase. They, therefore, produce interference fringe patterns when they are recombined. If they are recombined at an analyzer (shown in Fig. 3.11), then the amount of interference in the

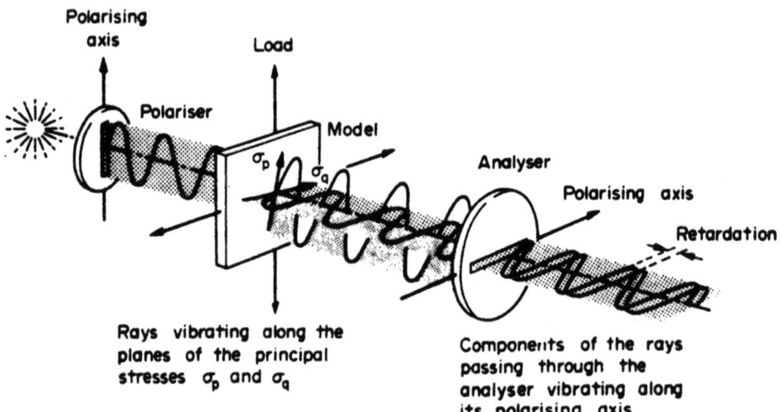

Figure 3.11 Interaction of polarized light with loaded specimen prior to recombination after passing through analyzer. (After Hearn, 1985. Courtesy of Elsevier Science.)

emerging rays is directly proportional to the difference between the principal stress levels σ_p and σ_q. Therefore, the amount of interference is related to the maximum shear, which is given by

$$\tau_{max} = \frac{1}{2}(\sigma_p - \sigma_q) \tag{3.44}$$

The fringe patterns therefore provide a visual indication of the variations in the maximum shear stress. However, in the case of stresses along a free unloaded boundary, one of the principal stresses is zero. The fringe patterns therefore correspond to half of the other principal stress. Quantitative information on local principal/maximum shear stress levels may be obtained from the following expression:

$$\sigma_p - \sigma_q = \frac{nf}{t} \tag{3.45}$$

where σ_p and σ_q are the principal stress levels, f is the material fringe coefficient, n is the fringe number or fringe order at a point, and t is the thickness of the model. The value of f may be determined from a stress calibration experiment in which known values of stress on an element are plotted against the fringe number (at that point) corresponding to various loads.

3.8 MECHANICAL TESTING

Displacements and loads are usually applied to laboratory specimens using closed-loop electromechanical (Fig. 3.12) and servohydraulic (Fig. 3.13) testing machines. Electromechanical testing machines are generally used for simple tests in which loads or displacements are increased at relatively slow rates, while servohydraulic testing machines are used for a wider variety of "slow" or "fast" tests. Both types of testing machines are usually controlled by feedback loops that enable loads or displacements to be applied to test specimens with reasonably high levels of precision. The loads are measured with load cells, which are essentially calibrated springs connected to load transducers. The latter generate electrical signals that are proportional to the applied loads.

Displacements are typically measured with extensometers (Figs 3.12 and 3.13) that are attached to the test specimen. Transducers attached to the extensometer generate voltage changes that are proportional to the relative displacements between the extensometer attachments. Alternatively, the composite displacement of the load train (a combination of the test specimen and all the loading fixtures) may be determined from the so-called

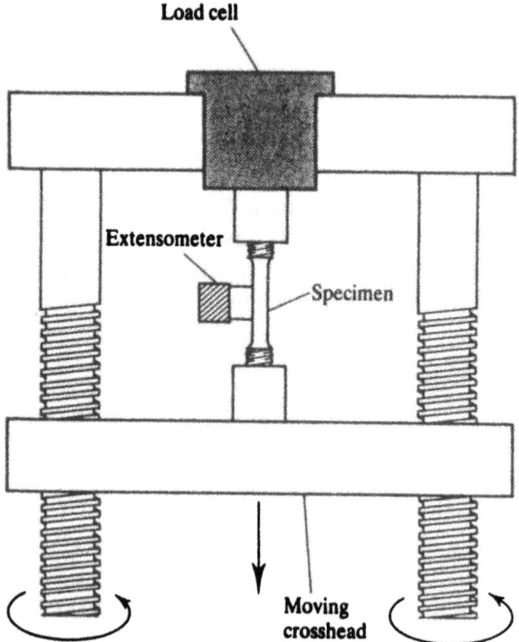

Figure 3.12 Schematic of screw-driven electromechanical testing machine. (After Courtney, 1990. Courtesy of McGraw-Hill.)

stroke reading on a electromechanical or servohydraulic testing machine. However, it is important to remember that the composite stroke reading includes the displacements of all the elements along the load train (Figs 3.12 and 3.13). Such stroke readings may, therefore, not provide a good measure of the displacements within the gauge section of the test specimen.

Electromechanical and servohydraulic test machines may also be controlled under strain control using signals from load cells, extensometers, strain gauges, or other strain transducers. However, it is important to note that strain gauges and extensometers have a resolution limit that is generally between 10^{-3} and 10^{-4}. Higher resolution strain gauges and laser-based techniques can be used to measure strain when better resolution is required. These can measure strains as low as $\sim 10^{-5}$ and 10^{-6}.

Furthermore, in many cases, it is informative to obtain plots of load versus displacement, or stress versus strain. The resolution of such plots often depends on the speed with which data can be collected by the electrical circuits in the data-acquisition units (mostly computers although chart recorders are still found in some laboratories) that are often attached to

FIGURE 3.13 Servohydraulic testing machine. (Courtesy of the MTS Systems Corporation, Eden Prarie, MN.)

mechanical testing systems. Some of the important details in the stress–strain plots may, therefore, be difficult to identify or interpret when the rate of data acquisition is slow. However, very fast data collection may also lead to problems with inadequate disk space for the storage of the acquired load–displacement data.

In any case, electromechanical and servohydraulic testing machines are generally suitable for the testing of all classes of materials. Electromechanical testing machines are particularly suitable for tests in which the loads are increased continuously (monotonic loading) or decreased continuously with time. Also, stiff electromechanical testing machines are suitable for the testing of brittle materials such as ceramics and intermetallics under monotonic loading, while servohydraulic testing machines are well suited to testing under monotonic or cyclic loading. Finally, it is important to note that the machines may be programmed to apply complex load/displacement spectra that mimick the conditions in engineering structures and components.

3.9 SUMMARY

An introduction to the fundamental concepts of stress and strain is presented in this chapter. Following some basic definitions, the geometrical relationships between the stress components (or strain components) on an element were described using Mohr's circle.

Polynomial expressions were then presented for the computation of principal stresses and principal strains for any generalized state of stress or strain on an element. Finally, hydrostatic and deviatoric stresses/strains were introduced before describing some simple experimental techniques for the measurement and application of strain and stress.

BIBLIOGRAPHY

Ashby, M. F. and D. R. H. Jones, D. R. H. (1996) Engineering Materials 1: An Introduction to Their Properties and Applications. 2nd ed. Butterworth-Heinemann, Oxford, UK.

Courtney, T. H. (1990) Mechanical Behavior of Materials, McGraw-Hill, New York.

Dieter, G. E. (1986) Mechanical Metallurgy. 3rd ed. McGraw-Hill, New York.

Dowling, N. (1993) Mechanical Behavior of Materials, Prentice Hall, Englewood Cliffs, NJ.

Hearn, E. J. (1985) Mechanics of Materials. vols. 1 and 2, 2nd ed. Pergamon Press, New York.

Hertzberg, R. W. (1996) Deformation and Fracture Mechanics of Engineering Materials. 4th ed. Wiley, New York.

McClintock, F. A. and A. S. Argon, A. S. (eds) (1965) Mechanical Behavior of Materials. Tech Books, Fairfax, VA.

4

Introduction to Elastic Behavior

4.1 INTRODUCTION

Elastic deformation, by definition, does not result in any permanent deformation upon removal of the applied loads. It is induced primarily by the stretching or bending of bonds in crystalline and noncrystalline solids. However, in the case of polymeric materials, elastic deformation may also involve the rotation of bonds in addition to the stretching and unwinding of polymer chains.

Elastic deformation may be linear or nonlinear in nature (Fig. 4.1). It may also be time dependent or time independent. When it is time independent, the strains are fully (instantaneously) recovered on removal of the applied loads, Fig. 4.1(a). However, in materials, e.g., polymeric materials, some time may be needed for the viscous flow of atoms or chain stretching/rotation to occur to return the atoms to their initial configurations. The elastic behavior of such materials is, therefore, time independent, i.e., viscoelastic. Also, the elastic strains in polymers can be very large (typically 5–1000%) compared to relatively low elastic strain limits (0.1–1.0%) in metallic and nonmetallic materials (Figs 4.1).

This chapter presents an introduction to basic concepts in elasticity. Following a brief description of the atomic displacements that are responsible for elastic deformation, the anisotropic elasticity of crystalline materi-

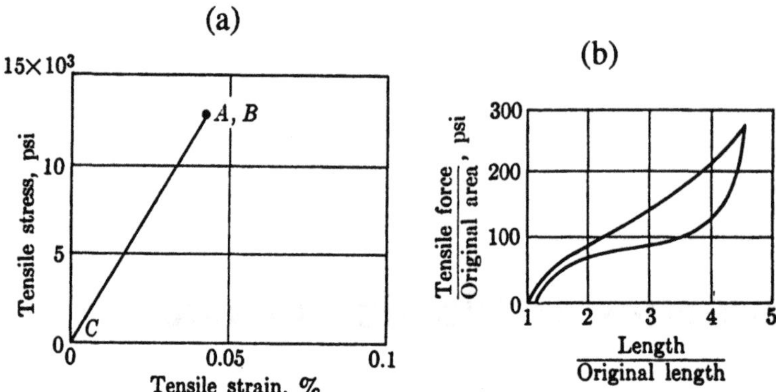

Figure 4.1 Schematic illustration of (a) linear elastic time-independent elastic deformation of 1020 steel at room temperature and (b) nonlinear elastic deformation of rubber at room temperature. (Adapted from McClintock and Argon, 1966. Courtesy of Addison-Wesley.)

als is described before presenting an overview of continuum elasticity theory. The final sections of the chapter contain advanced topics that may be omitted in an introductory course on mechanical properties. These include sections on tensors and generalized elasticity theory using shorthand tensor nomenclature.

4.2 REASONS FOR ELASTIC BEHAVIOR

Consider two atoms (A and B) that are chemically bonded together. Note that these bonds may be strong primary bonds (ionic, covalent, or metallic) or weaker secondary bonds (Van der Waals' forces or hydrogen bonds). During the earliest stages of deformation, the response of materials to applied loads is often controlled by the stretching of bonds (Fig. 4.2).

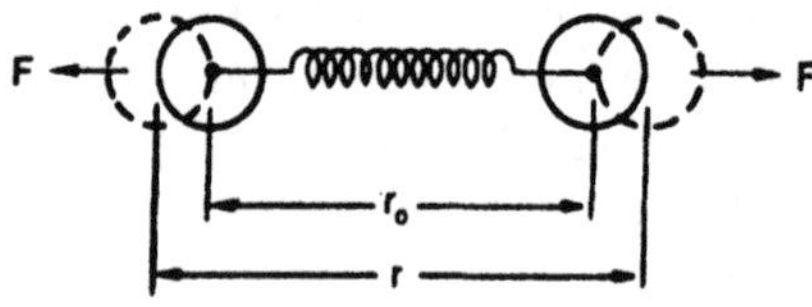

Figure 4.2 Stretching of chemical bonds between A and B. (Adapted From Ashby and Jones, 1996. Reprinted with permission from Butterworth Heinemann.)

This occurs before the onset of plasticity (irreversible/permanent damage). Under such conditions, the potential energy, U, of the bonded system depends on the attractive and repulsive components, U_A and U_R, respectively. These are given by

$$U_A = \frac{-A}{r^m} \tag{4.1}$$

and

$$U_R = \frac{B}{r^n} \tag{4.2}$$

where r is the separation between the atoms, A and B are material constants, and m and n are constants that depend on the type of chemical bonds. Combining Eqs (4.1) and (4.2), the total potential energy, U, is thus given by

$$U = U_A + U_R = \frac{-A}{r^m} + \frac{B}{r^n} \tag{4.3}$$

The relationship between U and r is shown schematically in Fig. 4.3 along with the corresponding plots for U_R versus r, and U_A versus r. Note that the repulsive term, U_R, is a short-range energy, while the attractive term, U_A, is a long-range energy. The force, F, between the two atoms is given by the first derivative of U. This gives

$$F = -\frac{dU}{dr} = -\frac{Am}{r^{m+1}} + \frac{Bn}{r^{n+1}} \tag{4.4}$$

Similarly, Young's modulus is proportional to the second derivative of U with respect to r. That is

$$E \sim \frac{d^2U}{dr^2} = \frac{dF}{dr} = \frac{Am(m+1)}{r^{m+2}} - \frac{Bn(n+1)}{r^{n+2}} \tag{4.5}$$

The relationship between F and r is shown schematically in Fig. 4.3(b). Note that the point where $F = 0$ corresponds to the equilibrium separation, r_0, between the two atoms where the potential energy is a minimum, Fig. 4.3(a). Also, the relationship between F and r is almost linear in the regime where $r \sim r_0$. Small displacements (by forces) of the atoms, therefore, result in a linear relationship between force and displacement, i.e, apparently linear elastic behavior. Note that large forces may also result in nonlinear elastic behavior since the force–separation curves are not linear for large deformations, Fig. 4.3(b). Elastic moduli for several engineering materials are listed in Table 4.1.

It should be readily apparent from the above discussion that Young's modulus is a measure of resistance to deformation. Also, since Young's modulus varies with the type of chemical bonding, it does not change

(a)

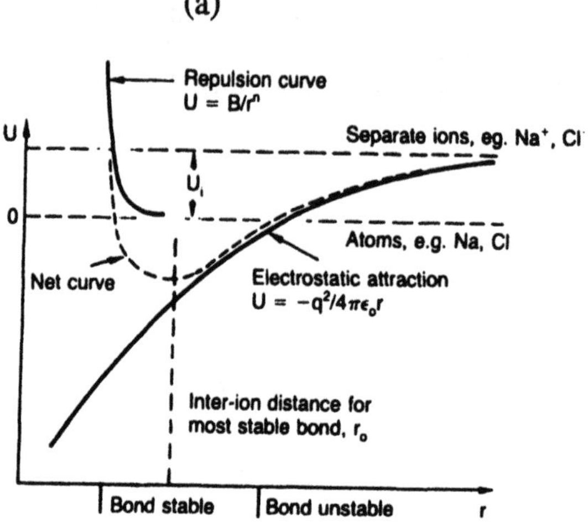

(b)

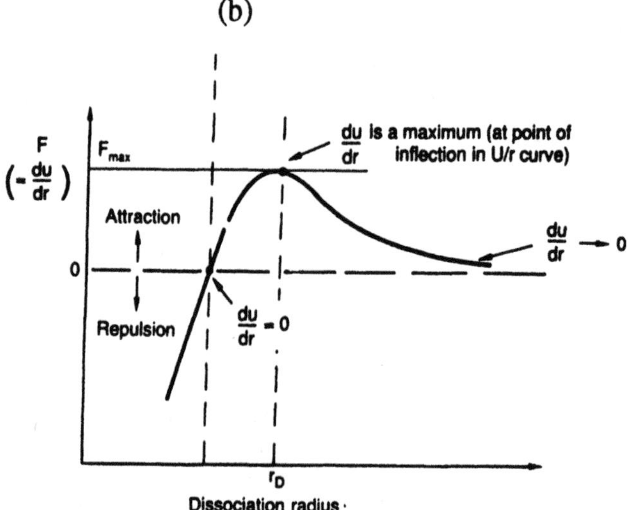

Figure 4.3 Formation of a chemical bond: relationship between U and r; (b) relationship between F and r. (Adapted from Ashby and Jones, 1996). Reprinted with permission from Butterworth Heinemann.)

much (~ ±5%) with processing/heat treatment variations or minor alloying additions that can have little effect on chemical bonding. In contrast, minor alloying, processing, and heat treatment can have very significant effects (as much as ±2000%) on strength. Young's modulus is, therefore, a material property that is generally independent of microstructure. However, since it depends strongly on the equilibrium atomic separation, which increases as a result of more vigorous atomic vibrations at elevated temperature, Young's modulus will generally decrease with increasing temperature.

Finally in this section, it is important to note that the slope of the force–distance curve in the linear regime close to r_0 is a measure of the resistance of a solid to small elastic deformation. However, the modulus of a solid may also depend strongly on the direction of loading, especially in crystals that are highly anisotropic. The modulus may also vary with direction, as discussed in the next section.

4.3 INTRODUCTION TO LINEAR ELASTICITY

The simple relationship between stress and strain was first proposed by Robert Hooke in 1678. For this reason, the basic relationship between stress and strain in the elastic regime is often referred to as Hooke's law. This law states simply that the strain, ε, in an elastic body is directly proportional to the applied stress, σ.

For axial loading [Figs 3.1(a) and 3.1(b)], the proportionality constant is commonly referred to as Young's modulus, E. The modulus for shear loading [Fig. 3.1(c)] is defined as the shear modulus, while the modulus for triaxial/pressure loading is (Figs. 3.5 and 3.9) generally referred to as the bulk modulus, K. The respective elasticity equations for isotropic tensile, shear, and bulk deformation are given by

$$\sigma = E\varepsilon \tag{4.6a}$$

$$\tau = G\gamma \tag{4.6b}$$

and

$$p = -K\frac{\Delta V}{V} = -K\Delta \tag{4.6c}$$

where σ is the applied axial stress, ε is the applied axial strain, τ is the applied shear stress, γ is the applied shear strain, p is the applied volumetric pressure/triaxial stress, and Δ is the volumetric strain (Equation 3.20).

It is also important to recall that axial extension is typically associated with lateral contraction, while axial compression often results in lateral extension. The extent of lateral contraction (or extension) may be represented by Poisson's ratio, ν, which is defined as (Fig. 4.4):

$$\nu = -\frac{\text{Lateral strain}}{\text{Longitudinal strain}} = -\frac{\varepsilon_{yy}}{\varepsilon_{xx}} \tag{4.7}$$

Table 4.1 Elastic Moduli For Different Materials

Material	E (GN m^{-2})	Material	E (GN m^{-2})
Diamond	1000	Palladium	124
Tungsten carbide (WC)	450–650	Brasses and bronzes	103–124
Osmium	551	Niobium and alloys	80–110
Cobalt/tungsten carbide cermets	400–530	Silicon	107
Borides of Ti, Zr, Hf	450–500	Zirconium and alloys	96
Silicon carbide, SiC	430–445	Silica glass, SiO_2 (quartz)	94
Boron	441	Zinc and alloys	43–96
Tungsten and alloys	380–411	Gold	82
Alumina (Al_2O_3)	385–392	Calcite (marble, limestone)	70–82
Beryllia (BeO)	375–385	Aluminium	69
Titanium carbide (TiC)	370–380	Aluminium and alloys	69–79
Titanium carbide (TaC)	360–375	Silver	76
Molybdenum and alloys	320–365	Soda glass	69
Niobium carbide (NbC)	320–340	Alkali halides (NaCl, LiF, etc.)	15–68
Silicon nitride (Si_3N_4)	280–310	Granite (Westerly granite)	62
Beryllium and alloys	290–318	Tin and alloys	41–53
Chromium	285–290	Concrete, cement	30–50
Magnesia (MgO)	240–275	Fiberglass (glass-fiber/epoxy)	35–45
Cobalt and alloys	200–248	Magnesium and alloys	41–45
Zirconia (ZrO_2)	160–241	GFRP	7–45
Nickel	214	Calcite (marble, limestone)	31
Nickel alloys	130–234	Graphite	27
CFRP	70–200	Shale (oil shale)	18
Iron	196	Common woods, ǁ to grain	9–16
Iron-based super-alloys	193–214	Lead and alloys	16–18
Ferritic steels, low-alloy steels	196–207	Alkyds	14–17
Stainless austenitic steels	190–200	Ice (H_2O)	9.1
Mild steel	200	Melamines	6–7
Cast irons	170–190	Polyimides	3–5
Tantalum and alloys	150–186	Polyesters	1.8–3.5
Platinum	172	Acrylics	1.6–3.4
Uranium	172	Nylon	2–4
Boron/epoxy composites	80–160	PMMA	3.4
Copper	124	Polystyrene	3–3.4
Copper alloys	120–150	Epoxies	2.6–3
Mullite	145		
Vanadium	130		
Titanium	116		
Titanium alloys	80–130		

TABLE 4.1 Continued

Material	E (GN m^{-2})	Material	E (GN m^{-2})
Polycarbonate	2.6	Foamed polyurethane	0.01–0.06
Common woods, ⊥ to grain	0.6–1.0	Polyethylene (low density)	0.2
Polypropylene	0.9	Rubbers	0.01–0.1
PVC	0.2–0.8	Foamed polymers	0.001–0.01
Polyethylene (high density)	0.7		

After Ashby and Jones, 1996. Reprinted with permission from Butterworth Heinemann.

Furthermore, the resistance of a crystal to deformation is strongly dependent on its orientation. It is, therefore, important to develop a more complete description of elastic behavior that includes possible crystal anisotropy effects. This can be achieved by rewriting Eqs (4.6a) and (4.6b) with their respective stress and strain components as the independent variables:

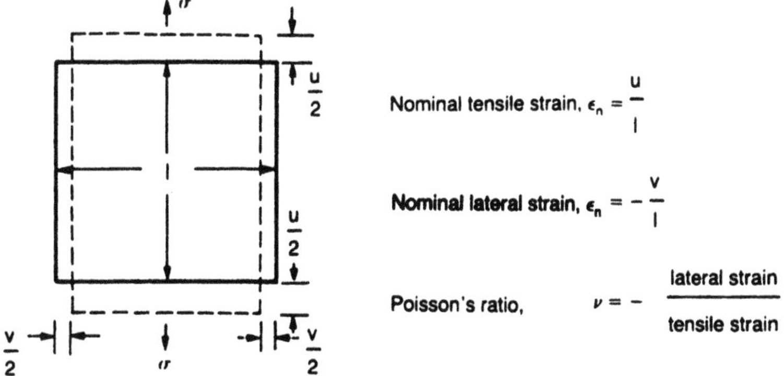

FIGURE 4.4 Schematic illustration of lateral contraction. (Adapted from Ashby and Jones, 1996—reprinted with permission from Butterworth Heinemann.)

$$\begin{bmatrix} \sigma_{11} \\ \sigma_{22} \\ \sigma_{33} \\ \tau_{23} \\ \tau_{13} \\ \tau_{12} \end{bmatrix} = \begin{bmatrix} C_{11} & C_{12} & C_{13} & C_{14} & C_{15} & C_{16} \\ C_{21} & C_{22} & C_{23} & C_{24} & C_{25} & C_{26} \\ C_{31} & C_{32} & C_{33} & C_{34} & C_{35} & C_{36} \\ C_{41} & C_{42} & C_{43} & C_{44} & C_{45} & C_{46} \\ C_{51} & C_{52} & C_{53} & C_{54} & C_{55} & C_{56} \\ C_{61} & C_{62} & C_{63} & C_{64} & C_{65} & C_{66} \end{bmatrix} \begin{bmatrix} \varepsilon_{11} \\ \varepsilon_{22} \\ \varepsilon_{33} \\ \gamma_{23} \\ \gamma_{13} \\ \gamma_{12} \end{bmatrix} \tag{4.8a}$$

where the terms in the C_{ij} matrix represent the elastic stiffness. Similarly, the three-dimensional strains can be expressed as

$$\begin{bmatrix} \varepsilon_{11} \\ \varepsilon_{22} \\ \varepsilon_{33} \\ \gamma_{23} \\ \gamma_{13} \\ \gamma_{12} \end{bmatrix} = \begin{bmatrix} S_{11} & S_{12} & S_{13} & S_{14} & S_{15} & S_{16} \\ S_{21} & S_{22} & S_{23} & S_{24} & S_{25} & S_{26} \\ S_{31} & S_{32} & S_{33} & S_{34} & S_{35} & S_{36} \\ S_{41} & S_{42} & S_{43} & S_{44} & S_{45} & S_{46} \\ S_{51} & S_{52} & S_{53} & S_{54} & S_{55} & S_{56} \\ S_{61} & S_{62} & S_{63} & S_{64} & S_{65} & S_{66} \end{bmatrix} \begin{bmatrix} \sigma_{11} \\ \sigma_{22} \\ \sigma_{33} \\ \tau_{23} \\ \tau_{13} \\ \tau_{12} \end{bmatrix} \tag{4.8b}$$

where the terms in the S_{ij} matrix are the so-called compliance coefficients. At first glance, both the C_{ij} and S_{ij} matrices contain 36 terms. However, due to the existence of a unique strain energy, only 21 of the terms are independent in each of matrices in Eqs (4.8) and (4.9). The number of independent terms in the C_{ij} and S_{ij} matrices also decreases with increasing crystal symmetry. The least symmetric triclinic crystals have 21 independent elastic constants; orthorhombic crystals have nine independent elastic constants, and tetragonal crystals have six. Hexagonal crystals have five independent elastic constants and cubic crystals have three. Hence, for cubic crystals, Eqs (4.8a) and (4.8b) reduce to Eqs (4.9a) and (4.9b):

$$\begin{bmatrix} \sigma_{11} \\ \sigma_{22} \\ \sigma_{33} \\ \tau_{23} \\ \tau_{13} \\ \tau_{12} \end{bmatrix} = \begin{bmatrix} C_{11} & C_{12} & C_{12} & 0 & 0 & 0 \\ C_{12} & C_{11} & C_{12} & 0 & 0 & 0 \\ C_{12} & C_{12} & C_{11} & 0 & 0 & 0 \\ 0 & 0 & 0 & C_{44} & 0 & 0 \\ 0 & 0 & 0 & 0 & C_{44} & 0 \\ 0 & 0 & 0 & 0 & 0 & C_{44} \end{bmatrix} \begin{bmatrix} \varepsilon_{11} \\ \varepsilon_{22} \\ \varepsilon_{33} \\ \gamma_{23} \\ \gamma_{13} \\ \gamma_{12} \end{bmatrix} \tag{4.9a}$$

$$\begin{bmatrix} \varepsilon_{11} \\ \varepsilon_{22} \\ \varepsilon_{33} \\ \gamma_{23} \\ \gamma_{13} \\ \gamma_{12} \end{bmatrix} = \begin{bmatrix} S_{11} & S_{12} & S_{12} & 0 & 0 & 0 \\ S_{12} & S_{11} & S_{12} & 0 & 0 & 0 \\ S_{12} & S_{12} & S_{11} & 0 & 0 & 0 \\ 0 & 0 & 0 & S_{44} & 0 & 0 \\ 0 & 0 & 0 & 0 & S_{44} & 0 \\ 0 & 0 & 0 & 0 & 0 & S_{44} \end{bmatrix} \begin{bmatrix} \sigma_{11} \\ \sigma_{22} \\ \sigma_{33} \\ \tau_{23} \\ \tau_{13} \\ \tau_{12} \end{bmatrix} \tag{4.9b}$$

The moduli in the different directions can be estimated from the stiffness coefficients. For cubic crystals, Young's modulus along a crystallographic [*hkl*] direction is given by

$$\frac{1}{E_{hkl}} = S_{11} - 2\left[(S_{11} - S_{12}) - \frac{S_{44}}{2}\right](l_1^2 l_2^2 + l_2^2 l_3^2 + l_3^2 l_2^2) \qquad (4.10)$$

where l_1, l_2, and l_3 are the direction cosines of the angle between the vector corresponding to the direction, and the x, y, z axes, respectively.

The room-temperature values of C_{ij} and S_{ij} for selected materials are listed in Table 4.2. The direction cosines for the most widely used principal directions in the cubic lattice are also given in Table 4.3. Note that the uniaxial elastic moduli of cubic crystals depend solely on their compliance/stiffness coefficients, and the magnitudes of the direction cosines. Also, the modulus in the [100] direction $1/E_{111} = S_{11} - 2/3[(S_{11} - S_{12}) - S_{44}/2]$. Hence, depending on the relative magnitudes of $(S_{11} - S_{12})$ and $S_{44/2}$, the modulus may be greatest in the [111] or [100] directions. The average modulus of a polycrystalline material in a given direction depends on the relative proportions of grains in the different orientations.

The modulus of a polycrystalline cubic material may be estimated from a simple mixture rule of the form: $E = \sum_{i=1}^{n} V_i E_i$, where V_i is the volume fraction of crystals with a particular crystallographic orientation, E_i is the modulus in that particular orientation, and n is the number of possible crystallographic orientations.

Finally in this section, it is important to note that the degree of anisotropy of a cubic crystal (anisotropy ratio) is given by:

$$\text{Anisotropy ratio} = \frac{2(S_{11} - S_{12})}{S_{44}} \qquad (4.11)$$

Anisotropy ratios are presented along with elastic constants for several cubic materials in Table 4.4. Note that with the exception of tungsten, all of the materials listed are anisotropic, i.e., their moduli depend strongly on direction. The assumption of isotropic elasticity in several mechanics and materials problems may therefore lead to errors. However, in many problems in linear elasticity, the assumption of isotropic elastic behavior is made to simplify the analysis of stress and strain.

4.4 THEORY OF ELASTICITY

4.4.1 Introduction

The rest of the chapter may be omitted in an undergraduate class on the mechanical behavior of materials. However, this section is recommended

TABLE 4.2 Summary of Elastic Stiffness and Compliance Coefficients

Material	$(10^{10}$ Pa)			$(10^{-11}$ $Pa^{-1})$		
	c_{11}	c_{12}	c_{44}	s_{11}	s_{12}	s_{44}
Cubic						
Aluminum	10.82	6.13	2.85	1.57	−0.57	3.51
Copper	16.84	12.14	7.54	1.50	−0.63	1.33
Gold	18.60	15.70	4.20	2.33	−1.07	2.38
Iron	23.70	14.10	11.60	0.80	−0.28	0.86
Lithium fluoride	11.2	4.56	6.32	1.16	−0.34	1.58
Magnesium oxide	29.3	9.2	15.5	0.401	−0.096	0.648
Molybdenum[b]	46.0	17.6	11.0	0.28	−0.08	0.91
Nickel	24.65	14.73	12.47	0.73	−0.27	0.80
Sodium chloride[b]	4.87	1.26	1.27	2.29	−0.47	7.85
Spinel ($MgAl_2O_4$)	27.9	15.3	14.3	0.585	−0.208	0.654
Titanium carbide[b]	51.3	10.6	17.8	0.21	−0.036	0.561
Tungsten	50.1	19.8	15.14	0.26	−0.07	0.66
Zinc sulfide	10.79	7.22	4.12	2.0	−0.802	2.43

	c_{11}	c_{12}	c_{13}	c_{33}	c_{44}	s_{11}	s_{12}	s_{13}	s_{33}	s_{44}
Hexagonal										
Cadmium	12.10	4.81	4.42	5.13	1.85	1.23	−0.15	−0.93	3.55	5.40
Cobalt	30.70	16.50	10.30	35.81	7.53	0.47	−0.23	−0.07	0.32	1.32
Magnesium	5.97	2.62	2.17	6.17	1.64	2.20	−0.79	−0.50	1.97	6.10
Titanium	16.0	9.0	6.6	18.1	4.65	0.97	−0.47	−0.18	0.69	2.15
Zinc	16.10	3.42	5.01	6.10	3.83	0.84	0.05	−0.73	2.84	2.61

After Hertzberg, 1996—reprinted with permission from John Wiley.
[a]*Sources*: Huntington (1958) and Hellwege (1969).
[b]Note that $E_{100} > E_{111}$.

TABLE 4.3 Summary of Direction Cosines For Cubic Lattices

Direction	l_1	l_2	l_3
$\langle 100 \rangle$	1	0	0
$\langle 110 \rangle$	$1/\sqrt{2}$	$1/\sqrt{2}$	0
$\langle 111 \rangle$	$1/\sqrt{3}$	$1/\sqrt{3}$	$1/\sqrt{3}$

After Hertzberg, 1996. Reprinted with permission from John Wiley.

for graduate students, or those who simply want to develop a deeper understanding of elasticity and plasticity concepts. Following a review of equilibrium equations and possible states of stress, compatibility conditions are described prior to a simple presentation of Airy stress functions. Short-hand tensor notation is also explained in a simple presentation that will enable the reader to interpret abbreviated versions of equations that are often used in the literature. A generalized form of Hooke's law is then presented along with a basic definition of the strain energy density function at the end of the chapter.

TABLE 4.4 Summary Anisotropy Ratios at Room Temperature

Metal	Relative degree of anisotropy $\left[\frac{2(s_{11} - s_{12})}{s_{44}}\right]$	E_{111} (10^6 psi)	E_{100} (10^6 psi)	$\left[\frac{E_{111}}{E_{100}}\right]$
Aluminum	1.219	11.0	9.2	1.19
Copper	3.203	27.7	9.7	2.87
Gold	2.857	16.9	6.2	2.72
Iron	2.512	39.6	18.1	2.18
Magnesium oxide	1.534	50.8	36.2	1.404
Spinel ($MgAl_2O_4$)	2.425	52.9	24.8	2.133
Titanium carbide	0.877	62.2	69.1	0.901
Tungsten	1.000	55.8	55.8	1.00

After Hertzberg, 1996. Reprinted with permission from John Wiley.

4.4.2 Equilibrium Equations

The relationships between stress and strain may be obtained by determining the conditions for equilibrium of an element. If we consider a magnified cubic element, as shown in Fig. 4.5, it is easy to imagine all the forces that must be applied to the cube to keep it suspended in space. These forces may consist of applied normal and shear stresses as described in Chap. 3 (Fig. 3.1). They also consist of body forces that have x, y, and z components. Body forces may be due to gravitational or centrifugal forces which act throughout the volume. They have the units of force per unit volume. Considering stress gradients across the cube and force equilibrium in the (x, y, z) directions, it is relatively easy to show that

$$\frac{\partial \sigma_{xx}}{\partial x} + \frac{\partial \tau_{xy}}{\partial y} + \frac{\partial \tau_{xz}}{\partial z} + F_x = \rho \frac{\partial^2 x}{\partial t^2} \tag{4.12a}$$

$$\frac{\partial \tau_{yx}}{\partial x} + \frac{\partial \sigma_{yy}}{\partial y} + \frac{\partial \tau_{yz}}{\partial z} + F_y = \rho \frac{\partial^2 y}{\partial t^2} \tag{4.12b}$$

$$\frac{\partial \tau_{zx}}{\partial x} + \frac{\partial \tau_{zy}}{\partial y} + \frac{\partial \sigma_{zz}}{\partial z} + F_z = \rho \frac{\partial_z^2}{\partial t^2} \tag{4.12c}$$

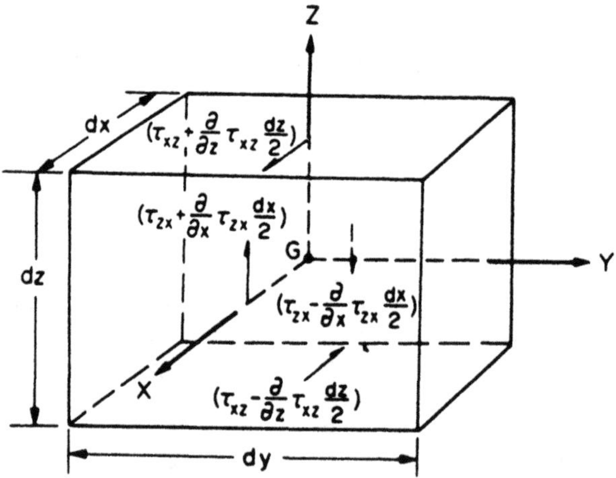

FIGURE 4.5 State of stress on an element in the Cartesian co-ordinate system. Only components affecting equilibrium are labeled. (Adapted from Hearn, 1985. Reprinted with permission from Pergamon Press.)

where F_x, F_y, and F_z are the body forces per unit volume, ρ is the density, x, y, and z are the Cartesian co-ordinates, and the acceleration terms in Newton's second law are given by the second derivatives on the right-hand side of Eqs 4.12(a)–(c). In the case of stationary bodies, the acceleration terms on the right-hand side of these equations are all equal to zero. Also, the body forces are usually small when compared to the applied forces. Hence, we have the usual forms of the equilibrium equations that are generally encountered in mechanics and materials problems. These are

$$\frac{\partial \sigma_{xx}}{\partial x} + \frac{\partial \tau_{xy}}{\partial y} + \frac{\partial \tau_{xz}}{\partial z} = 0 \tag{4.13a}$$

$$\frac{\partial \tau_{yx}}{\partial x} + \frac{\partial \sigma_{yy}}{\partial y} + \frac{\partial \tau_{yz}}{\partial z} = 0 \tag{4.13b}$$

$$\frac{\partial \tau_{zx}}{\partial x} + \frac{\partial \tau_{zy}}{\partial y} + \frac{\partial \sigma_{zz}}{\partial z} = 0 \tag{4.13c}$$

The equilibrium equations for cylindrical and spherical co-ordinate systems may also be derived from appropriate free-body diagrams (Fig. 4.6). These alternative co-ordinate systems are often selected when the geometrical shapes that are being analyzed have cylindrical or spherical symmetry. Under such conditions, considerable simplification may be achieved in the analysis by using cylindrical or spherical co-ordinate systems. Force components in the free-body diagrams can also be obtained by applying

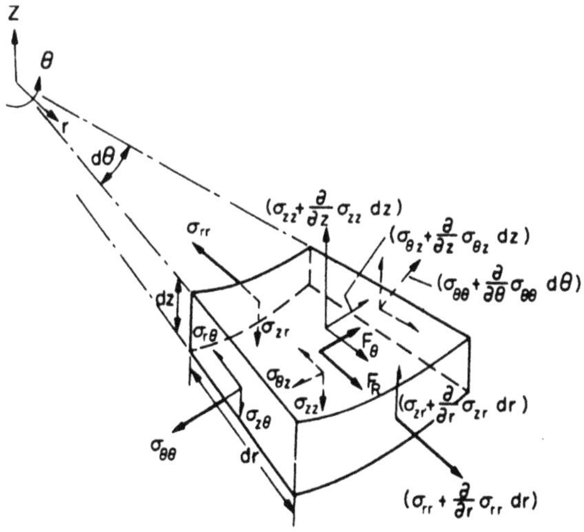

FIGURE 4.6 State of stress over an element in the cylindrical co-ordinate system.

Newton's second law to the special case of a stationary body. In the case of cylindrical co-ordinates, if we neglect body forces, the equations of equilibrium are given by

$$\frac{\partial \sigma_{rr}}{\partial r} + \frac{1}{r}\frac{\partial \tau_{r\theta}}{\partial \theta} + \frac{\partial \tau_{rz}}{\partial z} + \frac{\sigma_{rr} - \sigma_{\theta\theta}}{r} = 0 \tag{4.14a}$$

$$\frac{\partial \tau_{r\theta}}{\partial r} + \frac{1}{r}\frac{\partial \sigma_{\theta\theta}}{\partial \theta} + \frac{\partial \tau_{z\theta}}{\partial z} + \frac{2\tau_{r\theta}}{r} = 0 \tag{4.14b}$$

$$\frac{\partial \tau_{rz}}{\partial r} + \frac{1}{r}\frac{\partial \tau_{z\theta}}{\partial \theta} + \frac{\partial \sigma_{zz}}{\partial z} + \frac{\tau_{zr}}{r} = 0 \tag{4.14c}$$

Similarly, the equations of equilibrium for problems with spherical symmetry may be derived by summing the force components in the r, θ, and ϕ directions. Hence, the equilibrium equations are given by

$$\frac{\partial \sigma_{rr}}{\partial r} + \frac{1}{r\sin\phi}\frac{\partial \tau_{r\theta}}{\partial \theta} + \frac{1}{r}\frac{\partial \tau_{r\theta}}{\partial \phi} + \frac{2\sigma_{rr} - \sigma_{\theta\theta} - \sigma_{\phi\phi} + \tau_{r\phi}\cot\phi}{r} + F_r = 0 \tag{4.14d}$$

$$\frac{\partial \tau_{r\theta}}{\partial r} + \frac{1}{r}\frac{\partial \sigma_{\theta\theta}}{\partial \theta} + \frac{1}{r}\frac{\partial \tau_{\phi\theta}}{\partial \phi} + \frac{3\tau_{\phi\theta}}{r}\cot\phi + F_\theta = 0 \tag{4.14e}$$

$$\frac{\partial \tau_{r\theta}}{\partial r} + \frac{1}{r}\frac{\partial \tau_{\phi\theta}}{\partial \theta} + \frac{1}{r}\frac{\partial \sigma_{\phi\phi}}{\partial \phi} + \frac{\sigma_{\phi\phi} - \sigma_{\theta\theta}}{r}\cot\phi + \frac{3\tau_{r\phi}}{r} + F_\phi = 0 \tag{4.14f}$$

The above equations (three each for any three-dimensional coordinate system) are insufficient to solve for the six independent stress components. The remaining three components of stress can only be found from a simultaneous solution with the stress–strain relationships in most elasticity problems.

4.4.3 States of Stress

4.4.3.1 Plane Stress and Plane Strain Conditions

The above discussion has focused largely on the general three-dimensional state of stress on an element [Figs 4.5 and 4.6). However, in many problems in mechanics and materials, it is possible to achieve considerable simplification in the analysis of stress and strain by assuming biaxial stress (plane stress) or biaxial strain (plane strain) conditions. Such problems are often referred to as plane problems (Fig. 4.7).

In plane elastic problems, neither the stresses nor strains vary in the z direction. Furthermore, the loads on the sides and the body forces must be distributed uniformly across the thickness. Also, plane stress conditions often apply to problems in which the thickness is small, while plane strain conditions usually apply to problems in which the thickness is large with respect to the other dimensions.

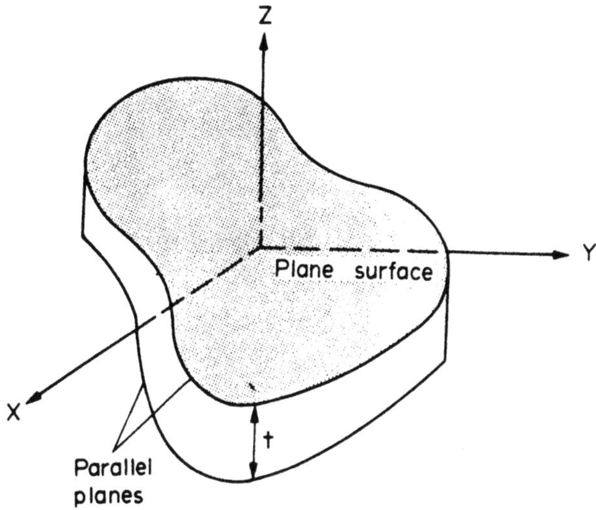

FIGURE 4.7 Schematic of a plane element.

In plane stress problems, all the z components of stress are assumed to be zero, i.e., σ_{zz}, τ_{xz}, and τ_{yz} are all equal to zero. Note also that the effective value of σ_{zz} may be reduced to zero by imposing an equal stress of opposite sign in the z direction. The equilibrium equations thus reduce to the following expressions:

$$\frac{\partial \sigma_{xx}}{\partial x} + \frac{\partial \tau_{xy}}{\partial y} + F_x = 0 \tag{4.15a}$$

$$\frac{\partial \tau_{xy}}{\partial x} + \frac{\partial \sigma_{yy}}{\partial y} + F_y = 0 \tag{4.15b}$$

The relationships between strain and stress are now given by

$$\varepsilon_{xx} = \frac{\sigma_{xx}}{E} - \frac{\nu \sigma_{yy}}{E} \tag{4.16a}$$

$$\varepsilon_{yy} = \frac{\sigma_{yy}}{E} - \frac{\nu \sigma_{xx}}{E} \tag{4.16b}$$

$$\gamma_{xy} = \frac{\tau_{xy}}{G} \tag{4.16c}$$

Under plane strain conditions, all of the strains in the z direction are zero. Hence, $\varepsilon_{zz} = \varepsilon_{xz} = \varepsilon_{yz} = 0$. Also, for all plane problems, $\tau_{xz} = \tau_{yz} = 0$. The equilibrium equations are thus given by

$$\frac{\partial \sigma_{xx}}{\partial x} + \frac{\partial \tau_{xy}}{\partial y} + F_x = 0 \tag{4.17a}$$

$$\frac{\partial \tau_{xy}}{\partial x} + \frac{\partial \sigma_{yy}}{\partial y} + F_y = 0 \tag{4.17b}$$

$$\frac{\partial \sigma_{zz}}{\partial z} + F_z = 0 \tag{4.17c}$$

The relationships between strain and stress are now given by

$$\varepsilon_{xx} = \frac{(1-\nu^2)}{E}\left[\sigma_{xx} - \frac{\nu}{(1-\nu)}\sigma_{yy}\right] \tag{4.18a}$$

$$\varepsilon_{yy} = \frac{(1-\nu^2)}{E}\left[\sigma_{yy} - \frac{\nu}{(1-\nu)}\sigma_{xx}\right] \tag{4.18b}$$

$$\gamma_{xy} = \frac{\tau_{xy}}{G} \tag{4.18c}$$

Note that the plane strain equations can be obtained from the plane stress equations simply by replacing ν with $\nu/(1-\nu)$ and E with $E/(1-\nu^2)$. Furthermore, the above equations may be rearranged to obtain stress components in terms of strain. For plane stress conditions, this gives

$$\sigma_{xx} = \frac{E}{(1-\nu^2)}[\varepsilon_{xx} + \nu\varepsilon_{yy}] \tag{4.19a}$$

$$\sigma_{yy} = \frac{E}{(1-\nu^2)}[\varepsilon_{yy} + \nu\varepsilon_{xx}] \tag{4.19b}$$

$$\tau_{xy} = G\gamma_{xy} \tag{4.19c}$$

Similarly, for plane strain conditions, we may rearrange Eqs 4.19(a–c) to obtain the following expressions for stress components in terms of strain:

$$\sigma_{xx} = \frac{E(1-\nu)}{(1+\nu)(1-2\nu)}\left[\varepsilon_{xx} + \frac{\nu}{(1-\nu)}\varepsilon_{yy}\right] \tag{4.20a}$$

$$\sigma_{yy} = \frac{E(1-\nu)}{(1+\nu)(1-2\nu)}\left[\varepsilon_{yy} + \frac{\nu}{(1-\nu)}\varepsilon_{xx}\right] \tag{4.20b}$$

$$\tau_{xy} = G\gamma_{xy} \tag{4.20c}$$

4.4.3.2 Generalized Three-Dimensional State of Stress

Thermal stresses may also be included in the equilibrium equations simply by treating them as body forces. For a generalized three-dimensional state of stress, the equations of equilibrium are given by

$$\frac{\partial \sigma_{xx}}{\partial x} + \frac{\partial \tau_{xy}}{\partial y} + \frac{\partial \tau_{xz}}{\partial z} + \alpha_x \Delta T = 0 \tag{4.21a}$$

$$\frac{\partial \tau_{yx}}{\partial x} + \frac{\partial \sigma_{yy}}{\partial y} + \frac{\partial \tau_{yz}}{\partial z} + \alpha_y \Delta T = 0 \tag{4.21b}$$

$$\frac{\partial \tau_{zx}}{\partial x} + \frac{\partial \tau_{zy}}{\partial y} + \frac{\partial \sigma_{zz}}{\partial z} + \alpha_z \Delta T = 0 \tag{4.21c}$$

Equations 4.21(a–c) are not applicable to plasticity problems. Also, similar expressions may be written for the equilibrium conditions in spherical and cylindrical co-ordinates under elastic conditions. However, in general, the above equations cannot be solved without satisfying the so-called compatibility conditions. These are discussed in the next section.

4.4.3.3 Compatibility Conditions and Stress Functions

To ensure that the solutions to the above equations are consistent with single valued displacements, the compatibility conditions must be satisfied. These are derived in most mechanics texts on elasticity. The compatibility conditions are given by three equations of the form:

$$\frac{\partial^2 \varepsilon_{ii}}{\partial x_j^2} + \frac{\partial^2 \varepsilon_{jj}}{\partial x_i^2} = \frac{\partial^2 \gamma_{ij}}{\partial x_i \partial x_j} \tag{4.22}$$

where subscripts i and j can have values between 1 and 3 corresponding to subscripts x, y, and z, respectively. Also, three other compatibility equations of the following form can be obtained from linear elasticity theory, which gives

$$\frac{2\partial^2 \varepsilon_{ii}}{\partial x_j \partial x_k} = \frac{\partial}{\partial x_i}\left(-\frac{\partial \gamma_{jk}}{\partial x_i} + \frac{\partial \gamma_{ki}}{\partial x_j} + \frac{\partial \gamma_{ij}}{\partial x_k}\right) \tag{4.23}$$

Three equations of the above form can be obtained by the cyclic permutation of i, j, and k in the above equation. As the reader can probably imagine, the above equations are difficult to solve using standard methods. It is, therefore, common to employ trial and error procedures in attempts to obtain solutions to elasticity problems.

One class of trial functions, known as Airy functions, are named after Sir George Airy, the British engineer who was the first person to introduce them in the 19th century. The Airy stress function, χ, defines the following relationships for σ_{xx}, σ_{yy}, and τ_{xy}:

$$\sigma_{xx} = \frac{\partial^2 \chi}{\partial y^2} \tag{4.24a}$$

$$\sigma_{yy} = \frac{\partial^2 \chi}{\partial x^2} \tag{4.24b}$$

$$\tau_{xy} = \frac{-\partial^2 \chi}{\partial x \partial y} \tag{4.24c}$$

The compatibility condition can also be expressed in terms of the Airy stress function. This gives

$$\frac{\partial^4 \chi}{\partial x^4} + \frac{2\partial^4 \chi}{\partial x^2 \partial y^2} + \frac{\partial^4 \chi}{\partial y^4} = 0 \tag{4.25}$$

For stress fields with polar symmetry, the Airy stress functions are given by

$$\sigma_{rr} = \frac{1}{r}\frac{\partial \chi}{\partial r} + \frac{1}{r^2}\frac{\partial^2 \chi}{\partial \theta^2} \tag{4.26a}$$

$$\sigma_{\theta\theta} = \frac{\partial^2 \chi}{\partial r^2} \tag{4.26b}$$

$$\sigma_{r\theta} = -\frac{\partial}{\partial r}\left(\frac{1}{r}\frac{\partial \chi}{\partial \theta}\right) \tag{4.26c}$$

The compatibility conditions are satisfied when the Airy stress function satisfies Laplace's equation and the biharmonic equation. Hence,

$$\nabla^2 c = 0 \tag{4.27}$$

and

$$\nabla^4 c = 0 \tag{4.28}$$

where the ∇^2 operator is given by

$$\nabla^2 \equiv \frac{\partial^2}{\partial r^2} + \frac{1}{r}\frac{\partial}{\partial r} + \frac{1}{r^2}\frac{\partial^2}{\partial \theta^2} \tag{4.29}$$

The compatibility condition is, therefore, satisfied when Laplace's equation and the biharmonic function are satisfied, i.e., when Eqs (4.27) and (4.28) are satisfied.

4.5 INTRODUCTION TO TENSOR NOTATION

So far, the description of linear elasticity concepts has used simple matrix notation that any reader with a basic knowledge of linear algebra can follow without too much difficulty. However, it is common in technical publications to formalize the use of matrix notation in a manner that is not self-evident to the untrained reader. The formal notation that is often used is generally referred to as tensor notation. The latter is commonly used because it facilitates the simplified/abbreviated presentation of groups of numbers.

Tensors are essentially groups of numbers that represent a physical quantity such as the state of stress on an element. The order, n, of a tensor determines the number of components of a tensor. The number of components of a tensor is given by 3^n. Hence, the simplest tensors are scalar quantities, which are a special class of tensors of order zero, i.e., $n = 0$ and the number of components is $3^n = 3^0 = 1$. Temperature is one example of a zeroth-order tensor. Vectors are tensor quantities of the first order. They, therefore, have three components given by $3^n = 3^1 = 3$. The three components are often referred to three independent axes, e.g., x, y, z or 1, 2, 3. Similarly, second rank tensors may be defined as tensors with $n = 2$. They have $3^n = 3^2 = 9$ components.

Tensors can be used to represent stress, strain, and physical properties such as electrical/thermal conductivity and diffusivity. Tensors may also be used to represent the anisotropy of stiffness and compliance. Stress and strain are examples of second rank tensors since they require specifications of the directions of the plane normal and applied force. Stress and strain are also examples of symmetric tensors since $\sigma_{ij} = \sigma_{ji}$ and $\varepsilon_{ij} = \varepsilon_{ji}$, i.e., their components are symmetric about their diagonals.

Tensor notation is particularly useful because it provides a short-hand notation for describing transformations between different orthogonal sets of axes. If we now consider the simple case of two sets of orthogonal coordinate axes (x_1, x_2, x_3) and (x'_1, x'_2, x'_3) that describe the same vector, A, then it is easy to express the vector A in terms of the unit vectors along the old axes (i, j, k) or the unit vectors across the new axes (i', j', k'). The components of A are transformed from one co-ordinate system to the other simply by multiplying them by the direction cosines between the old axes and the new axes (Fig. 4.8). Hence, the components of the the vector A in the new vector basis are given by the following expression:

$$A_i = \ell_{ik} A_k \tag{4.30}$$

where A_i are the vector components in the new basis, A_k are the vector components in the old basis, and ℓ_{ik} is the direction cosine of the angle between the ith axis in the new basis and the kth axis in the old basis.

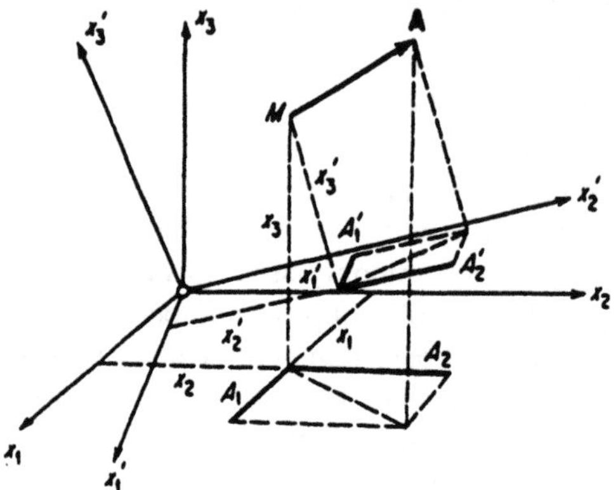

FIGURE 4.8 Definition of direction cosines between two sets of axes.

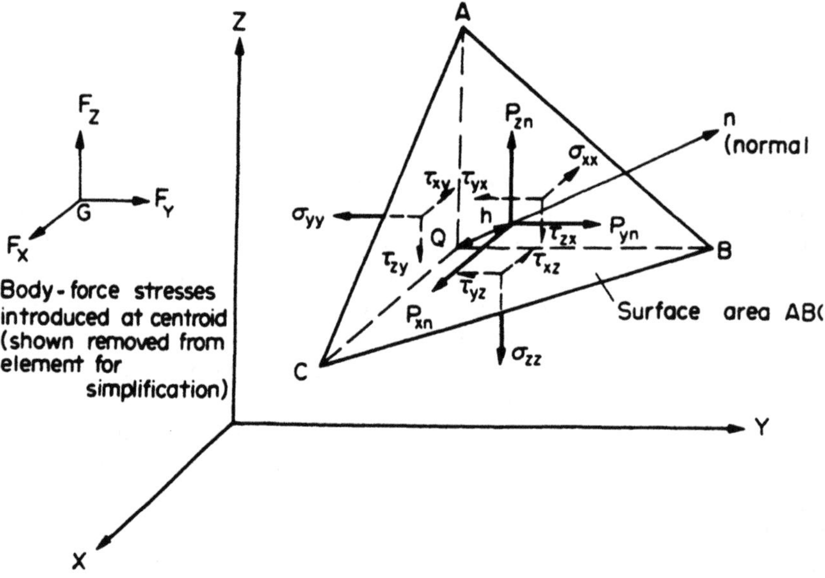

FIGURE 4.9 State of stress on an inclined plane through a given point in a three-dimensional Cartesian co-ordinate system.

We may now turn our attention to second-order tensors which are next in order of complexity after scalars and vectors. A second-order tensor consists of nine components. One example of a tensor quantity is the stress tensor, which the reader should be familiar with from Chap. 3.

The nine components of the second-order stress tensor are given by (Figs. 4.8 and 4.9):

$$[\sigma] = \begin{bmatrix} \sigma_{xx} & \tau_{xy} & \tau_{xz} \\ \tau_{yx} & \sigma_{yy} & \tau_{yz} \\ \tau_{zx} & \tau_{zy} & \sigma_{zz} \end{bmatrix} \tag{4.31}$$

Note that Eq. (4.31) presents the components of the stress tensor in the old co-ordinate system. However, by definition, the components of a second-order tensor transform from one co-ordinate system to another via the so-called transformation rule. This is given by (Figs. 4.8 and 4.9):

$$\sigma_{ij} = \sum_{k=1}^{3}\sum_{l=1}^{3} \ell_{ik}\ell_{jl}\sigma_{kl} \tag{4.32}$$

where the σ_{kl} terms represent the stress components referred to the old axes, ℓ_{ik} represents the direction cosines that transform the plane normal components to the new set of axes, while ℓ_{jl} represent the direction cosines that transform the force components to the new set of axes.

Similar expressions may also be written for the strain tensor and the transformation of the strain tensor. These are given by

$$\varepsilon_{ij} = \frac{1}{2}\left(\frac{\partial u_i}{\partial x_j} + \frac{\partial u_j}{\partial x_i}\right) \tag{4.33}$$

and

$$\varepsilon_{ij} = \sum_{k=1}^{3}\sum_{l=1}^{3} \ell_{ik}\ell_{jl}\varepsilon_{kl} \tag{4.34}$$

We note here that a more compact form of the above equations can be obtained by applying what is commonly known as the *Einstein notation* or the summation convention. This notation, which was first proposed by Albert Einstein, states that if a suffix occurs twice in the same term, then summation is automatically implied over values of i and j between 1 and 3. Hence, Eqs (4.32) and (4.34) can be expressed as

$$\sigma_{ij} = \ell_{ik}\ell_{jl}\sigma_{kl} \tag{4.35}$$

and

$$\varepsilon_{ij} = \ell_{ik}\ell_{jl}\varepsilon_{kl} \tag{4.36}$$

where the ℓ_{ik} and ℓ_{jl} terms represent the direction cosines of the angles between the new and old axes. Note that the subscripts k and l in the above equations can be replaced by any other letter (apart from i and j) without changing their functional forms. Hence, these subscripts are commonly referred to as "dummy suffixes" since they can be replaced by any letter. Similarly, the equilibrium equations [4.12(a–c)] may be expressed in abbreviated tensor notation, which gives

$$\sigma_{ij,j} + F_i = \rho_{ii,i} \tag{4.37}$$

It is also important to note here that the Kronecker delta function may also be defined as the unit tensor, δ_{ij}, with a value of 1 when $i = j$. This is related to the direction cosines via the following expression:

$$\ell_{ij}\ell_{il} = \delta_{jk} \tag{4.38}$$

So far, we have learned that zeroth, first and second order tensors transform, respectively, according to the following transformation laws: $\phi = \phi'$; $A'_i = \ell_{ik}A_k$, and $\sigma_{ij} = \ell_{ik}\ell_{jl}\sigma_{kl}$. Also, in general, a tensor of order n has 3^n components, as discussed at the start of this section.

In general, a tensor of order n may be defined as one that undergoes transformation from one co-ordinate system to another by the following transformation rule:

$$A_{i_1 i_2 \ldots, i_n} = \ell_{i_1 k_1}\ell_{i_2 k_2}\ell_{i_3 k_3} \cdots \ell_{i_n k_n} A_{k_1 k_2 k_3 \ldots, k_n} \tag{4.39}$$

where $A_{k_1 k_2 k_3 \ldots, k_n}$ are the components of the tensor in the old co-ordinate system, $A_{i_1 i_2 \ldots, i_n}$ are the components of the tensor of order n in the new co-ordinate system, and $\ell_{i_1 k_1}, \ell_{i_2 k_2}, \ell_{i_3 k_3} \ldots, \ell_{i_n k_n}$ are the direction cosines between the axes in the new and old co-ordinate systems.

For example, three vectors **A**, **B** and **C** would form a third-order tensor given by $A_{i_1 i_2 i_3} = \ell_{i_1 k_1}\ell_{i_2 k_2}\ell_{i_3 k_3} A_{k_1 k_2 k_3}$. Similarly, if a second-order tensor, $A_{ik} = \lambda_{iklm}B_{lm}$ is a linear function of another second-order function, they are related by a fourth-order tensor via:

$$A_{ik} = \lambda_{iklm}B_{lm} \tag{4.40}$$

Finally, in this section, it is important to note that the permutation tensor, ε_{ijk}, may be defined as having values of 1, −1, and 0, depending on whether the order of i, j, and k is cyclic (123, 231, 312) or repeated (112, 221, 331), etc. The permutation tensor is not commonly used, although it is useful in selected problems involving the application of couple stresses. The permutation tensor may also be used to represent the vector cross-product, which is given by

$$a \times b = \varepsilon_{ijk} n_i a_j b_k \tag{4.41}$$

where a and b are vectors, n_i is the normal to both vectors a and b, and ε_{ijk} is the permutation tensor. Furthermore, a small rotation of an element may be expressed as

$$\omega = \omega_k n_k = \frac{n_k \varepsilon_{ijk} u_{i,j}}{2} \tag{4.42}$$

This completes our brief introduction to tensor notation. The interested reader is referred to a number of excellent texts (listed in the bibliography) for further details on the subject. In particular, the classical text by Nye (1953) provides what is generally considered by many to be the clearest introduction to tensor notation.

4.6 GENERALIZED FORM OF LINEAR ELASTICITY

With the above introduction to tensor notation now complete, we will now return to Hooke's law of linear elasticity. This is simply an expression of the linear spring-like behavior of elastic solids (Fig. 4.2). Using tensor notation, Hooke's law may be expressed as

$$\sigma_{ij} = C_{ijkl}\varepsilon_{kl} \tag{4.43}$$

where C_{ijkl} is the fourth-order tensor that represents all the possible elastic constants. Expressed in terms of the direction cosines, C_{ijkl} is given by

$$C_{ijkl} = \sum_{i'=1}^{3}\sum_{j'=1}^{3}\sum_{k'=1}^{3}\sum_{l'=1}^{3} C_{i'j'k'l'}\ell_{i'i}\ell_{j'j}\ell_{k'k}\ell_{l'l} \tag{4.44}$$

The primed terms in Eq. (4.44) refer to the new co-ordinates, while the nonprimed terms refer to the old axes. It is also important to note that the order of ij and kl does not matter. Hence, $C_{ijkl} = C_{klji}$, $C_{ij} = C_{ji}$, and $C_{kl} = C_{lk}$. Similarly, the above elastic expressions can be expressed in terms of the fourth order elastic compliance tensor S_{ijkl} :

$$\varepsilon_{ij} = \sum_{k=1}^{3}\sum_{l=1}^{3} S_{ijkl}\sigma_{kl} + \alpha_{ij}\Delta T \tag{4.45}$$

where α_{ij} represents the thermal expansion coefficients, and ΔT is the temperature difference between the actual temperature and a stress-free temperature. The components of the compliance tensor transform in a manner similar to the elastic stiffness tensor, C_{ijkl}. For an isotropic material, i.e., a material with two independent elastic constants, C_{ijkl} is given by

$$C_{ijkl} = \lambda\delta_{ij}\delta_{kl} + \mu(\delta_{ik}\delta_{jl} + \delta_{il}\delta_{jk}) \tag{4.46}$$

where μ is the shear modulus, λ is Lame's constant, and δ_{ij} is the Kronecker delta for which $\delta_{ij} = 0$ when $i \neq j$ and $\delta_{ij} = 1$ when $i = j$. Young's modulus, E, and Poisson's ratio, υ, are given by

$$E = \frac{\mu(3\lambda + 2\mu)}{\lambda + \mu} \tag{4.47}$$

and

$$\upsilon = \frac{\lambda}{2(\mu + \lambda)} \tag{4.48}$$

As discussed earlier in this chapter, E represents the resistance to axial deformation, and υ represents the ratio of the transverse contractions to the axial elongation under axial loading. Values of E, G, and υ for selected materials are presented in Table 4.5. Note that these elastic properties are not significantly affected by minor alloying or microstructural changes.

TABLE 4.5 Summary of Elastic Properties of Assumed Isotropic Solids

Material at 68°F	E (10^6 psi)	G (10^6 psi)	ν
Metals			
Aluminum	10.2	3.8	0.345
Cadmium	7.2	2.8	0.300
Chromium	40.5	16.7	0.210
Copper	18.8	7.0	0.343
Gold	11.3	3.9	0.44
Iron	30.6	11.8	0.293
Magnesium	6.5	2.5	0.291
Nickel	28.9	11.0	0.312
Niobium	15.2	5.4	0.397
Silver	12.0	4.4	0.367
Tantalum	26.9	10.0	0.342
Titanium	16.8	6.35	0.321
Tungsten	59.6	23.3	0.280
Vanadium	18.5	6.8	0.365
Other materials			
Aluminum oxide (fully dense)	~ 60	—	—
Diamond	~ 140	—	—
Glass (heavy flint)	11.6	4.6	0.27
Nylon 66	0.17	—	—
Polycarbonate	0.35	—	—
Polyethylene (high density)	0.058–0.19	—	—
Poly(methyl methacrylate)	0.35–0.49	—	—
Polypropylene	0.16–0.39	—	—
Polystyrene	0.39–0.61	—	—
Quartz (fused)	10.6	4.5	0.170
Silicon carbide	~ 68	—	—
Tungsten carbide	77.5	31.8	0.22

Adapted from Hertzberg, 1996—reprinted with permission from John Wiley.
[a] *Source*: Kaye and Laby (1973).

TABLE 4.6 Relationships Between Elastic Properties of Isotropic Solids

	G	K	E	ν
G, E		$\frac{GE}{3(3G-E)}$		$\frac{E-2G}{2G}$
G, ν		$\frac{2G(1+\nu)}{3(1-2\nu)}$	$2G(1+\nu)$	
G, K			$\frac{9KG}{3K+G}$	$\frac{1}{2}\left[\frac{3K-2G}{3K+G}\right]$
E, ν	$\frac{E}{2(1+\nu)}$	$\frac{E}{3(1-2\nu)}$		
E, K	$\frac{3EK}{9K-E}$			$\frac{1}{2}\left[\frac{3K-E}{3K}\right]$
ν, K	$\frac{3K(1-2\nu)}{2(1+\nu)}$		$3K(1-2\nu)$	

Courtesy of L. Anand, MIT.

Also, typical values for Poisson's ratio, ν, are close to 0.3 for a large variety of materials.

It is important to note here that an isotropic material has only two independent elastic constants. Hence, if any two of the elastic constants (E, G, K, and υ) are known, then the other two elastic constants may be calculated from equations of isotropic linear elasticity. The expressions that relate the elastic constants for isotropic solids are summarized in Table 4.6. In reviewing the table, it is important to remember that the different moduli and elastic properties are associated with the chemical bonds between atoms. They are, therefore, intrinsic properties of a solid that do not vary significantly with microstructure or minor alloying additions.

The elastic constants may also be derived from energy potentials of the kind presented earlier in this chapter. However, detailed quantum mechanics derivations of the potentials are only now becoming available for selected materials. A summary of stiffness and compliance coefficients for a range of materials is presented in Table 4.2.

4.7 STRAIN ENERGY DENSITY FUNCTION

Under isothermal elastic conditions, the work done per unit volume in displacing the surfaces/boundaries of a system, $\mathrm{d}w = \mathrm{d}W/V$, can be expressed in terms of the incremental work done per unit volume. This yields:

$$dw = \sigma_{ij} d\varepsilon_{ij} = C_{ijkl} \varepsilon_{kl} d\varepsilon_{ij} \tag{4.49}$$

Equation (4.49) may also be applied to incremental plasticity problems, as discussed in Chap. 5. In any case, under isothermal incremental elastic loading conditions, the total work per unit volume is a single valued function of the form:

$$w = \frac{1}{2} C_{ijkl} \varepsilon_{ij} \varepsilon_{kl} = \phi(\varepsilon_{ij}) \tag{4.50}$$

where w is the strain energy density, which is given by

$$\sigma_{ij} = \frac{\partial \phi}{\partial \varepsilon_{ij}} \tag{4.51}$$

Differentiating the strain energy density gives

$$\frac{\partial^2 \phi}{\partial \varepsilon_{ij} \partial \varepsilon_{kl}} = \frac{\partial \sigma_{ij}}{\partial \varepsilon_{kl}} = C_{ijkl} \tag{4.52}$$

It is important to note that Eqs (4.43) and (4.52) suggest that there are 81 independent elastic constants. However, the equalities $\sigma_{ij} = \sigma_{ji}$ and $\varepsilon_{ij} = \varepsilon_{ji}$ reduce the number of independent elastic constants to 36. Also, the reversibility of elastic deformation leads to the result that the work done during elastic deformation is a unique function of strain that is independent of the loading path. Hence,

$$\frac{\partial^2 \phi}{\partial \varepsilon_{ij} \partial \varepsilon_{kl}} = \frac{\partial^2 \phi}{\partial \varepsilon_{kl} \partial \varepsilon_{ij}} = C_{ijkl} = C_{klij} \tag{4.53}$$

From Eq (4.53), it can be deduced that $C_{ij} = C_{ji}$, and hence there are only 21 independent elastic constants, as discussed in Sect 4.3. The concept of the strain energy density will be discussed further in subsequent sections on plasticity and fracture mechanics.

4.8 SUMMARY

An introduction to elasticity has been presented in this chapter. Following a brief description of the physical basis for elastic behavior, an introduction to anisotropic linear elasticity was presented. Equilibrium equations were then introduced for Cartesian, spherical, and cylindrical co-ordinate systems. An overview of the mathematical theory of elasticity was then presented before introducing tensor notation. Finally, the basic equations of elasticity were described in tensor form before concluding with a section on the strain energy density function.

BIBLIOGRAPHY

Ashby, M. F. and Jones, D. R. H. (1996) Engineering Materials I. 2nd ed., Butterworth Heinemann, Oxford, UK.

Courtney, T. H. (1990) Mechanical Behavior of Materials. McGraw-Hill, New York.

Crandall, S.H., Dahl, N. C., and Lardner, T. J., (eds), (1959) An Introduction to the Mechanics of Solids. McGraw-Hill, New York.

Hearn, E. J. (1985) Mechanics of Materials, 2nd ed., vol. 2. Pergamon Press, New York.

Hellwege, K. H. (1969) Elastic, Piezoelectric and Related Constants of Crystals. Springer-Verlag, Berlin, p 3.

Hertzberg, R. W. (1996) Deformation and Fracture Mechanics of Engineering Materials, 4th ed., John Wiley, New York.

Huntington, H. B. (1958) Solid State Physics. vol. 7. Academic Press, New York, p 213.

Kaye, G. W. C. and Laby, T. H. (1973) Tables of Physical and Chemical Constants. 14th ed. Longman, London, p. 31.

Love, A. E. H. (1944) Mathematical Theory of Elasticity. Dover, New York.

McClintock, F. and Argon, A. S. (eds) (1966) Mechanical Behavior of Materials. Addison-Wesley, Reading, MA.

Muskhelishvili, N. I. (1953) Some Basic Problems in the Mathematical Theory of Elasticity. Nordhoff, Groningen, The Netherlands.

Nye, J. F. (1957) Physical Properties of Crystals. Oxford University Press, London, UK.

Sokolnikoff, I. S. (1956) Mathematical Theory of Elasticity. McGraw-Hill, New York.

Timoshenko, S. and Goodier, J. N. (1951) Theory of Elasticity. McGraw-Hill, New York.

5

Introduction to Plasticity

5.1 INTRODUCTION

After a high enough stress is reached, the strain no longer disappears on the release of stress. The remaining permanent strain is called a "plastic" strain (Fig. 5.1). Additional incremental plastic strains may also be accumulated on subsequent loading and unloading, and these can lead ultimately to failure. In some cases, the dimensional and shape changes associated with plasticity may lead to loss of tolerance(s) and premature retirement of a structure or component from service. An understanding of plasticity is, therefore, important in the design and analysis of engineering structures and components.

This chapter presents a basic introduction to the mechanisms and mechanics of plasticity in monolithic materials. Following a simple review of the physical basis for plasticity in different classes of monolithic materials (ceramics, metals, intermetallics, and polymers), empirical plastic flow rules are introduced along with multiaxial yield criteria. Constitutive equations of plasticity are then presented in the final section of the chapter.

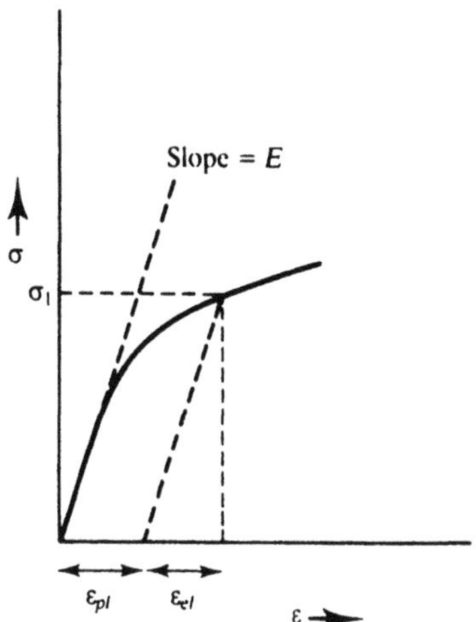

Figure 5.1 Schematic illustration of plastic strain after unloading.

5.2 PHYSICAL BASIS FOR PLASTICITY

5.2.1 Plasticity in Ceramics

Most ceramics only undergo only elastic deformation prior to the onset of catastrophic failure at room temperature. Hence, most reports on the mechanical properties of ceramics are often limited to elastic properties. Furthermore, most ceramists report flexural properties obtained under three- or four-point bending. Typical strength properties of selected ceramic materials are presented in Table 5.1. Note that ceramics are stronger (almost 15 times stronger) in compression than in tension. Also, the flexural strengths are intermediate between the compressive and tensile strength levels. Reasons for these load-dependent properties will be discussed in subsequent chapters. For now, it is simply sufficient to state that the trends are due largely to the effects of pre-existing defects such as cracks in the ceramic structures.

The limited capacity of ceramic materials for plastic deformation is due largely to the limited mobility of dislocations in ceramic structures. The latter may be attributed to their large Burgers (slip) vectors and unfavorable (for plastic deformation) ionically/covalently bonded crystal structures. Plastic deformation in ceramics is, therefore, limited to very small strains (typically < 0.1–1%), except at elevated temperatures where thermally activated dislocation motion and grain boundary sliding are possible. In fact, the extent of plasticity at elevated temperatures may be very significant in

TABLE 5.1 Strength Properties of Selected Ceramic Materials

Material	Compressive strength (MPa (ksi)]	Tensile strength [MPa (ksi)]	Flexural strength [MPa (ksi)]	Modulus of elasticity [GPa (10^6 psi)]
Alumina (85% dense)	1620 (235)	125 (18)	295 (42.5)	220 (32)
Alumina (99.8% dense)	2760 (400)	205 (30)	345 (60)	385 (56)
Alumina silicate	275 (40)	17 (2.5)	62 (9)	55 (8)
Transformation toughened zirconia	1760 (255)	350 (51)	635 (92)	200 (29)
Partially stabilized zirconia +9% MgO	1860 (270)	—	690 (100)	205 (30)
Cast Si_3N_4	138 (20)	24 (3.5)	69 (10)	115 (17)
Hot-pressed Si_3N_4	3450 (500)	—	860 (125)	—

Sources: After Hertzberg, 1996. Reprinted with permission from John Wiley.
[a] *Guide to Engineering Materials*. vol. 1(1). ASM, Metals Park, OH, 1986, pp 16, 64, 65.

ceramics deformed at elevated temperature, and superplasticity (strain levels up to 1000% plastic strain) has been shown to occur due to creep phenomena in some fine-grained ceramics produced.

However, in most ceramics, the plastic strains to failure are relatively small (< 1%), especially under tensile loading which tends to open up preexisting cracks that are generally present after processing. Also, since incipient cracks in ceramics tend to close up under compressive loading, the strength levels and the total strain to failure in compression are often greater than those in tension. Furthermore, very limited plasticity (permanent strains on removal of applied stresses) may occur in some ceramics or ceramic matrix composites by microcracking or stress-induced phase transformations.

Microcracking generally results in a reduction in Young's modulus, E, which may be used as a global/scalar measure of damage (Fig. 5.2). If we assume that the initial "undeformed" material has a damage state of zero, while the final state of damage at the point of catastrophic failure corresponds to a damage state of 1, we may estimate the state of damage using some simple damage rules. For an initial Young's modulus of E_0 and an intermediate damage state, the damage variable, D, is given simply by $D = 1 - E/E_0$. Damage tensors may also be used to obtain more rigorous descriptions of damage (Lemaitre, 1991).

Plasticity in ceramics may also occur by stress-induced phase transformations. This has been observed in partially stabilized zirconia (ZrO_2 alloyed with CaO, Y_2O_3, or CeO to stabilize the high-temperature tetragonal

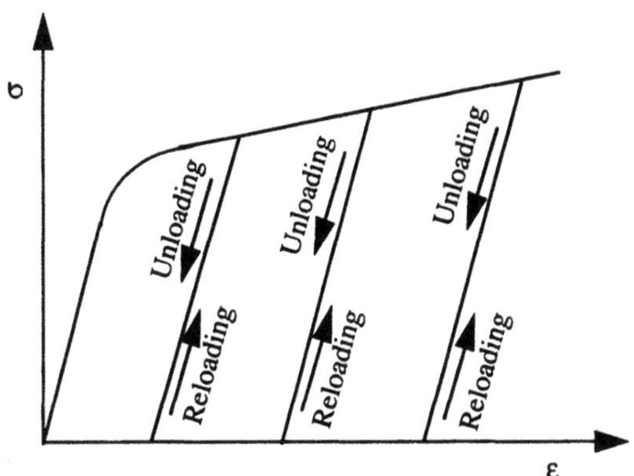

FIGURE 5.2 Schematic showing the change in modulus due to damage during loading and unloading sequences.

phase down to room temperature). Under monotonic loading, the metastable tetragonal phase can undergo stress-induced phase transformations from the tetragonal to the monoclinic phase. This stress-induced phase transformation is associated with a volume increase of ~ 4%, and can give rise to a form of toughening known as transformation toughening, which will be discussed in Ch. 13.

Stress-induced phase transformations occur gradually in partially stabilized zirconia, and they give rise to a gradual transition from linearity in the elastic regime, to the nonlinear second stage of the stress–strain curve shown in Fig. 5.3. The second stage ends when the stress-induced transformation spreads completely across the gauge section of the specimen. This is followed by the final stage in which rapid hardening occurs until failure. It is important to note that the total strain to failure is limited, even in partially stabilized zirconia polycrystals. Also, as in conventional plasticity, stress-induced transformation may be associated with increasing, level, or decreasing stress–strain behavior (Fig 5.4).

5.2.2 Plasticity in Metals

In contrast to ceramics, plastic deformation in metals is typically associated with relatively large strains before final failure. This is illustrated in Fig. 5.5 using data obtained for an aluminum alloy. In general, the total plastic strains can vary between 5 and 100% in ductile metals deformed to failure at room temperature. However, the elastic portion of the stress–strain curve is generally limited to strains below ~ 0.1 to 1%. Furthermore, metals and their alloys may exhibit stress–strain characteristics with rising, level, or

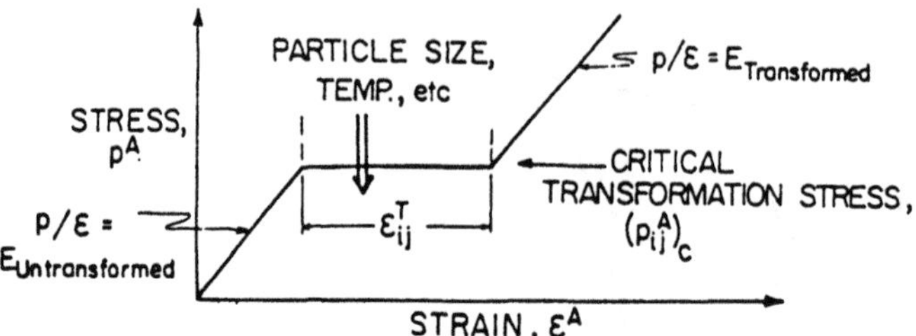

Figure 5.3 Schematic of the three stages of deformation in material undergoing stress-induced phase transformation. (After Evans et al., 1981.)

decreasing stress, as shown in Fig. 5.4. Materials in which the stress level remains constant with increasing strain [Fig. 5.4(b)] are known as elastic–perfectly plastic. Materials in which the stress level decreases with increasing strain are said to undergo strain softening [Fig. 5.4(c)], while those in which the stress level increases with increasing strain are described as strain hardening materials, Fig. 5.4(a).

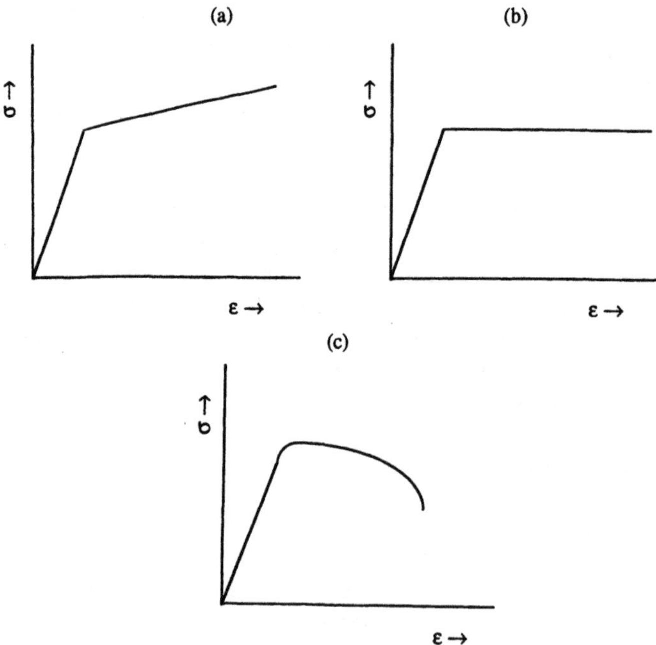

Figure 5.4 Types of stress–strain response: (a) strain hardening; (b) elastic–perfectly plastic deformation; (c) strain softening.

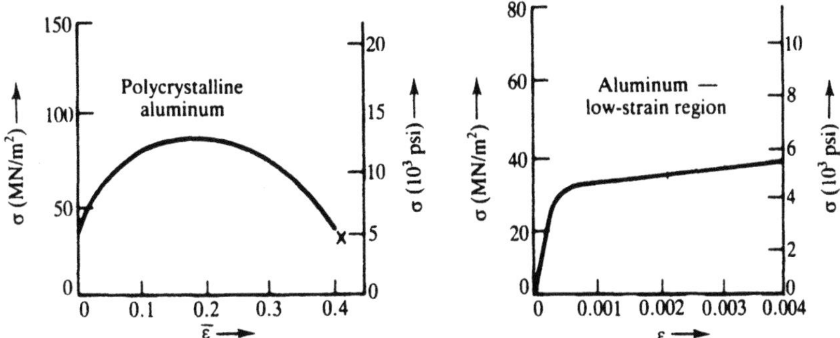

FIGURE 5.5 Stress–strain behavior in an aluminum alloy. (After Courtney, 1990. Reprinted with permission from McGraw-Hill.)

Strain hardening occurs as a result of dislocation interactions in the fully plastic regime. These may involve interactions with point defects (vacancies, interstitials, or solutes), line defects (screw, edge, or mixed dislocations), surface defects (grain boundaries, twin boundaries, or stacking faults), and volume defects (porosity, entrapped gases, and inclusions). The dislocation interactions may give rise to hardening when additional stresses must be applied to overcome the influence of defects that restrict dislocation motion. This may result in rising stress–strain curves that are characteristic of strain hardening behavior, Fig. 5.4(a).

As discussed earlier, the stress–strain curves may also remain level [Fig. 5.4(b)], or decrease or increase continuously with increasing strain, Fig. 5.4(c). The reasons for such behavior are generally complex, and not fully understood at present. However, there is some limited evidence that suggests that elastic–perfectly plastic behavior is associated with slip planarity, i.e., slip on a particular crystallographic plane, while strain softening tends to occur in cases where slip localizes on a particular microstructural feature such as a precipitate. The onset of macroscopic yielding, therefore, corresponds to the stress needed to shear the microstructural feature. Once the initial resistance to shear is overcome, the material may offer decreasing resistance to increasing displacement, giving rise ultimately to strain softening behavior, Fig. 5.4(c).

Since the moving dislocations interact with solute clouds, serrated yielding phenomena may be observed in the stress–strain behavior [Fig 5.6). Different types of serrated yielding phenomena have been reported due to the interactions of dislocations with internal defects such as solutes and interstitials. The phenomenon is generally referred to as the Portevin–

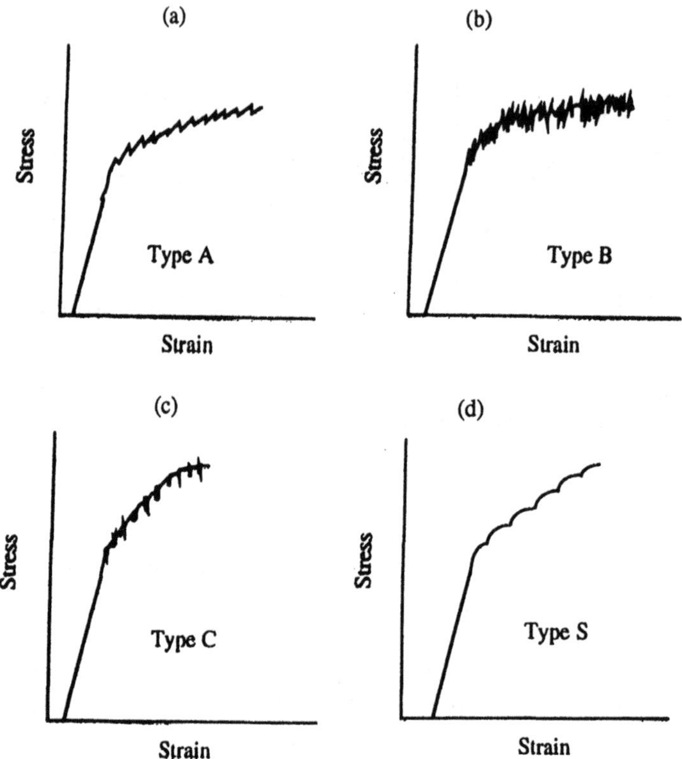

FIGURE 5.6 Types of serrated yielding phenomena: (a) Type A; (b) Type B; (c) Type C; (d) Type S. (Types A–C After Brindley and Worthington, 1970; Type S After Pink, 1994. Reprinted with permission from *Scripta Met.*)

LeChatelier effect, in honor of the two Frenchmen who first reported it (Portevin and LeChatelier, 1923). The serrations are caused by the pinning and unpinning of groups of dislocations from solutes that diffuse towards it as it moves through a lattice. The mechanisms is particularly effective at particular parametric ranges of strain-rate and temperature (Cottrell, 1958).

Finally in this section, it is important to discuss the so-called anomalous yield phenomena that has been reported in some plain carbon steels (Fig. 5.7). The stress–strain plots for such materials have been observed to exhibit double yield points in some annealed conditions, as shown in Fig. 5.7. The upper yield point (UYP) corresponds to the unpinning of dislocations from interstitial carbon clouds. Upon unpinning, the load drops to a lower yield point (LYP). Lüder's bands (shear bands inclined at $\sim 45^\circ$ degrees to the loading axis) are then observed to propagate across the

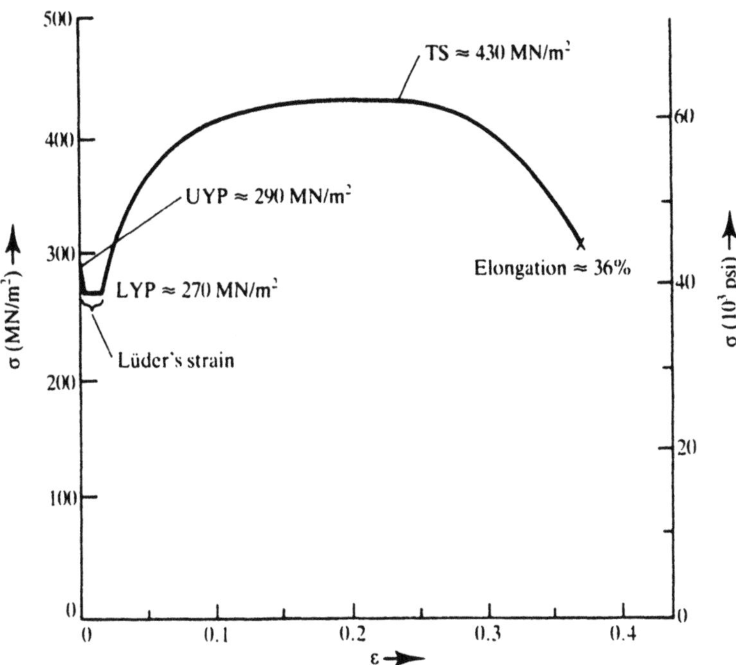

Figure 5.7 Anomalous yielding in 1018 plain carbon steel. (After Courtney, 1990. Reprinted with permission from McGraw-Hill.)

gauge sections of the tensile specimens, as the strain is increased further (Fig. 5.7). Note that the stress remains relatively constant in the so-called Lüder's strain regime, although serrations may be observed with sufficiently sensitive instrumentation. The strain at the end of this constant stress regime is known as the Lüders strain. This corresponds to the point at which the Lüder's bands have spread completely across the gauge section of the specimen. Beyond this point, the stress generally increases with increasing due to the multiple interactions between dislocations, as discussed earlier for conventional metallic materials (Fig. 5.5).

5.2.3 Plasticity in Intermetallics

As discussed in Chap. 1, intermetallics are compounds between metals and other metals. Due to their generally ordered structures, and partially covalently or ionically bonded structures, intermetallics generally exhibit only limited plasticity at room-temperature. Nevertheless, some ductility has been reported for ordered gamma-based titanium aluminide intermetallics with duplex $\alpha_2 + \gamma$ microstructures. These two phase intermetallics have room temperature plastic elongations to failure of about 1–2% due to deformation by slip and twinning (Kim and Dimiduk, 1991). Their limited room-temperature ductility has been attributed to the soaking up of interstitial oxygen by the α_2 phase. This results in a reduction in interstitial oxygen content in the gamma phase, and the increased dislocation mobility of dislocations in the latter which gives rise to the improved ductility in two-phase gamma titanium aluminides (Vasudevan et al., 1989).

Niobium aluminide intermetallics with plastic elongations of 10–30% have also been developed in recent years (Hou et al., 1994; Ye et. al., 1998). The ductility in these B2 (ordered body-centered cubic structures) intermetallics has been attributed to the partial order in their structures. Similar improvements in room-temperature (10–50%) ductility have been reported in Ni_3Al intermetallics that are alloyed with boron (Aoki and Izumi, 1979; Liu et al., 1983), and Fe_3Al intermetallics alloyed with boron (Liu and Kumar, 1993).

The improvements in the room-temperature ductilities of the nickel and iron aluminide intermetallics have been attributed to the cleaning up of the grain boundaries by the boron additions. However, the reasons for the improved ductility in ordered or partially ordered intermetallics are still not fully understood, and are under investigation. Similarly, anomalous yield-point phenomena (increasing yield stress with increasing temperature) and the transition from brittle behavior at room temperature to ductile behavior at elevated temperature are still under investigation.

5.2.4 Plasticity in Polymers

Plasticity in polymers is not controlled by dislocations, although dislocations may also exist in polymeric structures. Instead, plastic deformation in polymers occurs largely by chain sliding, rotation, and unkinking (Figs 1.7 and 1.8). Such chain sliding mechanisms do not occur so readily in three-dimensional (thermoset) polymers (Fig. 1.8). However, chain sliding may occur relatively easily in linear (thermoplastic) polymers when the sliding of polymer chains is not hindered significantly by radical side groups or other steric hindrances. The plastic deformation of polymers is also associated with significant changes in entropy, which can alter the local driving force for deformation.

Elasticity and plasticity [Fig. 5.8(a)] in rubbery polymers may result in strain levels that are between 100 and 1000% at fracture. Such large strains are associated with chain sliding, unkinking, and uncoiling mechanisms. Furthermore, unloading does not result in a sudden load drop. Instead, unloading follows a time-dependent path, as shown in Fig. 5.8(b).

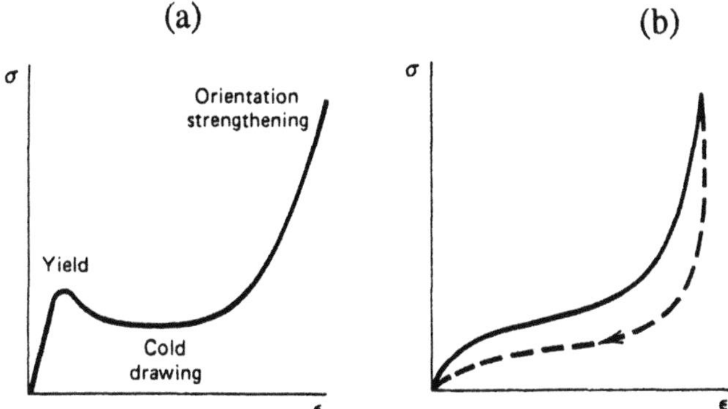

FIGURE 5.8 Elastic–plastic deformation in rubbery polymers. (a) Rubber rand deformed at room temperature. (After Argon and McClintock, 1990) (b) Viscoelasticity in a rubbery polymer. (After Hertzberg, 1996. Reprinted with permission from John Wiley.)

Elasticity and plasticity in rubbery polymers are, therefore, often time dependent, since time is often required for the polymer chains to flow to and from the deformed configurations. Cyclic deformation may result in hysterisis loops since the strain generally lags the stress (Fig. 5.9), and anomalous stress–strain behavior may also be associated with chain interactions with distributed side groups which are often referred to as steric hindrances.

Crystalline polymers (Fig. 1.9) may also exhibit interesting stress–strain behavior. The minimum in the stress–strain curve is due to cold drawing and the competition between the breakdown of the initial crystalline structure, and the reorganization into a highly oriented chain structure.

5.3 ELASTIC–PLASTIC BEHAVIOR

A generic plot of stress versus strain is presented in Fig. 5.10. This shows a transition from a linear "elastic" regime to a nonlinear "plastic regime." The linear elastic regime persists up to the proportional limit, at which the deviation from linear elastic behavior occurs. However, the onset of nonlinear stress–strain behavior is generally difficult to determine experimentally. An engineering offset yield strength is, therefore, defined by drawing a line parallel to the original linear elastic line, but offset by a given strain (usually an engineering strain level of 0.002 or 0.2%).

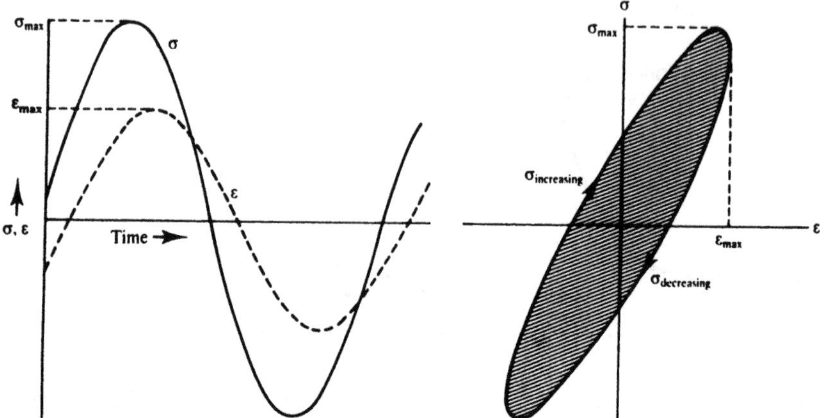

Figure 5.9 Hysterisis loop in a cyclically deformed polymer.

The arbitrary offset strain level of 0.002 is recommended by the ASTM E-8 code for tensile testing for the characterization of stresses required for bulk yielding. However, it is important to remember that the offset strain level is simply an arbitrary number selected by a group of experts with a considerable amount of combined experience in the area of tensile testing.

Above the offset yield strength, *A*, the stress may continue to increase with increasing applied strain. The slope of the stress–strain curve in the

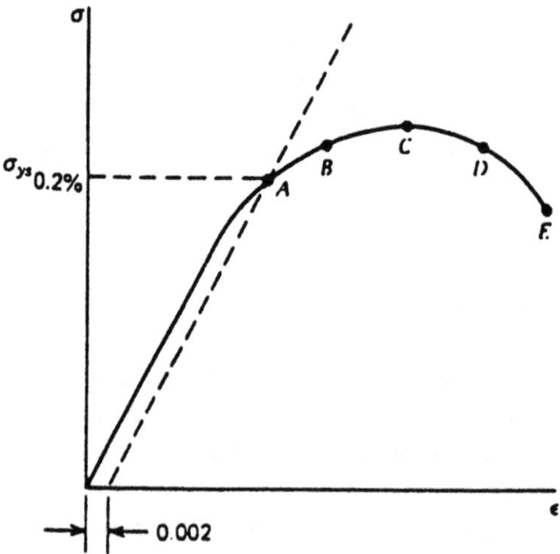

Figure 5.10 Schematic of stress–strain behavior in the elastic and plastic regimes.

plastic regime depends largely on the underlying dislocation interactions. The resulting shape changes in the gauge sections of tensile specimens are illustrated in Fig. 5.11. Stretching in the vertical direction is accompanied by Poisson contraction in the elastic regime. However, the contraction in the horizontal direction is countered by hardening during the initial stages of plastic deformation in which the gauge section deforms in a relatively uniform manner, Fig. 5.12(a). The rate of rate of hardening is, therefore, greater than the rate of horizontal contraction, and the total volume of deformed material remains constant, Fig. 5.12(a). This inequality persists until the ultimate tensile strength, M, is reached in Fig. 5.11. At this stress level, the rate of hardening is equal to the rate contraction of the gauge area, as shown in Fig. 5.12(b).

Beyond the point M, in the stress–strain plot, geometrical instabilities (internal microvoids and microcracks within the gauge section) dominate the plastic response, and the rate of horizontal contraction is greater than the rate of hardening, Fig. 5.12(c). The deformation is thus concentrated within regions with the highest crack/microvoid density, and a phenomenon known as "necking" [Figs 5.11 and 5.12(c)] occurs beyond the ultimate

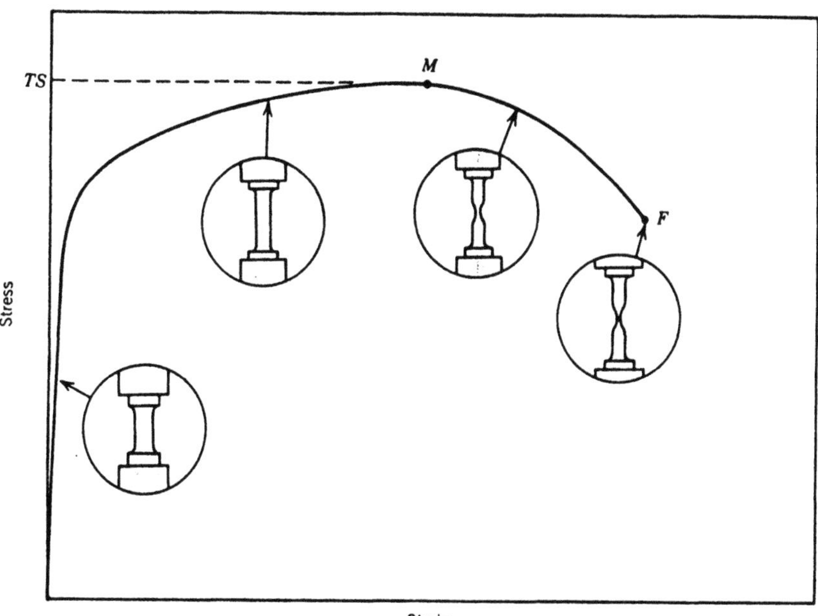

FIGURE 5.11 Schematic illustration of gauge deformation in the elastic and plastic regimes.

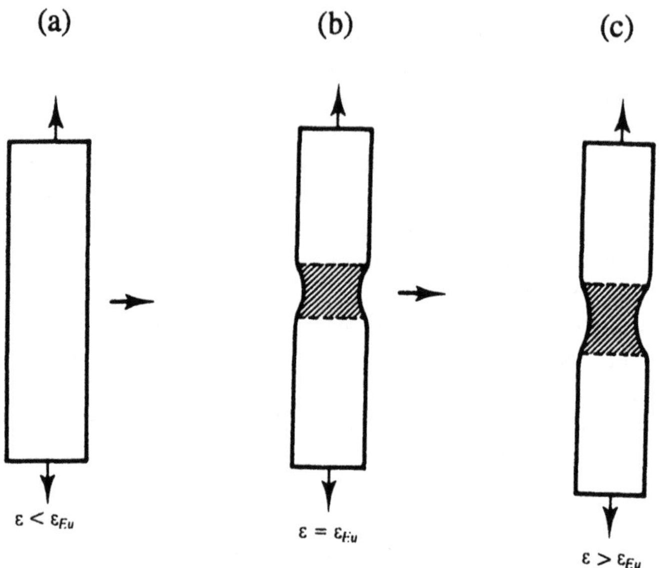

Figure 5.12 Hardening versus geometrical instability: (a) rate of hardening > rate of geometrical instability formation; (b) rate of hardening = rate of geometrical instability formation (onset of necking); (c) rate of hardening < rate of geometrical instability formation (necking down to failure) (After Courtney, 1990. Reprinted with permission from McGraw-Hill.)

tensile stress. This involves the gradual reduction in the cross-sectional area in the regime of concentrated deformation. This reduction occurs because of the rate of horizontal contraction is now greater than the rate of hardening, Fig. 5.12(c). Necking may continue until the geometrical instabilities coalesce. In any case, catastrophic failure occurs when a critical condition is reached.

It is important to note here that the onset of necking may be delayed by the application of hydrostatic stresses to the gauge section of a tensile specimen. This was first shown by Bridgman (1948) who demonstrated that the ductility of metals could be increased significantly with increasing hydrostatic stress. This is because the hydrostatic stresses tend to close up pores and voids that lead ultimately to necking and fracture.

The geometrical instabilities are, therefore, artifacts of the test conditions and specimen geometries that are used in tensile tests (Fig 5.13). Note that the tensile specimen geometries (usually dog-bone shapes) are typically designed to minimize stress concentrations in the region of transition from the grip to the gauge sections. This is done to avoid fracture outside the

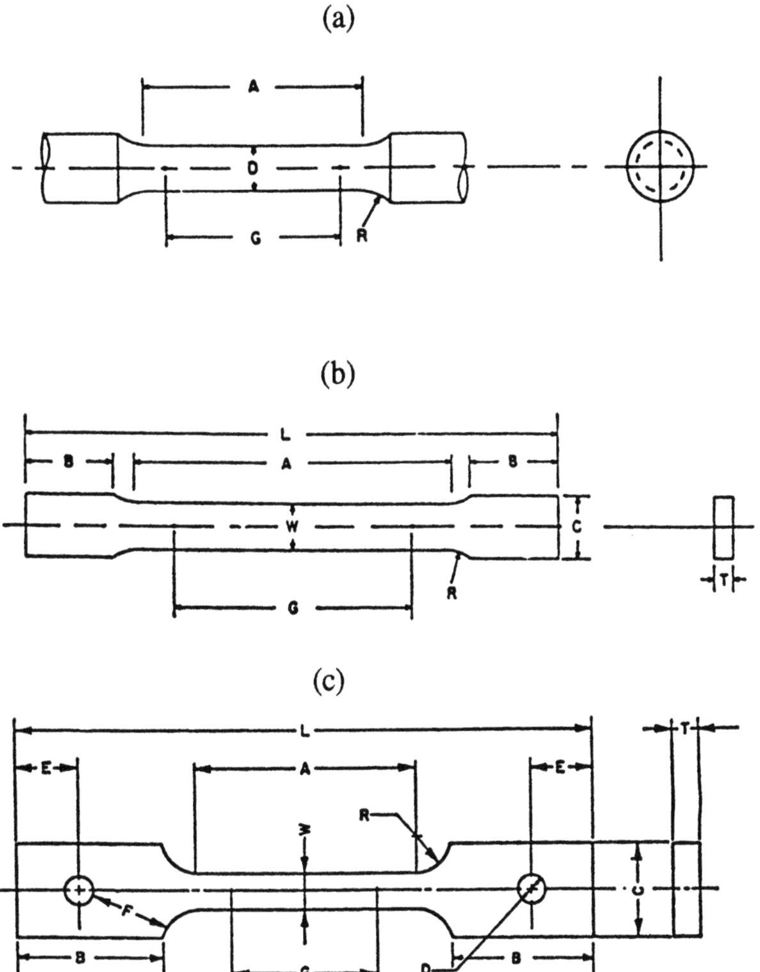

FIGURE 5.13 Types of tensile specimen geometries: (a) cylindrical cross--sections, (b) dog-bone specimen (wedge grips); (c) dog-bone specimen (pin loaded).

gauge section. Also, the engineering definitions of stress and strain may not be applicable to situations in which the cross-sectional area changes significantly during incremental plastic deformation to failure (Figs 5.11 and 5.12).

True stress and true strain levels must, therefore, be defined, especially in the plastic regime. The true engineering stress, σ_T, is given by the ratio of applied load, P, to the actual cross-sectional area, A. This gives

$$\sigma_T = \text{True stress} = \frac{\text{Applied load}}{\text{Actual cross-sectional area}} = \frac{P}{A} \tag{5.1}$$

In contrast, the engineering stress, σ_E, is given simply by the ratio of applied load, P, to the original cross-sectional area, A_0. This gives

$$\sigma_E = \text{Engineering stress} = \frac{\text{Applied load}}{\text{Original cross-sectional area}} = \frac{P}{A_0} \tag{5.2}$$

Similarly, the engineering strain levels are different from the true strain levels which are known to increase in an incremental manner. The true strain, ε_T, is obtained from the incremental theory of plasticity by separating the total displacement into incremental portions. This gives

$$\varepsilon_T = \sum_{i=1}^{n} \frac{d\ell_i}{\ell_i} \tag{5.3}$$

where $d\ell_i$ is the increment in specimen length that occurs during the *ith* deformation stage, and n is the number of incremental deformation stages. In the limit, Eq. (5.3) may be expressed in integral form, which gives

$$\varepsilon_T = \int_{\ell_0}^{\ell} \frac{d\ell}{\ell} = \ln\left(\frac{\ell}{\ell_0}\right) \tag{5.4}$$

where ℓ is the actual instantaneous (deformed) length, ℓ_0 is the initial (undeformed) length, and ln denotes natural logarithms. The true strain is, therefore, different from the engineering strain, ε_E, which is given simply by the ratio of the change in gauge length, $\delta\ell = \ell - \ell_0$, to the original (undeformed) length, ℓ_0:

$$\varepsilon_E = \frac{\delta\ell}{\ell_0} = \frac{\ell - \ell_0}{\ell_0} \tag{5.5}$$

Furthermore, since the deformed volume remains constant during plastic deformation, the initial volume of the gauge section before plastic deformation must be equal to the final volume of the gauge section during plastic deformation. If the initial (undeformed) gauge cross-sectional area is A_0 and the deformed cross-sectional area is A (during plastic deformation), then the initial volume ($A_0\ell_0$) and the deformed volume ($A\ell$) must be the same. Hence, the area ratio, A_0/A, must be equal to the length ratio, ℓ/ℓ_0. The equations for engineering strain and true strain (Equations (5.4) and (5.5)] may, therefore, be expressed as

$$\varepsilon_E = \frac{\ell}{\ell_0} - \ell = \frac{A_0}{A} - \ell \tag{5.6}$$

and

$$\varepsilon_T = 1\text{n}\left(\frac{\ell}{\ell_0}\right) = 1\text{n}\left(\frac{A_0}{A}\right) = 1\text{n}(\ell + \varepsilon_E) \tag{5.7}$$

Similarly, the true stress, σ_T, may be expressed in terms of the engineering stress since

$$\sigma_T = \frac{P}{A} = \frac{P}{A_0} \cdot \frac{A_0}{A} = \sigma_E \frac{A_0}{A} = \sigma_E \frac{1}{1_0} = \sigma_E(1 + \varepsilon_E) \tag{5.8}$$

It is important to note here that all of the above definitions of true stress and true strain are valid for stress levels below or equal to the ultimate tensile strength (the maximum engineering stress) in the plot of engineering stress versus engineering strain (Fig. 5.11). However, due to the effects of geometrical instabilities (Fig. 5.12), the expressions involving the gauge length terms should not be used in the regime beyond the ultimate tensile strength. This is because the cross-sectional areas decrease by necking in areas with the greatest concentrations of geometrical instabilities. Area ratios should, therefore, be used in the determination of true stress and true strain levels for deformation beyond the ultimate tensile strength.

A characteristic plot of true stress versus true strain is compared with a typical engineering stress–strain plot in Fig. 5.14. The "true" stress–strain plot obtained from length ratios is shown in solid lines, while the true stress–strain plot obtained from area ratios is represented by the dashed lines in Fig. 5.14. The two plots are coincident until the ultimate tensile stress is reached. Note also that the true stress–strain plots are shifted to the top and to the left of the original points on the engineering stress–strain plots (Fig. 5.14). Also, there is no indication of an ultimate tensile strength on the true stress–strain plot. This is because the ultimate tensile strength is purely a test artifact that is due to the presence of geometrical instabilities within a test specimen.

As discussed earlier, the smooth parabolic stress–strain curves observed in the plasticity regime are generally associated with the bulk/irreversible movement of dislocations in metals and intermetallics. The plastic stress–strain curves may also be associated with chain sliding and chain uncoiling/unkinking processes in noncrystalline polymers. In any case, the slope in the rising portion of the plastic stress–strain plots is a measure of resistance to plastic deformation (Fig. 5.11). The material is said to undergo strain hardening in this regime (Fig. 5.11).

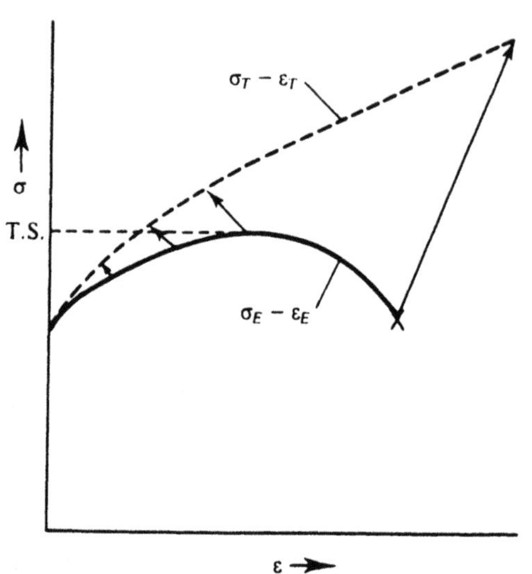

A schema showing the relationship between tensile true stress–true strain (dotted line) and engineering stress–engineering strain (solid line). For engineering strains less than ε_{Eu}, $\sigma_T > \sigma_E$ and $\varepsilon_T < \varepsilon_E$. The necking point ($\sigma_E$ = T.S.) has no particular significance in the true stress–true strain curve. At some strain greater than ε_{Eu}, ε_T (when calculated on the basis of neck area) becomes greater than ε_E, although σ_T remains greater than σ_E.

Figure 5.14 Comparison of true stress–strain behavior with engineering stress–strain behavior.

It is important to note here, however, that some materials do not undergo any strain hardening. In cases where constant stress levels are required to continue the plastic straining, the materials are described as perfectly plastic materials, Fig. 5.4(b). In contrast, strain softening occurs when the stress required for deformation decreases with increasing strain, Fig. 5.4(c). In most materials, however, the portion of the stress strain curve between the onset of bulk yielding (bulk yield stress) and the onset of necking (the ultimate tensile stress) tends to exhibit the type of rising stress–strain behavior shown in Fig. 5.4(a).

5.4 EMPIRICAL STRESS–STRAIN RELATIONSHIPS

It is currently impossible to develop *ab initio* methods for the prediction of the stress–strain behavior of materials from detailed descriptions of the underlying atomic and defect structures. However, some useful *empirical* relationships have been developed for the characterization of the true stress–strain behavior. The most popular empirical mathematical relationship is usually attributed to Hollomon (1945), although it was first proposed by Bülfinger (1735) about 200 years earlier. The so-called Hollomon equation is given by

$$\sigma = K(\varepsilon)^n \tag{5.9}$$

where σ is the true stress, K is a proportionality constant that represents the true stress at a true strain of 1.0, and n is the strain hardening/work hardening exponent which is a measure of the resistance to plastic deformation. In general, the strain hardening, n, is a number between 0 and 1. Also, n is sensitive to thermomechanical processing and heat treatment, and it is generally higher in materials tested in the annealed or hot worked conditions. Strain hardening exponents for selected materials are presented in Table 5.2. Such data may be obtained readily from log–log plots of true stress versus true strain. Taking logarithms of Eq. (5.9) gives

$$\log \sigma = \log k + n \log \varepsilon \tag{5.10}$$

Equation (5.10) is the equation of a straight line, with a slope of n, and a y axis intercept of log K. Hence, from Eq. (5.10), the material constants K and n may be obtained from the intercept and slope, respectively, of a plot of log σ versus log ε. However, it is important to note that the straight-line relationship suggested by Eq. (5.10) is not always followed by every material. Error analysis must, therefore, be performed to determine the applicability of the Hollomon equation. Also, in several materials, the hardening coefficient, n, is often found empirically to be approximately equal to the true strain at the ultimate tensile strain. This is referred to as the *Considere criterion*, and it may be derived simply by noting that $P = \sigma A$, and finding the condition for which $dP = \sigma dA + A d\sigma = 0$ (see Sect. 5.5). In any case, it is important to remember that the applicability of the Considere criterion must be verified by appropriate error analysis.

In general, however, the strain hardening exponent of a metallic material increases with increasing strength and decreasing dislocation mobility. The stress–strain behavior of a material may also be significantly affected by

Table 5.2 Strain Hardening Exponents of Selected Metallic Materials

Material	Strain hardening coefficient, n
Stainless steel	0.45–0.55
Brass	0.35–0.4
Copper	0.3–0.35
Aluminum	0.15–0.25
Iron	0.05–0.15

After Hertzberg, 1996. Reprinted with permission from John Wiley.

the strain-rate, $\dot{\varepsilon} = d\varepsilon/dt$. The effects of strain rate can be modeled using the following empirical power law equation:

$$\sigma = K'(\dot{\varepsilon})^m \tag{5.11}$$

where σ is the true stress, $\dot{\varepsilon}$ is the true strain-rate, K' is a proportionality constant corresponding to the stress for a strain-rate of 1 s^{-1}, and m is the strain-rate sensitivity factor which can have values between 0 and 1. As for the strain hardening exponent, the strain-rate sensitivity can be determined from log–log plots of stress versus $\dot{\varepsilon}$. Materials with strain-rate sensitivity factors between 0 and 0.1 are not strain-rate sensitive, while materials with strain-rate sensitivities between 0.5 and 1 are very strain-rate sensitive.

Most materials have strain-rate sensitivity values close to 0.2. However, very strain-rate sensitive materials such as "silly putty" may have strain-rate sensitivity values close to 1. Such materials are resistant to fracture due to necking. They may, therefore, deform extensively by necking down to a point. This is because the rate of hardening at high strain-rates tends to prevent the onset of necking. In the most extreme cases, this leads to superplastic behavior which is associated with extremely high plastic strain levels between 100 and 1000%. High values of the strain-rate sensitivity index, m, are, therefore, one indication of the potential for superplasticity.

Finally in this section, it is of interest to note that the combined effects of strain-rate sensitivity and strain hardening (on the true stress) can be assessed using the following equation which is obtained by combining Eqs (5.9) and (5.11). This gives

$$\sigma = K''(\varepsilon)^n(\dot{\varepsilon})^m \tag{5.12}$$

where σ is the true stress, and K'' is a proportionality constant which is related to the material constants, K and K'. The variable, ε, is the true plastic strain, n is the strain hardening exponent, $\dot{\varepsilon}$ is the strain-rate, and m is the strain-rate sensitivity. As before, the proportionality constants may be determined form appropriate log–log plots. Note that either strain, ε, or strain-rate, $\dot{\varepsilon}$, may be varied independently or simultaneously in Eq. (5.12). The applicability of Eq. (5.12) must also be established by comparing predicted true stresses with actual true stresses obtained for each material, i.e., error analysis must be performed to determine the applicability of Eq. (5.12).

Tensile properties for some common engineering materials are presented in Table 5.3. Note that the data presented in this latter table are very dependent on microstructure and composition. The effects of these variables on yield strength will be discussed in detail in Chap. 7. For now, it is simply sufficient to note that tensile strength generally increases with decreasing grain size.

5.5 CONSIDERE CRITERION

As stated in Sect. 5.4, the Considere criterion may be derived by considering the conditions that must be satisfied at the ultimate tensile strength (Fig. 5.11). The ultimate tensile strength corresponds to the maximum value of load, P, in the plot of load versus strain. Since $P = \sigma A$, the maximum value of P may be obtained by equating the first derivative, dP, to zero. This gives

$$dP = A d\sigma + \sigma dA = 0 \tag{5.13}$$

Rearranging Eq. (5.13) and separating variables gives

$$-\frac{dA}{A} = \frac{d\sigma}{\sigma} \tag{5.14}$$

Since there is no change in volume, $V = A\ell$, associated with plastic deformation, we may also assume that d$V = 0$. Hence,

$$dV = A d\ell + \ell dA = 0 \tag{5.15}$$

Rearranging Eq. (5.15) and separating variables, we obtain:

$$-\frac{dA}{A} = \frac{d\ell}{\ell} = d\varepsilon \tag{5.16}$$

If we now assume that the Hollomon equation can be used to describe the stress–strain response, i.e., $\sigma = K\varepsilon^n$, then dσ, obtained by differentiating Eq. (5.9), is given by

$$d\sigma = Kn\varepsilon^{n-1} d\varepsilon = n(K\varepsilon^n)\varepsilon^{-1} d\varepsilon = n\sigma\varepsilon^{-1} d\varepsilon \tag{5.17}$$

Also, substituting Eqs (5.14) and (5.16) into Eq. (5.17) yields:

$$d\varepsilon = \frac{n\sigma\varepsilon^{-1} d\varepsilon}{\sigma} \tag{5.18a}$$

or

$$n = \varepsilon \tag{5.18b}$$

Equation 5.18(b) is often referred to as the *Considere criterion.* It states that *the strain at the onset of necking is equal to the strain hardening exponent.* It is a very useful "rule-of-thumb." It is important to remember that the Considere criterion is only applicable when Hollomon's equation can be used to describe the stress–strain behavior of a material.

Table 5.3 Tensile Properties of Selected Engineering Materials

Material	Treatment	Yield strength (MPa)	Tensile strength (MPa)	Elongation in 5-cm gauge (%)	Reduction in area (1.28 cm diameter) (%)
Steel alloys					
1015	As-rolled	315	420	39	61
1050	As-rolled	415	725	20	40
1080	As-rolled	585	965	12	17
1340	$Q+T$ (205°C)	1590	1810	11	35
1340	$Q+T$(425°C)	1150	1260	14	51
1340	$Q+T$(650°C)	620	800	22	66
4340	$Q+T$ (205°C)	1675	1875	10	38
4340	$Q+T$(425°C)	1365	1470	10	44
4340	$Q+T$(650°C)	855	965	19	60
301	Annealed plate	275	725	55	—
304	Annealed plate	240	565	60	—
310	Annealed plate	310	655	50	—
316	Annealed plate	250	565	55	—
403	Annealed bar	275	515	35	—
410	Annealed bar	275	515	35	—
431	Annealed bar	655	860	20	—
AFC-77	Variable	560–1605	835–2140	10–26	32–74
PH 15-7Mo	Variable	380–1450	895–1515	2–35	—
Titanium alloys					
Ti-5Al-2.5Sn	Annealed	805	860	16	40
Ti-8Al-I Mo-1V	Duplex annealed	950	1000	15	28
Ti-6Al-4V	Annealed	925	995	14	30
Ti-13V-11Cr-3Al	Solution + age	1205	1275	8	—
Magnesium alloys					
AZ31B	Annealed	103–125	220	9–12	—
AZ80A	Extruded bar	185–195	290–295	4–9	—
ZK60A	Artificially aged	215–260	295–315	4–6	—
Aluminum alloys					
2219	-T31, -T351	250	360	17	—
2024	-T3	345	485	18	—
2024	-T6, -T651	395	475	10	—
2014	-T6, -T651	415	485	13	—
6061	-T4, -T451	145	240	23	—
7049	-T73	475	530	11	—

TABLE 5.3 Continued

Material	Treatment	Yield strength (MPa)	Tensile strength (MPa)	Elongation in 5-cm gauge (%)	Reduction in area (1.28 cm diameter) (%)
7075	-T6	505	570	11	—
7075	-T73	415	505	11	—
7178	-T6	540	605	11	—
Plastics					
ACBS	Medium impact	—	46	6–14	—
Acetal	Homopolymer	—	69	25–75	—
Poly(tetra fluorethylene)	—	—	14–48	100–450	—
Poly(vinylidene fluoride)	—	—	35–48	100–300	—
Nylon 66	—	—	59–83	60–300	—
Polycarbonate	—	—	55–69	130	—
Polyethylene	Low density	—	7–21	50–800	—
Polystyrene	—	—	41–54	1.5–2.4	—
Polysulfone	—	69	—	50–1000	—

Sources: After Hertzberg, 1996. Reprinted with permission from John Wiley
[a]*Datebook 1974, Metal Progress* (mid-June 1974).

5.6 YIELDING UNDER MULTIAXIAL LOADING

5.6.1 Introduction

So far, we have considered only yielding under uniaxial loading conditions. However, in several engineering problems of practical interest, yielding may occur under multiaxial loading conditions that include both axial and shear components (Fig. 3.2). The yielding conditions under multiaxial loading will clearly depend on the magnitudes and directions of the local axial and shear components of stress. To avoid unnecessary dependence on the choice of co-ordinate systems, stress invariants of the stress tensor are often defined for the local stress states. These stress invariants are independent of the choice of co-ordinate system, and they can be used to develop yielding criteria that are independent of co-ordinate system. Multiaxial yielding criteria will be presented in this section for monolithic materials. The reader should review the section on the invariants of the stress tensor in Chap. 3 before proceeding with the rest of this chapter.

5.6.2 Multiaxial Yield Criteria

Unlike uniaxial loading, yielding under multiaxial loading may be induced by an almost infinite combination of stresses. It is, therefore, very difficult to develop first principles models for the prediction of the combinations of stresses that are required for the initiation of bulk plastic flow under multiaxial loading conditions. Instead of first principles models, empirical flow rules have been used for the prediction of the combined stresses required to cause yielding under multiaxial loading conditions. It is important to remember that these flow rules are *empirical* in nature. They are, therefore, approximate solutions. However, extensive work has been done to verify their general applicability to a wide range of engineering materials.

5.6.2.1 Tresca Yield Criterion

The simplest and most commonly used flow rule was first proposed by Tresca (1869). It is, therefore, called the *Tresca yield criterion*. This criterion states that yielding will occur under multiaxial loading when the shear stress at a point is a maximum, i.e., when the shear stress is equal to half the uniaxial yield stress, Fig. 5.15(a). The Tresca yield criterion is given by

$$\frac{\sigma_1 - \sigma_3}{3} = \frac{\sigma_y}{2} = \tau_y \tag{5.19}$$

where σ_1 and σ_3 are the maximum and minimum principal stress values, is the uniaxial yield stress, and τ_y is the shear yield stress. Note that the yield locus for the Tresca yield criterion is a hexagon in two dimensions. It is a useful exercise to try to construct this locus from Eq. (5.19). The trick is to note that the signs of the principal stresses change from quadrant to quadrant.

The Tresca yield criterion is often used in industry because of its simplicity. However, it neglects the possible contributions from shear components. Significant errors may, therefore, be associated with the Tresca yield criterion, especially in cases where the terms in the deviatoric stress tensor are significant.

5.6.2.2 Von Mises Yield Criterion

The Von Mises yield criterion is used in many cases where improved accuracy is required. It was proposed independently by Von Mises (1913), although an equivalent expression was suggested in earlier work by Huber (1904). Both Huber and Von Mises equate the yield stress to the distortional

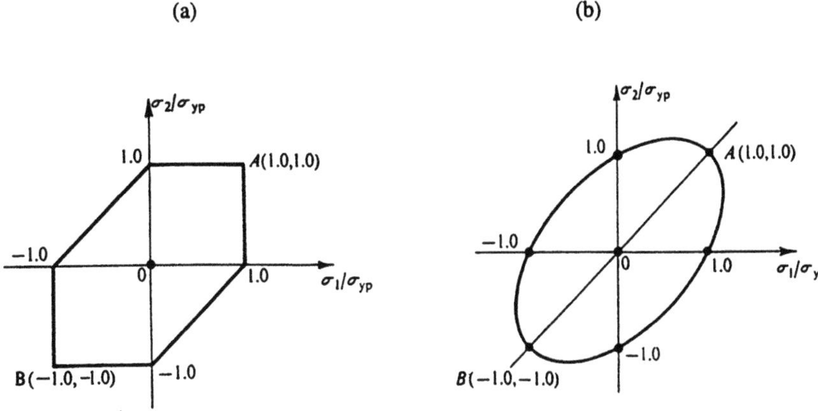

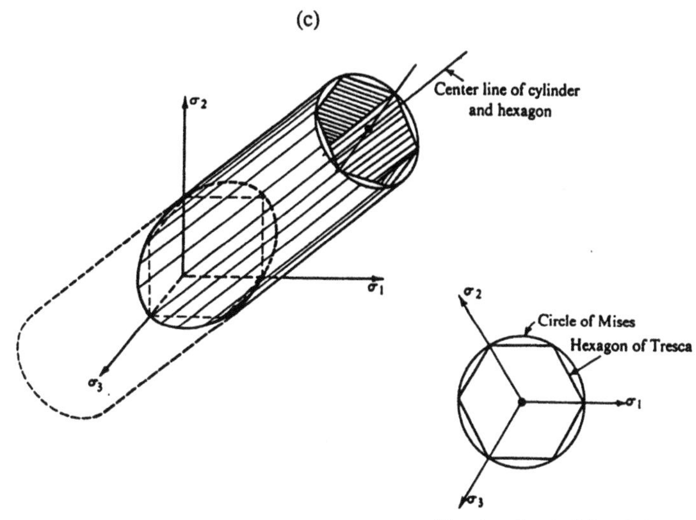

FIGURE 5.15 Loci of yield criteria: (a) two-dimensional Tresca yield criterion; (b) two-dimensional Von Mises yield criterion; (c) three-dimensional Tresca and Von Mises yield criteria.

energy, U. This gives the following empirical condition for plastic flow under multiaxial loading:

$$\sigma_y = \frac{1}{\sqrt{2}}\left\{(\sigma_{xx}-\sigma_{yy})^2+(\sigma_{yy}-\sigma_{zz})^2+(\sigma_{zz}-\sigma_{xx})^2 + 6(\tau_{xy}^2+\tau_{yz}^2+\tau_{zx}^2\right\}^{\frac{1}{2}} \tag{5.20a}$$

or

$$\sigma_y = \frac{1}{6}\left[(\sigma_1-\sigma_2)^2+(\sigma_2-\sigma_3)^2+(\sigma_3-\sigma_1)^2\right]^{\frac{1}{2}} \tag{5.20b}$$

For consistency, we will refer to the above yield criterion as the Von Mises yield criterion. We note here that the Von Mises yield criterion includes all the six independent stress components (Fig. 3.2). Also, in the case of two-dimensional stress states, the Von Mises yield locus is an ellipse, Fig. 5.15(b). The Von Mises ellipse can be constructed from the hexagonal yield locus of the Tresca yield locus. However, unlike the Tresca yield criterion discussed earlier, the Von Mises yield criterion accounts for the effects of the shear stress components on yielding under multiaxial loading conditions. Furthermore, the three-dimensional yield locus for the Von Mises yield criterion is a cylinder, as shown in Fig. 5.15(c). Similarly, the three-dimensional yield locus for the Tresca yield criterion is a hexagonal prism, Fig. 5.15(c). Finally, it is important to recognize that yielding in several metallic materials is generally observed to occur at stress levels that are intermediate between those predicted by the Tresca and Von Mises yield criteria (Fig. 5.16).

5.7 INTRODUCTION TO J_2 DEFORMATION THEORY

It is useful at this stage to identify some general relationships between stress state and yield criteria under multiaxial loading. First, it is important to note that yielding generally occurs when the combinations of J_2 [see Eqs (3.38) and (3.39)] and the yield stress in pure shear, τ_y, reach a critical value. Hence, yielding occurs when

$$f(J_2, \tau_y) = 0 \tag{5.21}$$

where J_2 is the second invariant of the deviatoric stress tensor. In tensor notation, J_2 is given by

$$J_2 = \frac{1}{2}\sigma'_{ij}\sigma'_{ij} \tag{5.22}$$

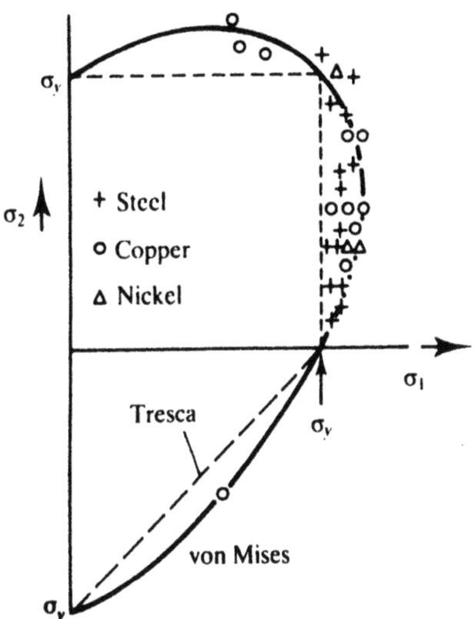

Figure 5.16 Comparison of experimental and empirical multiaxial yield criteria. (After Courtney, 1990. Reprinted with permission from McGraw-Hill.)

where σ'_{ij} is the deviatoric stress described in Chap. 3. As discussed, J_2 and τ_y are related to both the Tresca and Von Mises yield criteria. The Tresca yield criterion may thus be expressed as

$$\tau_y = \frac{\sigma_1 - \sigma_3}{2} \tag{5.23}$$

where σ_1 and σ_3 are the maximum and minimum principal stresses, respectively. The Von Mises yield criterion may also be expressed in terms of J_2 and τ_y, which give

$$J_2 = \tau_y^2 \tag{5.24}$$

However, it is important to remember that the value of τ_y depends on the yield criterion used. Appropriate values of τ_y for the Tresca and Von Mises yield criteria are given below:

$$\tau_y = \begin{cases} \dfrac{\sigma_y}{2} \text{(Tresca yield criterion)} \\ \dfrac{\sigma_y}{\sqrt{3}} \text{(Von Mises yield criterion)} \end{cases} \tag{5.25}$$

where σ_y is the yield stress under uniaxial loading conditions.

5.8 FLOW AND EVOLUTIONARY EQUATIONS (CONSTITUTIVE EQUATIONS OF PLASTICITY)

Finally in this chapter, constitutive equations will be presented for the prediction of plastic flow in monolithic materials. The most commonly used equations were first proposed by Prandtl and Reuss. The so-called Prandtl–Reuss equations are constitutive equations that describe the elastic–plastic response of work hardening and non-work hardening material. For non-work hardening materials, i.e., elastic–perfectly plastic materials, the incremental plastic strains and stresses are given by

$$d\varepsilon'_{ij} = \frac{3\sigma'_{ij}}{2\sigma_y} d\bar{\varepsilon}_p + \frac{d\sigma'_{ij}}{2G} \tag{5.26}$$

$$d\varepsilon_{ii} = \frac{(1-2\nu)}{E} d\sigma_{ii} \tag{5.27}$$

$$\sigma'_{ij}\sigma'_{ij} = 2\tau_y^2 = \text{constant} \tag{5.28}$$

For work hardening materials, the hydrostatic and deviatoric strain increments are given by

$$d\varepsilon_{ii} = \frac{(1-2\nu)}{E} d\sigma_{ii} \tag{5.29}$$

$$d\varepsilon'_{ij} = \frac{3\sigma'_{ij} d\bar{\sigma}}{2\bar{\sigma}H} + \frac{d\sigma'_{ij}}{2G} \tag{5.30}$$

where $\bar{\sigma}$ is the effective stress given by $3\sigma_{ij}\sigma_{ij}/2$, H is the slope of the uniaxial/effective stress versus plastic strain, ε_p, plot, E is Young's modulus, G is the shear modulus, ν is Poisson's ratio and $d\varepsilon_p$ is the equivalent plastic strain defined by $3d\varepsilon_{ij}^p d\varepsilon_{ij}^p/2$. Upon unloading, the above equations can be reduced to an elastic equation given by

$$d\varepsilon_{ii} = \frac{(1-2\nu)}{E} d\sigma_{ii} \tag{5.31}$$

$$d\varepsilon'_{ij} = \frac{d\sigma_{ij}}{2G} \tag{5.32}$$

The above equations are useful in incremental simulations of plastic flow processes such as metal forming. Such simulations often involve incremental changes in plastic strains. Further details on the application of incremental plasticity theories may be found in texts on plasticity theory by Nadai (1950), Hill (1950), and Prager (1951). It is interesting to note that both Nadai and Prager were students of Prandtl in Gottingen. Other students of Prandtl include Von Karman, Professor and Mrs. Flügge, and Den Hartog, who all went on to make important contributions in the field of mechanics. In recent years, the original ideas of Prandtl have also been extended by Anand and Kalidindi (1994) to include microscopic details on crystal plasticity models within a finite element framework.

5.9 SUMMARY

An introduction to plasticity is presented in this chapter. Following a brief description of the physical basis of plasticity in ceramics, metals, intermetallics, and polymers, empirical plastic flow rules were introduced along with the Considere criterion. A simple review of J_2 deformation theory was also presented along with the flow and evolutionary (constitutive) equations of plasticity (Prandtl–Reuss equations).

BIBLIOGRAPHY

Anand, L. and Kalidindi, S. R. (1994) The Process of Shear Band formation in Plane Strain Compression in FCC Metals: Effects of Crystallographic Texture, Mechanics and Materials. Vol. 17, pp 223–243.

Aoki, K. and Izumi, O. (1979) Nippon Kinzoku Gakkaishi. vol. 43, p 1190.

ASTM Handbook For Mechanical Testing. E-8 Code. American Society for Testing and Materials, Philadelphia, PA.

Bridgman, P. W. (1948) Fracturing of Metals. ASM International, Materials Park, OH, p 246.

Brindley and Worthington (1970) Metall Rev. vol. 15, p 101.

Bülfinger, G. B. (1735) Common Acad Petrop. vol. 4, p 164.

Cottrell, A. H. (1958) Vacancies and Other Point Defects in Metals and Alloys. Institute of Metals, London, p 1.

Courtney, T. (1990) Mechanical Behavior of Materials. McGraw-Hill, New York.

Hertzberg, R. W. (1996) Deformation and Fracture Mechanics of Engineering Materials. 4th ed., John Wiley, New York.

Hill, R. (1950) The Mathematical Theory of Plasticity, Oxford University Press, Oxford, UK.

Hollomon, J. H. (1945) Trans AIME. vol. 162, p 248.

Hou, D. H., Shyue, J., Yang, S. S. and Fraser, H. L. (1994) Alloy Modeling and Design. In: G. M. Stocks and P. Z. A. Turch, eds. TMS, Warrendale, PA, pp 291–298.

Huber, M. T. (1904) Czasopismo Tech. vol. 15, Lwov.

Kim, Y.-W. and Dimiduk, D. (1991) Journal of Metals. vol. 43, pp 40–47.

Lemaitre, J. (1991) A Course on Damage Mechanics, Springer-Verlag, New York.

Liu, C. T. and Kumar, K. (1993) Metals. vol. 45, p 38.

Liu, C. T., White, C. L., Koch, C. C., and Lee, E. H. (1983) Proceedings of Symposium on High Temperature Materials Chemistry II. In: Munir et al., eds). Electrochemical Society, Inc.

McClintock, F. and Argon, A. S., (eds) (1963) Mechanical Behavior of Materials, Addison Wesley, MA.

Nadai, A. (1950) Theory of Flow and Fracture of Solids. 2nd ed.

Pink (1994) Scripta Met. vol. 30, p 767.

Portevin, A. and LeChatelier, F. (1923) Sur un Phenomene Observe lors de Pessai de traction d'alliages en cours de Transformation, Acad Sci Compt Rend. vol. 176, pp 507–510.

Prager, W. and Hodge, P.G., Jr. (1951) Theory of Perfectly Plastic Solids. John Wiley, New York.

Tresca (1869) Mémoires Sur l'Écoulement des Corps Solides, Mém. Présentés par Divers Savants. vol. 20.

Vasudevan, V. K., Court, S. A., Kurath, P. and Fraser, H. L. (1989) Scripta Metall. vol. 23, pp 467–469.

Von Mises, R. (1913) Gottinger Nachr. p 582.

Ye, F., Mercer, C. and Soboyejo, W. O. (1998) Metall Mater Trans. vol. 29A, pp 2361–2374.

6

Introduction to Dislocation Mechanics

6.1 INTRODUCTION

Early in the 20th century, a number of scientists tried to predict the theoretical strength of a crystalline solid by estimating the shear stress required to move one plane of atoms over another (Fig. 6.1). They found that the predicted theoretical strengths were much greater than the measured strengths of crystalline solids. The large discrepancy (an order of magnitude or two) between the theoretical and measured shear strengths puzzled many scientists until Orowan, Polanyi, and Taylor (1934) independently published their separate classical papers on dislocations (line defects).

The measured strengths were found to be lower than the predicted theoretical levels because plasticity occurred primarily by the movement of line defects called dislocations. The stress levels required to induce dislocation motion were lower than those required to shear complete atomic planes over each other (Fig. 6.1). Hence, the movement of dislocations occurred prior to the shear of atomic planes that was postulated by earlier workers such as Frenkel (1926).

Since 1934, numerous papers have been published on the role of dislocations in crystalline plasticity. A number of books (Hirth and Lothe, 1982; Hull and Bacon, 1984; Weertman and Weertman, 1992) have also been written on the subject. This chapter will, therefore, not attempt to

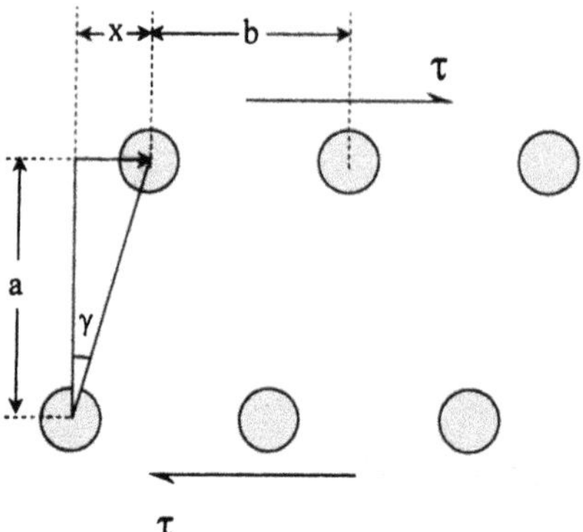

FIGURE 6.1 Shear of one row of atoms over another in a perfect crystal.

present a comprehensive overview of dislocations. Instead, the fundamental ideas in dislocation mechanics required for a basic understanding of crystalline plasticity will be presented at an introductory level. The interested reader is referred to papers and more advanced texts that are listed in the bibliography at the end of the chapter.

6.2 THEORETICAL SHEAR STRENGTH OF A CRYSTALLINE SOLID

Frenkel (1926) obtained a useful estimate of the theoretical shear strength of a crystalline solid. He considered the shear stress required to cause shear of one row of crystals over the other (Fig. 6.1). The shear strain, γ, associated with a small displacement, x, is given by

$$\gamma = \frac{x}{a} \tag{6.1}$$

Hence, for small strains, the shear stress, τ, may be obtained from

$$\tau = G\gamma = G\frac{x}{a} \tag{6.2}$$

where G is the shear modulus. Similarly, we may also use an approximate sinusoidal potential function to obtain an expression for the variation in the applied shear stress, τ, as a function of displacement, x (Fig. 6.2). This gives

$$\tau = \tau_{max} \sin\left(\frac{2\pi x}{b}\right) \tag{6.3}$$

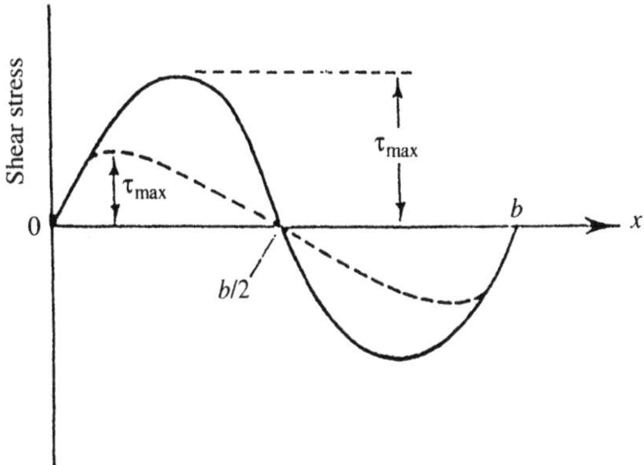

FIGURE 6.2 Schematic illustration of shear stress variations. The dashed curve corresponds to more precise shear stress – displacement function.

where τ_{max} is the maximum shear stress in the approximately sinusoidal τ versus x curve shown in Fig. 6.2; x is the displacement, and b is the interatomic spacing (Fig. 6.1). For small displacements, $\sin(2\pi x/b) \sim 2\pi x/b$. Hence, τ is given by

$$\tau = \tau_{max}\left(\frac{2\pi x}{b}\right) \tag{6.4}$$

We may now equate Eqs (6.4) and (6.2) to obtain an expression for τ_{max}. Noting that for cubic crystals b this gives

$$\tau_{max} = \frac{G}{2\pi} \tag{6.5}$$

Equation 6.5 provides an approximate measure of the theoretical shear strength of a crystalline solid. More rigorous analysis using more representative interatomic potentials (Fig. 6.2) gives estimates of the theoretical shear strength to be $\sim G/30$. However, most estimates of the theoretical shear strength are about one or two orders of magnitude greater than the measured values obtained from actual crystalline solids.

This discrepancy between the measured and theoretical strengths led Orowan, Polanyi, and Taylor (1934) to recognize the role of line defects (dislocations) in crystal plasticity. However, these authors were not the first to propose the idea of dislocations. Dislocation structures were first proposed by Volterra (1907), whose purely mathematical work was unknown to Orowan, Polanyi, and Taylor in 1934 when they published their original papers on dislocations.

Since the early ideas on dislocations, considerable experimental and analytical work has been done to establish the role of dislocations in crystal plasticity. Materials with low dislocation/defect content (whiskers and fibers) have also been produced by special processing techniques. Such materials have been shown to have strength levels that are closer to theoretical strength levels discussed earlier (Kelly, 1986). The concept of dislocations has also been used to guide the development of stronger alloys since much of what we perceive as strengthening is due largely to the restriction of dislocation motion by defects in crystalline solids.

6.3 TYPES OF DISLOCATIONS

There are basically two types of dislocations. The first type of dislocation that was proposed in 1934 is the edge dislocation. The other type of dislocation is the screw dislocation, which was proposed later by Burgers (1939). Both types of dislocations will be introduced in this section before discussing the idea of mixed dislocations, i.e., dislocations with both edge and screw components.

6.3.1 Edge Dislocations

The structure of an edge dislocation is illustrated schematically in Fig. 6.3. This shows columns of atoms in a crystalline solid. Note the line of atoms at which the half-filled column terminates. This line represents a discontinuity in the otherwise perfect stacking of atoms. It is a line defect that is generally referred to as an edge dislocation. The character of an edge dislocation may also be described by drawing a so-called right-handed Burgers circuit around the dislocation, as shown in Fig. 6.4(a). Note that S in Fig. 6.4 corresponds to the *start* of the Burgers circuit, while F corresponds to the *finish*. The direction of the circuit in this case is also chosen to be right-handed, although there is no general agreement on the sign convention in the open literature. In any case, we may now proceed to draw the same Burgers circuit in a perfect reference crystal, Fig. 6.4(b). Note that the finish position, F, is different from the start position, S, due to the absence of the edge dislocation in the perfect reference crystal.

We may, therefore, define a vector to connect the finish position, F, to the start position, S, in Fig. 6.4(b). This vector is called the *Burgers vector*. It is often denoted by the letter, b, and it corresponds to one atomic spacing for a single edge dislocation. It is important to remember that we have used a right-handed finish-to-start definition in the above discussion. However, this is not always used in the open literature. For consistency, however, we

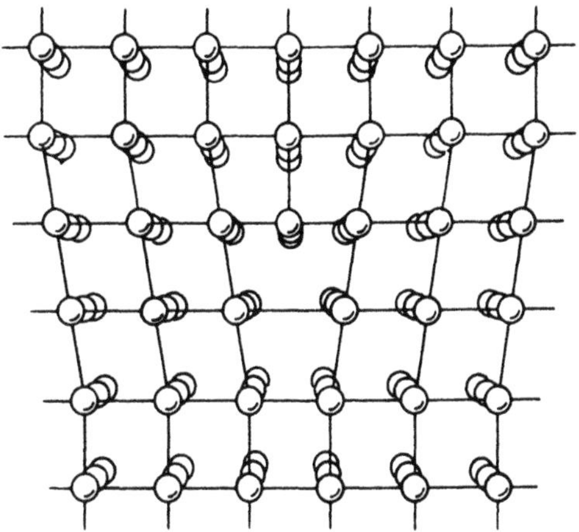

Figure 6.3 Schematic of edge dislocation. (Taken from Hirthe and Lothe, 1982. Reprinted with permission from John Wiley.)

will retain the current sign convention, i.e., the finish-to-start (F/S) right-handed rule.

Finally in this section, it is important to note that we may define the sense vector, **s**, of an edge dislocation in a direction along the dislocation line (into the page). The sense of an edge dislocation, **s**, is therefore perpendicular to the Burgers vector, **b**. Hence, we may describe an edge dislocation

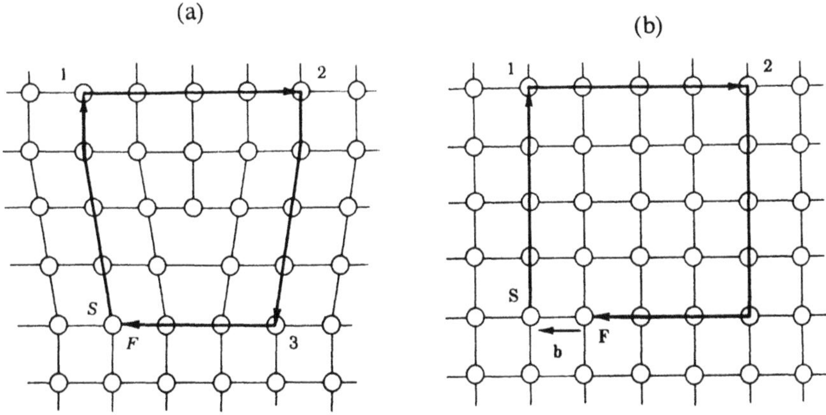

Figure 6.4 Finish to start (F/S) right-handed Burgers circuits: (a) around edge dislocation; (b) in a perfect reference crystal. (Taken from Hirthe and Lothe, 1982. Reprinted with permission from John Wiley.)

as a line defect with a sense vector, **s**, that is perpendicular to the Burgers vector, **b**, i.e. **b.s** = 0.

6.3.2 Screw Dislocations

The structure of a screw dislocation may be visualized by considering the shear displacement of the upper half of a crystal over the lower half, as shown in Fig. 6.5(a). If the atoms in the upper half of the crystal are denoted as open circles, while those in the lower half are denoted as filled circles [Fig. 6.5(a)], then the relative displacements between the open and filled circles may be used to describe the structure of a screw dislocation. The arrangement of the atoms around the dislocation line AB follows a spiral path that is somewhat similar to the path that one might follow along a spiral staircase. This is illustrated clearly in Fig. 6.5(b) for a right-handed screw dislocation.

As before, we may also define a Burgers vector for a screw dislocation using a finish-to-start right-handed screw rule. This is shown schematically in Fig. 6.6. Note that the Burgers vector is now parallel to the sense vector, **s**, along the dislocation line. This is in contrast with the edge dislocation for which the Burgers vector is perpendicular to the sense vector. In any case, we may now formally describe a right-handed screw dislocation as one with **b.s** = **b**. A left-handed screw dislocation is one with **b.s** = −**b**.

6.3.3 Mixed Dislocations

In reality, most dislocations have both edge and screw components. It is, therefore, necessary to introduce the idea of a mixed dislocation (one with both edge and screw components). A typical mixed dislocation structure is shown in Fig. 6.7(a). Note that this dislocation structure is completely screw in character at A, and completely edge in character at B. The segments of the dislocation line between A and B have both edge and screw components. They are, therefore, mixed dislocation segments.

Other examples of mixed dislocation structures are presented in Figs 6.7(b) and 6.7(c). The screw components of the mixed dislocation segments, $\mathbf{b}_s$, may be obtained from the following expression:

$$\mathbf{b}_s = (\mathbf{b}.\mathbf{s})\mathbf{s} \tag{6.6}$$

Similarly, the edge components, $\mathbf{b}_e$, of the mixed dislocation segments may be obtained from

$$\mathbf{b}_e = \mathbf{s} \times (\mathbf{b} \times \mathbf{s}) \tag{6.7}$$

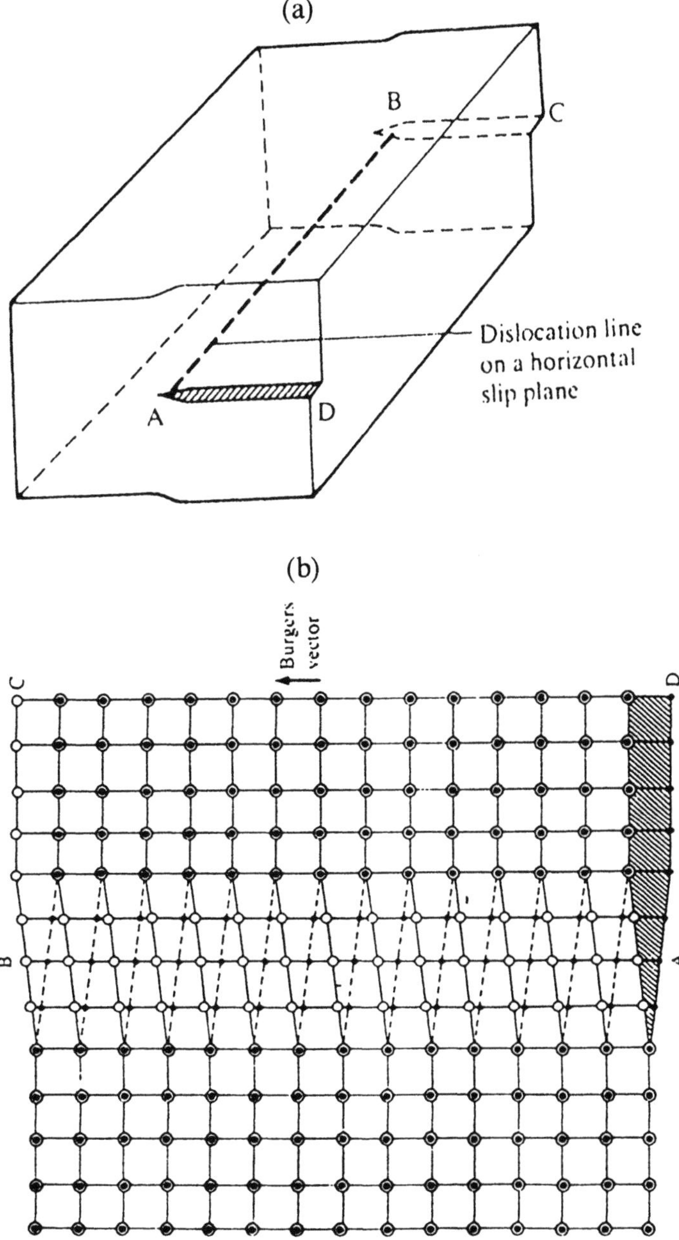

Figure 6.5 Structure of a screw dislocation: (a) displacement of upper half of crystal over lower half; (b) spiral path along the dislocation line. (From Read, 1953. Reprinted with permission from McGraw-Hill.)

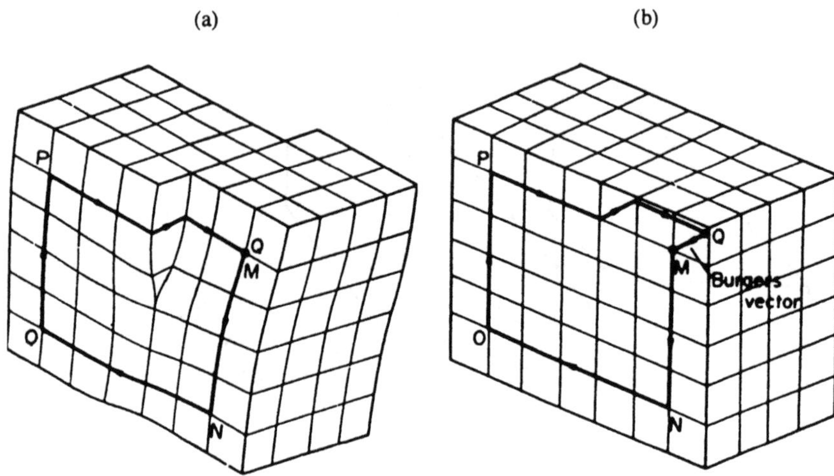

Figure 6.6 Right-handed Burgers circuits: (a) around screw dislocation; (b) in perfect reference crystal. (From Hull and Bacon, 1984. Reprinted with permission from Pergamon Press.)

6.4 MOVEMENT OF DISLOCATIONS

As discussed earlier, crystal plasticity is caused largely by the movement of dislocations. It is, therefore, important to develop a clear understanding of how dislocations move through a crystal. However, dislocations also encounter lattice friction as they move through a lattice. Estimates of the lattice friction stress were first obtained by Peierls (1940) and Nabarro (1947). Considering the motion of a dislocation in a lattice with lattice parameters a and b (Fig. 6.1), they obtained a simple expression for the lattice friction stress, τ. The so-called Peierls–Nabarro lattice friction stress is given by

$$\tau_f = G \exp\left[\frac{-2\pi a}{b(1-\nu}\right] \tag{6.8a}$$

or

$$\tau_f = G \exp\left[\frac{-2\pi w}{b}\right] \tag{6.8b}$$

where a is the vertical spacing between slip planes, b is the slip distance or Burgers vector, G is the shear modulus, w is the dislocation width (Fig 6.8),

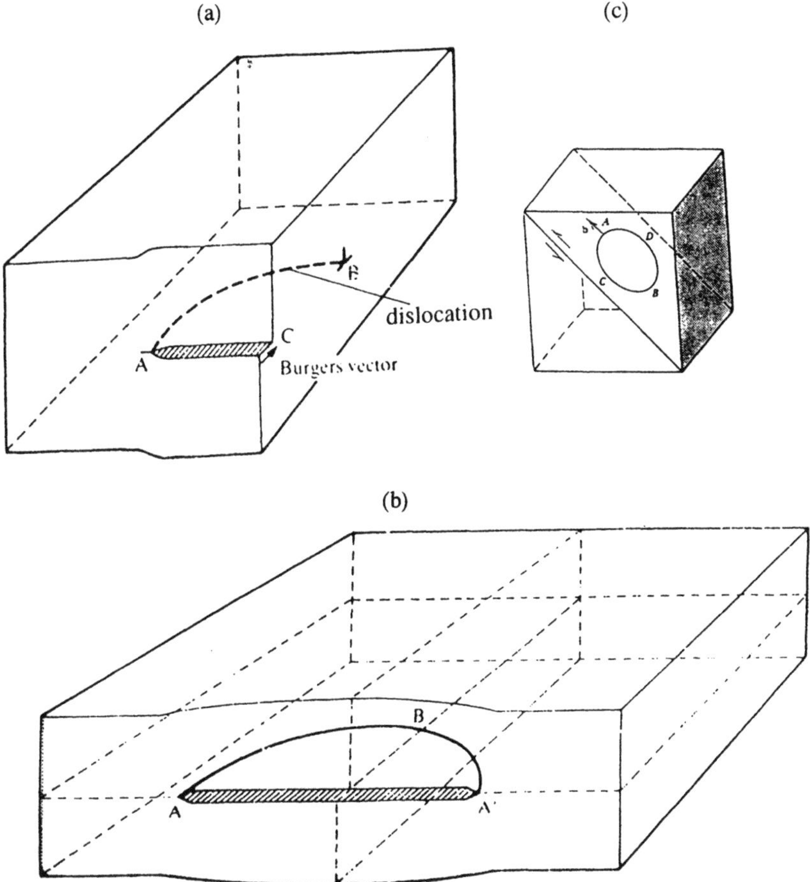

FIGURE 6.7 Structure of a mixed dislocation: (a) quarter loop; (b) half loop; (c) full loop.

and ν is Poisson's ratio. The lattice fraction stress is associated with the energy or the stress that is needed to move the edge dislocation from position A to position D (Fig. 6.9). Note that the dislocation line energy [Fig. 6.10(a)] and the applied shear stress [Fig. 6.10(b)] vary in a sinusoidal manner. Also, the shear stress increases to a peak value corresponding to τ_f [Fig. 6.10(b)], the friction stress. The latter may, therefore, be considered as the lattice resistance that must be overcome to enable dislocation motion to occur between A and D (Fig. 6.9). It is important to note that τ_f is generally much less than the theoretical shear strength of a perfect lattice, which is

(a)

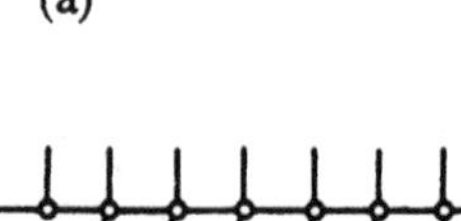
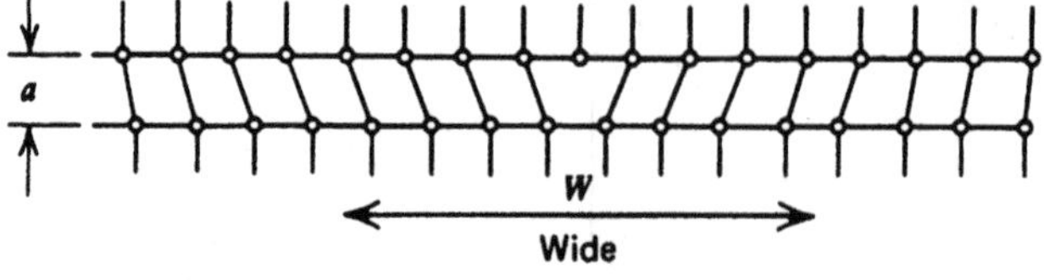

(b)

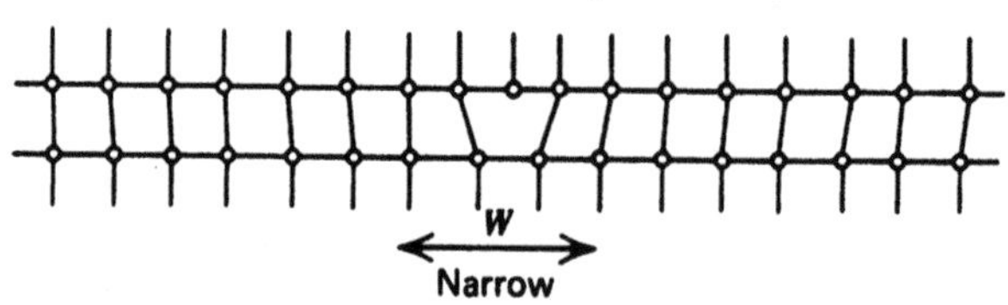

Figure 6.8 Schematic of (a) wide and (b) narrow dislocations. (From Cottrell, 1957. Reprinted with permission from Institute of Mechanical Engineering.)

given by Eq. (6.5) for a cubic lattice. Slip is, therefore, more likely to occur by the exchange of bonds, than the complete shear of atomic planes over each other, as suggested by Fig. 6.1.

The reader should examine Eqs (6.8a) and (6.8b) carefully since the dependence of the lattice friction stress, τ_f, on lattice parameters a and b has some important implications. It should be readily apparent that the friction stress is minimized on planes with large vertical spacings, a, and small horizontal spacings, b. Dislocation motion is, therefore, most likely to occur on close-packed planes which generally have the largest values of a and the smallest values of b. Dislocation motion is also most likely to occur along close-packed directions with small values of b. Hence, close-packed materials are more likely to be ductile, while less close-packed materials such as ceramics are more likely to be brittle.

We are now prepared to tackle the problem of dislocation motion in crystalline materials. First, we will consider the movement of edge dislocations on close-packed planes in close-packed directions. Such movement is generally described as conservative motion since the total number of atoms on the slip plane is conserved, i.e., constant. However, we will also consider the nonconservative motion of edge dislocations which is often described as

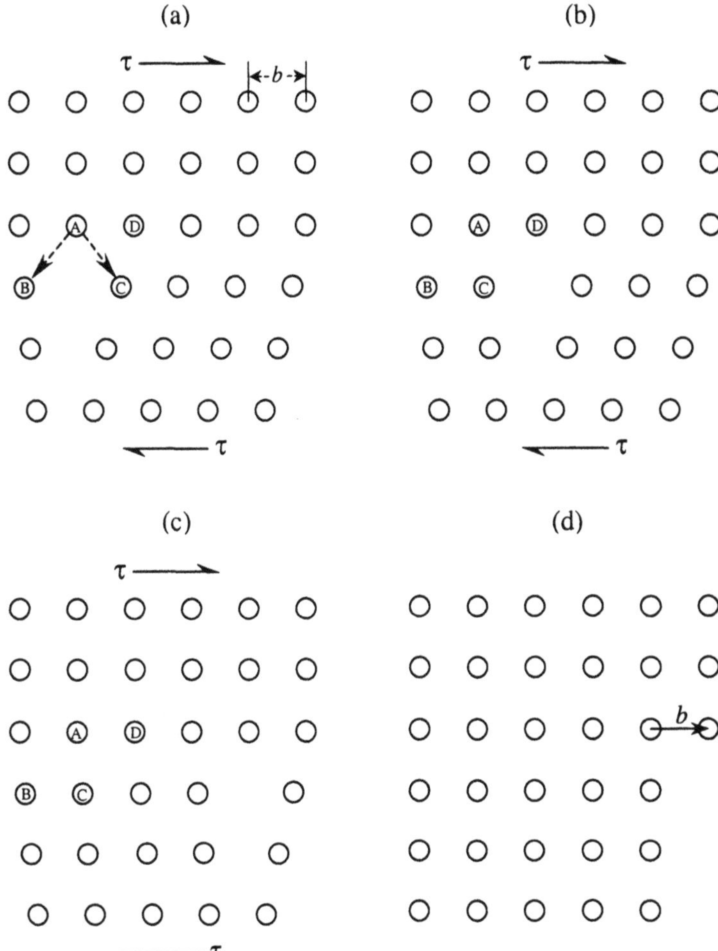

FIGURE 6.9 Schematic of atomic rearrangements associated with edge dislocation motion: (a) atoms B and C equidistant from atom A along edge dislocation line at start of deformation; (b) greater attraction of C towards A as crystal is sheared; (c) subsequent motion of edge dislocation to the right; (d) formation of step of Burgers vector when dislocation reaches the edge of the crystal.

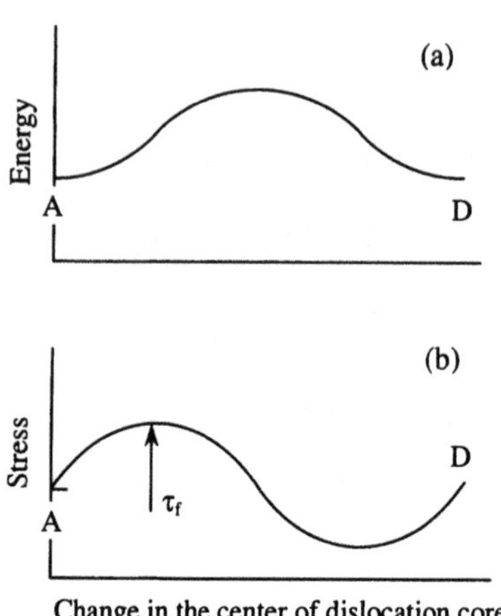

FIGURE 6.10 Variation of (a) dislocation line energy and (b) stress with the position of the dislocation core.

dislocation climb.* Since dislocation climb involves the exchange of atoms and vacancies outside the slip plane, the total number of atoms in the slip plane is generally not conserved by dislocation climb mechanisms. Following the discussion of edge dislocation motion by slip and climb, we will then discuss the conventional movement of screw dislocations, and the cross-slip of screw dislocations.

6.4.1 Movement of Edge Dislocations

The movement of edge dislocations is relatively easy to visualize. Let us start by considering the movement of the positive edge dislocation shown schematically in Fig. 6.9. Prior to the application of shear stress to the crystal, the atom A at the center of the edge dislocation is equidistant from atoms B and C, Fig. 6.9(a). It is, therefore, equally attracted to atoms B and C. However, on the application of a small shear stress, τ, to the top and bottom faces of the crystal, atom A is displaced slightly to the right. The slight asymmetry develops in a greater attraction between A and C, compared to that between A and B. If the applied shear stress is increased, the increased attraction between atoms A and C may be sufficient to cause the displacement of atom C and surrounding atoms to the left by one atomic spacing, b, Fig. 6.9(b). The half column of atoms (positive edge dislocation),

*Note that conservative climb may also occur by the movement of prismatic loops. The interested reader is referred to Hull and Bacon (1984).

therefore, appears to move to the right by a distance of one lattice spacing, b [Fig. 6.9(b)].

If we continue to apply a sufficiently high shear stress to the crystal, dislocation motion will continue [Fig. 6.9(c)] until the edge dislocation reaches the edge of the crystal, Fig. 6.9(d). This result in slip steps with step dimensions that are proportional to the total number of dislocations that have moved across to the edge of the crystal. The slip sites may actually be large enough to resolve under an optical or scanning electron microscope when the number of dislocations that reach the boundary is relatively large. However, in many cases, the slip steps may only be resolved by high magnification scanning electron or transmission electron microscopy techniques.

In addition to the conservative motion of edge dislocations on close-packed planes along close-packed directions, edge dislocation motion may also occur by nonconservative dislocation climb mechanisms. These involve the exchange of atoms and vacancies, shown schematically in Fig. 6.11. The exchange of atoms and vacancies may be activated by stress and/or temperature and is diffusion controlled. Hence, dislocation climb is most often observed to occur at elevated temperature.

6.4.2 Movement of Screw Dislocations

The movement of screw dislocations is a little more difficult to visualize. Let us start by considering the effects of an applied shear stress on the screw dislocation shown in Fig. 6.12. The shear stress on the upper part of the crystal displaces the atoms on one half of the crystal over the other, as shown in Fig. 6.12. However, in this case, the Burgers vector is parallel to the dislocation line. The direction of screw dislocation motion is perpendicular to the direction of the applied shear stress (Fig. 6.12).

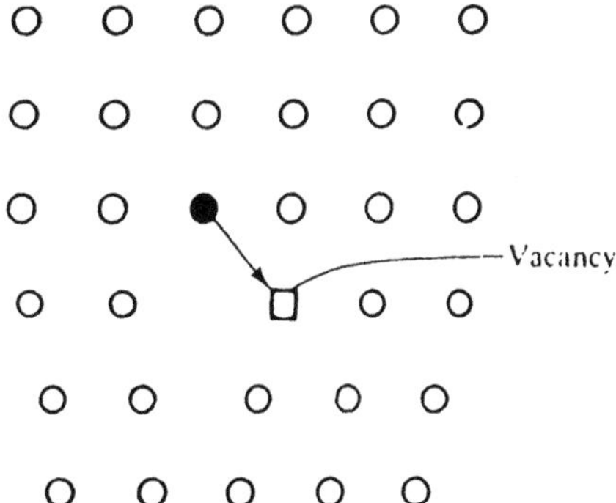

FIGURE 6.11 Climb by the exchange of atoms and vacancies.

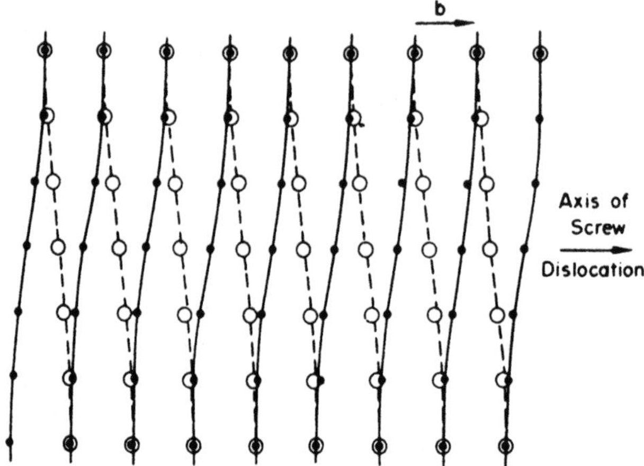

FIGURE 6.12 Arrangement of atoms around a screw dislocation—open circles above and closed circles below plane of diagram. (Taken from Hull and Bacon, 1984. Reprinted with permission from Pergamon Press.)

Unlike the edge dislocation, the screw dislocation can glide on a large number of slip planes since the Burgers vector, *b*, and the sense vector, *s*, are parallel. However, in most cases, screw dislocation motion will tend to occur on close-packed planes in close-packed directions. Screw dislocations also generally tend to have greater mobility than edge dislocations.

Nevertheless, unlike edge dislocations, screw dislocations cannot avoid obstacles by nonconservative dislocation climb processes. Instead, screw dislocations may avoid obstacles by cross-slip on to intersecting slip planes, as shown in Fig. 6.13. Note that the Burgers vector is unaffected by cross-slip process. The screw dislocation may also cross-slip back on to a parallel slip plane, or the original slip plane, after avoiding an obstacle.

6.4.3 Movement of Mixed Dislocations

In reality, most dislocations in crystalline materials are mixed dislocations, with both edge and screw components. Such dislocations will, therefore, exhibit aspects of screw and edge dislocation characteristics, depending on the proportions of screw and edge components. Mixed dislocations are generally curved, as shown in Fig. 6.7. Also, the curved dislocation loops may have pure edge, pure screw, and mixed dislocation segments.

It is a useful exercise to identify the above segments of the mixed dislocation loops shown in Fig. 6.7. It is also important to note that dislocation loops may be circular or elliptical, depending on the applied stress

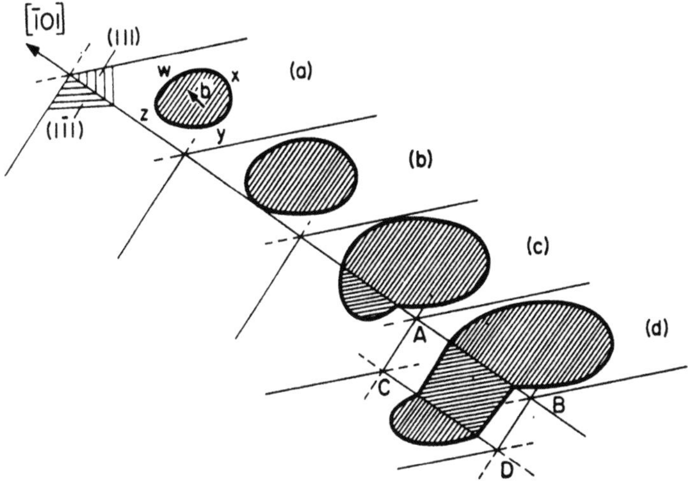

Figure 6.13 Schematic illustration of the cross-slip of a screw dislocation in a face-centered cubic structure. Note that since [$\bar{1}$01] direction is common to both the (111) and (1$\bar{1}$1) closed packed planes, the screw dislocation can glide on either of these planes: (a,b) before cross-slip; (c) during cross-slip; (d) double cross-slip. (Taken from Hull and Bacon, 1984. Reprinted with permission from Pergamon Press.)

levels. Furthermore, dislocation loops tend to develop semielliptical shapes in an attempt to minimize their strain energies. This will become apparent later after the concept of the dislocation line/strain energy is introduced.

Finally, it is important to note that dislocations cannot terminate inside a crystal. They must, therefore, either form loops or terminate at other dislocations, grain boundaries, or free surfaces. This concept is illustrated in Fig. 6.14 using a schematic of the so-called Frank net. Note that when three dislocations meet at a point (often called a dislocation node), the algebraic sum of the Burgers vectors, b_1, b_2, and b_3 (Fig. 6.14) is zero. Hence,

$$\sum_{i=1}^{3} b_1 + b_2 + b_3 = 0 \tag{6.9}$$

When the sense vectors of the dislocations are as shown in Fig. 6.14, then Eq. (6.9) may be expressed as

$$b_1 = b_2 + b_3 \tag{6.10}$$

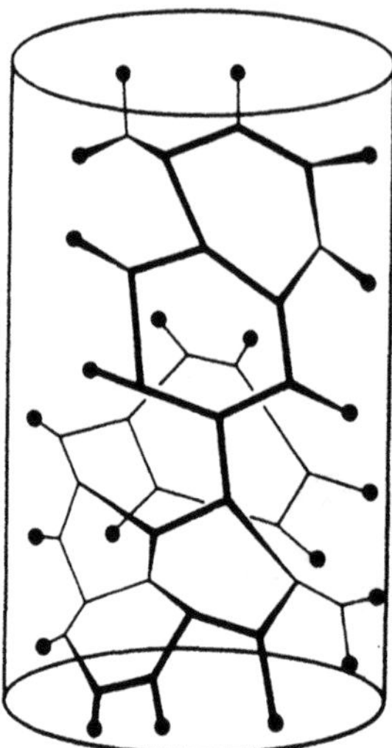

FIGURE 6.14 The Frank net. (Taken from Cottrell, 1957. Reprinted with permission from Institute of Mechanical Engineering.)

The above expressions are, therefore, analogous to Kirchoff's equations for current flow in electrical circuits.

6.5 EXPERIMENTAL OBSERVATIONS OF DISLOCATIONS

A large number of experimental techniques have been used to confirm the existence of dislocations. They include:

1. Etch-pit techniques
2. Dislocation decoration techniques
3. X-ray techniques
4. Transmission electron microscopy
5. Field ion microscopy

Other specialized techniques have also been used to reveal the existence of dislocations. However, these will not be discussed in this section. The interested reader is referred to the text by Hull and Bacon (1984) that is listed in the bibliography at the end of the chapter. This section will, therefore, present only a brief summary of experimental techniques that have been used to confirm the existence of dislocations.

The most widely used tool for the characterization of dislocation substructures is the transmission electron microscope (TEM). It was first used by Hirsch (1956) to study dislocation substructures. Images in the TEM are produced by the diffraction of electron beams that are transmitted through thin films ($\sim$ 1–2 μm) of the material that are prepared using special specimen preparation techniques. Images of dislocations are actually produced as a result of the strain fields associated with the presence of dislocations. In most cases, dislocations appear as dark lines such as those shown in Fig. 6.15(a). It is important also to note that the dark lines are actually horizontal projections of dislocation structures that are inclined at an angle with respect to the image plane, as shown in Fig. 6.15(b).

Dislocations have also been studied extensively using etch-pit techniques. These rely on the high chemical reactivity of a dislocation due to its strain energy. This gives rise to preferential surface etching in the presence of certain chemical reagents. Etch-pit techniques have been used notably by Gilman and Johnston (1957) to study dislocation motion LiF crystals (Fig. 6.16). Note that the etch pit with the flat bottom corresponds to the position of the dislocation prior to motion to the right. The strain energy associated with the presence of dislocations has also been used to promote precipitation reactions around dislocations. However, such reactions are generally limited to the observation of relatively low dislocation densities.

At higher dislocation densities, dislocations are more likely to interact with other defects such as solutes/interstitials, other dislocations and precipitates (Fig. 6.17). These interactions are typically studied using the TEM. Dislocation substructures have also been studied using special x-ray diffraction and field ion microscopy techniques. However, by far the most commonly used technique today for the study of dislocation substructures is the TEM. The most modern TEMs may be used today to achieve remarkable images that are close to atomic resolution. Recently developed two-gun focused ion beam may also be used to extract TEM foils with minimal damage to the material in the foils. These also offer some unique opportunities to do combined microscopy and chemical analyses during TEM analyses.

6.6 STRESS FIELDS AROUND DISLOCATIONS

The stress fields around dislocations have been derived from basic elasticity theory. These fields are valid for the region outside the dislocation core (which is a region close to the center of the dislocation where linear elasticity theory breaks down). The radius of the dislocation core will be denoted by R in subsequent discussion. Note that R is approximately

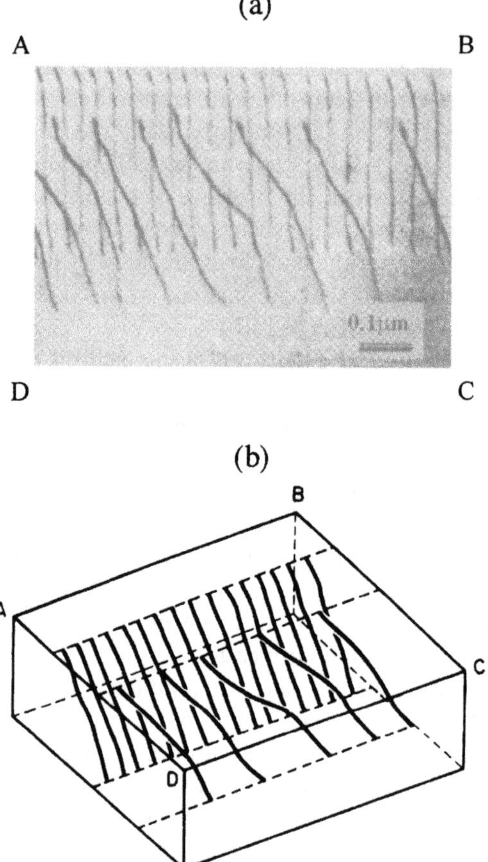

FIGURE 6.15 Images of dislocations: (a) thin-film transmission microscopy micrograph showing parallel rows of dislocations; (b) line diagram demonstrating that thin-foil image is a line projection of a three-dimensional configuration of dislocations. (Taken from Hull and Bacon, 1984. Reprinted with permission from Pergamon Press.)

equal to $5b$, where b is the Burgers vector. Linear elasticity theory may be used to estimate the stress fields around individual edge and screw dislocations in regions where $r > 5b$. Detailed derivations of the elastic stress fields around dislocations are beyond the scope of this book. However, the interested reader may refer to Hirth and Lothe (1982) for the derivations. Stress fields around individual stationary edge and screw dislocations are presented in this section.

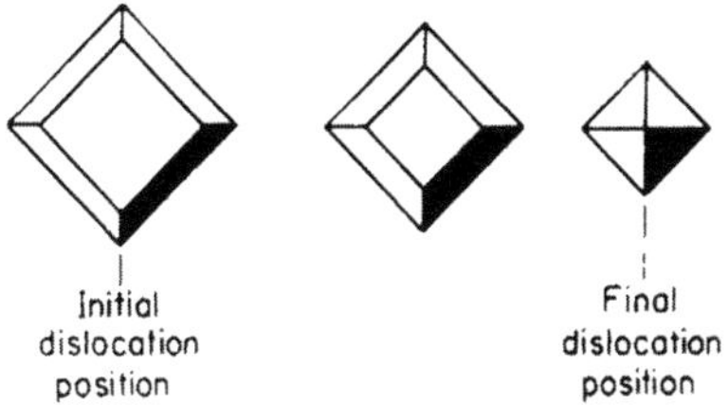

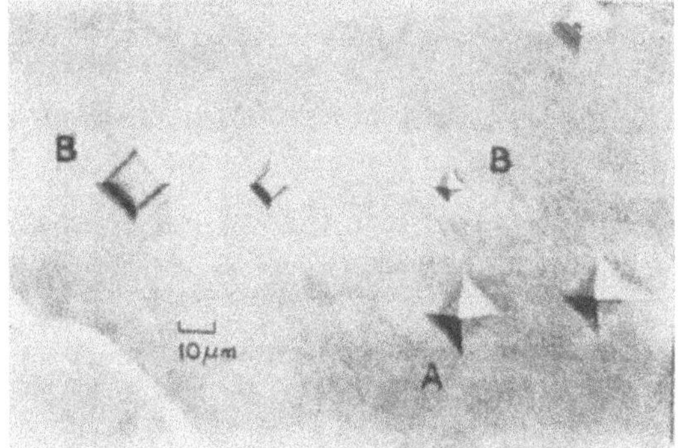

FIGURE 6.16 Etch pits in a lithium fluoride crystal after etching three times to reveal the motion of dislocations. The dislocation at A has not moved between each etching treatment since the etch pit still extends to a point. However, dislocation B has moved from the left to the right. The current position of dislocation corresponds to the sharp pit labeled B. Subsequent etching produces flat-bottomed pits at prior positions of dislocation B. (From Gilman and Johnston, 1957. Reprinted with permission from John Wiley.)

6.6.1 Stress Field Around a Screw Dislocation

The stress field around a single dislocation may be derived by considering the shear displacements (in the z direction) around a right-handed screw dislocation (Fig. 6.18). The displacement in the z direction, w, increases with the angle, θ, as shown in Fig. 6.18. This gives:

$$w = \frac{b\theta}{2\pi} = \frac{b}{2\pi}\tan^{-1}\left(\frac{y}{x}\right) \tag{6.11}$$

where b is the Burgers vector. The elastic strains and stresses (away from the dislocation core, $r > 5b$) around the dislocation may be calculated using elasticity theory. Noting that $x = r\cos\theta$ and $y = r\sin\theta$, the elastic strains around the screw dislocation are thus given by

$$\gamma_{zx} = \gamma_{xz} = -\frac{b}{2\pi}\frac{y}{(x^2+y^2)} = -\frac{b}{2\pi}\frac{\sin\theta}{r} \tag{6.12}$$

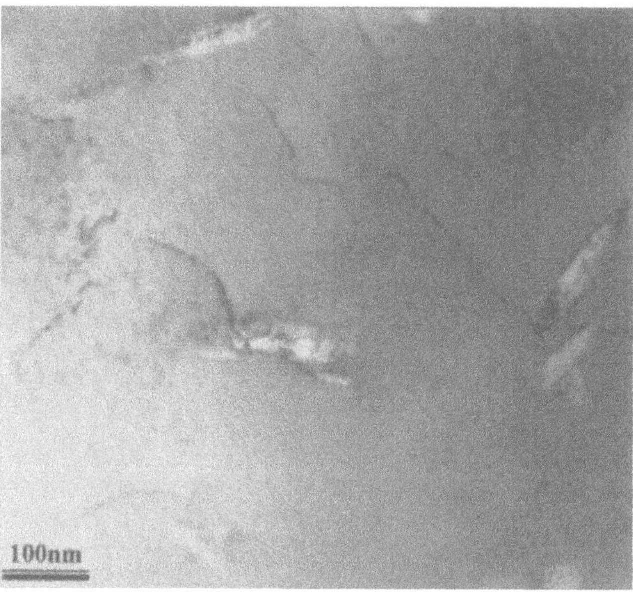

FIGURE 6.17 High-resolution transmission electron microscopy image of a dislocation at the interface between a titanium carbide precipitate in a niobium alloy (Courtesy of Dr. Seyed Allameh.)

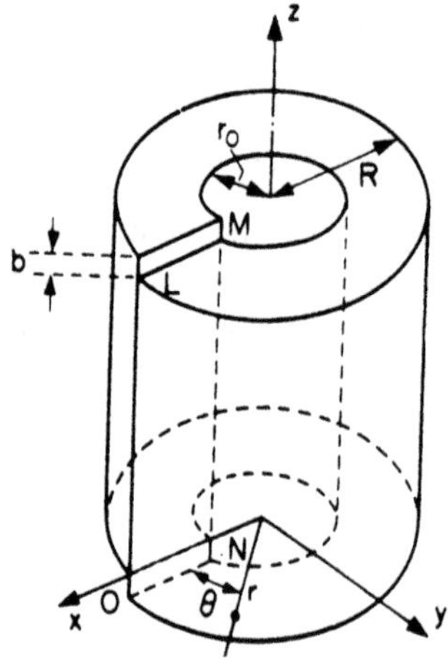

FIGURE 6.18 Stress field around a screw dislocation. (From Hull and Bacon, 1984—reprinted with permission from Pergamon Press.)

and

$$\gamma_{zy} = \gamma_{yz} = \frac{b}{2\pi} \frac{x}{(x^2 + y^2)} = \frac{b}{2\pi}\left(\frac{\cos\theta}{r}\right) \tag{6.13}$$

All the other strain terms are zero in the case of a pure screw dislocation. Hence, $\varepsilon_{xx} = \varepsilon_{yy} = \varepsilon_{zz} = \gamma_{xy} = 0$. The stress field around a screw dislocation is thus given by

$$\tau_{zx} = \tau_{xz} = -\frac{Gb}{2\pi} \frac{y}{(x^2 + y^2)} = \frac{-Gb \sin\theta}{2\pi \quad r} \tag{6.14}$$

and

$$\tau_{yz} = \tau_{yz} = \frac{Gb}{2\pi} \frac{x}{(x^2 + y^2)} = \frac{Gb \cos\theta}{2\pi \quad r} \tag{6.15}$$

Using the appropriate definitions of stress and strain and cylindrical co-ordinates, we may show that

$$\tau_{zr} = \tau_{xz} \cos\theta + \tau_{yz} \sin\theta \tag{6.16a}$$

and

$$\tau_{\theta z} = -\tau_{xz} \sin\theta + \tau_{yz} \cos\theta \tag{6.16b}$$

Substituting appropriate expressions for τ_{xz} and τ_{yz} from Eqs. (6.14), (6.15), and (6.16) gives the only non-zero stress term as:

$$\tau_{\theta z} = \frac{Gb}{2\pi r} \tag{6.17}$$

As before, the expression for the shear strain $\gamma_{\theta z} = \gamma_{z\theta} = b/(2\pi r)$.

It should be readily apparent from the above equations that the stresses and strains around a screw dislocation approach infinity as $r \to 0$. Since the stress levels in the dislocation core may not exceed the theoretical shear strength of a solid $(G/2\pi)$, we may estimate the size of the dislocation core by introducing a cut-off at the point where $\tau_{\theta z} \sim G/2\pi$. This gives an estimate of the size of the dislocation core to be on the order of $R \sim 5b$. Finally, it is important to note here that the elastic stress fields surrounding a screw dislocation generally decay as a function of $1/r$.

6.6.2 Stress and Strain Fields Around an Edge Dislocation

The elastic strain fields around single edge dislocations have also been determined using linear elasticity theory. For regions outside the dislocation core, if we adopt the sign convention shown in Fig. 6.19, the strain components are given by

$$\varepsilon_{xx} = -\frac{b}{2\pi(1-\nu)}\frac{y(3x^2+y^2)}{(x^2+y^2)^2} \tag{6.18}$$

$$\varepsilon_{yy} = \frac{b}{2\pi(1-\nu)}\frac{y(x^2-y^2)}{(x^2+y^2)^2} \tag{6.19}$$

$$\gamma_{xy} = \frac{b}{2\pi(1-\nu)}\frac{x(x^2-y^2)}{(x^2+y^2)^2} \tag{6.20}$$

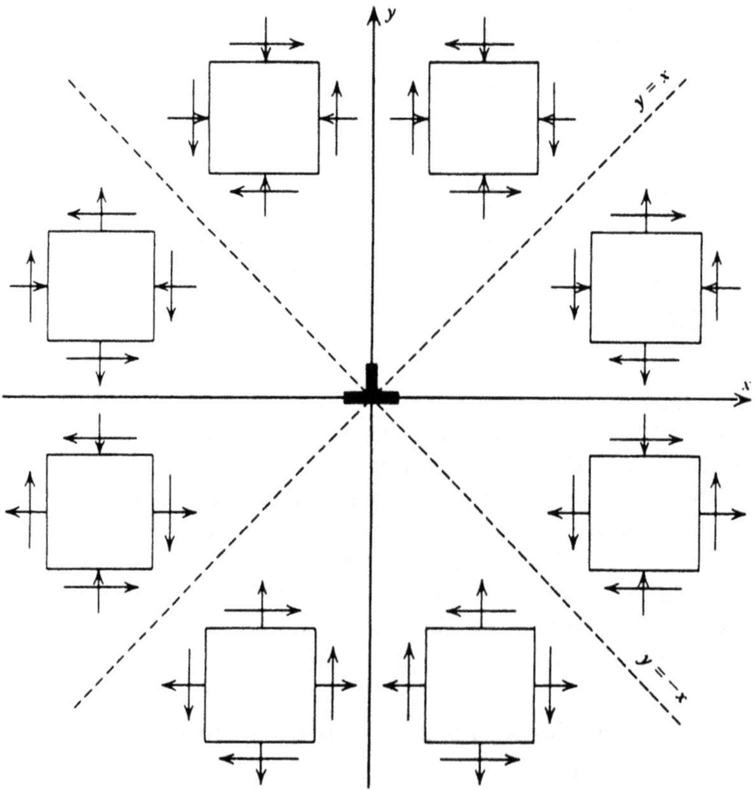

FIGURE 6.19 Stress field around an edge dislocation. (From Read, 1953. Reprinted with permission from McGraw-Hill.)

Note that $\tau_{zx} = \tau_{xz} = \tau_{yz} = \tau_{zy} = 0$ for an edge dislocation. The elastic stress field around a stationary edge dislocation is thus given by

$$\sigma_{xx} = -\frac{Gb}{2\pi(1-\nu)}\frac{y(3x^2+y^2)}{(x^2+y^2)^2} \tag{6.21}$$

$$\sigma_{yy} = \frac{Gb}{2\pi(1-\nu)}\frac{y(x^2-y^2)}{(x^2+y^2)^2} \tag{6.22}$$

$$\tau_{xy} = \frac{Gb}{2\pi(1-\nu)}\frac{x(x^2-y^2)}{(x^2+y^2)^2} \tag{6.23}$$

$$\sigma_{zz} = \nu(\sigma_{xx}+\sigma_{yy}) = -\frac{Gb\nu}{\pi(1-\nu)}\frac{y}{x^2+y^2)} \tag{6.24}$$

The above stress fields may also be expressed in cylindrical co-ordinates. This gives

$$\sigma_{rr} = \sigma_{\theta\theta} = -\frac{Gb\sin\theta}{2\pi(1-\nu)r} \tag{6.25}$$

$$\tau_{r\theta} = \frac{Gb\cos\theta}{2\pi(1-\nu)r} \tag{6.26}$$

$$\sigma_{zz} = \nu(\sigma_{rr}+\sigma_{\theta\theta}) = -\frac{Gb\nu\sin\theta}{\pi(1-\nu)r} \tag{6.27}$$

Also, $\tau_{rz} = \tau_{\theta z} = 0$ for an edge dislocation. Once again, the elastic stress fields exhibit a $1/r$ dependence. The self-stresses around an edge dislocation also approach infinity as $r \to 0$. The elastic stress fields are, therefore, valid only for regions outside the dislocation core, i.e., $r > 5b$.

6.7 STRAIN ENERGIES

The stress and strain fields surrounding individual dislocations give rise to internal strain energies that depend on the dislocation type. The magnitudes of the internal strain energies are particularly important since dislocations generally try to reduce their overall line energies by minimizing their lengths. The elastic strain energy per unit length of a screw dislocation may be estimated by considering the cylindrical domain around the screw dislocation shown in Fig. 6.20. If we now unroll the cylindrical element of radius, r, and thickness, dr, it is easy to see that the shear strain, γ, must be given by [Fig. 6.20]:

$$\gamma_{\theta z} = \frac{b}{2\pi r} \tag{6.28}$$

The increment in the strain energy per unit volume, dU'_s, may be determined by integrating the relationship between shear stress and shear strain in the elastic regime. This gives

$$dU'_s = \int_0^{\gamma} \tau_{\theta z} d\gamma_{\theta z} = \int_0^{\gamma} G\gamma_{\theta z}\, d\gamma_{\theta z} = \frac{G\gamma_{\theta z}^2}{2} \tag{6.29}$$

hence, dU''_s, the increment in the strain energy in Fig. 6.20(a) is given by the volume of the element multiplied by dU'_s. This gives

$$dU''_s = (2\pi r l\, dr)\left(\frac{Gb^2}{8\pi^2 r^2}\right) = \frac{Gb^2 l}{4\pi}\frac{dr}{r} \tag{6.30}$$

Since the core energy, U_{core}, is generally considered to be a small fraction of the overall energy, $U_{tot} = U_{core} + U_s$, the core energy, U_{core}, may be neglected. Hence, neglecting the core energy, we may integrate Eq. (6.30) between $r = r_0$ and $r = R$ to find an expression for the elastic strain energy per unit length, U_s. This is given by

$$U_s = \frac{U'_s}{l} = \frac{Gb^2}{4\pi}\ln\left(\frac{R}{r}\right) \tag{6.31}$$

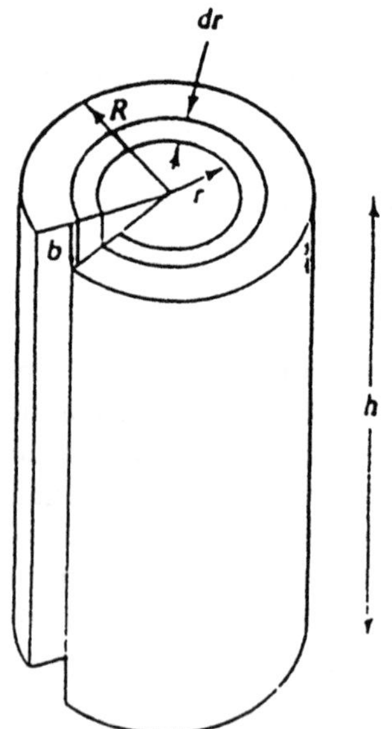

Figure 6.20 Strain energy around a screw dislocation.

For single crystals, R is the crystal radius. However, it is much more difficult to determine R for polycrystals that contain several dislocations. Nevertheless, the coefficient of Gb^2 in Eq. (6.31) may be approximated by unity for order-of-magnitude comparisons. This gives the strain energy per unit length of screw dislocations as

$$U_s \sim Gb^2 \tag{6.32}$$

Similar calculations of the elastic strain energy per unit length of an edge dislocation may also be carried out. However, the calculations are more complex since edge dislocations have more complex stress and strain fields associated with them. Using similar assumptions to those used in the screw dislocation derivation above, we may show that the strain energy per unit length of an edge dislocation, U_e, is given by

$$U_e \sim \frac{Gb^2}{1-\nu} \tag{6.33}$$

where ν is Poisson's ratio, and the other terms have their usual meaning. Since ν is generally close to 0.3 in most elastic solids, comparison of Eqs (6.32) and (6.33) shows that edge dislocations have higher strain energy per unit length than screw dislocations. Mixed dislocations (consisting of edge and screw components) will, therefore, try to minimize the lengths of their edge components, while maximizing the lengths of their screw components which have lower energies per unit length. By so doing, they can minimize their overall line energies. For this reason, a circular dislocation loop would tend to evolve into an elliptical shape that maximizes the length of the screw segments. Similarly, an initially straight edge dislocation may become curved in an effort to minimize the overall line energy per unit length.

6.8 FORCES ON DISLOCATIONS

When a crystal is subjected to external stresses, the resulting motion of dislocations may be considered to arise from the effects of "virtual" internal forces that act in directions that are perpendicular to segments on the dislocation line. Let us start by considering the motion of the right-handed screw dislocation in the crystal shown in Fig. 6.21. The external force applied to the surface of the crystal is given by the product of stress, τ_{xz}, multiplied by surface area, $L\,dx$, where dx is the distance the dislocation moves and L is the dislocation length. This causes a Burgers vector displacement, b, of the upper half of the crystal relative to the lower half. Hence, the external work done, W, by the applied stress is given by

$$W = \tau_{xz} L \Delta x \tag{6.34}$$

If the magnitude of the "virtual" force per unit length is f, then the internal work done in moving the dislocation of length, L, through a distance Δx is $fL\Delta x$. Hence, equating the internal and external work gives

$$fL\Delta x = \tau_{xz} L dx b \tag{6.35a}$$

or

$$f = \tau_{xz} b \tag{6.35b}$$

The force per unit length, f, is, therefore, the product of the applied shear stress and the Burgers vector, b. It is also important to note that this force acts along the slip plane in a direction that is perpendicular to the dislocation line.

Similarly, we may consider the motion of a positive edge dislocation in a crystal subjected to an external shear stress (Fig. 6.22). As before, if we assume that the dislocation extends across the width of the crystal, then the external force on the surface of the crystal is given by the product of the surface shear stress, τ_{xy}, and the area, Ldy. The external work, W, is now the product of this force and the Burgers vector, b. This gives

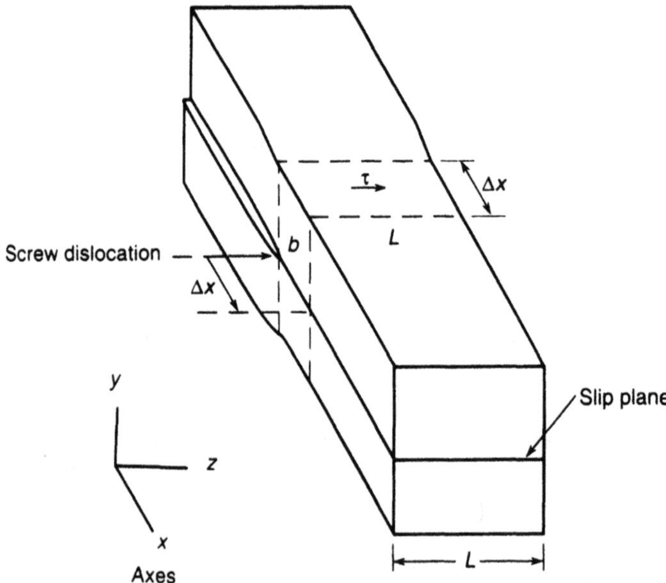

Figure 6.21 Effect of applied shear stress on the glide of a screw dislocation. (Taken from Read-Hill and Abbaschian, 1994. Reprinted with permission from PWS Publishing Co.)

$$W = \tau_{xy} L \, dyb \tag{6.36}$$

Furthermore, the internal "virtual" force is given by fL, and the internal work done by this force is $fL\Delta y$. Equating the internal and external work now gives

$$f \, L \, \Delta y = \tau_{xy} L \, \Delta yb \tag{6.37a}$$

or

$$f = \tau_{xy} b \tag{6.37b}$$

The fictitious force per unit length f on an edge dislocation is, therefore, the product of the applied shear stress, τ_{xy}, and the Burgers vector, b. As in the case of the screw dislocation, the force acts along the slip plane in a direction that is perpendicular to the dislocation line.

Furthermore, when a normal stress is applied to a crystal, the resulting virtual force on an edge dislocation may cause it to move up or down, depending on the sign of the applied stress (tensile or compressive). This is shown schematically in Fig. 6.23 for an applied tensile stress. Once again,

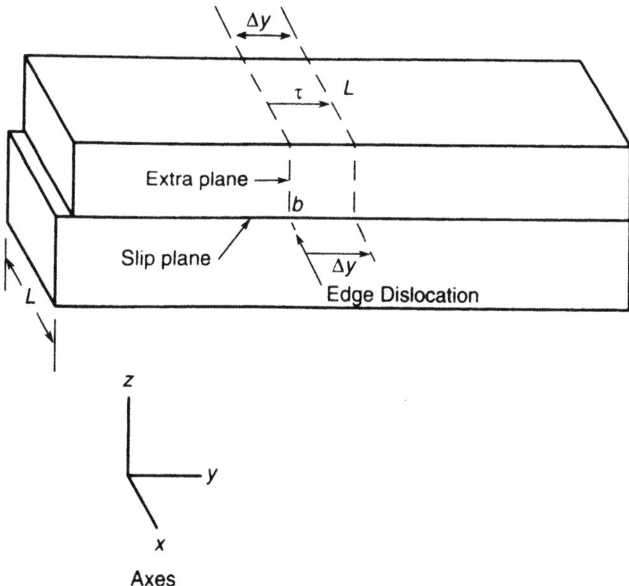

Figure 6.22 Effect of applied shear stress on the glide of an edge dislocation. (Taken from Read-Hill and Abbaschian, 1994. Reprinted with permission from PWS Publishing Co.)

the external force is the product of the stress, σ_{xx}, and the area, $L\Delta z$. Hence, the external work, W, required to move the dislocation down through a distance of one Burgers vector is given by

$$W = -\sigma_{xx} L \Delta z b \tag{6.38}$$

The corresponding internal work due to the virtual force arising from the imposed external stress is now $fL\Delta z$. Hence, equating the internal and external work gives

$$fL \Delta z = -\sigma_{xx} L \Delta z b \tag{6.39a}$$

or

$$f = -\sigma_{xx} b \tag{6.39b}$$

The climb force per unit length, f, is, therefore, a product of the applied tensile stress, σ_{xx}, and the Burgers vector, b. Note that a tensile stress will cause the dislocation to move down, while an applied compressive stress will cause the dislocation to move up. Furthermore, unlike applied shear stresses, which induce dislocation motion along the slip plane, applied normal

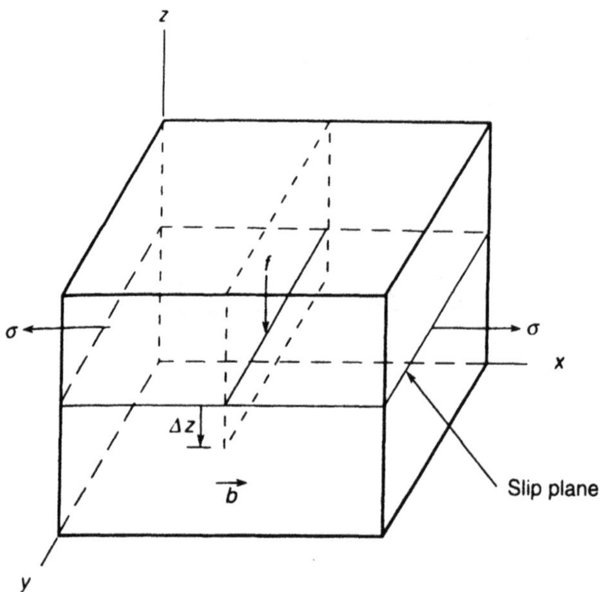

Figure 6.23 Effect of applied tensile stress on the climb of an edge dislocation. (Taken from Read-Hill and Abbaschian, 1994. Reprinted with permission from PWS Publishing Co.)

(tensile or compressive) stresses induce vertical dislocation motion out of the slip plane. However, in both cases (applied normal or shear stresses), the motion of the dislocation is always in a direction that is perpendicular to the dislocation line (Figs 6.21–6.23).

Since most dislocations are mixed dislocations (that contain both screw and edge components), it is of interest to examine how mixed dislocations move. However, before doing this, it is necessary to recall our definitions of the sense and Burgers vectors. The sense vector, **s**, defines the unit vector that is tangential to the dislocation line. It has components, s_x, s_y and s_z. Similarly, the Burgers vector, **b**, has components b_x, b_y and b_z. If we denote a general state of stress, F, by components F_x, F_y and F_z, then the virtual force per unit length of dislocation, f, is given by the vector cross product of F and s:

$$f = F \times s = \begin{vmatrix} i & j & k \\ F_x & F_y & F_z \\ s_x & s_y & s_z \end{vmatrix} \tag{6.40}$$

where the components of $F(F_x, F_y, F_z)$ are obtained from

$$\begin{aligned} F_x &= b_x \sigma_{xx} + b_y \tau_{xy} + b_z \tau_{xz} \\ F_y &= b_x \tau_{yx} + b_y \sigma_{yy} + b_z \tau_{yz} \\ F_z &= b_x \tau_{zx} + b_y \tau_{zy} + b_z \sigma_{zz} \end{aligned} \tag{6.41}$$

The relationship between the force, Burgers vector, stress, and the sense vector may now be expressed as

$$f = (b \cdot \sigma) \times s \tag{6.42}$$

This is the so-called *Peach–Kohler equation.* It is an extremely useful expression for calculating the force on a dislocation. Furthermore, since the fictitious forces on the edge and screw dislocation segments are perpendicular, then the virtual force on a mixed dislocation segment must also be perpendicular to the dislocation segment.

6.9 FORCES BETWEEN DISLOCATIONS

The total energy of two dislocations, 1 and 2, U_{tot}, may be obtained from the sum of the self-energies of dislocations 1 and 2, U_1 and U_2, and an interaction term, U_{int}, due to the interactions between the stress fields of dislocations 1 and 2. This gives

$$U_{tot} = U_1 + U_2 + U_{int} \tag{6.43}$$

The forces between the two dislocations are obtained from the derivatives of U_{int} with respect to the appropriate axes. The interaction energy is determined from the work done in displacing the faces of the cut that creates dislocation 2 in the presence of the stress field due to dislocation 1. This may be done by cutting in a direction parallel to the x or y axes. The resulting two alternative expressions for U_{int} are

$$U_{int} = \int_{x}^{\infty} (b_x \tau_{xy} + b_y \sigma_{yy} + b_z \sigma_{zy}) dx \tag{6.44a}$$

or

$$U_{int} = \int_{y}^{\infty} (b_x \sigma_{xx} + b_y \tau_{yx} + b_z \tau_{zx}) dy \tag{6.44b}$$

The components of force, F_x and F_y, on dislocation 2 (Fig. 6.24) are obtained from the first derivatives of the interaction energy, U_{int}, with respect to the appropriate axes. These are given by

$$F_x = -\frac{\partial U_{int}}{\partial x} \tag{6.45a}$$

$$F_y = -\frac{\partial U_{int}}{\partial y} \tag{6.45b}$$

For two parallel edge dislocations (Fig. 6.24) with parallel Burgers vectors, $b = b_x$, $b_y = b_z = 0$, the forces F_x and F_y on dislocation 2 (Fig. 6.24) are given by

$$F_x = \tau_{xy} b \tag{6.46}$$

and

$$F_y = -\sigma_{xx} b \tag{6.47}$$

Note that the expressions for the force between two edge dislocations are identical to those obtained earlier for forces due to an applied shear stress on a crystal. Furthermore, the expressions for σ_{xx} and τ_{xy} are the self-stresses acting at (x, y). Also, σ_{xx} and τ_{xy} are given by the expressions for the self-stresses in Eqs (6.21) and (6.23). Note that the signs of the forces are reversed if the sign of either dislocation 1 or 2 is reversed. Also, equal and opposite forces (from those at 2) act on dislocation 1.

The force F_x is particularly important since it determines the horizontal separation between the two dislocations in Fig. 6.24. Plots of F_x versus separation are presented in Fig. 6.25. The solid lines correspond to forces

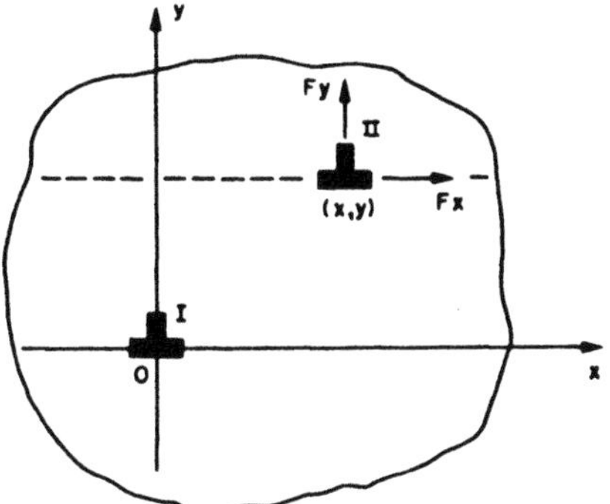

Figure 6.24 Schematic of interaction forces between two edges. (From Cottrell, 1953.)

between edge dislocations of the same kind, while the dashed line corresponds to edge dislocations of opposite sign. Note that the force is zero at $x = y$. Hence, stable dislocation configurations (tilt boundaries) tend to occur when like edge dislocations lie above each other, Fig. 6.26(a). Also, edge dislocations of opposite sign tend to glide past each other and form edge dislocation dipoles, as shown in Fig. 6.26(b).

The component of force, F_y, does not promote conservative dislocation glide on the slip plane. Instead, it acts to promote nonconservative dislocation climb out of the slip plane. This occurs by the exchange of atoms and vacancies, as shown schematically in Fig. 6.11. Dislocation climb is, therefore, both thermally and stress assisted since it involves atom/vacancy exchanges. In any case, when the two edge dislocations are of opposite sign, dislocation climb occurs in the opposite direction. This results ultimately in the annihilation of the two edge dislocations.

Let us now consider the force between two screw dislocations, with dislocation 1 lying on the z axis, and dislocation 2 on a parallel axis. In this case, the components of force are given by

$$F_r = \tau_{\theta z} b \tag{6.48}$$

and

$$F_\theta = \tau_{zr} b \tag{6.49}$$

where τ_{zr} and $\tau_{\theta z}$ are given by Eqs (6.16a) and (6.16b), respectively. Substituting Eqs (6.14)–(6.16) into Eqs (6.48) and (6.49), it is easy to show that the force components F_r and F_θ are given by

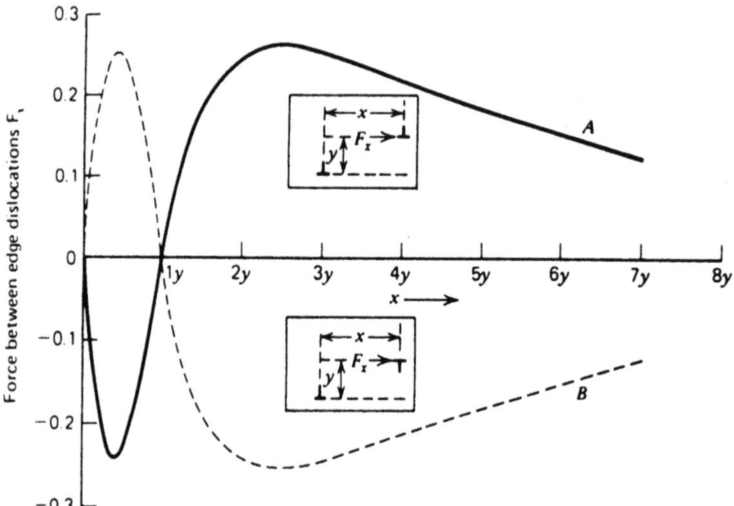

FIGURE 6.25 Forces between parallel edge dislocations. The solid curve corresponds to dislocations of the same sign. Note that the attractive force that occurs for $x < y$ causes sub-boundaries to form. (From Cottrell, 1957. Reprinted with permission from Institute of Mechanical Engineering.)

$$F_r = \frac{Gb^2}{2\pi r} \tag{6.50}$$

and

$$F_\theta = 0 \tag{6.51}$$

The forces between two parallel screw dislocations are, therefore, much simpler than those between two edge dislocations due to the radial symmetry of the dislocation field around a screw dislocation. As before, the signs of

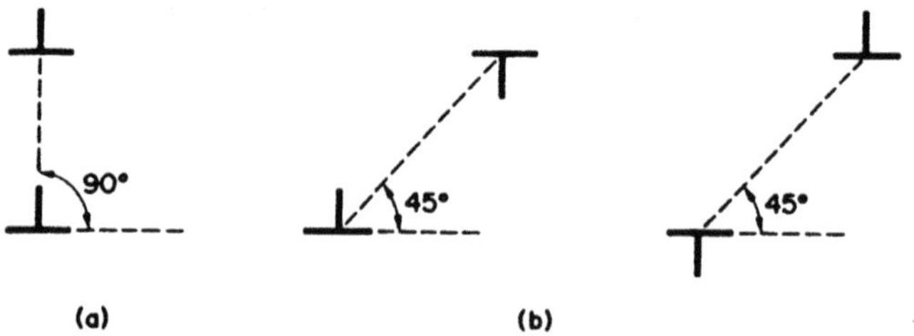

FIGURE 6.26 Stable positions for two dislocations: (a) same sign; (b) opposite sign. (Taken from Hull and Bacon, 1984. Reprinted with permission from Pergamon Press.)

the forces depend on whether the dislocations are parallel or antiparallel. Finally in this section, it is important to note that there are no forces between parallel edge and screw dislocations.

6.10 FORCES BETWEEN DISLOCATIONS AND FREE SURFACES

Since a compliant free surface offers no stress in opposition to the displacements of an approaching dislocation, the strain energy of a crystal will decrease as a dislocation approaches a free surface. This tends to pull the dislocation towards the free surface to form a step of one interatomic distance. The reduction in the strain energy of the crystal may also be expressed in terms of a "force" that pulls the dislocation out of a crystal. This has been shown by Koehler (1941) and Head (1953) to correspond to a force that would be exerted by an image dislocation of opposite sign on the opposite side of the surface (Fig. 6.27).

If the surface contains a thin layer, e.g., a surface film, that prevents the dislocation from being pulled out, some limited surface hardening may occur due to pile up of dislocations against the surface layer. Image forces are often used to model the effects of grain boundaries on dislocations in a crystal. However, it is important to recognize that the boundary conditions for free surfaces are much simpler, and significantly different from those for grain boundaries.

For a screw dislocation approaching a surface, the components of stress are (Fig. 6.28a):

$$\tau_{zx} = -\frac{Gb}{2\pi}\frac{y}{[(x-d)^2+y^2]} - \frac{Gb}{2\pi}\frac{y}{[(x+d)^2+y^2]} \tag{6.52}$$

and

$$\tau_{zy} = \frac{Gb}{2\pi}\frac{(x-d)}{[(x-d)^2+y^2]} - \frac{Gb}{2\pi}\frac{y}{[(x+d)^2+y^2]} \tag{6.53}$$

The force, F_x, on the screw dislocation at $x = d$, $y = 0$ is thus given by

$$F_x = \tau_{zy} b = -\frac{Gb^2}{4\pi d} \tag{6.54}$$

Note that this force diminishes with increasing distance from the free surface. Similarly, for an edge dislocation near a surface (Fig. 6.28b), we may superpose the stress fields of the dislocation at $x = d$ on those of the image dislocation. The σ_{xx} terms cancel out, but the τ_{xy} terms do not. The resulting stress field is given by

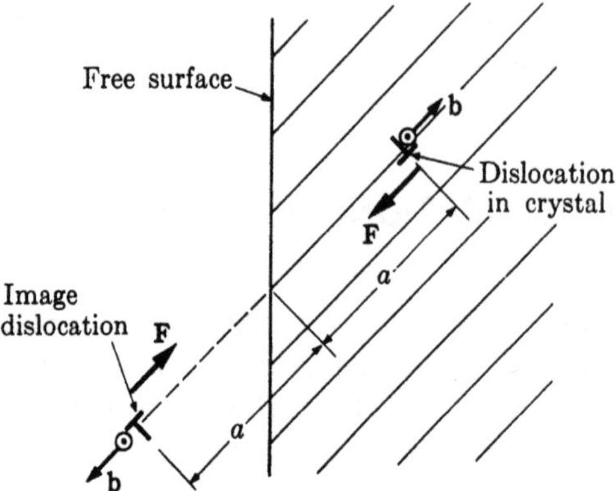

FIGURE 6.27 Image dislocation tending to pull a dislocation near a surface out of a crystal by glide. (Taken from Argon and McClintock, 1966: Addison Wesley.)

$$\tau_{xy} = \frac{D(x-d)\left[(x-d)^2 - y^2\right]}{\left[(x-d)^2 + y^2\right]^2} - \frac{D(x+d)\left[(x+d)^2 - y^2\right]}{\left[(x+d)^2 + y^2\right]}$$

$$- \frac{2Dd\left[(x-d)(x+d)^3 - 6x(x+d) + y^2 + y^4\right]}{\left[((x+d)^2 + y^2)^3\right]} \tag{6.55}$$

where $D = Gb/[2\pi(1-\nu)]$. The first term in the above equation corresponds to the self-field of dislocation 1 in the absence of the second dislocation. The

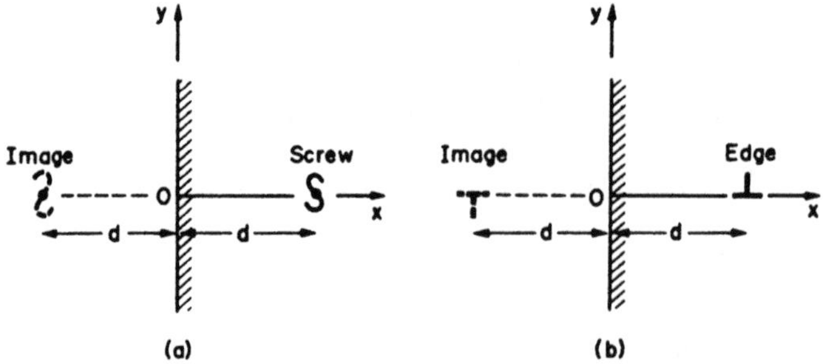

FIGURE 6.28 Image dislocations at a distance d from the surface: (a) screw dislocation approaching a surface; (b) edge dislocation approaching a surface. (Taken from Hull and Bacon, 1984. Reprinted with permission from Pergamon Press.

second term is the interaction term, and the third term is required to keep $\tau_{xy} = 0$ at the free surface. The force F_x on the dislocation at $x = d$, $y = 0$ is now given by

$$F_x = b\,\tau_{xy} = \frac{-Gb^2}{4\pi(1-\nu)d} \tag{6.56}$$

Once again, the image force decreases with increasing distance from the free surface. However, edge dislocations that are sufficiently close to the surface will be attracted to the surface, and some may be removed from the surface by such image forces.

6.11 SUMMARY

This chapter has presented an introduction to dislocation mechanics. Following a brief derivation of the theoretical shear strength of a solid, the discrepancy between the measured and theoretical strengths was attributed to the presence of dislocations. The different types of dislocations (screw, edge, and mixed) were then introduced before describing the atomic rearrangements associated with dislocation motion and lattice friction. The stress fields around individual dislocations were also presented, along with expressions for dislocation line energies, forces per unit length on dislocations, and forces between dislocations. The chapter concluded with a brief description of image forces between dislocations and free surfaces. The dislocation mechanics topics covered in this chapter should provide the foundation for the development of a basic understanding of the plastic deformation of metals in Chap. 7.

BIBLIOGRAPHY

The treatment of dislocation mechanics in this chapter is brief. It is intended to serve as an introduction to the subject. More detailed treatment of dislocation mechanics may be found in texts by Hirthe and Lothe (1982), Weertman and Weertman (1992), and Nabarro (1967). The books by Nabarro (1947) and Hirthe and Lothe (1982) provide comprehensive descriptions of almost all aspects of advanced dislocation theory. The books by Hull and Bacon (1984) and Weertman and Weertman provide more of an introduction to basic concepts. They are also lucid, and relatively easy to read.

Argon, A. and McClintock, F.A. (1966) Mechanical Behavior of Materials. Addison Wesley, Reading, MA.

Burgers, J.M. (1939) Proc Kon Ned Akad Wetenschap. vol. 42, p 293.

Cottrell, A.H. (1953) Dislocations and Plastic Flow in Crystals. Oxford, UK.

Cottrell, A.H. (1957) The Properties of Materials at High Rates of Strain. Institute of Mechanical Engineering, London, UK.

Courtney, T.H. (1990) Mechanical Behavior of Materials, John Wiley, New York.

Frenkel, J. (1926) Z Phys. vol. 37, p 572.

Gilman, J.J. and Johnston, W.G. (1957) Dislocations and Mechanical Properties of Crystals. In: J.C. Fisher, W.G. Johnston, R. Thomson and T. Vreeland, Jr., ed. John Wiley, New York.

Head, A.K. (1953) Edge dislocations in homogeneous Media. Proc Phys Soc. vol. B66, pp 793–801.

Hirsch, P. (1956) Prog Met Phys. vol. 6, p 236.

Hirthe, J.P. and Lothe, J. (1982) Theory of Dislocations. 2nd ed. John Wiley, New York.

Hull, D. and Bacon, D.J. (1984) Introduction to Dislocations, 3rd ed. Pergamon Press, Oxford, UK.

Kelly, A. (1986) Strong Solids. Clarendon Press, Oxford, UK.

Koehler, J.S. (1941) On the dislocation theory of plastic deformation. Phys Rev. vol 60, p 397.

Nabarro, F.R.N. (1947) Dislocations in a simple cubic lattice. Proc Phys Soc. vol. 59, p 256.

Orowan, E. (1934) Crystal plasticity III. On the mechanism of the glide process. Z Phys. vol. 89, p 634.

Peierls, R. (1940) The size of a dislocation. Proc Phys Soc. vol. 52, p 34.

Polanyi, M. (1934) On a kind of glide disturbance that could make a crystal plastic. Z Phys. vol. 89, p 660.

Read, W.T. (1953) Dislocations in Crystals. McGraw-Hill, New York.

Read-Hill, R.E. and Abbaschian, R. (1991) Physical Metallurgy Principles. 3rd ed. PWS Publishing Co., Boston, MA.

Taylor, G.I. (1934) The mechanisms of plastic deformation of crystals. Part I, Theoretical. Proc R Soc, London. vol. A145, p 362.

Volterra, V. (1907) Ann Ecole Normale Super. vol. 24, p 400.

Weertman, J. and Weertman, J.R. (1992) Elementary Dislocation Theory. Oxford University Press, Oxford, UK.

7

Dislocations and Plastic Deformation

7.1 INTRODUCTION

Let us begin this chapter by performing the following thought experiment. Imagine picking up a piece of copper tubing that can be bent easily, at least the first time you try to bend it. Now think about what really happens when you bend the piece of copper a few times. You will probably remember from past experience that it becomes progressively harder to bend the piece of copper tubing after each bend. However, you have probably never asked yourself why.

Upon some reflection, you will probably come to the conclusion that the response of the copper must be associated with internal changes that occur in the metal during bending. In fact, the strength of the copper, and the progressive hardening of the copper, are associated with the movement of dislocations, and their interactions with defects in the crystalline copper lattice. This is hard to imagine. However, it is the basis for crystalline plasticity in most metallic materials and their alloys.

This chapter presents an overview of how dislocation motion and dislocation interactions contribute to plastic deformation in crystalline materials. We begin with a qualitative description of how individual dislocations move, interact, and multiply. The contributions of individual dislocations to bulk plastic strain are then considered within a simple continuum framework. This is followed by an introduction to the crystallography of slip in hexagonal and cubic materials. The role that dislocations play in the deformation of single crystals and polycrystals is then explained.

7.2 DISLOCATION MOTION IN CRYSTALS

As discussed in Chap. 6, dislocations tend to glide on close-packed planes along close-packed directions. This is due to the relatively low lattice friction stresses in these directions, Eq. (6.8a) or (6.8b). Furthermore, the motion of dislocations along a glide plane is commonly referred to as conservative motion. This is because the total number of atoms across the glide plane remains constant (conserved) in spite of the atomic interactions associated with dislocation glide (Fig. 6.9). In contrast to conservative dislocation motion by glide, nonconservative dislocation motion may also occur by climb mechanisms (Fig. 6.11). These often involve the exchange of atoms with vacancies. Since the atom/vacancy exchanges may be assisted by both stress and temperature, dislocation climb is more likely to occur during loading at elevated temperature.

So far, our discussion of dislocation motion has focused mostly on straight dislocations. Furthermore, it is presumed that the dislocations lie in the positions of lowest energy within the lattice, i.e., energy valleys/troughs (Fig. 6.10). However, in many cases, kinked dislocations are observed (Fig. 7.1). These have inclined straight or curved line segments that all lie on the same glide plane (Fig. 7.1). The shape of the kinked dislocation segment is dependent on the magnitude of the energy difference between the energy peaks and energy valleys in the crystalline lattice. In cases where the energy difference is large, dislocations can minimize their overall line energies by minimizing their line lengths in the higher energy peak regime. This gives rise to sharp kinks (A in Fig. 7.1) that enable dislocations to minimize their line lengths in the high-energy regions. It also maximizes the dislocation line lengths in the low-energy valley regions.

In contrast, when the energy difference (between the peaks and the valleys) is small, a diffuse kink is formed (C in Fig. 7.1). The diffuse kink has significant fractions of its length in the low-energy valleys and high-energy peak regions. In this way, a diffuse kink can also minimize the overall line energy of the dislocation.

The motion of kinked dislocations is somewhat complex, and will only be discussed briefly in this section. In general, the higher energy regions along the kink tend to move faster than those along the low-energy valleys which have to overcome a larger energy barrier. Once the barriers are overcome, kink nucleation and propagation mechanisms may be likened to the

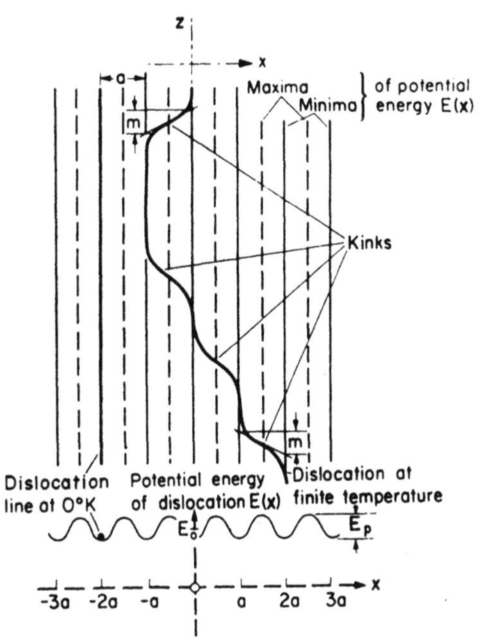

Schematic illustration of the potential energy surface of a dislocation line due to the Peierls stress τ_p. From (Seeger, Donth and Pfaff, *Disc. Faraday Soc.* **23**, 19, 1957.)

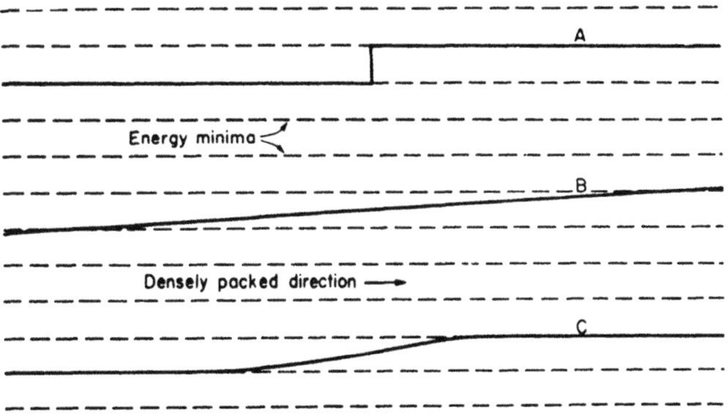

Shape of dislocations running almost parallel to a densely packed direction. The energy per unit length of the dislocation is a minimum along the dashed lines and varies periodically at right angles to the lines. The shape of the dislocation (curve *C*) is somewhere between the extremes *A* and *B*. (From Read (1953), *Dislocations in Crystals*, McGraw-Hill.)

Figure 7.1 Schematic of kinked dislocation configurations between peaks and valleys in a crystalline lattice. Note that sharp kink is formed when energy difference is large, diffuse kink is formed when energy difference is small (B), and most kinks are between the two extremes. (From Hull and Bacon, 1984. Reprinted with permission from Pergamon Press.)

snapping motion of a whip. As in the case of a snapped whip, this may give rise to faster kink propagation than that of a straight dislocation. The overall mobility of a kink will also depend on the energy difference between the peaks and the valleys, and the orientation of the dislocation with respect to the lattice.

Before concluding this section, it is important to note here that there is a difference between a sharp kink [A in Fig. 7.1 and Figs 7.2(a) and 7.2(b)] and a jog, Figs 7.2(a) and 7.2(b). A kink has all its segments on the same plane as the glide plane (Fig. 7.1). In contrast, a jog is produced by dislocation motion out of the glide plane as the rest of the dislocation line. Kinks and jogs may exist in edge and screw dislocations, Figs 7.2(a)–7.2(d). However, kinked dislocations tend to move in a direction that is perpendicular to the dislocation line, from one valley position to the other. Furthermore, kinks may also move faster than straight dislocation segments, while jogged dislocation segments are generally *not* faster than the rest of the dislocation line. In fact, they may be less mobile than the rest of the dislocation line, depending on the directions of their Burgers vectors relative to those of the unjogged segments.

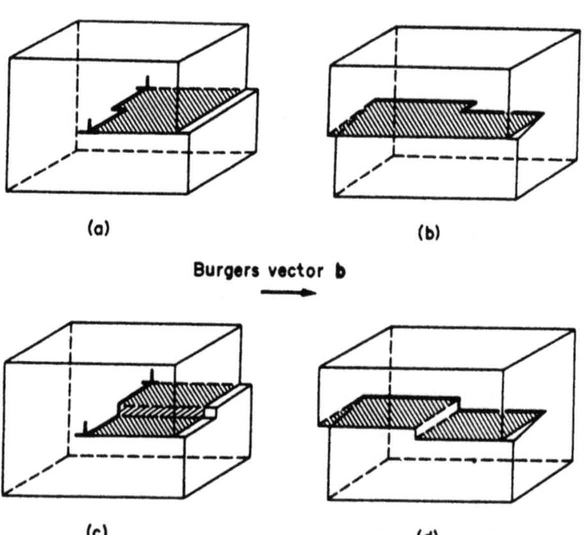

FIGURE 7.2 (a), (b) Kinks in edge and screw dislocations; (c), (d) jogs in edge and screw dislocations. The slip planes are shaded. (From Hull and Bacon, 1984. Reprinted with permission from Pergamon Press.)

7.3 DISLOCATION VELOCITY

When the shear stress that is applied to a crystal exceeds the lattice friction stress, dislocations move at a velocity that is dependent on the magnitude of the applied shear stress. This has been demonstrated for LiF crystals by Johnston and Gilman (1959). By measuring the displacement of etch pits in crystals with low dislocation densities, they were able to show that the dislocation velocity is proportional to the applied shear stress. Their results are presented in Fig. 7.3 for both screw and edge dislocations.

Note that, at the same stress level, edge dislocations move at faster speeds (up to 50 times faster) than screw dislocations. Also, the velocities of dislocations extend over 12 orders of magnitude on the log–log plot shown

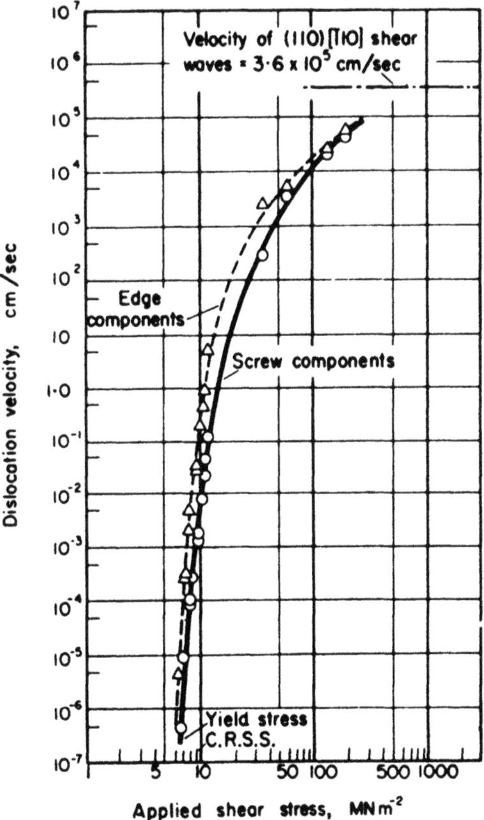

FIGURE 7.3 Dependence of dislocation velocity on applied shear stress. (From Johnston and Gilman, 1959. Reprinted with permission from J. Appl. Phys.)

in Fig. 7.3. However, for uniform dislocation motion, the limiting velocity for both screw and edge dislocations corresponds to the velocity of transverse shear waves. Also, damping forces increasingly oppose the motion of dislocations at velocities above 10^3 cm/s.

Dislocation velocities in a wide range of crystals have been shown to be strongly dependent on the magnitude of the applied shear stress (Fig. 7.4), although the detailed shapes of the dislocation velocity versus stress curves may vary significantly, as shown in Fig. 7.4. For the straight sections of the dislocation–velocity curves, it is possible to fit the measured velocity data to power-law equations of the form:

$$v = A(\tau)^m \tag{7.1}$$

where v is the dislocation velocity, τ is the applied shear stress, A is a material constant, and m is a constant that increases with decreasing temperature. An increase in dislocation velocity with decreasing temperature has also been demonstrated by Stein and Low (1960) in experiments on Fe–3.25Si crystals (Fig. 7.5). This increase is associated with the reduced damping forces due to the reduced scattering (phonons) of less frequent lattice vibrations at lower temperatures.

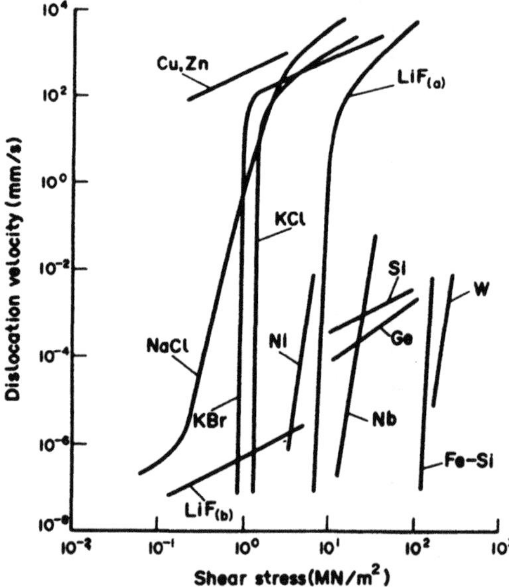

FIGURE 7.4 Dependence of dislocation velocity on applied shear stress. The data are for 20°C except for Ge (450°C) and Si (850°C). (From Haasen, 1988. Reprinted with permission from Cambridge University Press.)

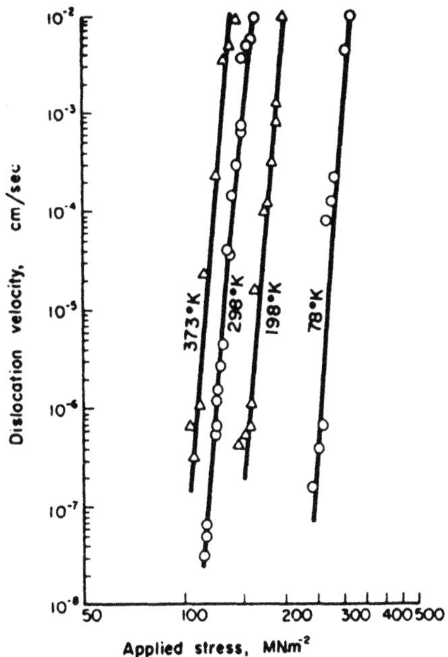

FIGURE 7.5 Dependence of dislocation velocity on temperature and applied shear stress in Fe–3.25Si Crystals. (From Stein and Low, 1960. Reprinted with permission from J. Appl. Phys.)

7.4 DISLOCATION INTERACTIONS

The possible interactions between screw and edge dislocations will be discussed in this section. Consider the edge dislocations (Burgers vectors perpendicular to the dislocation lines) AB and XY with perpendicular Burgers vectors, b_1 and b_2, shown in Fig 7.6. The moving dislocation XY [Fig. 7.6(a)] glides on a slip plane that is a stationary dislocation AB. During the intersection, a jog PP′ corresponding to one lattice spacing is produced as dislocation XY cuts dislocation AB, Fig. 7.6(b). Since the jog has a Burgers vector that is perpendicular to PP′, it is an edge jog. Also, since the Burgers vector of PP′ is the same as that of the original dislocation, AB, the jog will continue to glide along with the rest of the dislocation, if there is a large enough component of stress to drive it along the slip plane, which is perpendicular to that of line segments AP or P′ B, Fig. 7.6(b).

Let us now consider the interactions between two edge dislocations (XY and AB) with parallel Burgers vectors, Fig. 7.7(a). In this case, disloca-

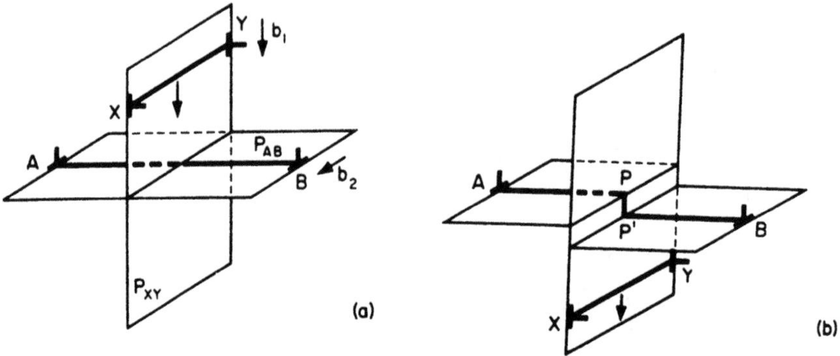

FIGURE 7.6 Interactions between two edge dislocations with perpendicular Burgers vectors: (a) before intersection; (b) after intersection. (From Hull and Bacon, 1984. Reprinted with permission from Pergamon Press.)

tion XY intersects dislocation AB, and produces two screw jogs PP′ and QQ′. The jogs PP′ and QQ′ are screw in nature because they are parallel to the Burgers vectors b_1 and b_2, respectively, Figs 7.7(a) and 7.7(b). Since the jogged screw dislocation segments have greater mobility than the edge dislocations to which they belong, they will not impede the overall dislocation motion. Hence, interactions between edge dislocations do not significantly affect dislocation mobility.

This is not true for interactions involving screw dislocations. For example, in the case of a right-handed screw dislocation that intersects a moving edge dislocation [Fig. 7.8(a)], the dislocation segment PP′ glides

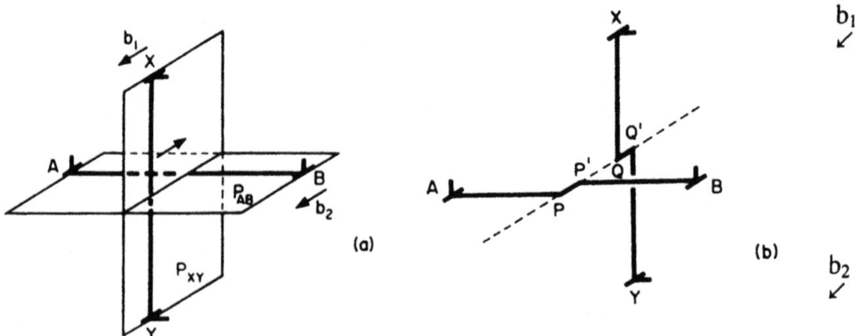

FIGURE 7.7 Interactions between two edge dislocations with parallel Burgers vectors: (a) before intersection; (b) after intersection. (From Hull and Bacon, 1984. Reprinted with permission from Pergamon Press.)

down one level (from one atomic plane to the other) following a spiral path (staircase) along the dislocation line XY, as it cuts the screw dislocation XY, Fig. 7.8(b). This produces a jog PP′ in AB, and a jog QQ′ in XY. Hence, the segments AP′ and PB lie on different planes, Fig. 7.8(b). Furthermore, since the Burgers vectors of the dislocation line segments PP′ and QQ′ are perpendicular to their line segments, the jogs are edge in character. Therefore, the only way the jog can move conservatively is along the axis of the screw dislocation, as shown in Fig. 7.9. This does not impede the motion of the screw dislocation, provided the jog glides on the plane (PP′RR′).

However, since edge dislocation components can only move conservatively by glide on planes containing their Burgers vectors and line segments, the movement of the edge dislocation to A′QQ′B (Fig. 7.9) would require nonconservative climb mechanisms that involve stress- and thermally assisted processes. This will leave behind a trail of vacancies or interstitials, depending on the direction of motion, and the sign of the dislocation. This is illustrated in Fig. 7.10 for a jogged screw dislocation that produces a trail of vacancies. Note that the dislocation segments between the jogs are bowed due to the effects of line tension. Bowing of dislocations due to line tension effects will be discussed in the next section. In closing, however, it is important to note here that the interactions between two screw dislocations (Fig. 7.11) can give rise to similar phenomena to those discussed above. It is a useful exercise to try to work out the effects of such interactions.

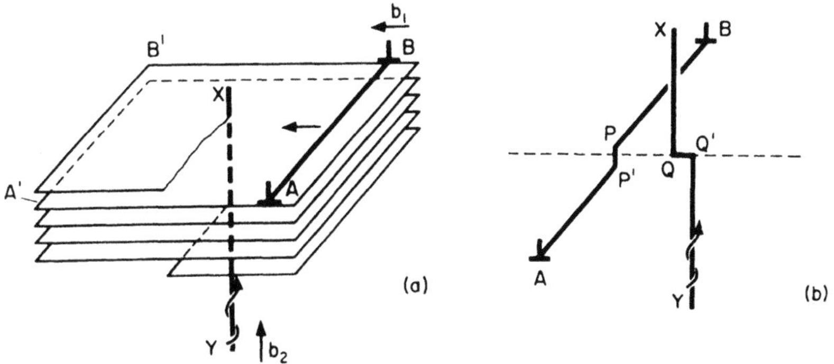

FIGURE 7.8 Interactions between right-handed screw dislocation and edge dislocations: (a) before intersection; (b) after intersection. (From Hull and Bacon, 1984. Reprinted with permission from Pergamon Press.)

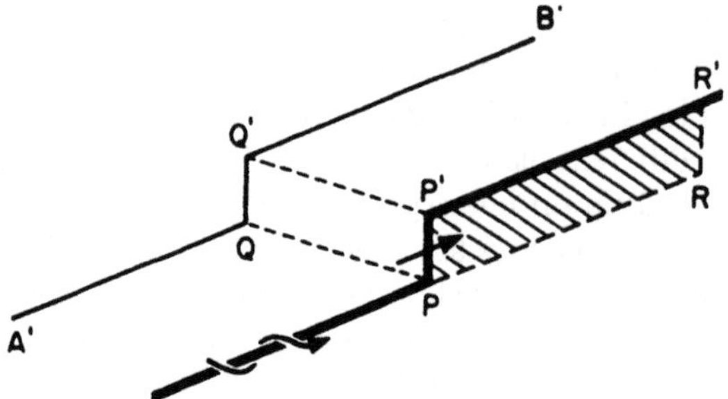

Figure 7.9 Movement of edge jog on a screw dislocation; conservative motion of jog only possible on plane PP′RR. Motion of screw dislocation to A′QQ′B would require climb of the jog along plane PQQ′P. (From Hull and Bacon, 1984. Reprinted with permission from Pergamon Press.)

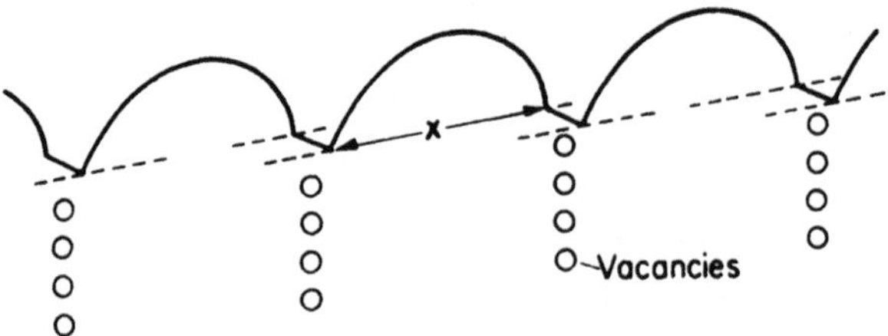

Figure 7.10 Schematic illustration of trail of vacancies produced by glide of screw dislocation. (From Hull and Bacon, 1984—reprinted with permission from Pergamon Press.)

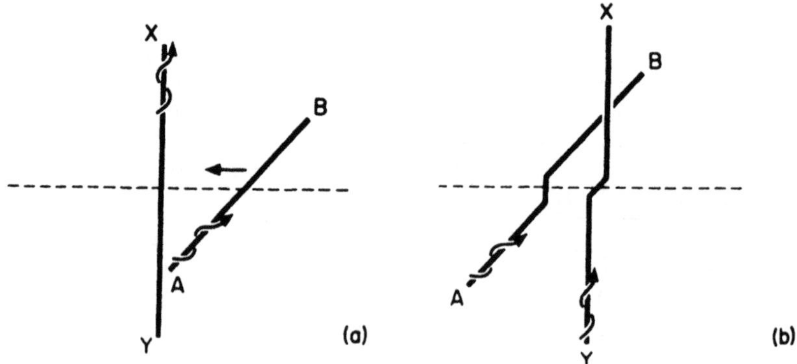

Figure 7.11 Interactions between two screw dislocations: (a) before intersection; (b) after intersection. (From Hull and Bacon, 1984. Reprinted with permission from Pergamon Press.)

7.5 DISLOCATION BOWING DUE TO LINE TENSION

It should be clear from the above discussion that interactions between dislocations can give rise to pinned dislocation segments, e.g., dislocation line segments that are pinned by jogs, solutes, interstitials, or precipitates. When a crystal is subjected to a shear stress, the so-called line tension that develops in a pinned dislocation segment can give rise to a form of dislocation bowing that is somewhat analogous to the bowing of a string subjected to line tension, T. In the case of a dislocation, the line tension has a magnitude $\sim Gb^2$. The bowing of dislocation is illustrated schematically in Fig 7.12.

Let us now consider the free body diagram of the bowed dislocation configuration in Fig. 7.12(b). For force equilibrium in the y direction, we may write:

$$2T\sin\left(\frac{\delta\theta}{2}\right) = \tau bL \tag{7.2}$$

where τ is the applied shear stress, b is the Burgers vector, L is the dislocation line length, and the other parameters are shown in Fig. 7.12. For small curvatures, $\sin(\delta\theta/2) \sim \delta\theta/2$, and so Eq. (7.2) reduces to

$$T\,\delta\theta \cong \tau bL \tag{7.3}$$

Recalling that $T \sim Gb^2$ and that $\delta\theta = L/R$ from Fig. 7.12, we may simplify Eq. (7.3) to give

$$Gb^2\frac{L}{R} = \tau bL \tag{7.4a}$$

or

$$\tau = \frac{Gb}{R} \tag{7.4b}$$

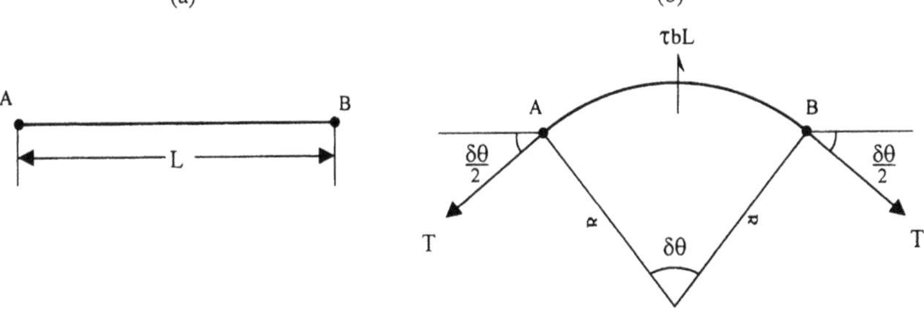

FIGURE 7.12 Schematics of (a) pinned dislocation segment and (b) bowed dislocation configuration due to applied shear stress, τ. Note that T is the line tension $\sim Gb^2$.

The critical shear stress described by Eq. (7.4b) is sufficient to cause the pinned dislocation to continue to bow in a stable manner until it reaches a semicircular configuration with $r = L/2$. This bowing forms the basis of one of the most potent mechanisms for dislocation multiplication, which is discussed in the next section.

7.6 DISLOCATION MULTIPLICATION

The discerning reader is probably wondering how plastic deformation can actually continue in spite of the numerous interactions between dislocations that are likely to give rise to a reduction in the density of mobile dislocations. This question will be addressed in this section. However, before answering the question, let us start by considering the simplest case of a well-annealed crystal. Such crystals can have dislocation densities as low as 10^8–10^{12} m/m^3. When annealed crystals are deformed, their dislocation densities are known to increase to $\sim 10^{16}$–10^{18} m/m^3. How can this happen when the interactions between dislocations are reducing the density of mobile dislocations?

This question was first answered by Frank and Read in a discussion that was held in a pub in Pittsburgh. Their conversation led to the mechanism of dislocation breeding that is illustrated schematically in Fig. 7.13. The schematics show one possible mechanism by which dislocations can multiply when a shear stress is applied to a dislocation that is pinned at both ends. Under an applied shear stress, the pinned dislocation [Fig. 7.13(a)] bows into a circular arc with radius of curvature, $r = L/2$, shown in Fig. 7.13(b). The bowing of the curved dislocation is caused by the line tension, T, as discussed in Sect. 7.5 (Fig. 7.12). This causes the dislocation to bow in a stable manner until it reaches the circular configuration illustrated schematically in Fig. 7.13(b). From Eq. (7.4b), the critical shear stress required for this to occur is $\sim Gb/L$.

Beyond the circular configuration of Fig. 7.13(b), the dislocation bows around the pinned ends, as shown in Fig. 7.13(c). This continues until the points labeled X and X′ come into contact, Fig. 7.13(d). Since these dislocation segments are opposite in sign, they annihilate each other. A new loop is, therefore, produced as the cusped dislocation [Fig. 7.13(e)] snaps back to the original straight configuration, Fig. 7.13(a).

Note that the shaded areas in Fig. 7.13 correspond to the regions of the crystal that have been sheared by the above process. They have, therefore, been deformed plastically. Furthermore, subsequent bowing of the pinned dislocation AB may continue, and the newly formed dislocation loop will continue to sweep through the crystal, thereby causing further

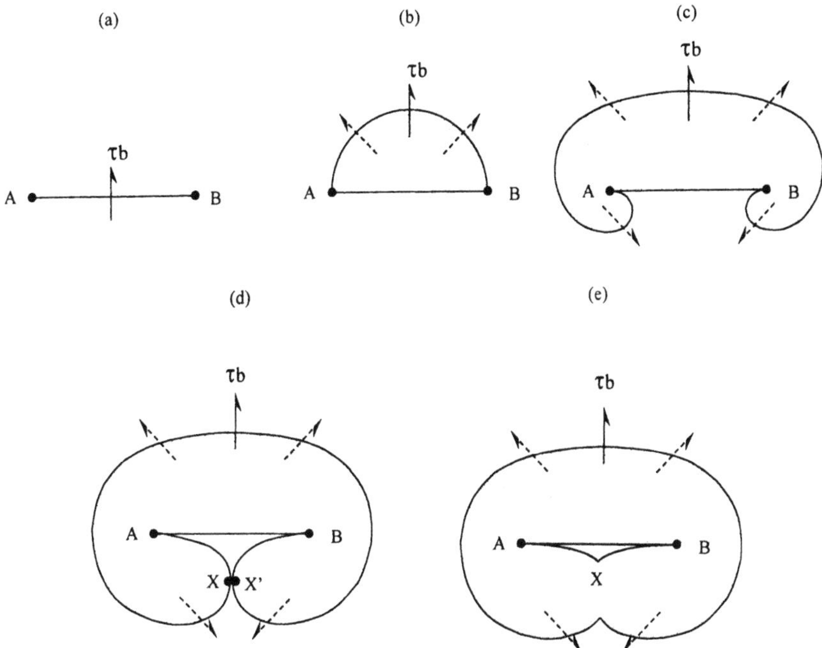

Figure 7.13 Breeding of dislocation at a Frank–Read source: (a) initial pinned dislocation segment; (b) dislocation bows to circular configuration due to applied shear stress; (c) bowing around pinned segments beyond loop instability condition; (d) annihilation of opposite dislocation segments X and X′, (e) loop expands out and cusped dislocation AXB returns to initial configuration to repeat cycle. (Adapted from Read, 1953. Reprinted with permission from McGraw-Hill.)

plastic deformation. New loops are also formed, as the dislocation AB repeats the above process under the application of a shear stress. This leads ultimately to a large increase in dislocation density (Read, 1953). However, since the dislocation loops produced by the Frank–Read sources interact with each other, or other lattice defects, back stresses are soon set up. These back stresses eventually shut down the Frank–Read sources. Experimental evidence of the operation of Frank–Read sources has been presented by Dash (1957) for slip in silicon crystals (Fig. 7.14).

A second mechanism that can be used to account for the increase in dislocation density is illustrated in Fig. 7.15. This involves the initial activation of a Frank–Read source on a given slip plane. Screw dislocation segments then cross-slip on to a different slip plane where a new Frank–Read

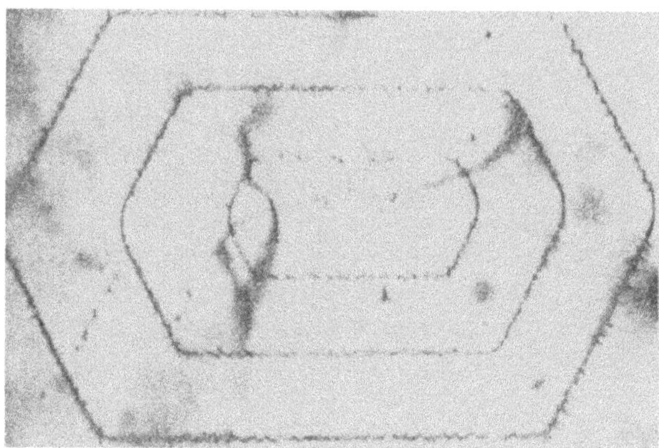

Figure 7.14 Photograph of Frank–Read source in a silicon crystal. (From Dash 1957. Reprinted with permission from John Wiley.)

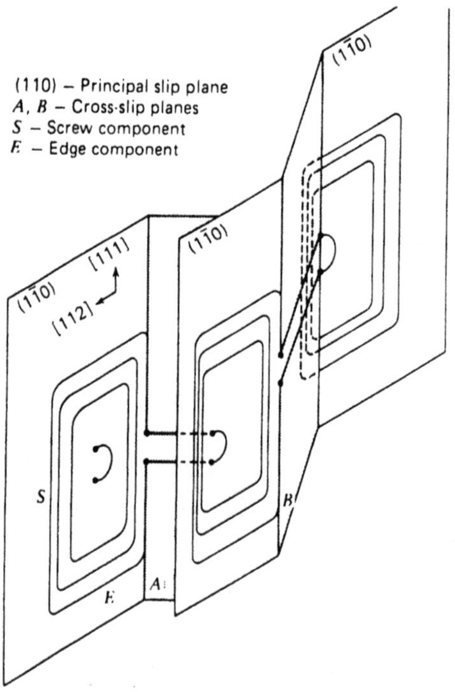

Figure 7.15 Dislocation multiplication by multiple cross-slip mechanism. (From Low and Guard, 1959. Reprinted with permission from Acta Metall.)

source is initiated. The above process may continue by subsequent cross-slip and Frank–Read source creation, giving rise to a large increase in the dislocation density on different slip planes. The high dislocation density (10^{15}–10^{18} m/m^3) that generally results from the plastic deformation of annealed crystals (with initial dislocation densities of $\sim 10^8$–10^{10} m/m^3) may, therefore, be explained by the breeding of dislocations at single Frank–Read sources (Fig. 7.12), or multiple Frank–Read sources produced by cross-slip processes (Fig. 7.15).

7.7 CONTRIBUTIONS FROM DISLOCATION DENSITY TO MACROSCOPIC STRAIN

The macroscopic strain that is developed due to dislocation motion occurs as a result of the combined effects of several dislocations that glide on multiple slip planes. For simplicity, let us consider the glide of a single dislocation, as illustrated schematically in Fig. 7.16. The crystal of height, h, is displaced by a horizontal distance, b, the Burgers vector, due to the glide of a single dislocation across distance, L, on the glide plane. Hence, partial slip across a distance, x, along the glide plane results in a displacement that is a fraction, x/L, of the Burgers vector, b. Therefore, the overall displacement due to N dislocations shearing different glide planes is given

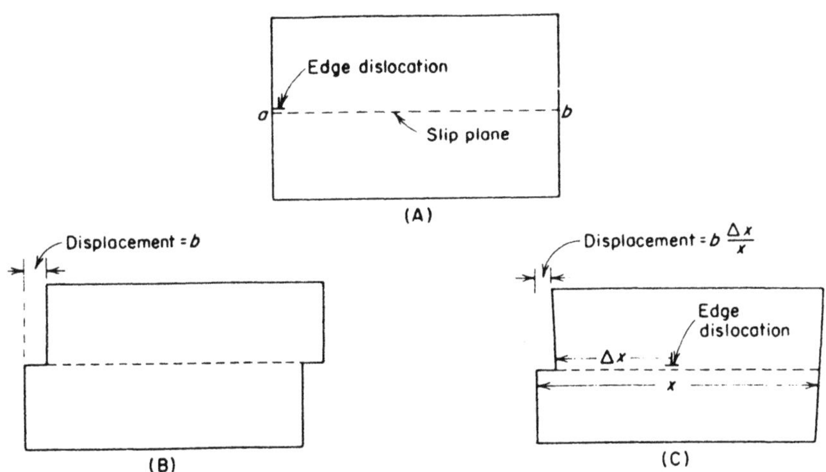

FIGURE 7.16 Macroscopic strain from dislocation motion: (A) before slip; (B) slip steps of one Burgers vector formed after slip; (C) displacement due to glide through distance Δx. (From Read-Hill and Abbaschian, 1994. Reprinted with permission from McGraw-Hill.)

by

$$\Delta = \sum_{i=1}^{N} \left(\frac{x_i}{h_i} \right) b \tag{7.5}$$

For small displacements, we may assume that the shear strain, γ, is $\sim \Delta/h$. Hence, from Eq. (7.4), we may write:

$$\gamma = \frac{\Delta}{h} = \sum_{i=1}^{N} \left(\frac{x_i}{L} \right) \frac{b}{h} \tag{7.6}$$

If we also note that the dislocation density, ρ, is given by $Nz/(hzL)$, we may rewrite Eq. (7.6) as

$$\gamma = \sum_{i=1}^{h} \frac{\rho b\, x_i}{N} \tag{7.7}$$

Assuming that the average displacement of each dislocation is $\overline{x}$, Eq. (7.7) may now be written as

$$\gamma = \rho b \overline{x} \tag{7.8}$$

The shear strain rate, $\dot{\gamma}$, may also be obtained from the time derivative of Eq. (7.8). This gives

$$\dot{\gamma} = \frac{d\gamma}{dt} = \rho b \overline{v} \tag{7.9}$$

where $\overline{v}$ is the average velocity of dislocations, which is given by dx/dt. It is important to note here that ρ in the above equations may correspond either to the overall dislocation density, ρ_{tot}, or to the density of mobile dislocations, ρ_m, provided that $\overline{x}$ and $\overline{v}$ apply to the appropriate dislocation configurations (mobile or total). Hence, $\rho_m \overline{x}_m = \rho_{tot} \overline{x}_{tot}$ or $\rho_m \overline{v}_m = \rho_{tot} \overline{v}_{tot}$ in Eqs (7.8) and (7.9).

Finally in this section, it is important to note that Eqs (7.8) and (7.9) have been obtained for straight dislocations that extend completely across the crystal width, z. However, the same results may be derived for curved dislocations with arbitrary configurations across multiple slip planes. This may be easily realized by recognizing that the sheared area fraction of the glide plane, dA/A, corresponds to the fraction of the Burgers vector, b, in the expression for the displacement due to glide of curved dislocations. Hence, for glide by curved dislocations, the overall displacement, Δ, is now given by

$$\Delta = \sum_{i=1}^{N} \frac{\Delta A_i}{A} b \tag{7.10}$$

As before [Eq. (7.6)], the shear strain, γ, due to the glide of curved dislocations is also given by Δ/h. Hence,

$$\gamma = \frac{\Delta}{h} \sum_{i=1}^{N} \left(\frac{\Delta A_i}{A}\right) \frac{b}{h} \tag{7.11}$$

Similarly, the shear strain rate may be obtained from the time derivative of Eq. (7.11).

7.8 CRYSTAL STRUCTURE AND DISLOCATION MOTION

In Chap. 6, we learned that the Peierl's (lattice friction) stress [Eq. (6.8)] is minimized by small Burgers vectors, b, and large lattice spacings, a. Hence, dislocation motion in cubic crystals tends to occur on closed packed (or closest packed) planes in which the magnitudes of the Burgers vectors, b, are minimized, and the vertical lattice spacings, a, are maximized. Since the lattice friction stresses are minimized on such planes, dislocation motion is most likely to occur on closed packed planes along closed packed directions (Table 7.1).

7.8.1 Slip in Face-Centered Cubic Structures

Close-packed planes in face-centered cubic (f.c.c.) materials are of the $\{111\}$ type. An example of a (111) plane is shown in Fig. 7.17(a). All the atoms touch within the closed packed (111) plane. Also, the possible $\{111\}$ slip planes form an octahedron if all the $\{111\}$ planes in the eight possible quadrants are considered. Furthermore, the closed packed directions correspond to the $\langle 110\rangle$ directions along the sides of a $\{111\}$ triangle in the octahedron. Hence, in the case of f.c.c. materials, slip is most likely to occur on octahedral $\{111\}$ planes along $\langle 110\rangle$ directions. Since there are four slip planes with three slip directions in the f.c.c. structure, this indicates that there are 12 (four slip lanes $\times$ three slip directions) possible $\{111\}$ $\langle 110\rangle$ slip systems (Table 7.1).

7.8.2 Slip in Body-Centered Cubic Structures

In the case of body-centered cubic (b.c.c.) structures, there are no close-packed planes in which all the atoms touch, although the $\{101\}$ planes are the closest packed. The close-packed directions in b.c.c. structures are the

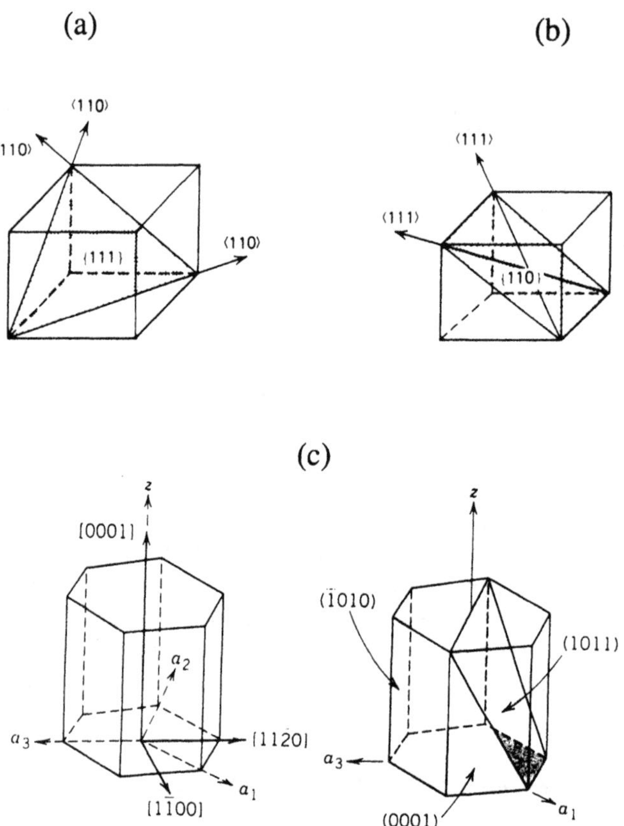

FIGURE 7.17 Closed packed planes and directions in (a) face-centered cubic structure, (b) body-centered cubic crystal, and (c) hexagonal closed packed structure. (From Hertzberg, 1996. Reprinted with permission from John Wiley.)

⟨111⟩ directions. Slip in b.c.c. structures is most likely to occur on {101} planes along ⟨111⟩ directions, Fig. 7.17(b). However, slip may also occur on {110}, {112}, and {123} planes along ⟨111⟩ directions. When all the possible slip systems are counted, there are 48 such systems in b.c.c. structures (Table 7.1). This rather large number gives rise to wavy slip in b.c.c. structures. Nevertheless, the large number of possible slip systems in b.c.c. crystals (four times more than those in f.c.c. materials) do not necessarily promote improved ductility since the lattice friction (Peierls–Nabarro) stresses are generally higher in b.c.c. structures.

7.8.3 Slip in Hexagonal Closed Packed Structures

The basal (0001) plane is the closed packed plane in hexagonal closed packed (h.c.p.) structures. Within this plane, slip may occur along closed packed $\langle 11\bar{2}0\rangle$ directions, Fig. 7.17(c). Depending on the c/a ratios, slip may also occur on nonbasal $(10\bar{1}0)$ and $(10\bar{1}1)$ planes along $\langle 11\bar{2}0\rangle$ directions (Table 7.1). This is also illustrated schematically in Fig. 7.18. Nonbasal slip is more likely to occur in h.c.p. metals with c/a ratios close to 1.63, which is the expected value for ideal close packing. Also, pyramidal $(10\bar{1}1)$ slip may be represented by equivalent combinations of basal (0001) and prismatic (1010) slip.

7.8.4 Condition for Homogeneous Plastic Deformation

The ability of a material to undergo plastic deformation (permanent shape change) depends strongly on the number of independent slip systems that can operate during deformation. A necessary (but not sufficient) condition for homogeneous plastic deformation was first proposed by Von Mises (1928). Noting that six independent components of strain would require six independent slip components for grain boundary compatibility between two adjacent crystals (Fig. 7.19), he suggested that since plastic deformation occurs at constant volume, then $\Delta V/V = \varepsilon_{xx} + \varepsilon_{yy} + \varepsilon_{zz} = 0$. This reduces the number of grain boundary compatibility equations by one. Hence, only five independent slip systems are required for homogeneous plastic defor-

TABLE 7.1 Summary of Slip Systems in Cubic and Hexagonal Crystals

Crystal Structure	Slip Plane	Slip direction	Number of nonparallel planes	Slip directions per plane	Number of slip systems
Face-centered cubic	{111}	$\langle 1\bar{1}0\rangle$	4	3	12 = (4 × 3)
Body-centered cubic	{110}	$\langle \bar{1}11\rangle$	6	2	12 = (6 × 2)
	{112}	$\langle 11\bar{1}\rangle$	12	1	12 = (12 × 1)
	{123}	$\langle 11\bar{1}\rangle$	24	1	24 = (24 × 1)
Hexagonal close-packed	{0001}	$\langle 11\bar{2}0\rangle$	1	3	3 = (1 × 3)
	$\{10\bar{1}0\}$	$\langle 11\bar{2}0\rangle$	3	1	3 = (3 × 1)
	$\{10\bar{1}1\}$	$\langle 11\bar{2}0\rangle$	6	1	6 = (6 × 1)

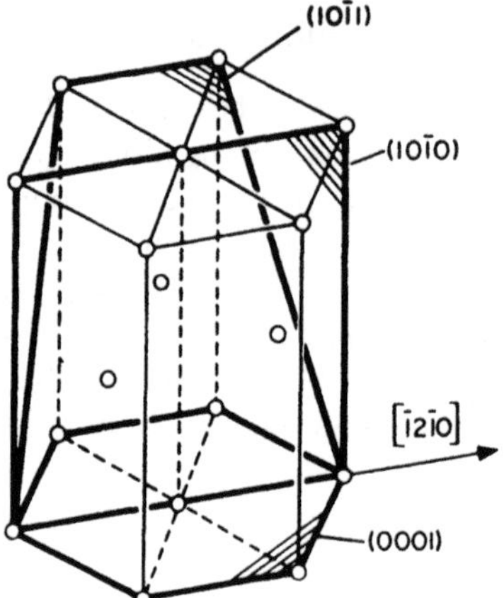

FIGURE 7.18 Planes in a hexagonal closed packed structure, with a common [11$\bar{2}$0] direction. (From Hull and Bacon, 1984. Reprinted with permission from Pergamon Press.)

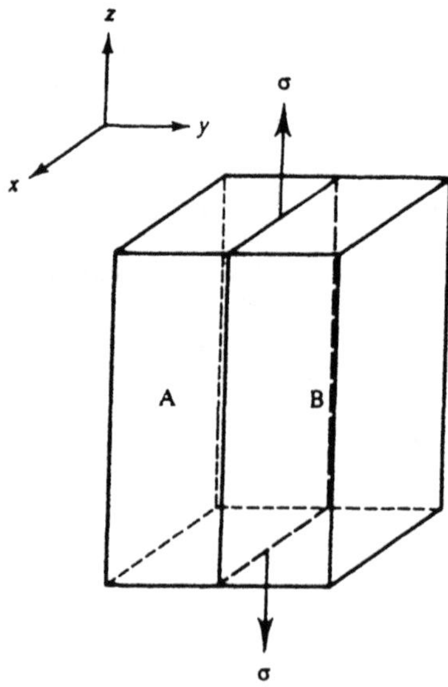

FIGURE 7.19 Strain conditions for slip compatibility at adjacent crystals. (From Courtney, 1990. Reprinted with permission from McGraw-Hill.)

mation. This is a necessary (but not sufficient) condition for homogeneous plastic deformation in polycrystals.

The so-called Von Mises condition for homogeneous plastic deformation is satisfied readily by f.c.c. and b.c.c. crystals. In the case of f.c.c. crystals, Taylor (1938) has shown that only five of the 12 possible $\{111\}$ $\langle 110\rangle$ slip systems are independent, although there are 384 combinations of five slip systems that can result in any given strain. Similar results have been reported by Groves and Kelly (1963) for b.c.c. crystals in which 384 sets of five $\{110\}$ $\langle 111\rangle$ slip systems can be used to account for the same strain. A much larger number of independent slip systems is observed in b.c.c. structures when possible slips in the $\{112\}$ $\langle 111\rangle$ and $\{123\}$ $\langle 111\rangle$ systems are considered. The large number of possible slip systems in this case have been identified using computer simulations by Chin and coworkers (1967, 1969).

In contrast to b.c.c. and f.c.c. crystals, it is difficult to show the existence of five independent slip systems in h.c.p. metals/alloys in which slip may occur on basal, prismatic, and pyramidal planes, Fig. 7.17(c). However, only two of the $\{0001\}$ $\langle 11\bar{2}0\rangle$ slip systems in the basal plane are independent. Similarly, only two of the prismatic $\{10\bar{2}0\}$ $\langle 11\bar{2}0\rangle$ type systems are independent. Furthermore, all the pyramidal slip systems can be reproduced by combinations of basal and prismatic slip. There are, therefore, only four independent slip systems in h.c.p. metals. So, how then can homogeneous plastic deformation occur in h.c.p. metals such as titanium? Well, the answer to this question remains an unsolved puzzle in the field of crystal plasticity.

One possible mechanism by which the fifth strain component may be accommodated involves a mechanism of deformation-induced twinning. This occurs by the co-ordinated movement of several dislocations (Fig. 7.18). However, further work is still needed to develop a fundamental understanding of the role of twinning in titanium and other h.c.p. metals/alloys.

7.8.5 Partial or Extended Dislocations

In f.c.c. crystals, the zig-zag motion of atoms required for slip in the $\langle 110\rangle$ directions may not be energetically favorable since the movement of dislocations requires somewhat difficult motion of the "white" atoms over the "shaded" atoms in Fig. 7.20. The ordinary $\langle 110\rangle$ dislocations may, therefore, dissociate into partial dislocations with lower overall energies than those of the original $\langle 110\rangle$ type dislocations.

The partial dislocations may be determined simply by vector addition, as shown schematically in Fig. 7.21. Note that $B_1C = b_2 = 1/6[\bar{1}2\bar{1}]$. Similarly, CB_2 can be shown to be given by $CB_2 = b_3 = 1/6[\bar{2}11]$. The ordinary dislocation $b_1 = 1/2[\bar{1}10]$ may, therefore, be shown by vector addition to be given by

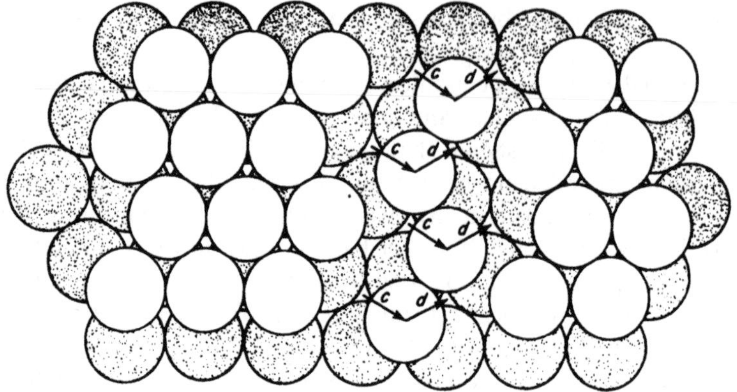

FIGURE 7.20 Zig-zag motion of atoms required for slip in face-centered cubic crystals. Note that "white" atoms are in a row above the "shaded" atoms. (From Read-Hill and Abbaschian, 1991. Reprinted with permission from McGraw-Hill.)

$$b_1 = b_2 + b_3 \tag{7.12a}$$

or

$$\frac{1}{2}[\bar{1}10] = \frac{1}{6}[\bar{2}11] + \frac{1}{6}[\bar{1}2\bar{1}] \tag{7.12b}$$

The partial dislocations, b_2 and b_3, are generally referred to as Shockley partials. They are formed because the elastic energies of the ordinary dislocations of type b_1 are greater than the sum of the line energies of the Shockley partials. Hence,

$$G(b_1)^2 > G(b_2)^2 + G(b_3)^2 \tag{7.13a}$$

or

$$\frac{G}{4}\left[(-1)^2 + (1)^2 + (0)^2\right] > \frac{G}{36}\left[(-1)^2 + (2)^2 + (-1)^2\right] + \frac{G}{36}\left[(-2)^2 + (1)^2 + (1)^2\right] \tag{7.13b}$$

or

$$\frac{G}{2} > \frac{G}{3} \tag{7.13c}$$

The above dislocation reaction is, therefore, likely to proceed since it is energetically favorable. This separation occurs because the net force on

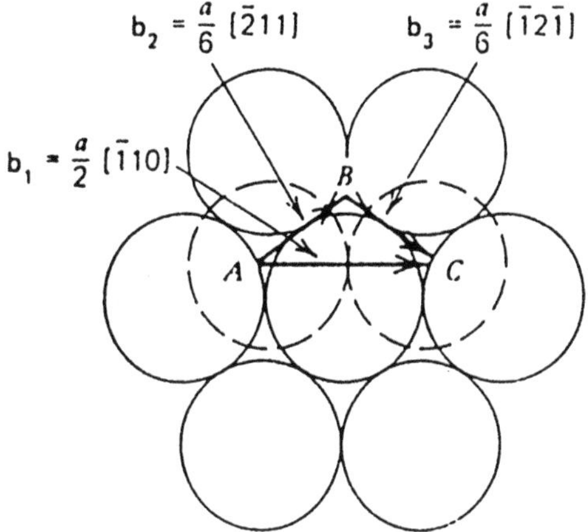

FIGURE 7.21 Path of whole (ordinary) and partial (Shockley) dislocations. (From Courtney, 1990. Reprinted with permission from McGraw-Hill.)

the partials is repulsive. As the partials separate, the regular ABC stacking of the f.c.c. lattice is disturbed. The separation continues until an equilibrium condition is reached where the net repulsive force is balanced by the stacking fault energy (Fig. 7.22). The equilibrium separation, d, between the two partials has been shown by Cottrell (1953) to be

$$d = \frac{Gb_2 b_3}{2\pi\gamma} \tag{7.14}$$

where G is the shear modulus, γ is the stacking fault energy, and b_2 and b_3 correspond to the Burgers vectors of the partial dislocations. Stacking faults ribbons corresponding to bands of partial dislocations are presented in Fig. 7.22(b). Typical values of the stacking fault energies for various metals are also summarized in Table 7.2. Note that the stacking fault energies vary widely for different elements and their alloys. The separations of the partial dislocations may, therefore, vary significantly, depending on alloy composition, atomic structure, and electronic structure.

The variations in stacking fault energy have been found to have a strong effect on slip planarity, or conversely, the waviness of slip in metals and their alloys that contain partial (extended) dislocations. This is because the movement extended dislocations is generally confined to the plane of the stacking fault. The partial dislocations must, therefore, recombine before cross-slip can occur. For this reason, metals/alloys with higher stacking fault energies will have narrow stacking faults [Eq. (7.14)], thus making recombi-

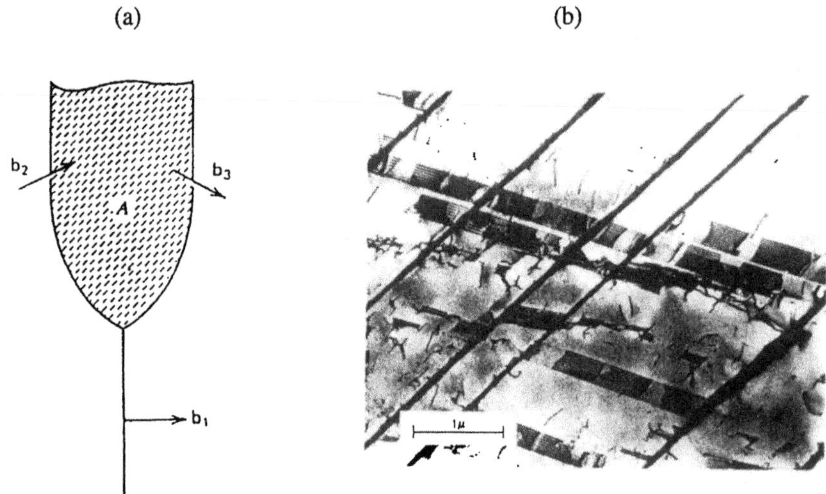

FIGURE 7.22 (a) Shockley b_2 and b_3 surrounding stacking fault region A; (b) stacking fault ribbons in a stainless steel. (From Michelak, 1976. Reprinted with permission from John Wiley.)

nation and cross-slip easier. This reduces the stress required for recombination.

Conversely, metals and alloys with low stacking fault energies have wide separations (stacking faults) between the partial dislocations. It is, therefore, difficult for cross-slip to occur, since the recombination of partial dislocations is difficult. Furthermore, because the movement of uncombined

TABLE 7.2 Stacking Fault Energies for Face Centered Cubic Metals and Alloys

Metal	Stacking faulty energy ($mJ/m^2 = ergs/cm^2$)
Brass	< 10
Stainless steel	< 10
Ag	~ 25
Au	~ 75
Cu	~ 90
Ni	~ 200
Al	~ 250

Source: Hertzberg (1996). Reprinted with permission from John Wiley.

TABLE 7.3 Stacking Faults and Strain Hardening Exponents

Metal	Stacking faulty energy (mJ/M^2)	Strain-hardening coefficient	Slip character
Stainless steel	<10	−0.45	Planar
Cu	~ 90	~ 0.3	Planar/wavy
Al	~ 250	~ 0.15	Wavy

Source: Hertzberg (1996). Reprinted with permission from John Wiley.

partial dislocations is confined to planes containing the stacking faults, materials with lower stacking fault energies (wide separations of partials) tend to exhibit higher levels of strain hardening (Table 7.3).

7.8.6 Superdislocations

So far, our discussion has focused on dislocation motion in disordered structures in which the solute atoms can occupy any position within the crystal structure. However, in some intermetallic systems (intermetallics are compounds between metals and metals), ordered crystal structures are formed in which the atoms must occupy specific sites within the crystal structure. One example of an ordered f.c.c. structure is the Ni_3Al crystal shown in Fig. 7.23. The nickel and aluminum atoms occupy specific positions in the structure, which must be retained after dislocation glide through {111} planes. However, the movement of a single dislocation on the glide plane disturbs the ordered arrangement of atoms, giving rise to an energetically unfavorable arrangement of atoms, Fig. 7.24(a).

A favorable arrangement is restored by the passage of a second dislocation, which restores the lower energy ordered crystal structure, Fig.

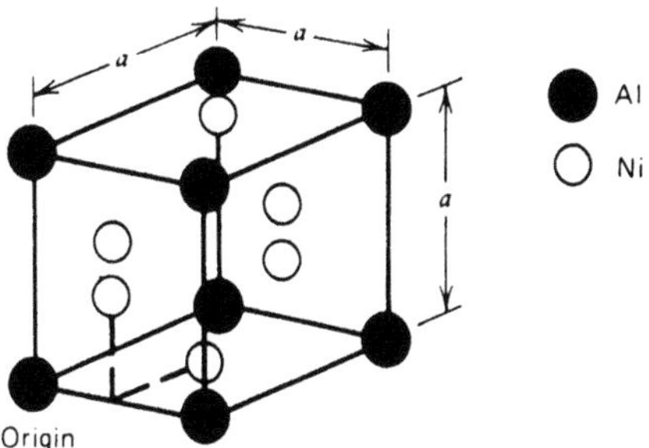

FIGURE 7.23 Ordered face-centered cubic structure of Ni_3Al. (From Hertzberg, 1996. Reprinted with permission from John Wiley.)

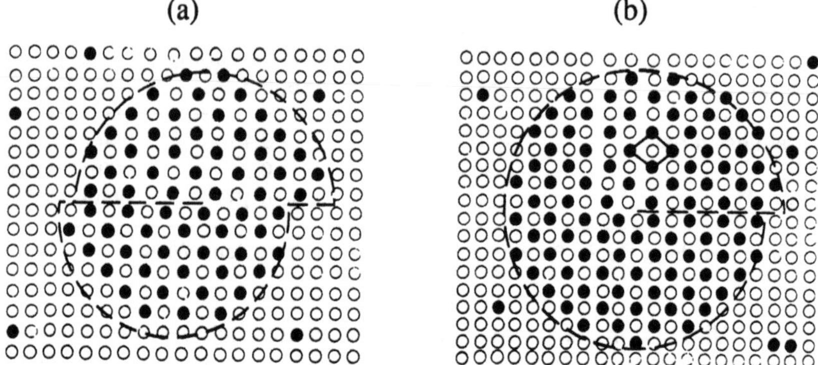

FIGURE 7.24 Effects of dislocation motion on atomic arrangements in Ni_3Al: (a) unfavorable atomic arrangement associated with passage of a single dislocation; (b) favorable ordered arrangement restored by passage of a second dislocation. (From Hertzberg, 1996. Reprinted with permission from John Wiley.)

7.24(b). The two dislocations are referred to as superlattice dislocations or superdislocations. Like superdislocations, superdislocations maintain an equilibrium separation that corresponds to the equilibrium separation between the repulsive force (between two like dislocations) and the antiphase boundary (APB). Examples of superdislocation pairs in Ni_3Al are shown in Fig. 7.25.

Finally in this section, it is of interest to note that the individual dislocations may in turn dissociate into partial dislocations that are separated by stacking faults and APBs, as shown schematically in Fig. 7.26.

7.9 CRITICAL RESOLVED SHEAR STRESS AND SLIP IN SINGLE CRYSTALS

Let us now consider a general case of slip in a single crystal that is subjected to axial loading, as shown in Fig. 7.27. Whether or not a dislocation will move on the slip plane in a given slip direction depends on the magnitude of the resolved shear stress in the direction of slip. Slip will only occur when the resolved shear stress (due to the applied load) is sufficient to cause dislocation motion. Noting that the magnitude of the resolved load along the slip plane is $P \cos \lambda$, and the inclined cross-sectional area is $A_o/\cos\phi$, we may then write the following expression for the resolved stress, τ_{RSS}:

$$\tau_{RSS} = \frac{P \cos \lambda}{(A_0 \cos \phi)} \frac{P}{A_0} \tag{7.15a}$$

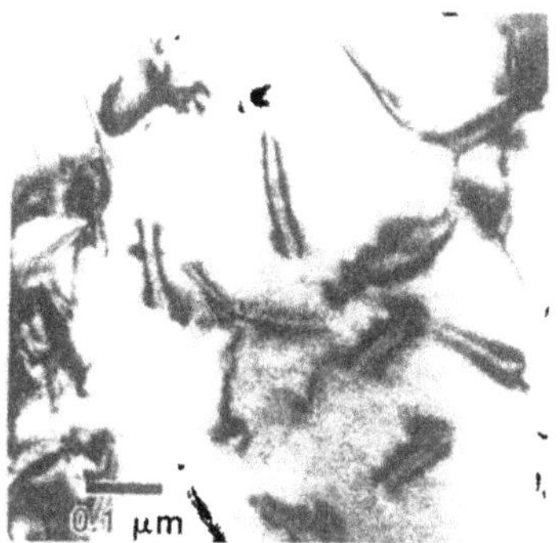

Figure 7.25 Superdislocation pairs in Ni_3Al. (Courtesy of Dr. Mohammed Khobaib and reprinted from Hertzberg, 1996. Reprinted with permission from John Wiley.)

where P is the applied axial load, A_0 is the cross-sectional area perpendicular to the applied load, and angles ϕ and λ are shown schematically in Fig. 7.27. The product cos λ cos ϕ is known as the *Schmid factor, m.* It is important because it represents a geometrical/orientation factor that determines the extent to which the applied load can induce shear stresses that may eventually cause dislocation motion to occur on possible slip lanes. Furthermore,

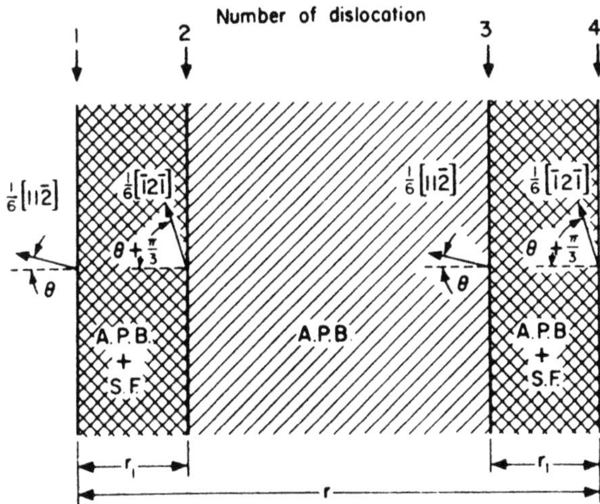

Figure 7.26 Schematic of stacking fault and antiphase boundaries bounded by partial dislocation pairs: A $\{01\bar{1}\}$ superlattice dislocation in an AB_3 superlattice. (From Marcinkowski et al., 1961.)

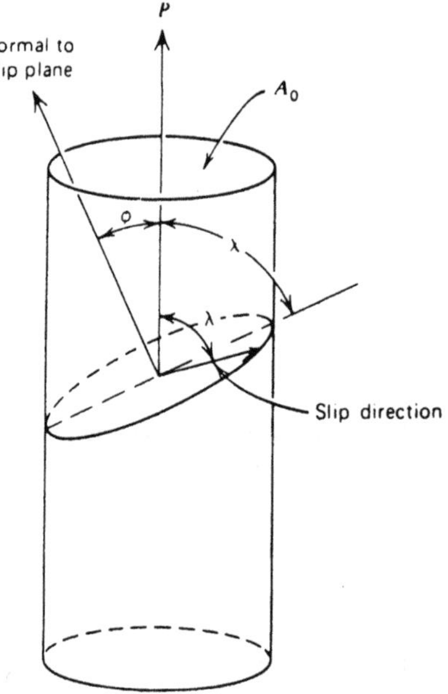

FIGURE 7.27 Slip plane and slip direction in a cylindrical single crystal subjected to axial deformation.

the onset of plasticity by dislocation motion corresponds to the critical shear stress (for a given slip system) that is just sufficient to induce dislocation motion. Hence, from Eq. (7.15a), yielding will initiate on the plane with the highest Schmid factor.

For the different combinations of slip planes and slip directions associated with any given slip system, yielding is generally found to occur at the same value of the critical resolved shear stress. The yield stress, σ_{ys}, of a single crystal may, therefore, be found by rearranging Eq. (7.15a) and noting that $\sigma_y = P_y/A_0$, where P_y is the load at the onset of yielding. This gives

$$\sigma_y = \frac{\tau_{CRSS}}{\cos\lambda\cos\phi} = \frac{\tau_{CRSS}}{m} \tag{7.15b}$$

where τ_{CRSS} is the critical resolved shear stress, m is the Schmid factor, and the other variables have their usual meaning. The unaxial yield strength of a single crystal will, therefore, depend on the slip system that has the highest Schmid factor, since this will result in the lowest value of σ_y. Consequently, the uniaxial yield strength of a single crystal may vary significantly with crystal orientation, even though the critical resolved shear stress does not generally change with crystal orientation, for yielding in a given slip system.

A schematic of a typical shear stress versus shear strain response of a single crystal is shown in Fig. 7.28. During the early stages of deformation, Stage I slip occurs by easy glide in a single slip direction along a single slip plane. There is limited interaction between dislocations, and the extent of hardening is limited. However, due to the constraints imposed by the specimen grips, the slipped segments of the single crystal experience close to pure rotation in the middle of the crystals, and pure bending near the grips (Fig. 7.29). The rotation of the slip plane gradually changes the Schmid factor until slip is induced in other slip systems. The interactions between dislocations gliding on multiple slip systems then results in hardening in Stage II, as shown schematically in Fig. 7.28. Stage II hardening is associated with a characteristic slope of $\sim G/300$ in several metals. Stage II hardening continues until Stage III is reached (Fig. 7.28), where the hardening is relaxed by cross-slip of screw dislocation segments. A cell structure is also likely to develop during Stage III in which the dislocation substructures closely resemble those observed in polycrystalline metals and their alloys.

The stress–strain behavior discussed above may vary significantly with test temperature and impurities. Furthermore, depending on crystal orientation and initial dislocation density, Stage I deformation may be absent in crystals in which two or more slip systems are initiated at the same stress level (Fig. 7.30). Since the dislocations can interact, rapid Stage II hardening may be observed in some crystals at the onset of plastic deformation.

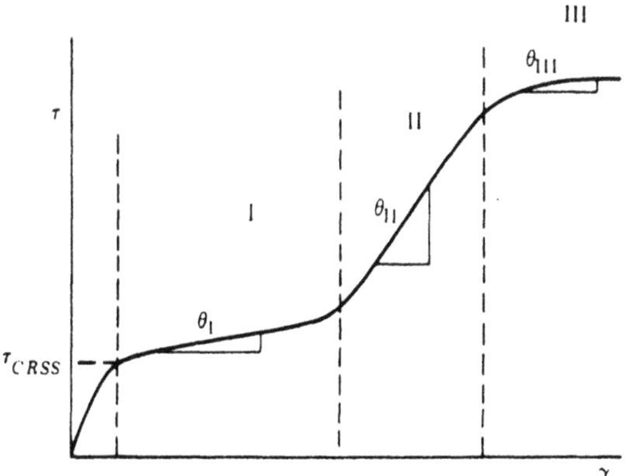

FIGURE 7.28 Three stages of plastic deformation in a single crystal.

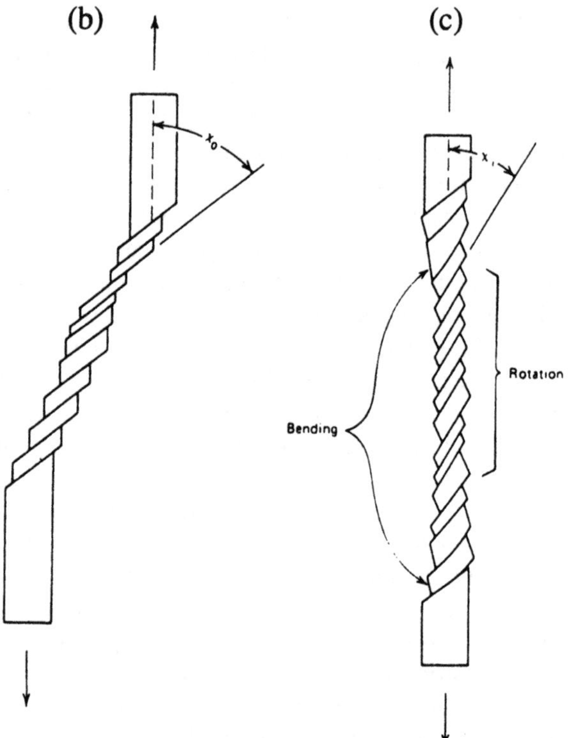

FIGURE 7.29 Schematic illustration of the effects of deformation constraint on the deformation behavior of a single crystal: (a) before deformation; (b) deformation without grip constraint; (c) deformation with group constraint.

Finally in this section, it is important to note that an alternative explanation of the above hardening behavior has been presented by Kuhlmann-Wilsdorf (1962, 1968). She attributes the low levels of Stage I hardening to heterogeneous slip of a low density of dislocations. In this theory, Stage II slip corresponds to the onset of significant dislocation–dislocation interactions, but not necessarily the onset of multiple slip. This results ultimately in the formation of a dislocation cell structure with a characteristic mesh length that remains stable from the onset of Stage III deformation.

7.10 SLIP IN POLYCRYSTALS

In Sect. 7.8.4, we showed that the homogeneous plastic deformation of polycrystals requires dislocation motion to occur on five independent slip

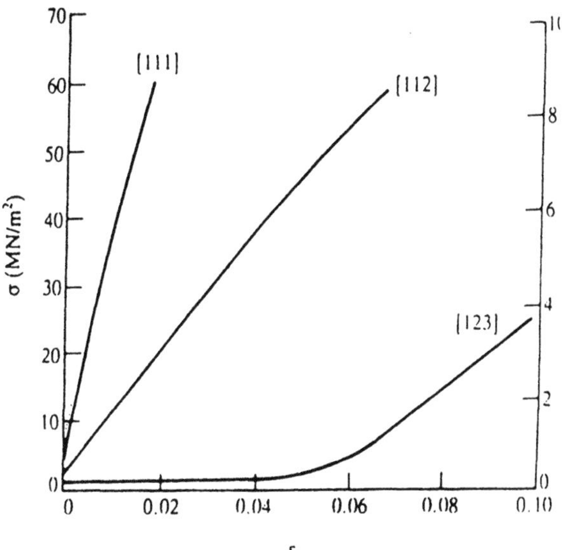

FIGURE 7.30 Different types of stress–strain behavior in copper single crystals with different initial orientations with respect to the tensile axis. Although the critical resolved shear stress is the same for all the crystals, only the [123] exhibits the easy glide Stage I regime. Duplex slip occurs initially in the [112] oriented crystal, giving rise to greater hardening at the onset of plastic deformation. More pronounced hardening to observed in the [111] oriented crystals in which six slip systems are activated initially. (From Diehl, 1956.)

systems (the Von Mises condition). This occurs relatively easily in f.c.c. and b.c.c. crystals. However, homogeneous plastic deformation is difficult to explain in h.c.p. crystals, which require the activation of additional deformation modes such as deformation-induced twinning. The simple picture of slip in single crystals developed in the previous section will be extended to the more general case of slip in polycrystals in this section.

Let us start by recalling from Sect. 7.9 that slip in a single crystal occurs when a critical resolved shear shear stress is reached. Furthermore, for slip without relative sliding of the boundary between adjacent grains A and B (Fig. 7.19), the strain components on either side of the boundary must be equal, i.e., $\varepsilon_{xx,\mathrm{A}} = \varepsilon_{xx,\mathrm{B}}$, $\varepsilon_{yy,\mathrm{A}} = \varepsilon_{yy,\mathrm{B}}$, $\varepsilon_{zz,\mathrm{A}} = \varepsilon_{zz,\mathrm{B}}$, $\gamma_{xy,\mathrm{A}} = \gamma_{xy,\mathrm{B}}$, $\gamma_{yz,\mathrm{A}} = \gamma_{yz,\mathrm{B}}$, and $\gamma_{zx,\mathrm{A}} = \gamma_{zx,\mathrm{B}}$. Since the volume does not increase during plastic deformation, it follows that $\Delta V/V = \varepsilon_{xx} + \varepsilon_{yy} + \varepsilon_{zz}$ is equal to zero. Hence, only five independent slip systems are needed for homogeneous plastic deformation.

However, the matching of strain components at the boundaries between grains imposes significant restrictions on the possible slip systems that can occur within polycrystals. The total number of possible slip systems in actual grains within polycrystals may, therefore, be significantly less than those in single crystals that are favorably oriented for slip. Furthermore, each grain has its own characteristic Schmid factor, and grains with the lowest Schmid factor deform last. The average Schmid factor is also more strongly affected by grains with unfavorable orientations (for plastic deformation).

In any case, the stress–strain behavior of polycrystals may be understood by considering an effective orientation factor, m, that is somewhat analogous to the Schmid factor, $\overline{m}$, that was introduced in Sect 7.9. However, the effective orientation factor, $\overline{m}$, is a more complex parameter than the Schmid factor, m, because it must somehow account for the numerous orientations of crystals that are possible between grains in a polycrystal. The effective orientation factor must also account for the stronger effects of less favorably oriented grains. This was first considered by Taylor (1938) for the deformation of f.c.c. crystals, which were shown to have values of $\bar{m}$ of ~ 3.1. More recent simulations have also shown that the values of $\bar{m}$ are close to 3.0 for b.c.c. crystals. However, due to the large number of possible slip systems in b.c.c. metals, the simulations are much more complex than those required for f.c.c. crystals which have fewer slip systems. The b.c.c. simulations have, therefore, required the use of computers, as discussed by Chin and coworkers (1967, 1969).

We may now apply this concept of the equivalent orientation factor to a polycrystal. Using similar concepts to those presented earlier for single crystals in Sect. 7.8, we may express the effective stress and strain in a polycrystal by the following relationships:

$$\bar{\tau} = \bar{m}\tau \tag{7.16}$$

and

$$\bar{\gamma} = \frac{\gamma}{\bar{m}} \tag{7.17}$$

where τ and γ are the resolved shear stress and shear strain, respectively, that would apply to an equivalent single crystal. Hence, differentiating Eqs (7.16) and (7.17) gives

$$\frac{d\bar{\tau}}{d\bar{\gamma}} = (\overline{m})^2 \frac{\tau}{\gamma} \tag{7.18}$$

Plots of shear stress versus shear strain will, therefore, be expected to exhibit stress levels that are a factor of $(\overline{m})^2$ greater than the equivalent plots for

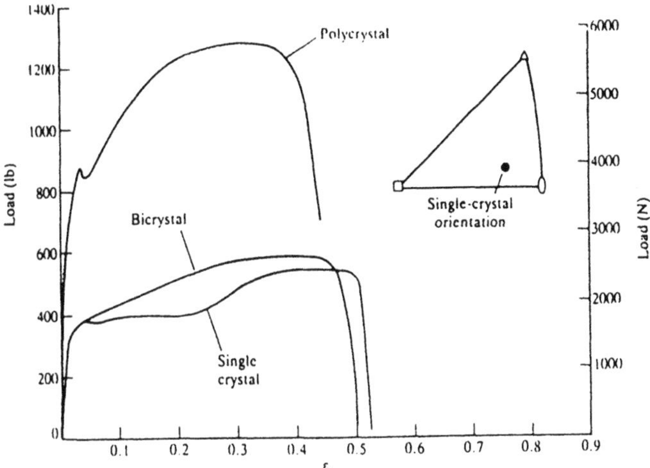

Figure 7.31 Comparison of stress–strain behavior in single crystal and polycrystalline niobium. (From Courtney, 1990. Reprinted with permission from McGraw-Hill.)

single crystals. This is shown to be the case for b.c.c. crystals of niobium (with $\overline{m}^2 \sim 3^2 = 9$) in Fig. 7.31 in which plots of shear stress versus shear strain are elevated by approximately one order of magnitude. Similar elevations are also observed in plots of axial flow stress versus axial strain for polycrystals.

7.11 GEOMETRICALLY NECESSARY AND STATISTICALLY STORED DISLOCATIONS

The discussion so far has focused largely on the role of initial dislocation substructures formed during processing, and those produced by dislocation breeding mechanisms, e.g., Frank–Read sources (Figs 7.13 and 7.14) and multiple cross-slip (Fig. 7.15) mechanisms. Such dislocations are produced by "chance" events and are generally termed *statistically stored dislocations (SSDs)*. The increase in the density of SSDs may be used qualitatively to account for the contributions of dislocations to plastic strain in the absence of high-stress gradients, as was done in Sect. 7.7.

However, yet another group of dislocations must be considered in cases where high-stress gradients are encountered, e.g., near grain boundaries or in metallic structures with thicknesses that are comparable to their grain sizes. This second group of dislocations are referred to as *geometrically necessary dislocations (GNDs)*. They were first proposed by Ashby (1970). GNDs are accumulated in regions of high-stress gradients, and are needed to avoid overlap or void formation during plastic deformation in such regions. They are accumulated in addition to the SSDs discussed earlier.

The need for GNDs may be visualized by considering the plastic bending of a rod, as shown schematically in Fig. 7.32. The initial configuration of the rod of length l and width t is shown in Fig. 7.32(a). On the application of a bending moment, a curved profile with a radius of curvature, r, is produced. The length of the outer surface is increased from l to $l + \delta l$, and the length of the inner surface is decreased from l to $l - \delta l$. Hence, the deformation of the outer surface is tensile, while that in the inner surface is compressive. There is, therefore, a stress gradient from the outer surface to the inner surface, which corresponds to the gradient of the bending stress field across the thickness, t, in Fig. 7.32(b).

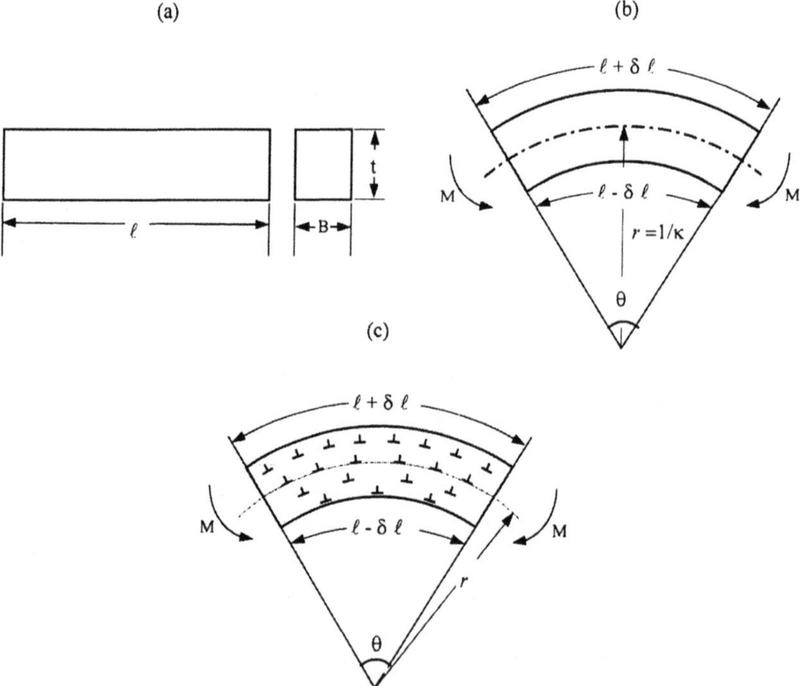

FIGURE 7.32 Schematics of plastic bending of a rod to highlight the need for geometrically necessary dislocations: (a) before deformation; (b) curvature after bending; (c) geometrically necessary dislocations. (From Ashby, 1970. Reprinted with permission from Phil. Mag.)

From basic geometry, $l = r\theta$, $l + \delta l = ((r + t)/2)\theta$ and $l\text{-}\delta l = ((r\text{-}t)/2)\theta$. Hence, $\theta = t\theta/2$. Furthermore, the strain on the outer surface is $l + \delta l$, while the strain in the inner surface is $-\delta l/l$. Since the strain gradient, $\mathrm{d}\varepsilon/\mathrm{d}t$, is linear across the length t, we may write:

$$\frac{\mathrm{d}\varepsilon}{\mathrm{d}t} = \frac{2}{t}\frac{\delta l}{l} = \frac{2}{t}\left(\frac{t\theta}{2l}\right) = \frac{\theta}{l} = \frac{\ell}{r} \tag{7.19}$$

To appreciate the next few steps, it is important to re-examine Fig. 7.32(a), and note that the number of atomic rows on the original surface is equal to the length, l, divided by the atomic separation, b. From Fig. 7.32(b), it should also be clear that the total number of rows on the outer surface of the crystal is $(l + \delta l)/b$. The difference between the total number of planes on the outer and inner surface can be accommodated by the introduction of edge dislocations of the same sign, as shown schematically in Fig. 7.32(c). These dislocations are GNDs. They are needed to maintain compatibility in the presence of stress gradients, e.g., near grain boundaries.

The total number of GNDs is given simply by $2\delta l/b$. The density of GNDs, ρ_G, may also be expressed as the ratio of the number of GNDs divided by the area (lt). This gives

$$\rho_G = \frac{2\Delta l}{b(lt)} = \frac{1}{rb} = \frac{\left(\frac{\mathrm{d}\varepsilon}{\mathrm{d}t}\right)}{b} \tag{7.20}$$

The total density of dislocations, ρ_{tot}, in a polycrystal is, therefore, given by the sum of the statistically stored dislocation dislocations, ρ_S, and the geometrically necessary dislocation density, ρ_G. This gives

$$\rho_{tot} = \rho_S + \rho_G \tag{7.21}$$

It is important to remember that the role of GNDs is only important in cases where stress gradients are present. Hence, in the absence of strain gradients, $\rho_{tot} \sim \rho_S$. The importance of GNDs has been highlighted in recent years by the development of strain gradient plasticity theories by Fleck et al. (1994). These theories include phenomenological models that attempt to predict length scale effects in crystal plasticity. These length scale effects are associated with the role of GNDs.

For example, the torsional stress–strain behavior of fine copper wires (Fleck et al., 1994) has been shown to exhibit a size scale dependence, with thinner wires having higher flow stresses than thicker wires (Fig. 7.33). Indentation tests (Stelmashenko et al., 1993; Ma and Clarke, 1995; Poole et al., 1996) on different metals have also revealed a dependence of hardness on indenter size, with smaller indenters (< 10–20 μm) resulting in higher

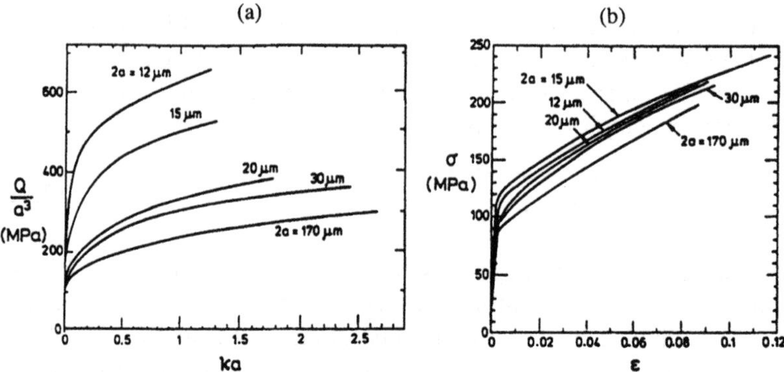

FIGURE 7.33 Effects of length scale on the flow stress of copper wires: (a) torque versus twist per unit length normalized in a way that material with no internal length scales would fall on to one another; (b) uniaxial stress versus strain for the same material shows almost no size effect in tension. (From Fleck et al., 1994. Reprinted with permission from Acta Metall. Mater.)

(2–3 times) hardness levels than the size-independent values that are typically observed for larger indenters.

In cases where the density of geometrically necessary dislocations, ρ_G, is significant, the flow stress, τ_y, is given by the following modified Taylor expression:

$$\tau_y = cGb\sqrt{\rho_s + \rho_G} \tag{7.22}$$

where c is a number that depends on the crystal type, G is the shear modulus, and b is the Burgers vector. Furthermore, in the vicinity of strain gradients of magnitude $d\varepsilon/dt$, the increment in plastic strain due to GNDs is $\sim \rho_G bL$, where L is the distance over which the strain gradients affect plastic flow. Similarly, the increment of plastic strain associated with the generation of new SSDs that travel through a distance L is $\rho_s\, bL$. An alternative derivation by Nix and Gao (1998) has shown that the length scale parameter, L, is given by

$$L = b\left(\frac{G}{\sigma_y}\right)^2 \tag{7.23}$$

where b is the Burgers vector, G is the shear modulus, and σ_y is the yield stress. It is important to note here that the magnitude of L is generally on the order of a few micrometers ($\sim$ 0.25–1 μm for stretch gradients and 4–5 μm for rotational gradients).

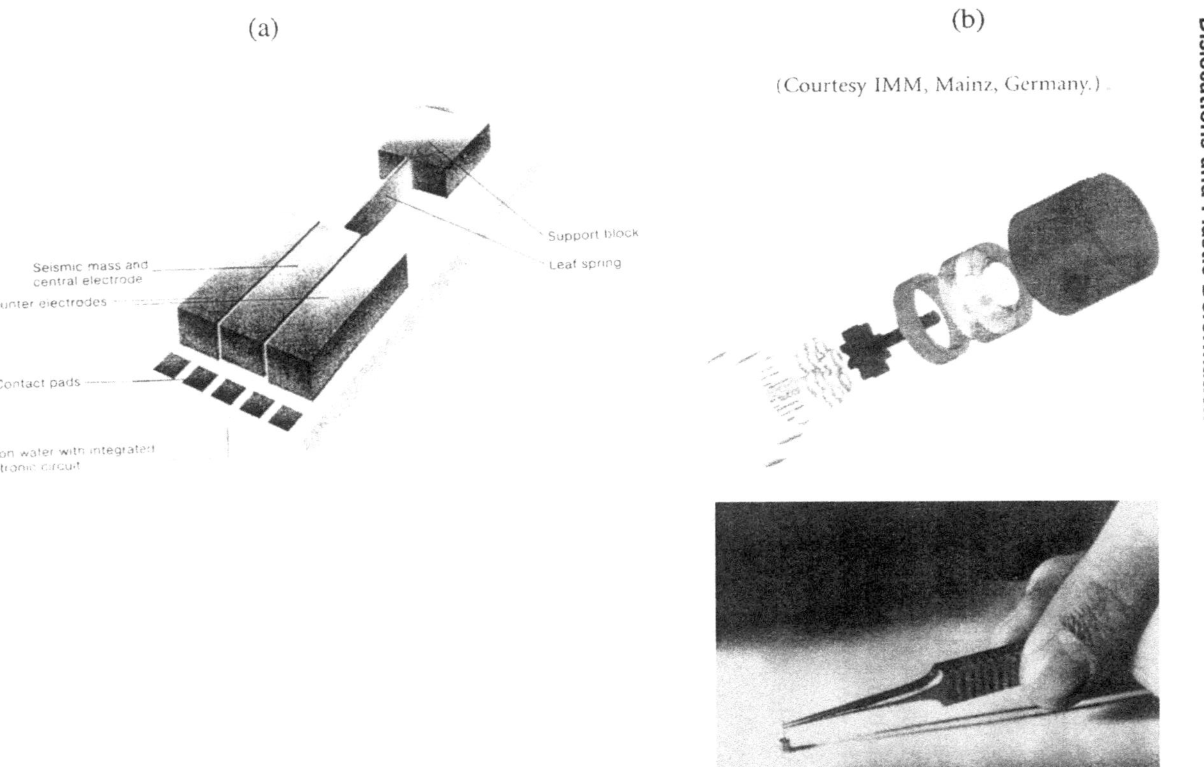

FIGURE 7.34 Examples of MEMS structures: (a) accelerometer; (b) magnetic motor. (From Madou, 1997. Reprinted with permission from CRC Press.)

Two approaches have been proposed for the estimation of the increment in plastic strain due to the combined effects of SSNs and GNDs. The first is by Fleck et al. (1994) who, guided by the Taylor relation, suggest that the dislocation density, ρ, increases in proportion to

$$\rho \propto \left[(\varepsilon^{\mathrm{p}})^{\lambda} + (\ell \partial \varepsilon^{\mathrm{p}}/\partial t)^{\lambda}\right]^{1/\lambda} \tag{7.24}$$

where ℓ is the length scale parameter, ε^{p} is the plastic strain, and λ is generally between 1 and 2.

The second approach proposed by Nix and Gao (1998) and Gao et al. (1998, 1999) assumes that GNDs have no direct effects on the accumulation of SSDs. Hence, from the Taylor relation, the density of SSDs is proportional to $f(\varepsilon^{\mathrm{p}})^2$. Strain gradients are introduced into this theory which gives the flow stress dependence as

$$\sigma_{\mathrm{y}} \propto \sqrt{f(\varepsilon_{\mathrm{p}})^2 + \ell \mathrm{d}\varepsilon^{\mathrm{p}}/\partial t} \tag{7.25}$$

The strain gradient plasticity (SGP) theory incorporates the length scale parameter, ℓ, into the J_2 deformation theory. As in the conventional J_2 theory, the J_2 SGP theory has both a small strain deformation version and an incremental version with a yield surface.

The deformation theory version gives the effective strain, E_{e}, as

$$E_{\mathrm{e}}^2 = \frac{2}{3}\varepsilon'_{ij}\varepsilon'_{ij} + \ell_1^2 \eta'^{(1)}_{ijk}\eta'^{(1)}_{ijk} + \ell_2^2 \eta'^{(2)}_{ijk}\eta'^{(2)}_{ijk} + \ell_3^2 \eta'^{(3)}_{ijk}\eta'^{(3)}_{ijk} \tag{7.26}$$

where ε_{ij} is the strain tensor, ε'_{ij} is the deviatoric strain, $\eta_{ijk} = u_{k,ij}$ is the strain gradient, and η'_{ijk} is the deviatoric strain gradient. The three deviatoric strain gradients are mutually ortogonal. Hence,

$$\eta'^{(I)}_{ijk}\eta'^{(J)}_{ijk} = 0 \text{ for } I \neq J \tag{7.27}$$

Furthermore, any strain gradient deviator η'_{ijk} can be expressed as a sum of three mutually orthogonal strain gradient tensors (Smyshlyaev and Fleck, 1996).

The deformation theory proposes the use of an energy density, $W(E_e, \varepsilon_{kk})$, which is determined by fitting monotonic shear or axial stress–strain data, and assuming elastic compressibility. The stress and higher order stress terms are obtained from

$$\sigma_{ij} = \frac{\partial W}{\partial \varepsilon_{ij}} \tag{7.28a}$$

and

$$\tau_{ijk} = \frac{\partial W}{\partial \eta_{ijk}} \tag{7.28b}$$

where the virtual work term, W, is given by Toupin (1962) and Mindlin (1965) to be:

$$W = \int_V [\sigma_{ij}\delta\varepsilon_{ij} + \tau_{ijk}\delta\eta_{ijk}]\mathrm{d}V = \int_A [t_i\delta u_i + r_i n_i \delta u_{i,j}]\,\mathrm{d}A \tag{7.29}$$

where η_I is the outward normal to the surface, $r_I = n_j n_k \tau_{ijk}$ is the double stress acting on the surface, and the surface traction is given by

$$t_k = n_i(\sigma_{ik} - \tau_{ijkj,}) + n_i n_j \tau_{ijk}(D_p\, n_p) - D_j(n_i\, \tau_{ijk}) \tag{7.30}$$

where D_j is the surface gradient, which is given by

$$D_j = (\delta_{jk} - n_j\, n_k)\,\partial_k \tag{7.31}$$

It is important to note here that the second and third invariants of the strain gradient depend only on the rotation gradient $X_{ij} = \theta_{ij}, = e_{ipk}\,\varepsilon_{kj,p}{}'$, where the rotation is given by $\theta_i = \varepsilon_{ijk}\, u_{k,j}/2$. The equivalent strain expression of Eq. (7.26) may thus be expressed as

$$\begin{aligned} E_e^2 &= \frac{2}{3}\varepsilon'_{ij}\varepsilon'_{ij} + \ell_1^2 \eta'^{(1)}_{ijk}\eta'^{(1)}_{ijk} + \frac{2}{3}\left(2\ell_2^2 + \frac{12}{5}\ell_3^2\right)\chi_{ij}\chi_{ij} \\ &+ \frac{2}{3}\left(2\ell_2^2 - \frac{12}{5}\ell_3^2\right)\chi_{ij}\chi_{ji} \end{aligned} \tag{7.32}$$

However, in most problems the relative contributions of the $\chi_{ij}\ \chi_{ji}$ term is relatively small. Hence, it is common to ignore this term and express the effective strain term as

$$E_e^2 = \frac{2}{3}\varepsilon'_{ij}\varepsilon'_{ij} + \ell_{SG}^2 \eta'(1)_{ijk}\eta'(1)_{ijk} + \frac{2}{3}\ell_{RG}^2 \chi_{ij}\chi_{ij} \tag{7.33}$$

where ℓ_{SG} and ℓ_{RG} are the stretch and rotation gradients, respectively, which are given by

$$\ell_{SG}^2 = \ell_1^2 \tag{7.34a}$$

and

$$\ell_{RG}^2 = 2\ell_2^2 + \frac{12}{5}\ell_3^2 \tag{7.34b}$$

Experimental measurements of ℓ_{SR} and ℓ_{RG} have been obtained by Fleck et al. (1994), Begley and Hutchinson (1998), and Stölken and Evans (1998). For annealed copper wires, Fleck et al. (1994) have shown that

$\ell_{RG} \sim 4\ \mu m$. Stölken and Evans (1998) have also shown, by measuring the amount of elastic springback in bending experiments, that nickel thin films have $\sqrt{\ell_{RG}^2 + 8/5\ell_{SG}^2} \sim 5\ \mu m$. Since the bending stress field is dominated by rotational gradients, $\ell_{RG} \sim 5\ \mu m$. Similarly, Begley and Hutchinson (1998) have analyzed indentation results from a number of metals, for which the length scale parameter is associated primarily with stretch gradients. Their analysis suggests that ℓ_{SG} is between 0.25 and 1.0 for various metals.

The above size effects are important in the modeling of plasticity in the microscale regime. For example, in the case of microelectromechanical systems (MEMS), machines are being fabricated on the microscale regime between ~ 1 and 750 μm. Most MEMS structures on a length scale between ~ 1 and 50 μm are being produced from silicon micromachining technology (Madou, 1997 (Fig. 7.34)). The emerging products include actuators, sensors, gears, microsatellites, and micromirrors. For thicker and larger devices, electroplated nickel structures are being used in applications such as microswitches and accelerometers in modern air bags. Aluminum MEMS structures are also being used in micromirror applications. In all the metal MEMS applications, it is likely that the phenomenological SGP J_2 theory could be useful in the modeling of length scale effects (Hutchinson, 2000). However, it is possible that higher order strain gradient theories may be needed to model the effects of constrained deformation due to dislocation pile-ups at the interfaces between brittle ceramic layers and ductile layers.

7.12 DISLOCATION PILE-UPS AND BAUSCHINGER EFFECT

It has been shown by numerous workers that the thermally induced nucleation of dislocations is energetically unfavorable (Argon and McClintock, 1963). However, dislocations may be produced within a grain by Frank–Read sources or by multiple cross-slip mechanisms. If we now consider the movement of dislocations from such sources within a grain, it is easy to envisage a dislocation pile-up that can result as the dislocations glide on a slip plane towards a barrier such as a grain boundary (Fig. 7.35). Since the dislocation loops are of the same sign, they will pile up without annihilating each other. The elastic interactions between the self-fields of the individual dislocations will result in an equilibrium separation of dislocations that decreases as the grain boundary is approached (Fig. 7.35).

Now consider the case of n dislocations approaching a boundary. The leading dislocation experiences a force due to the applied shear stress, τ. However, the remaining "trailing" $(n-1)$ dislocations experience a back stress, τ_b, due to the effect of the barrier. Hence, if the leading and trailing

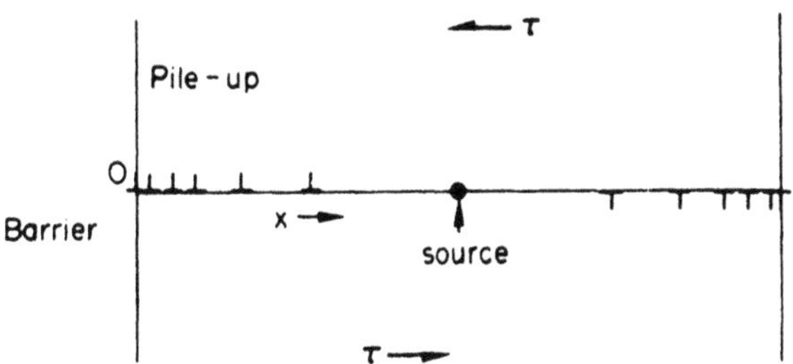

Figure 7.35 Schematic illustration of dislocation pile-up at grain boundaries. (From Hull and Bacon, 1984. Reprinted with permission from Pergamon Press.)

dislocations move forward by a small distance, δx, then the increase in the interaction energy between the loading dislocation and the barrier is $\tau_b\ \delta x$. Similarly, the work done per unit length of dislocation is $\tau(n\,\delta x)$. Since work and energy terms must be equal, and $\tau_l = \tau$, then the stress at the pile-up is given by

$$\tau = n\,\tau_b \tag{7.35}$$

The stress at a pile-up is, therefore, amplified by the total number of dislocations involved in the pile-up. This may result ultimately in the nucleation of slip or deformation-induced twinning, or crack nucleation in adjacent grains. The total number of dislocations, n, has been shown by Eshelby et al. (1951) to be given by

$$m = \frac{L\tau}{A} \tag{7.36}$$

where $A = Gb/\pi$ for screw dislocations and $A = Gb/\pi/(1 - \nu)$ for edge dislocations. Dislocation pile-ups, therefore, produce long range stresses, which is why they can exert some influence on the nucleation of yielding or cracking adjacent grains.

One of the major consequences of these long-range stresses are back stresses that can affect the yield strengths of crystalline solids deformed under cyclic tension–compression loading. During forward loading to a prestress, σ_p, forward dislocation motion results in plastic deformation and yielding at a yield stress, σ_y. However, if the loading is reversed, yielding is observed to occur in compression at a magnitude of stress that is lower than that in tension. This is illustrated schematically in Fig. 7.36. It was first observed by Bauschinger (1886) in experiments on wrought iron. It is, therefore, known as the *Bauschinger effect*.

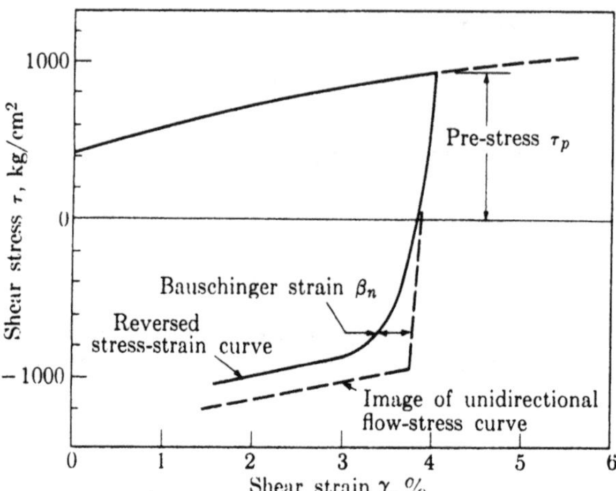

Figure 7.36 Bauschinger effect in a decarbonized tubular steel in torsion. (Data by Deak, 1961. From McClintock and Argon, 1966. Reprinted with permission from Addison Wesley.)

The difference between the flow strain in the tensile and compressive loading conditions (Fig. 7.36) is known as the Bauschinger strain. It is generally a function of the prestress, although it may also be a function of the prestrain. The compressive stress–strain curve, therefore, never reaches the image of the tensile stress strain curve shown by the dashed lines in Fig. 7.36. The Bauschinger effect must, therefore, involve some mechanisms of permanent softening.

The premature yielding that occurs on load reversal can be avoided by the use of stress relief heat treatments. However, its effects are often ignored, although they are known to be important in several engineering problems that involve the fatigue (damage due to cyclic loading in tension and/or compression) and creep (high-temperature deformation under static loads) of metals and their alloys. Work by Li et al. (2000) has resulted in the development of numerical finite element schemes for the modeling of the role of the Bauschinger effect in sheet metal forming processes. Such models are critical to the optimization of processing schemes that are used in the fabrication of smooth automotive car body panels.

7.13 MECHANICAL INSTABILITIES AND ANOMALOUS/SERRATED YIELDING

So far, our discussion of plastic deformation has ignored the possible effects of the dislocation interactions with solutes or interstitials that are present in all crystalline solids. These interactions can give rise to some interesting mechanical instabilities that are considered in this section.

7.13.1 Anomalous Yielding Phenomena

Anomalous yielding phenomena have been observed in mild steels and single crystal iron during tensile deformation. An example of anomalous yielding in a plain carbon steel is presented in Fig. 7.37. This shows a typical plot of stress versus strain obtained from a tensile test. Note that the stress rises initially to an upper yielding point (UYP) before dropping to a lower yield point (LYP). The initiation of deformation at the UYP is localized to a region within the gauge section of the tensile specimen. At the UYP, evidence of the localized deformation may be seen in the form of Lüders bands that are aligned at an angle of ~ 45° to the loading axis. The localized deformation (Lüders bands) then spreads across the gauge section until the gauge section is completely filled with Lüders bands at the so-called Lüders strain. Homogeneous deformation and hardening then continues (Fig. 7.37) to failure, as would be expected from a typical metallic material.

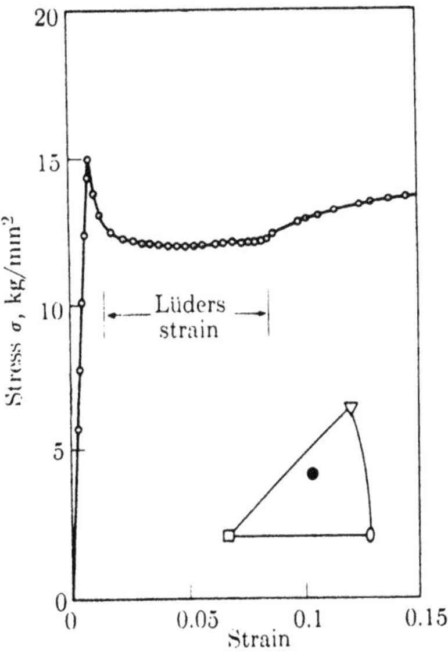

FIGURE 7.37 Anomalous yielding in a single crystal of iron containing 0.003% carbon and deformed continuously to failure at a strain rate of $10^5\ s^{-1}$ at 195 K. (Data by Paxton and Bear, 1955. Adapted from McClintock and Argon, 1963. Reprinted with permission from Addison Wesley.)

Two theories have been proposed to explain the anomalous yielding phenomena. One theory is by Cottrell and Bilby (1949) and the other is by Hahn (1962). The theory of Cottrell and Bilby (1949) attributes the yield phenomena to the effects of dislocation interactions with interstitials in b.c.c. metals. Since these cause large unsymmetrical distortions in the lattice structure, they interact strongly with edge and screw dislocations. At sufficiently high temperatures, the interstitials diffuse towards the dislocation cores, and thus impede the motion of dislocations. Higher stresses are, therefore, needed to break the dislocations free from the solute/interstitial clouds and move them through the lattice.

The theory of Hahn (1962) attributes the observed instabilities to the strong stress dependence of the velocity of dislocations. Hence, when the dislocations break away from the solute atmospheres, they must move at faster velocities to enable the imposed strain rate to be achieved. The high velocity requires high stress, which in turn results in rapid dislocation multiplication by double cross-slip (Fig. 6.13). Hence, as the dislocation density increases, both the velocity and the stress decrease. Both theories appear to be plausible. However, further research is needed to explain the observed dependence of the UYP and LYP on grain size.

7.13.2 Portevin–LeChatelier Effect

The interactions between dislocations and impurities (solutes and interstitials) or vacancies can give rise to serrated yielding (strain aging) phenomena in metallic and nonmetallic materials. The serrations were first observed by Portevin and LeChatelier (1923) in experiments on Duralumin (an Al–Cu alloy). For this reason, the occurrence of serrated yielding is often known as the *Portevin–LeChatelier effect*. The instabilities are associated with the interactions of groups of dislocations with solute/interstitial atoms. These result in jerky dislocation motion, and serrations in the stress–strain curves, as shown schematically in Fig 5.6.

The serrations are an indication of discontinuous yielding that occurs due to groups of dislocations breaking free from the pinning of dislocations which is caused by the (dislocation core stress-assisted) diffusion of interstitials and solutes towards the gliding dislocations (Fig. 7.38). These can give rise to different types of serrations (Fig. 5.6), depending on the nature of the dislocation/solute or dislocation/interstitial interactions. The periodicity and characteristic shapes of the serrations have been attributed by Cottrell (1958) to the thermally activated release of dislocations, and their subsequent pinning by interstitial and solute clouds. Serrated yielding (strain aging) has been observed in solid solutions of zinc and aluminum. It has also

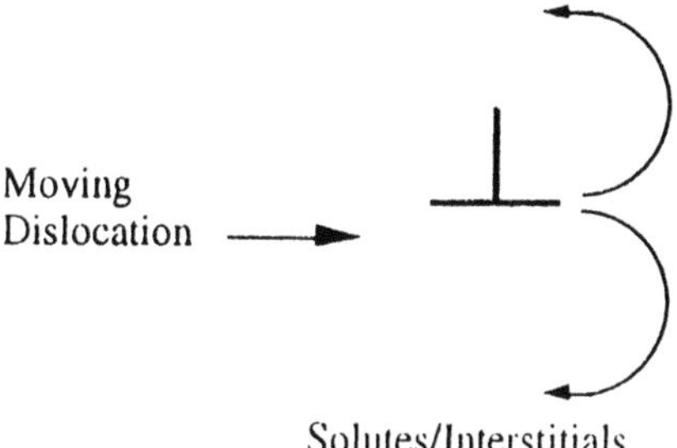

Figure 7.38 Pinning of dislocations by diffusing solutes and interstitials.

been reported to occur in some intermetallics such as gamma-based titanium aluminides and nickel aluminides.

7.14 SUMMARY

The role of dislocations in the plasticity of crystalline metals and their alloys has been examined in this chapter. Following a brief description of the motion of dislocations by glide, climb, and kink nucleation/propagation mechanisms, the factors that control dislocation velocity were discussed briefly for edge and screw dislocations. The bowing of dislocations due to line tension forces was then explored before introducing the concept of dislocation breeding from Frank–Read sources and multiple cross-slip sources. The contributions from dislocations to plastic strain were then elucidated within a simple continuum framework. The crystallography of slip was introduced for f.c.c., b.c.c., and h.c.p. structures, before describing dislocation dissociation mechanisms, partial/extended dislocations, stacking faults, superdislocations, and APBs. The concept of a critical resolved shear stress was also examined before describing the contributions from slip to plastic deformation in single crystals and polycrystals. Finally, SSNs and GNDs were described briefly before concluding with sections on the Bauschinger effect and mechanical instabilities.

BIBLIOGRAPHY

Ashby, M.F. (1970) Phil Mag. vol. 21, p 399.

Ashby, M.F. and Jones, D.R.H. (1996) Engineering Materials: An Introduction to the Properties and Applications. vol. 1. Pergamon Press, New York.

Bauschinger, J. (1886) On the Change of the Elastic Limit and Hardness of Iron and Steels Through Extension and Compression, Through Heating and Cooling, and Through Cycling. Mitteilung aus dem Mechanisch, Technischen Laboratorium der K. Technische Hochschule in Munchen. vol. 14, part 5, p 31.

Begley and Hutchinson (1998) The Mechanics of Size-Dependent Indentation, J. Mech. Phys. Solids. vol. 40, pp 2049–2068.

Chin, G.Y. and Mammel, W.L. (1967) Trans Met Soc, AIME. vol. 239, p 1400.

Chin, G.Y., Mammel, W.L., and Dolan, M.T. (1969) Trans Met Soc, AIME. vol. 245, p 383.

Cottrell, A.H. (1953) Dislocations and Plastic Flow in Crystals. Oxford University Press, Oxford, England.

Cottrell, A.H. (1958) Point Defects and the Mechanical Properties of Metals and Alloys at Low Temperatures, Vacancies and Other Point Defects in Metals and Alloys. Institute of Metals. London, pp 1–40.

Cottrell, A.H. and Bilby, B.A. (1949) Dislocation theory of yielding and strain aging of iron. Proc Phys Soc. vol. A62, pp 49–62.

Courtney, T.H. (1990) Mechanical Behavior of Materials. McGraw-Hill, New York.

Dash, W.C. (1957) Dislocations and Mechanical Properties of Crystals. In: J.C. Fisher, ed John Wiley, New York.

Eshelby, J.D., Frank, F.C., and Nabarro, F.R.N. (1951) The equilibrium of linear arrays of dislocations. Phil Mag. vol. 42, p 351.

Fleck, N.A., Muller, G.M., Ashby, M.F., and Hutchinson, J.W. (1994). Strain gradient plasticity: theory and experiments. Acta Metall Mater. vol. 42, pp 475–487.

Gao, H., Huang, Y., Nix, W.D., and Hutchinson, J.W. (1999) J Mech Phys Solids. vol. 41, pp 1239–1263.

Hertzberg, R.W. (1996) Deformation and Fracture Mechanics of Engineering Materials. 4th ed. John Wiley, New York.

Hull, D. and Bacon, D.J. (1984) Introduction to Dislocations. 3rd ed. Pergamon Press, New York.

Johnston, W.G. and Gilman, J.J. (1959) Dislocation velocities, dislocation densities, and plastic flow in lithium fluoride crystals. J Appl Phys. vol. 30, pp 129–144.

Li, K., Carden, W., and Wagoner, R. (2000) Numerical simulation of springback with the draw-bend test. Int J Mech Sci (in press).

Low, J.R. and Guard, R. W. (1959) Acta Metall. vol. 7, p 171.

Ma, Q. and Clarke, D.R. (1995) Size dependent hardness of silver single crystals. J Mater Res. vol. 10, pp 853–863.

Madou, M. (1997) Fundamentals of Microfabrication. CRC Press, New York.

McClintock, F.A. and Argon, A. (eds) (1966) Mechanical Behavior of Materials. Addison Wesley, MA.

Mindlin, R.D. (1965) Int J Solids Struct. vol. 1, pp 417–438.

Nix, W.D. and Gao, H. (1998) J Mech Phys Solids. vol. 46, pp 411–425.

Paxton, H. W. and Bear, I.J. (1955) Metall Trans. vol. 7, pp 989–994.

Poole, W.J., Ashby, M. J., and Fleck, N.A. (1996) Microhardness of annealed work hardened copper polycrystals. Scripta Metall Mater. vol. 34, pp 559–564.

Portevin, A. and LeChatelier, F. (1923) Acad Sci Compt Read. vol. 176, pp 507–510.

Read, W.T., Jr. (1953) Dislocations in Crystals. McGraw-Hill, New York.
Seeger, Donth, and Pfaff (1957) Disc Faraday Soc. vol. 23, p 19.
Smyshlyaev, V.P. and Fleck, N.A. (1966) J Mech Phys Solids. vol. 44, pp 465–496.
Stein, D.F. and Low, J.R. (1960) J Appl Phys. vol. 31, p 362.
Stelmashenko, N.A., Walls, M.G. Brown, L.M., and Milman, Y.V. (1993) Microindentations on W and Mo oriented single crystals: an STM study. Acta Metall Mater. vol 41, pp 2855–2865.
Stölken, J.S. and Evans, A.G. (1998) Acta Mater. vol. 46, pp 5109–5115.
Taylor, G.I. (1938) J. Inst Met. vol. 62, p 307.
Toupin, R.A. (1962) Arch Rational Math Anal. vol. 11, pp 385–414.
von Mises, R. (1928) Z. Ang Math. vol. 8, p 161.
Weeks, R.W. Pati, S.R., Ashby, M.F., and Barrand, P. (1969) Acta Metall. vol. 17, p 1403.

8

Dislocation Strengthening Mechanisms

8.1 INTRODUCTION

The dislocation strengthening of metals and their alloys is perhaps one of the major technological accomplishments of the last 100 years. For example, the strength of pure metals such as aluminum and nickel have been improved by factors of 10–50 by the use of defects that restrict dislocation motion in a crystal subjected to stress. The defects may be point defects (solutes or interstitials), line defects (dislocations), surface defects (grain boundaries or twin boundaries), and volume defects (precipitates or dispersions). The strain fields that surround such defects can impede the motion of dislocations, thus making it necessary to apply higher stresses to promote the movement of dislocations. Since yielding and plastic flow are associated primarily with the movement of dislocations, the restrictions give rise ultimately to intrinsic strengthening.

The basic mechanisms of intrinsic strengthening are reviewed in this chapter, and examples of technologically significant materials that have been strengthened by the use of the strengthening concepts are presented. The strengthening mechanisms that will be considered include:

1. Solid solution strengthening (dislocation interactions with solutes or interstitials).

2. Dislocation strengthening which is also known as work/strain hardening (dislocation interactions with other dislocations).
3. Boundary strengthening (dislocation interactions with grain boundaries or stacking faults).
4. Precipitation strengthening (dislocation interactions with precipitates).
5. Dispersion strengthening (dislocation interactions with dispersed phases).

Note the above sequence of dislocation interactions with: zero-dimensional point defects (solutes or interstitials); one-dimensional line defects (other dislocations; two-dimensional defects (grain boundaries or stacking faults), and three-dimensional defects (precipitates or dispersoids).

8.2 DISLOCATION INTERACTIONS WITH OBSTACLES

Before presenting the specific details of individual dislocation strengthening mechanisms, it is important to examine the interactions of dislocations with arrays of obstacles such as solutes/interstitials and particles/precipitates (Fig. 8.1). When dislocations encounter such arrays as they glide through a lattice under an applied stress, they are bent through an angle, ϕ, before they can move on beyond the cluster of obstacles (note that $0 < \phi < 180°$). The angle, ϕ, is a measure of the strength of the obstacle, with weak obstacles having values of ϕ close to 180°, and strong obstacles having obstacles close to 0°.

It is also common to define the strength of a dislocation interaction by the angle, $\phi' = 180 - \phi$ through which the interaction turns the dislocation (Fig. 8.1). Furthermore, the number of obstacles per unit length (along the dislocation) depends strongly on ϕ. For weak obstacles with $\phi \sim 180°$, the

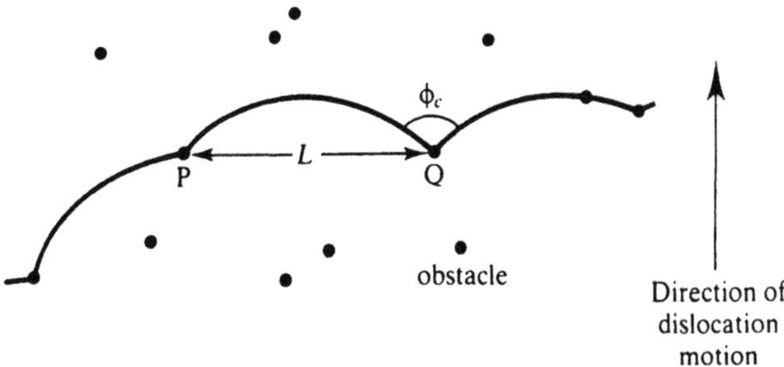

FIGURE 8.1 Dislocation interactions with a random array of particles.

number of obstacles per unit length may be found by calculating the number of particle intersections with a random straight line. Also, as ϕ decreases, the dislocations sweep over a larger area, and hence interact with more particles. Finally, in the limit, the number of intersections is close to the square root of the number of particles that intersect a random plane.

The critical stress, τ_c, required for a dislocation to break away from a cluster of obstacles depends on the particle size, the number of particles per unit volume, and the nature of the interaction. If the critical breakaway angle is ϕ_c, then the critical stress at which breakaway occurs is given by

$$\tau_c = \frac{Gb}{L}\cos\left(\frac{\phi_c}{2}\right) \tag{8.1}$$

Equation (8.1) may be derived by applying force balance to the geometry of Fig. 8.1. However, for strong obstacles, breakaway may not occur, even for $\phi = 0$. Hence, in such cases, the dislocation bows to the semi-circular Frank–Read configuration and dislocation multiplication occurs, leaving a small loop (Orowan loop) around the unbroken obstacle. The critical stress required for this to occur is obtained by substituting $r = L/2$ and $\phi = 0$ into Eq. (8.1). This gives

$$\tau_c = \frac{Gb}{L} \tag{8.2}$$

Hence, the maximum strength that can be achieved by dislocation interactions is independent of obstacle strength. This was first shown by Orowan (1948). The above expressions [Eqs (8.1) and (8.2)] provide simple order-of-magnitude estimates of the strengthening that can be achieved by dislocation interactions with strong or weak obstacles. They also provide a qualitative understanding of the ways in which obstacles of different types can affect a range of strengthening levels in crystalline materials.

8.3 SOLID SOLUTION STRENGTHENING

When foreign atoms are dissolved in a crystalline lattice, they may reside in either interstitial or substitutional sites (Fig. 8.2). Depending on their sizes relative to those of the parent atoms. Foreign atoms with radii up to 57% of the parent atoms may reside in interstitial sites, while those that are within ±15% of the host atom radii substitute for solvent atoms, i.e. they form solid solutions. The rules governing the formation of solid solutions are called the Hume–Rothery rules. These state that solid solutions are most likely to form between atoms with similar radii, valence, electronegativity, and chemical bonding type.

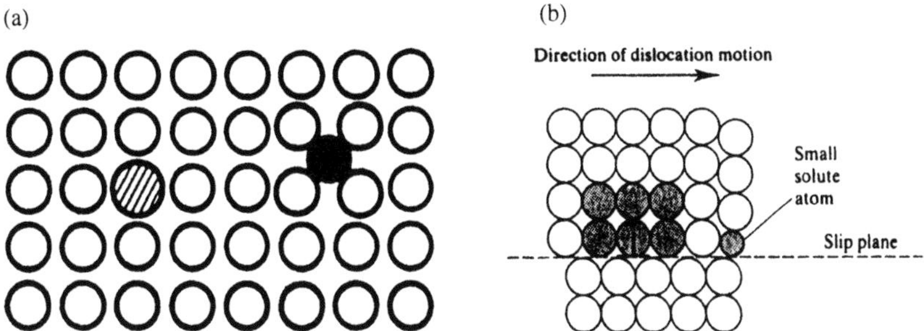

FIGURE 8.2 Interstitial and solute atoms in a crystalline lattice: (a) schematic of interstitial and solute atoms; (b) effects on dislocation motion. [(a) Adapted from Hull and Bacon (1984) and (b) adapted from Courtney (1990). Reprinted with permission from Pergamon Press.]

Since the foreign atoms have different shear moduli and sizes from the parent atoms, they impose additional strain fields on the lattice of the surrounding matrix. These strain fields have the overall effect of restricting dislocation motion through the parent lattice, Fig. 8.2(b). Additional applied stresses must, therefore, be applied to the dislocations to enable them to overcome the solute stress fields. These additional stresses represent what is commonly known as solid solution strengthening.

The effectiveness of solid solution strengthening depends on the size and modulus mismatch between the foreign and parent atoms. The size mismatch gives rise to misfit (hydrostatic) strains that may be symmetric or asymmetric (Fleischer, 1961, 1962). The resulting misfit strains, are proportional to the change in the lattice parameter, a, per unit concentration, c. This gives

$$\varepsilon_b = \frac{1}{a}\frac{da}{dc} \tag{8.3}$$

Similarly, because the solute/interstitial atoms have different moduli from the parent/host atoms, a modulus mismatch strain, ε_G, may be defined as

$$\varepsilon_G = \frac{1}{G}\frac{dG}{dc} \tag{8.4}$$

In general, however, the overall strain, ε_s, due to the combined effect of the misfit and modulus mismatch, may be estimated from

$$\varepsilon_s = |\varepsilon'_G - \beta\varepsilon_b| \tag{8.5}$$

where β is a constant close to 3 $\varepsilon'_G = \varepsilon_G/(1 + (1/2)|\varepsilon_G|)$, and ε_b is given by equation 8.3. The increase in the shear yield strength, $\Delta\tau_s$, due to the solid solution strengthening may now be estimated from

$$\Delta\tau_s = \frac{G\varepsilon_s^{3/2}c^{1/2}}{700} \tag{8.6}$$

where G is the shear modulus, ε_s is given by Eq. (8.5), and c is the solute concentration specified in atomic fractions. Also, $\Delta\tau_s$ may be converted into $\Delta\sigma_s$ by multiplying by the appropriate Schmid factor.

Several models have been proposed for the estimation of solid solution strengthening. The most widely accepted models are those of Fleischer (1961 and 1962). They include the effects of Burgers vector mismatch and size mismatch. However, in many cases, it is useful to obtain simple order-of-magnitude estimates of solid solution strengthening, $\Delta\sigma_s$, from expressions of the form:

$$\Delta\sigma_s = k_s c^{1/2} \tag{8.7}$$

where k_s is a solid solution strengthening coefficient, and c is the concentration of solute in atomic fractions. Equation (8.7) has been shown to provide reasonable fits to experimental data for numerous alloys. Examples of the $c^{1/2}$ dependence of yield strength are presented in Fig. 8.3.

In summary, the extent of solid solution strengthening depends on the nature of the foreign atom (interstitial or solute) and the symmetry of the stress field that surrounds the foreign atoms. Since symmetrical stress fields

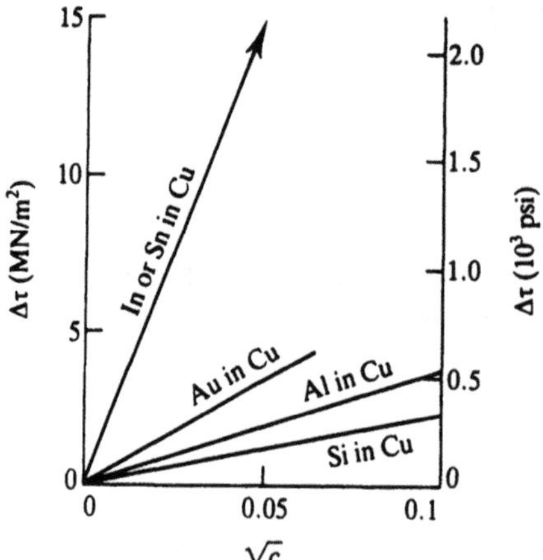

FIGURE 8.3 Dependence of solid solution strengthening on $c^{1/2}$. (Data taken from Fleischer (1963). Reprinted with permission from Acta Metall.)

interact only with edge dislocations, the amount of strengthening that can be achieved with solutes with symmetrical stress fields is very limited (between $G/100$ and $G/10$). In contrast, asymmetric stress fields around solutes interact with both edge and screw dislocations, and their interactions give rise to very significant levels of strengthening ($\sim 2G - 9G$), where G is the shear modulus. However, dislocation/solute interactions may also be associated with strain softening, especially at elevated temperature.

8.4 DISLOCATION STRENGTHENING

Strengthening can also occur as a result of dislocation interactions with each other. These may be associated with the interactions of individual dislocations with each other, or dislocation tangles that impede subsequent dislocation motion (Fig. 8.4). The actual overall levels of strengthening will also depend on the spreading of the dislocation core, and possible dislocation reactions that can occur during plastic deformation. Nevertheless, simple estimates of the dislocation strengthening may be obtained by considering the effects of the overall dislocation density, ρ, which is the line length, ℓ, of dislocation per unit volume , ℓ^3.

The dislocation density, ρ, therefore scales with ℓ/ℓ^3. Conversely, the average separation, $\bar{\ell}$, between dislocations may be estimated from

$$\bar{\ell} = \frac{1}{\rho^{-1/2}} \tag{8.8}$$

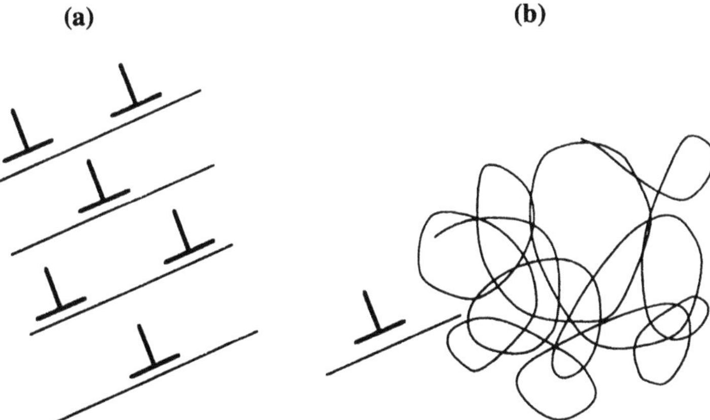

FIGURE 8.4 Strain hardening due to interactions between multiple dislocations: (a) interactions between single dislocations; (b) interactions with forest dislocations.

The shear strengthening associated with the pinned dislocation segments is given by

$$\Delta\tau_d = \frac{\alpha G b}{\ell} \tag{8.9}$$

where α is a proportionality constant, and all the other variables have their usual meaning. We may also substitute Eq. (8.8) into Eq. (8.9) to obtain the following expression for the shear strengthening due to dislocation interactions with each other:

$$\Delta\tau_d = \alpha G b \rho^{1/2} \tag{8.10}$$

Once again, we may convert from shear stress increments into axial stress increments by multiplying by the appropriate Schmid factor, m. This gives the strength increment, $\Delta\sigma_d$, as (Taylor, 1934):

$$\Delta\sigma_d = m\alpha G b \rho^{1/2} = k_d \rho^{1/2} \tag{8.11}$$

where $k_d = m\alpha Gb$ and the other variables have their usual meaning. Equations (8.10) and (8.11) have been shown to apply to a large number of metallic materials. Typical results are presented in Fig. 8.5. These show

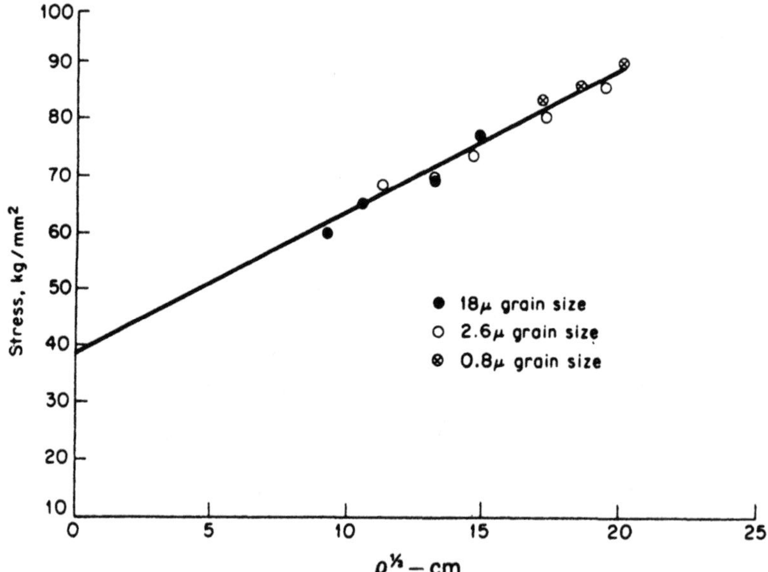

Figure 8.5 Dependence of shear yield strength on dislocation density. (From Jones and Conrad, 1969. Reprinted with permission from TMS-AIME.)

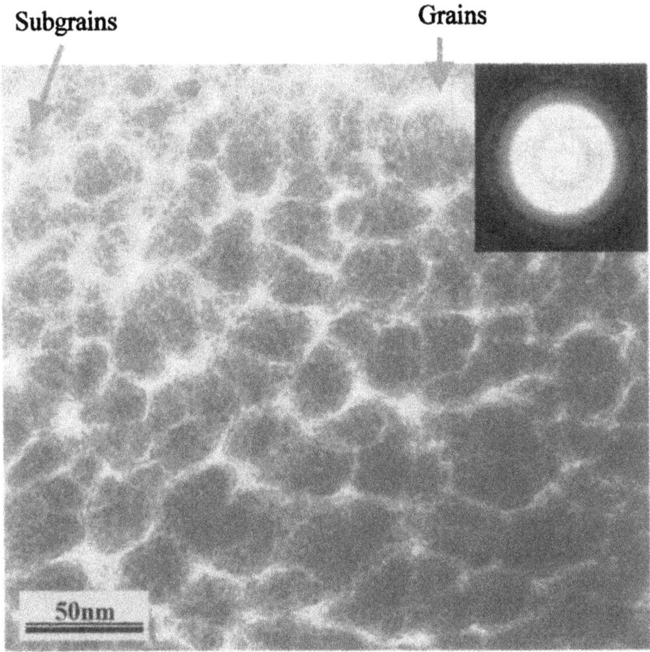

Figure 8.6 Dislocation cell structure in a Nb–Al–Ti based alloy. (Courtesy of Dr. Seyed Allameh.)

that the linear dependence of strengthening on the square root of dislocation density provides a reasonable fit to the experimental data.

It is important to note here that Eq. (8.11) does not apply to dislocation strengthening when cell structures are formed during the deformation process (Fig. 8.6). In such cases, the average cell size, s, is the length scale that controls the overall strengthening level. This gives

$$\Delta\sigma'_d = k'_d(s)^{-1/2} \tag{8.12}$$

where $\Delta\sigma'_d$ is the strengthening due to dislocation cell walls, k'_d is the dislocation strengthening coefficient for the cell structure, and s is the average size of the dislocation cells.

8.5 GRAIN BOUNDARY STRENGTHENING

Grain boundaries also impede dislocation motion, and thus contribute to the strengthening of polycrystalline materials (Fig. 8.7). However, the strengthening provided by grain boundaries depends on grain boundary

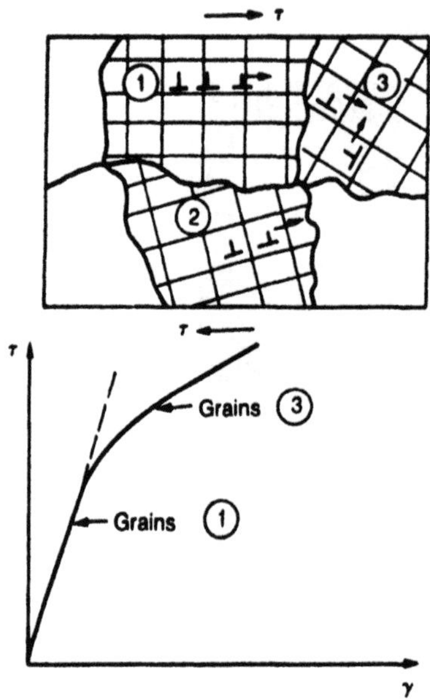

Figure 8.7 Dislocation interactions with grain boundaries. (From Ashby and Jones, 1996. Reprinted with permission from Pergamon Press.)

structure and the misorientation between individual grains. This may be understood by considering the sequence of events involved in the initiation of plastic flow from a point source (within a grain) in the polycrystalline aggregate shown schematically in Fig. 8.8.

Due to an applied shear stress, τ_{app} dislocations are emitted from a point source (possibly a Frank–Read source) in one of the grains in Fig. 8.8. These dislocations encounter a lattice friction stress, τ_i, as they glide on a slip plane towards the grain boundaries. The effective shear stress, τ_{eff}, that contributes to the glide process is, therefore, given by

$$\tau_{eff} = \tau_{app} - \tau_i \tag{8.13}$$

However, since the motion of the dislocations is impeded by grain boundaries, dislocations will generally tend to pile-up at grain boundaries. The stress concentration associated with this pile-up has been shown by Eshelby et al. (1951) to be $\sim (d/4r)^{1/2}$, where d is the grain size and r is the distance from the source. The effective shear stress is, therefore, scaled by this stress concentration factor. This results in a shear stress, τ_{12}, at the grain boundaries, that is given by

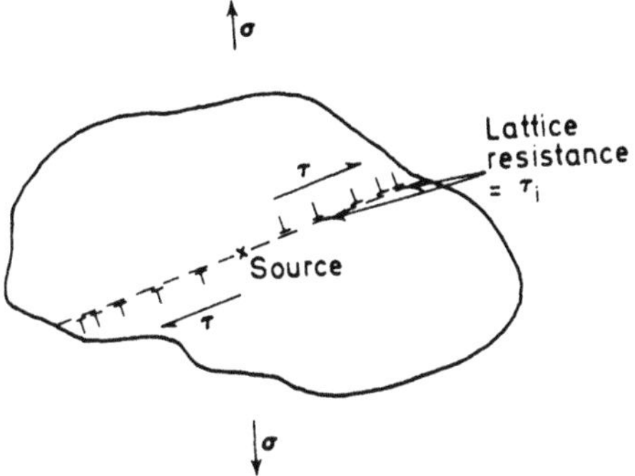

FIGURE 8.8 Schematic illustration of dislocation emission from a source. (From Knott, 1973. Reprinted with permission from Butterworth.)

$$\tau_{12} = (\tau_{app} - \tau_i)\left(\frac{d}{4r}\right)^{1/2} \tag{8.14}$$

If we now consider bulk yielding to correspond to the condition slip for transmission to adjacent grains when a critical τ_{12} is reached, then we may rearrange Eq. (8.14) to obtain the following expression for τ_{app} at the onset of bulk yielding:

$$\tau_{app} = \tau_i + \left(\frac{4r}{d}\right)^{1/2} \tau_{12} \tag{8.15}$$

The magnitude of the critical shear stress, τ_{12}, required for slip transmission to adjacent grains may be considered as a constant. Also, the average distance, r, for the dislocations in the pile-up is approximately constant. Hence, $(4r)^{1/2}\ \tau_{12}$ is a constant, k'_y, and Eq. (8.15) reduces to

$$\tau_y = \tau_i + k'_y d^{-1/2} \tag{8.16}$$

Once again, we may convert from shear stress into axial stress by multiplying Eq. (8.16) by the appropriate Schmid factor, m. This gives the following relationship, which was first proposed by Hall (1951) and Petch (1953):

$$\sigma_y = \sigma_0 + k_y d^{-1/2} \tag{8.17}$$

where σ_0 is the yield strength of a single crystal, k_y is a microstructure/grain boundary strengthening parameter, and d is the grain size. The reader should note that Eq. (8.17) shows that yield strength increases with decreasing grain size. Furthermore, (σ_o may be affected by solid solution alloying effects and dislocation substructures.

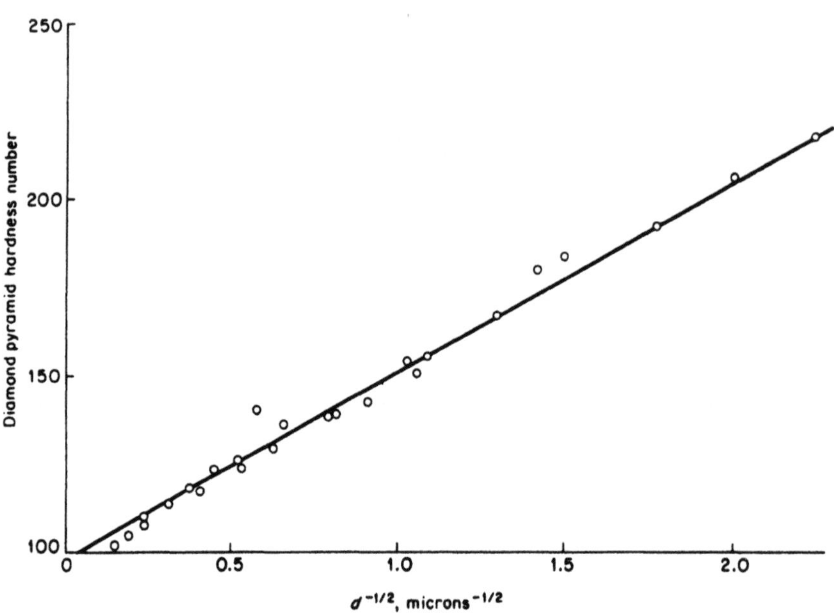

FIGURE 8.9 Hall–Petch dependence of yield strength. (From Hu and Cline, 1968. Original data presented by Armstrong and Jindal, 1968. Reprinted with permission from TMS-AIME.)

Evidence of Hall–Petch behavior has been reported in a large number of crystalline materials. An example is presented in Fig. 8.9. Note that the microstructural strengthening term, k_y, may vary significantly for different materials. Furthermore, for a single-phase solid solution alloy with a dislocation density, ρ, the overall strength may be estimated by applying the principle of linear superposition. This gives

$$\sigma_y = \sigma_0 + k_s c^{1/2} + k_d \rho^{1/2} + k_y d^{1/2} \tag{8.18}$$

Note that Eq. (8.18) neglects possible interactions between the individual strengthening mechanisms. It also ignores possible contributions from precipitation strengthening mechanisms that are discussed in the next section.

8.6 PRECIPITATION STRENGTHENING

Precipitates within a crystalline lattice can promote strengthening by impeding the motion of dislocations. Such strengthening may occur due to the additional stresses that are needed to enable the dislocations to shear the precipitates (Fig. 8.10), or avoid the precipitates by looping/extruding in

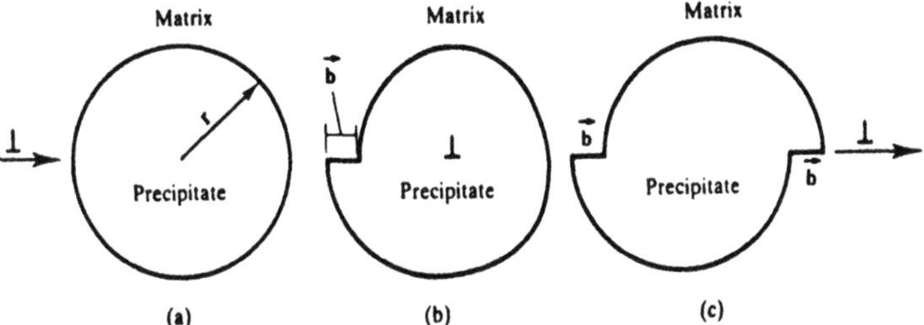

Figure 8.10 Schematic illustration of ledge formation and precipitation strengthening due to dislocation cutting of precipitates: (a) before cutting; (b) during cutting; (c) after cutting.

between the spaces that separate the precipitates (Fig. 8.11). The favored mechanism depends largely on the size, coherence, and distribution of the precipitates.

The different ways in which dislocations can interact with particles make the explanation of precipitation strengthening somewhat complicated. However, we will attempt to simplify the explanation by describing the mechanisms in different sub-sections. We will begin by considering the strengthening due to looping of dislocations around precipitates (Fig. 8.1). This will be followed by brief descriptions of particle shearing that can give rise to ledge formation (Fig. 8.10) in disordered materials, and complex association phenomena in ordered materials. The applications of precipitation strengthening to the strengthening of aluminum alloys will then be discussed after exploring the transitions that can occur between dislocation looping and particle cutting mechanisms.

8.6.1 Dislocation/Orowan Strengthening

Precipitation strengthening by dislocation looping (Fig. 8.11) occurs when sub-micrometer precipitates pin two segments of a dislocation. The rest of the dislocation line is then extruded between the two pinning points due to the additional applied shear stress $\Delta\tau$ (Fig. 8.11). The strengthening resulting from this mechanisms was first modeled by Orowan, and is commonly known as *Orowan strengthening*. This gives

$$\Delta\tau = \frac{Gb}{L - 2r} = \frac{Gb}{L'} \tag{8.19}$$

where G is the shear modulus, b is the Burger's vector, L is the center-to-center separation between the precipitates, r is the particle radius, and L' is the effective particle separation, $L' = L - 2r$. Note that the simplest, one-

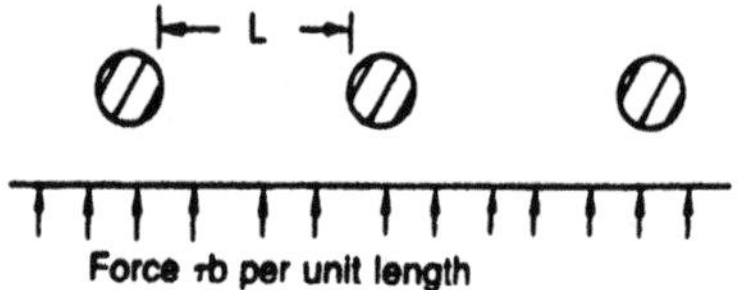

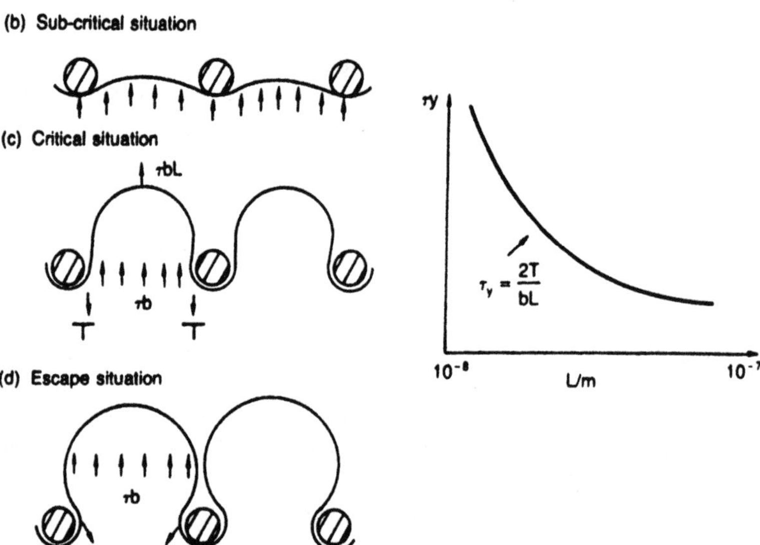

FIGURE 8.11 Schematic of Orowan strengthening due to looping of dislocations between precipitates: (a) dislocation approaching particles; (b) dislocation extruding through particles; (c) critical situation when extruded dislocation reaches semicircular configuration; (d) escape situation. (From Ashby and Jones, 1996. Reprinted with permission from Pergamon Press.)

dimensional estimate of the particle volume fraction, f, for the above configuration is equal to r/L. The shear strengthening term may also be converted into an axial strengthening term by premultiplying by an effective Schmid factor.

Equation (8.19) neglects changes in dislocation character along the line length of the dislocation. The critical stress, τ_c, for dislocations bowing through two pinning segments (Fig. 8.11) may be estimated from expressions of the form:

$$\tau_c = A(\theta)\frac{Gb}{2\pi L'}\ln\left(\frac{L'}{r} + B(\theta)\right) \tag{8.20}$$

where θ is the angle between the dislocation line and the Burgers vector, $A(\theta)$ and $B(\theta)$ are both functions of θ, L' is the effective particle separation, $L' = L - 2r$, b is the Burgers vector, and r is the particle radius. The function $A(\theta)$ has been determined by Weeks et al. (1969) for critical conditions corresponding to the instability condition in the Frank–Read mechanism. The special result for this condition is

$$\tau_c = A(\theta)\frac{Gb}{2\pi L'}\ln\left(\frac{L'}{r}\right) \tag{8.21}$$

where the function $A(\theta) = 1$ for initial edge dislocation segments or $A(\theta) = 1/(1 - \nu)$ for initial screw dislocation segments. Critical stresses have been calculated for different types of bowing dislocation configurations (Bacon, 1967; Foreman, 1967; Mitchell and Smialek, 1968).

Average effective values of A and B have been computed for the dislocation configurations since the values of θ vary along the dislocation lines. For screw dislocations with horizontal side arms, Fig. 8.11(a), $B = -1.38$, while $A = -0.92$ for corresponding edge dislocation configurations. Similarly, for bowing screw and edge dislocations with vertical side arms, $B = 0.83$ and 0.32, respectively. As the reader can imagine, different effective values of A and B have been obtained for a wide range of dislocation configurations. These are discussed in detail in papers by Foreman (1967) and Brown and Ham (1967).

8.6.2 Strengthening by Dislocation Shearing or Cutting of Precipitates

In addition to bowing between precipitates, dislocations may shear or cut through precipitates. This may result in the formation of ledges at the interfaces between the particle and the matrix, in the regions where dislocation entry or exit occurs (Fig. 8.10). Alternatively, since dislocation cutting of ordered precipitates by single dislocations will result in the disruption of the ordered structure, the passage of a second dislocation is often needed to restore the ordered structure (Fig. 8.12). Such pairs of dislocations are generally referred to as *superdislocations*. For energetic reasons, the superdislocations will often dissociate into superpartials that are bounded by stacking faults (SFs) and (APBs), as shown in Fig. 8.12.

Hence, the shearing of ordered precipitates results in the creation of new surfaces (APBs), while the shearing of disordered precipitates results in ledge formation, as shown schematically in Figs 8.10 and 8.12.

Furthermore, since an applied shear stress is needed to overcome the precipitate resistance to shear by APB or ledge formation, significant strengthening may be accomplished by dislocation shear/cutting mechan-

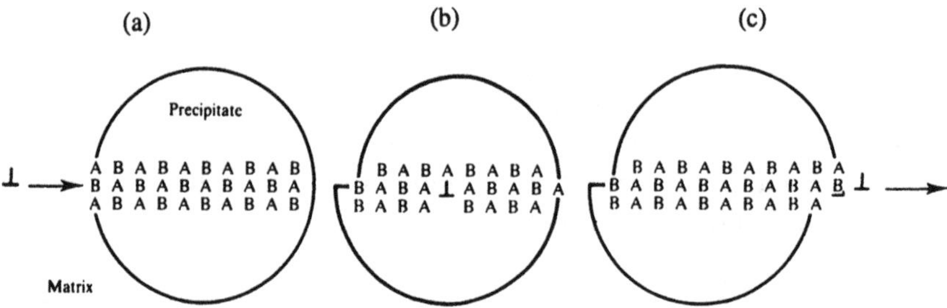

Figure 8.12 View of an edge dislocation penetrating an ordered precipitate (the crystal structure of the precipitate is simple cubic and its composition is AB). In (a) the dislocation has not yet entered the precipitate. In (b) it is partially through. Slip in the precipitate is accompanied by the formation of an antiphase domain boundary (A–A and B–B bonds) across the slip plane. After the dislocation exits the particle, the antiphase domain surface occupies the whole of the slip plane area of the precipitate and the energy increase is $\sim \pi r^2$ (APBE). The increase in energy is linear with the position of the dislocation in the particle. Thus, $F_{max} = \pi r^2$ (APBE)/2$r = \pi r$ (APBE)/2. (From Courtney, 1990. Reprinted with permission from McGraw-Hill.)

isms, especially when the nature of the particle boundaries permit dislocation entry into the particles, as shown schematically in Figs 8.13(a) and 8.13(b) for coherent and semicoherent interfaces (note that coherent interfaces have matching precipitate and matrix atoms at the interfaces, while semicoherent interfaces have only partial matching of atoms). In contrast, dislocation entry (into the precipitate) is difficult when the interfaces are incoherent, i.e., there is little or no matching between the matrix and precipitate atoms at the interfaces, Fig. 8.13(c). Dislocation entry into, or exit from, particles may also be difficult when the misfit strain, ε, induced as a result of lattice mismatch (between the matrix and precipitate atoms) is significant in semicoherent or coherent interfaces. This is because of the need to apply additional stresses to overcome the coherency strains/stresses associated with lattice mismatch.

As discussed earlier, particle shearing of disordered particles results in the formation of a slip step on entry, and another slip step on exit from the particle. For a particle volume fraction, f, and similar crystal structures in the matrix and particle, it can be shown that the shear strengthening provided by particle shearing of disordered precipitates is given by (Gleiter and Hornbogen, 1967):

$$\Delta\tau_{ps} = \frac{3\overline{G}(b_p - b_m)}{b}\left(\frac{r}{d}\right) \tag{8.22}$$

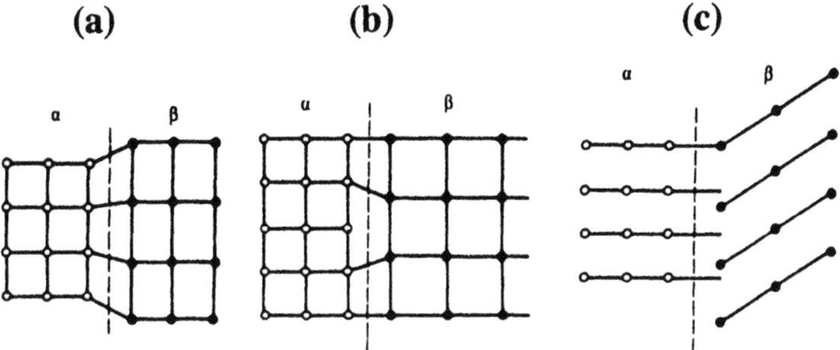

FIGURE 8.13 Schematic of the different types of interfaces: (a) coherent interface; (b) semicoherent interface; (c) incoherent interface. (From Courtney, 1990. Reprinted with permission from McGraw-Hill.)

where $\overline{G}$ is the average shear modulus, b_p is the particle Burgers vector, b_m is the matrix Burgers vector, b is the average Burgers vector, d is the distance traveled by the dislocation along the particle, and r is the particle radius. In cases where the misfit strain (due to lattice mismatch between the matrix and particle) is significant, the overall strengthening is the stress required to move the dislocations through the stress/strain fields at the particle boundaries.

The increase in shear strength is then given by

$$\Delta\tau_m = \sqrt{\frac{27 \cdot 4 \cdot b \cdot \varepsilon^3 E^3}{\pi T(1+\nu)^3}} f^{4/6} r^{1/2} \tag{8.23}$$

where E is the Young's modulus, b is the Burgers vector, ε is the misfit strain, T is the line energy of the dislocation, ν is Poisson's ratio, f is the precipitate volume fraction, and r is the precipitate radius. The mismatch strain, ε, is now given by

$$\varepsilon = \frac{3K(\Delta a/a)}{3K + 2/1 + \nu)} \tag{8.24}$$

where K is the bulk modulus, ν is Poisson's ratio, and the other constants have their usual meanings.

It is important to note here that Eq. (8.23) may be used generally in precipitation strengthening due to any type of misfit strain. Hence, for example, ε may represent misfit strains due to thermal expansion mismatch. In cases where the sheared particles are ordered, e.g., intermetallic compounds between metals and other metals (Fig. 8.12), the shearing of the particles often results in the creation of SFs and APBs. The shear strengthening term then becomes quite significant, and is given by (Gleiter and Hornbogen, 1965):

$$\Delta\tau_{pc} = \frac{0.28\gamma^{3/2} f^{1/3}}{b^2} \left(\frac{r}{G}\right)^{1/2} \tag{8.25}$$

where γ is the anti-phase boundary energy, f is the particle volume fraction, r is the particle radius, b is the Burgers vector, and G is the shear modulus. As before, the above strengthening equations can be multiplied by the appropriate Schmid factor to obtain expressions for axial strengthening.

8.6.3 Dislocation Looping Versus Shear

It is important at this stage to examine forms of the precipitation strengthening equations, (8.22), (8.23) and (8.25). Two of the expressions for dislocation cutting [Eqs (8.23) and (8.25)] show a strength dependence that varies as $r^{1/2}$, while the simplest cutting expression [Eq. (8.22)] shows a dependence on r. However, the strengthening due to dislocation looping [Equation (8.19)] exhibits dependence on r^{-1}. Hence, as particle size increases (for the same volume fraction of particles), the strength dependence due to dislocation looping and dislocation shearing will be of forms shown schematically in Fig. 8.14.

It should be clear from Fig. 8.14 that the stresses required for particle shearing are lower than those required for particle looping when the particle

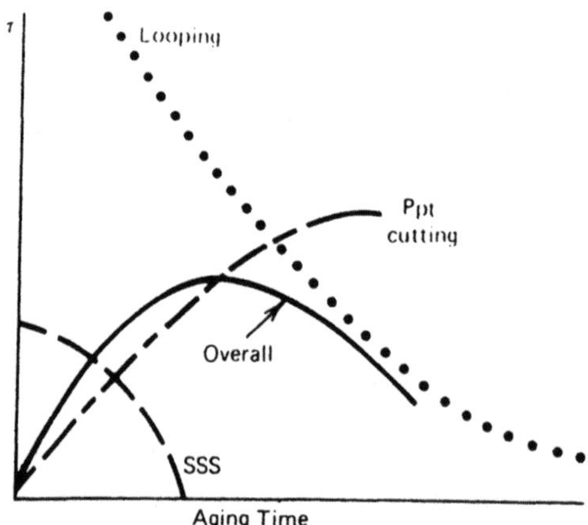

FIGURE 8.14 Schematic of the role of precipitation hardening mechanisms in the overall aging response. (From Hertzberg, 1996. Reprinted with permission from John Wiley.)

sizes are below the critical size, r_c. Hence, particle shearing will dominate when the particle sizes are below r_c (usually between 10 and 100 nm). However, above the critical size, r_c, the stress increment required for dislocation cutting (Figs 8.10 and 8.12) is greater than that required for dislocation looping (Fig. 8.11). Hence, it is easier to loop or extrude around the precipitates, and the effective strengthening mechanism is dislocation looping (Orowan strengthening) when the average particle size is above the critical particle radius r_c (Fig. 8.14).

The favored strengthening mechanism should, therefore, change from particle shearing to dislocation looping, as the precipitate size increases. The increase in the average particle size may be due to heat treatment, which can give rise to the diffusion-controlled coarsening of precipitates. Since heat treatments can be used to control the sizes and distributions of precipitates, it is common in industry to use aging heat treatments to achieve the desired amounts of precipitation strengthening.

It is important to note here that the above discussion has been based largely on idealized microstructures with uniform microstructures (precipitate sizes and spacings). This is clearly not the case in real microstructures, which generally exhibit statistical variations in precipitate size, distribution, and shape. A statistical treatment of the possible effects of these variables is, therefore, needed to develop a more complete understanding of precipitation strengthening. Nevertheless, the idealized presentation (based on average particle sizes and distribution and simple particle geometries) is an essential first step in the development of a basic understanding of the physics of precipitation strengthening.

In any case, when the average particle sizes that result from aging heat treatment schedules have radii that are less than r_c, the strengths are less than the peak strength values corresponding to $r = r_c$, and the material is said to be under-aged. If heat treatment results in precipitates with average radii, $r = r_c$, the material has the highest strength, and is described as peak aged. Aging heat treatment for even longer durations (or higher temperatures) will promote the formation of precipitates with average radii, $r > r_c$; hence, lower strengths that are associated with overaged conditions. The relative strengths in the underaged, peak aged, and overaged conditions may also be estimated easily by performing hardness tests.

8.6.4. Precipitation Strengthening of Aluminum Alloys

One example of the practical use of precipitation strengthening is in the age hardening of aluminum alloys. Such alloys are typically solution treated (to dissolve all second phases) and aged (Fig. 8.15) for different durations to precipitate out second phase particles with the desired sizes and coherence. The initial strengths of such alloys increase with increasing particle size, until they reach peak levels (peak aged condition) where their hardnesses are maximum (Fig. 8.16). This typically corresponds to the critical particle radius, r_c, described above. Beyond r_c, strengthening occurs by dislocation looping, and the strength decreases with increasing particle size, as the annealing duration is increased (Fig. 8.16). The alloys are said to be overaged in these conditions. Similarly, aging durations that result in particle radii below r_c correspond to underaged conditions (Figs 8.15 and 8.16).

One classical example of an age-hardened system is the Al–Cu system (Fig. 8.15). This was first studied during the first half of the 20th century, and is perhaps the best understood aluminum alloy system. Following a solution treatment (to remove any cold work or prior precipitates) and quenching, the Al–Cu system is supersaturated with Cu in solid solution (Fig. 8.15). Hence, subsequent aging results in the precipitation of Al–Cu platelets/precipitates, provided that the aging temperatures and durations are sufficient to promote the nucleation and growth of these new phases within reasonable periods of time. The nature of the precipitates that form also depend on the aging temperatures.

The first set of nano-scale particles that form are known as Guinier–Preston (G-P) zones. These consist of copper atoms that are arranged into plate-like structures. The G-P zones (Fig. 8.17) are named after the two scientists (Guinier and Preston) who first discovered them. The first set of

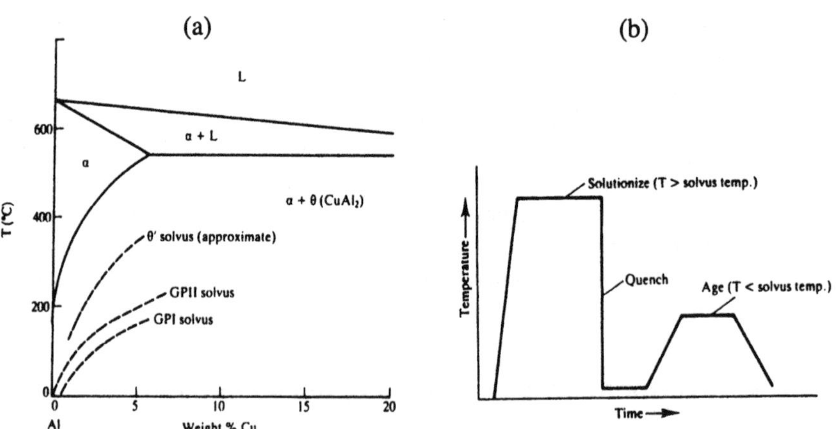

FIGURE 8.15 (a) Left-hand section of the Al–Cu phase diagram; (b) aging heat treatment schedule. (From Courtney, 1990. Reprinted with permission from Pergamon Press.)

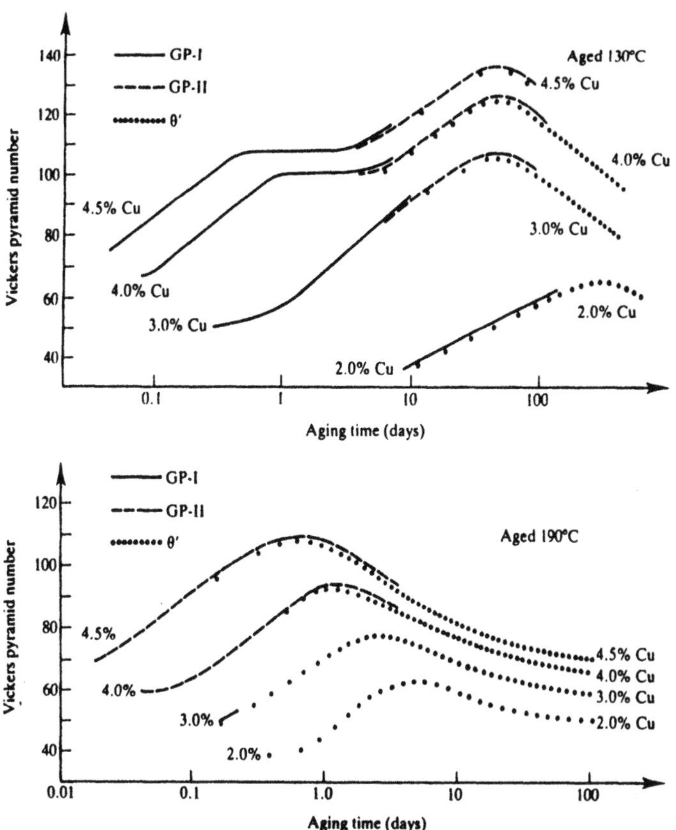

FIGURE 8.16 Role of Guinier–Preston zones and θ precipitates in the strengthening of Al–Cu alloys aged at 130° and 190°C. (From Courtney, 1990. Reprinted with permission from Pergamon Press.)

G-P zones that form are known as G-P I zones. They are ~ 25 atoms in diameter and are oriented parallel to {100} planes in the face-centered cubic aluminum solid solution matrix. The G-P I zones are coherent and they provide moderate strengthening by dislocation shearing. The second set of G-P zones are ~ 75 atoms wide and ~ 10 atoms thick. They contain an almost stoichiometric ratio of Al to Cu, and are known as G-P II zones.

Further aging results in the formation of metastable partially coherent or coherent θ' precipitates. The θ' phase corresponds to the equilibrium $CuAl_2$ phase, but it also has a different lattice structure. Furthermore, the maximum (peak) hardness is associated with the presence of both θ' and G-P II zones (Fig. 8.16). Eventually, only θ' precipitates are present when the

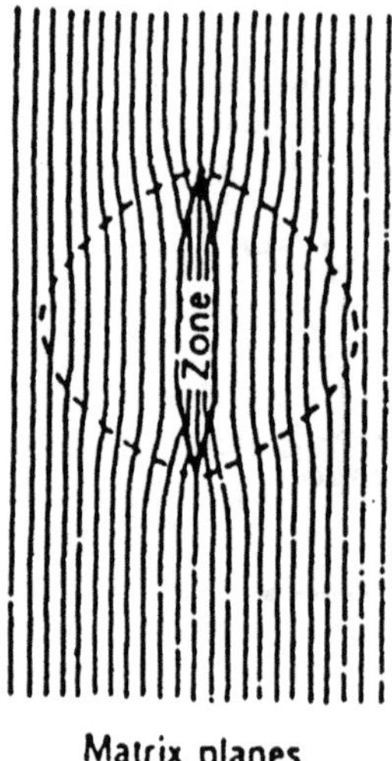

Figure 8.17 Schematic of Guinier–Preston zone.

alloys reach the overaged condition in which hardness, tensile strength, and ductility are reduced on further aging.

Subsequent aging may result in the formation of coarse incoherent θ precipitates, if aging is carried out between 170° and 300°C. The resulting material now contains microscale coarse $CuAl_2$ precipitates that are visible under a light microscope. The material is also softer than in the quenched state, because of the loss of the solid solution strengthening by Cu atoms, which have now largely diffused from the aluminum matrix to form the coarse $CuAl_2$ precipitates.

The above trends are summarized in Fig. 8.16. Note that the figure also indicates that the extent of hardening/strengthening increases with increasing volume fraction of precipitate/increasing copper content, and that the highest strengths are achieved at lower aging temperatures for longer aging temperatures and durations.

8.7 DISPERSION STRENGTHENING

Before concluding this section on strengthening, it is important to discuss a form of particulate strengthening that is known as dispersion strengthening. Dispersion strengthening is usually associated with incoherent precipitates that are larger in size than those encountered during Orowan strengthening or dislocation cutting. Since the elastic energies of gliding dislocations can be lowered by interactions with the boundaries of stiff elastic particles at high temperatures, the dislocations are attracted and pinned by the boundaries. This problem was first analyzed by Dundurs and Gundagarajan (1969) for time-independent deformation. Subsequent work by Srolovitz et al. (1983, 1984) extended the models to the analyses of diffusion-assisted relaxation of interfacial stresses that can occur during high-temperature creep deformation.

Details of the above elasticity models are beyond the scope of the current text. The interested reader is, therefore, referred to the relevant literature, which is cited at the end of this chapter. Nevertheless, it is important to note here that dispersion strengthening can be engineered to be as effective as precipitation strengthening.

One important example of dispersion strengthening is that provided by oxide particles. Oxide dispersion strengthening has been used recently to strengthen aluminum alloys and nickel-base superalloys. This has been achieved largely by the addition of Al_2O_3 flakes and ThO_2 particles to aluminum and nickel matrices. The overall strengthening from such relatively large incoherent particles is less than that from precipitation strengthened alloys at lower temperatures. However, oxide dispersion strengthened (ODS) alloys retain their strengths at very high temperatures (approaching the melting points of nickel- and aluminum-base alloys). This is because the Al_2O_3 and ThO_2 particles are morphologically stable at very high temperatures. It is, therefore, possible to design alloys for very high-temperature applications where intermetallic precipitates would coarsen. This has led to turbine blade and turbine vane applications of ODS nickel-base superalloys at temperatures greater than $\sim 950°C$.

8.8 OVERALL SUPERPOSITION

The above discussion has considered strengthening mostly within an athermal framework. However, it is important to realize that phase changes (coarsening/transformations) may change the overall strengthening contributions from different mechanisms. Dislocation pile-ups may also be relaxed by thermally assisted climb mechanisms. Furthermore, cutting and dislocation looping may occur simultaneously within a given alloy.

In any case, superposition concepts may be used to obtain order-of-magnitude estimates of the overall strengthening levels when the fractional contributions from each of the strengthening mechanisms are known approximately. In such cases, the contributions of the different strengthening mechanisms (to the overall strength of an alloy) may be estimated from expressions of the form:

$$\tau_y = \tau_0 + \Delta\tau_s + \Delta\tau_d + \Delta\tau_{gb} + \Delta\tau_p \tag{8.26}$$

or

$$\sigma_y = \sigma_0 + \Delta\sigma_s + \Delta\sigma_d + \Delta\sigma_{gb} + \Delta\sigma_p \tag{8.27}$$

where subscripts 0, s, d, gb, and p denote single crystal, solid solution, dislocation, grain boundary and particle strengthening components, respectively. It is also important to remember that the above expressions neglect possible interactions between strengthening mechanisms. They are intended only to provide insights into the sources of dislocation strengthening in metals and their alloys, i.e., they are not sufficiently well developed to be fully predictive tools.

8.9 SUMMARY

An introduction to dislocation strengthening mechanisms has been presented in this chapter. Strengthening was shown to occur by the use of defects (point defects, surface defects and volume defects in the restriction of dislocation motion in engineering alloys. A semiqualitative account of the different types of strengthening mechanisms include: (1) solid solution strengthening; (2) dislocation strengthening; (3) grain boundary strengthening; (4) precipitation strengthening, and (5) dispersion strengthening. The chapter concluded with a brief description of factors that can contribute to the overall strengths of engineering alloys. It is particularly important to note that the expressions presented in this chapter are only intended to serve as semiquantitative guides for the estimation of order-of-magnitude values of strengthening due to dislocation/defect interactions.

BIBLIOGRAPHY

Armstrong, R.W. and Jindal, P.C. (1968) TMS-AIME. vol. 242, p 2513.

Ashby, M.F. and Jones, D.R.H. (1996) Engineering Materials: An Introduction to the Properties and Applications. vol. 1. Pergamon Press, New York.

Bacon, D.J. (1967) Phys Stat Sol. vol. 23, p 527.

Brown, L.M. and Ham, R.K. (1967) Dislocation–particle interactions. In: A. Kelly and R.B. Nicholson, eds. Strengthening Methods in Crystals. Applied Science, London, pp 9–135.

Courtney, T.H. (1990) Mechanical Behavior of Materials. McGraw-Hill, New York.

Dundurs, J. and Mura, T. (1969) J Mech Phys Solids. vol. 17, p 459.

Eshelby, J.D., Frank, F.C., and Nabarro, F.R.N. (1951) The equilibrium of linear arrays of dislocations. Phil Mag. vol. 42, p 351.

Fleischer, R.L. (1961) Solution hardening. Acta Metall. vol. 9, pp 996–1000.

Fleischer, R.L. (1962) Solution hardening by tetragonal distortions: application to irradiation hardening in face centered cubic crystals. Acta Metall. vol. 10, pp 835–842.

Fleischer, R.L. (1963) Acta Metall., Vol. II, p. 203.

Foreman, A.J.E. (1967) Phil Mag. vol. 15, p 1011.

Gleiter, H. and Hornbogen, E. (1967) Precipitation hardening by coherent particles. Mater. Sci Eng. vol. 2, pp 285–302.

Hall, E.O. (1951) Proc Phys Soc B. vol. 64, p 747.

Hertzberg, R.W. (1996) Deformation and Fracture Mechanics of Engineering Materials. 4th ed. John Wiley, New York.

Hu, H. and Cline, R.S. (1968) TMS-AIME. vol. 242, p 1013.

Hull, D. and Bacon, D.J. (1984) Introduction to Dislocations. 3rd ed. Pergamon Press, New York.

Jones, R.L. and Conrad, H. (1969) TMS-AIME. vol. 245, p 779.

Knott, J.F. (1973) Fundamentals of Fracture Mechanics. Butterworth, London.

Mitchell, T.E. and Smialek, R.L. (1968) Proceedings of Chicago Work Hardening Symposium (AIME).

Orowan, E. (1948) Discussion in the Symposium on Internal Stresses in Metals and Alloys. Institute of Metals, London, p 451.

Petch, N.J. (1953) Iron Steel Inst. vol. 173, p 25.

Srolovitz, D.J., Petrovic-Luton, R.A., and Luton, M.J. (1983) Edge dislocation–circular inclusion interactions at elevated temperatures. Acta Metall. vol. 31, pp 2151–2159.

Srolovitz, D.J., Luton, M.J., Petrovic-Luton, R., Barnett, D.M., and Nix, W.D. (1984) Diffusionally modified dislocation–particle elastic interactions. Act Metall. vol. 32, pp 1079–1088.

Taylor, G.I. (1934) Proc Roy Soc. vol. A145, pp 362–388.

Weeks, R.W., Pati, S.R., Ashby, M.J., and Barrand, P. (1969) Acta Metall. vol. 17, p 1403.

9

Introduction to Composites

9.1 INTRODUCTION

Two approaches can be used to engineer improved mechanical properties of materials. One involves the modification of the internal structure of a given material system (intrinsic modification) by minor alloying, processing, and/or heat treatment variations. However, after a number of iterations, an asymptotic limit will soon be reached by this approach, as the properties come close to the intrinsic limits for any given system. In contrast, an almost infinite array of properties may be engineered by the second approach which involves extrinsic modification by the introduction of additional (external) phases.

For example, the strength of a system may be improved by reinforcement with a second phase that has higher strength than the intrinsic limit of the "host" material which is commonly known as the "matrix." The resulting system that is produced by the *mixture* of *two or more phases* is known as a *composite material.*

Note that this rather *general definition* of a composite applies to both synthetic (man-made) and natural (existing in nature) composite materials. Hence, concrete is a synthetic composite that consists of sand, cement and stone, and wood is a natural composite that consists primarily of hemicellulose fibers in a matrix of lignin. More commonly, however, most of us are familiar with polymer matrix composites that are often used in modern tennis racquets and pole vaults. We also know, from watching athletic events, that these so-called advanced composite materials promote significant improvements in performance.

This chapter introduces the concepts that are required for a basic understanding of the effects of composite reinforcement on composite strength and modulus. Following a brief description of the different types of composite materials, mixture rules are presented for composite systems reinforced with continuous and discontinuous fibers. This is followed by an introduction to composite deformation, and a discussion on the effects of fiber orientation on composite failure modes. The effects of statistical variations in fiber properties on the composite properties are then examined at the end of the chapter. Further topics in composite deformation will be presented in Chap. 10.

9.2 TYPES OF COMPOSITE MATERIALS

Synthetic composites are often reinforced with high-strength fibers or whiskers (short fibers). Such reinforcements are obtained via special processing schemes that generally result in low flaw/defect contents. Due to their low flaw/defect contents, the strength levels of whiskers and fibers are generally much greater than those of conventional bulk materials in which higher volume fractions of defects are present. This is shown in Table 9.1 in which the strengths of monolithic and fiber/whisker materials are compared. The higher strengths of the whisker/fiber materials allow for the development of composite materials with intermediate strength levels, i.e., between those of the matrix and reinforcement materials. Similarly, intermediate values of modulus and other mechanical/physical properties can be achieved by the use of composite materials.

The actual balance of properties of a given composite system depends on the combinations of materials that are actually used. Since we are generally restricted to mixtures of metals, polymers, or ceramics, most synthetic composites consist of mixtures of the different classes of materials that are shown in Fig. 9.1(a). However, during composite processing, interfacial reactions can occur between the matrix and reinforcement materials. These result in the formation of interfacial phases and interfaces (boundaries), as shown schematically in Fig. 9.1(b).

One example of a composite that contains easily observed interfacial phases is presented in Fig. 9.2. This shows a transverse cross-section from a titanium matrix (Ti–15V–3Cr–3Al–3Sn) composite reinforced with carbon-coated SiC (SCS-6) fibers. The interfacial phases in this composite

TABLE 9.1 Summary of Basic Mechanical/Physical Properties of Selected Composite Constituents: Fiber Versus Bulk Properties

	Young's modulus (GPa)	Strength[a] (MPa)
Alumina: fiber (Saffil RF)	300	2000
monolithic	382	332
Carbon: fiber (IM)	290	3100
monolithic	10	20
Glass, fiber (E)	76	1700
monolithic	76	100
Polyethylene: fiber (S 1000)	172	2964
monolithic (HD)	0.4	26
Silicon carbide: fiber (MF)	406	3920
monolithic	410	500

[a]Tensile and flexural strengths for fiber and monolithic, respectively.

have been studied using a combination of scanning and transmission electron microscopy. The multilayered interfacial phases in the Ti–15V–3Cr–3Al–3Sn/SCS-6 composite (Fig. 9.2) have been identified to contain predominantly TiC. However, some Ti_2C and Ti_5Si_3 phases have also been shown to be present in some of the interfacial layers (Shyue et al., 1995).

The properties of a composite can be tailored by the judicious control of interfacial properties. For example, this can be achieved in the Ti–15V–3Cr–3Al–3Sn/SCS-6 composite by the use of carbon coatings on the SiC/

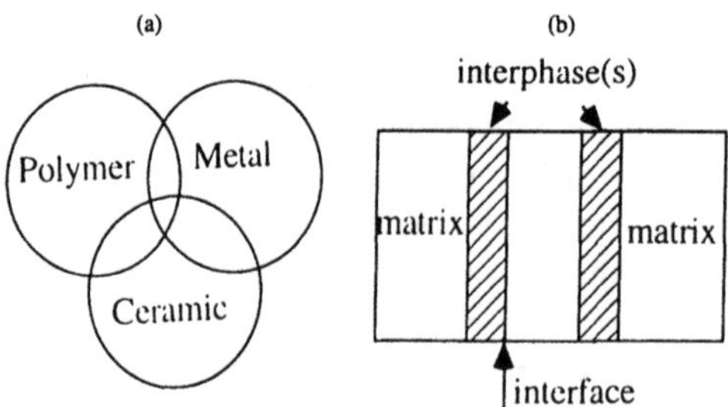

FIGURE 9.1 Schematic illustration of (a) the different types of composites and (b) interfaces and interfacial phases formed between the matrix and reinforcement materials.

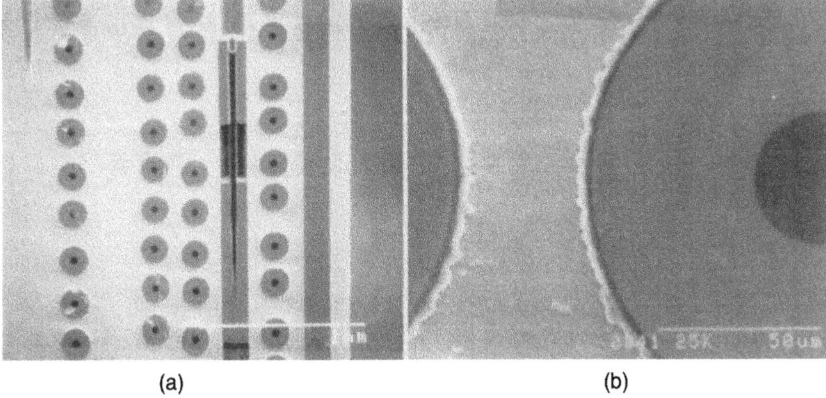

(a) (b)

Figure 9.2 (a) Transverse cross-section of Ti–15V–3Cr–3Al–3Sn composite reinforced with 35 vol% carbon-coated SiC (SCS-6) fibers and (b) Interfacial Phases in Ti-15-3/SCS6 composite.

SCS-6 fibers. The hexagonal graphite layers in the carbon coatings tend to align with axial stress, thus making easy shear possible in the direction of interfacial shear stress. Hence, the interfacial shear strengths of silicon carbide fiber-reinforced composites can be controlled by the use of carbon coatings that make interfacial sliding relatively easy. Such interfacial sliding is critical in the accommodation of strain during mechanical loading or thermal cycling.

Composite properties are also controlled by the selection of constituents with the appropriate mix of mechanical and physical properties (Tables 9.1 and 9.2). Since light weight is often of importance in a large number of structural applications, especially in transportation vehicles such as cars, boats, airplanes, etc., specific mechanical properties are often considered in the selection of composite materials. Specific properties are given by the ratio of a property (such as Young's modulus and strength) to the density. For example, the specific modulus is the ratio of Young's modulus to density, while specific strength is the ratio of absolute strength to density.

It is a useful exercise to compare the absolute and specific properties in Table 9.2. This shows that ceramics and metals tend to have higher absolute and specific moduli and strength, while polymers tend to have lower absolute properties and moderate specific properties. In contrast, polymer matrix composites can be designed with attractive combinations of absolute specific

Table 9.2 Summary of Basic Mechanical/Physical Properties of Selected Composite Constituents: Constituent Properties

	Density (mg/m^3)	Young's modulus (GPa)	Strength[a] (MPa)	Ductility (%)	Toughness, K_{IC} (M Pa $m^{1/2}$)	Specific modulus [(GPa)/(mg/m^3)]	Specific strength [(MPa)/(mg/m^3)]
Ceramics							
Alumina (Al_2O_3)	3.87	382	332	0	4.9	99	86
Magnesia (MgO)	3.60	207	230	0	1.2	58	64
Silicon nitride (Si_3N_4)		166	210	0	4.0		
Zirconia (ZrC_2)	5.92	170	900	0	8.6	29	152
β-Sialon	3.25	300	945	0	7.7	92	291
Glass–ceramic Silceram	2.90	121	174	0	2.1	42	60
Metals							
Aluminum	2.70	69	77	47		26	29
Aluminum–3%Zn–0.7%Zr	2.83	72	325	18		25	115
Brass (Cu-30%Zn)	8.50	100	550	70		12	65
Nickel–20%Cr–15%Co	8.18	204	1200	26		25	147
Mild steel	7.86	210	460	35		27	59
Titanium–2.5% Sn	4.56	112	792	20		24	174
Polymers							
Epoxy	1.12	4	50	4	1.5	4	36
Melamine formaldehyde	1.50	9	70			6	47
Nylon 6.6	1.14	2	70	60		18	61
Poly(ether ether ketone)	1.30	4	70			3	54
Poly(methyl methacrylate)	1.19	3	50	3	1.5	3	42
Polystyrene	1.05	3	50	2	1.0	3	48
Poly(vinyl chloride) rigid	1.70	3	60	15	4.0	2	35

[a]Strength values are obtained from the test appropriate for the material, e.g., flexural and tensile for ceramics and metals, respectively.

strength and stiffness. These are generally engineered by the judicious selection of polymer matrices (usually epoxy matrices) and strong and stiff (usually glass, carbon, or kevlar) fibers in engineering composites, which are usually polymer composites.

The specific properties of different materials can be easily compared using materials selection charts such as the plots of E versus ρ, or σ_f versus ρ in Figs 9.3 and 9.4, respectively. Note that the dashed lines in these figures correspond to different "merit" indices. For example, the minimum weight design of stiff ties, for which the merit index is E/ρ, could be achieved by selecting the materials with the highest E/ρ from Fig. 9.3. These are clearly the materials that lie on dashed lines at the top left-hand corner of Fig. 9.3.

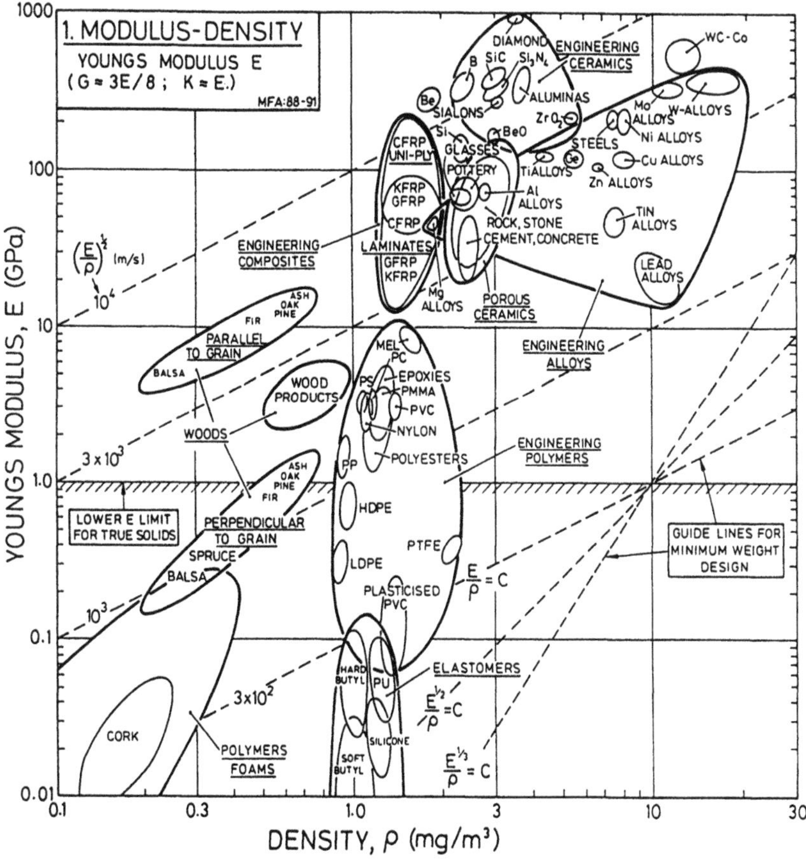

FIGURE 9.3 Materials selection charts showing attractive combinations of specific modulus that can be obtained from engineering composites.

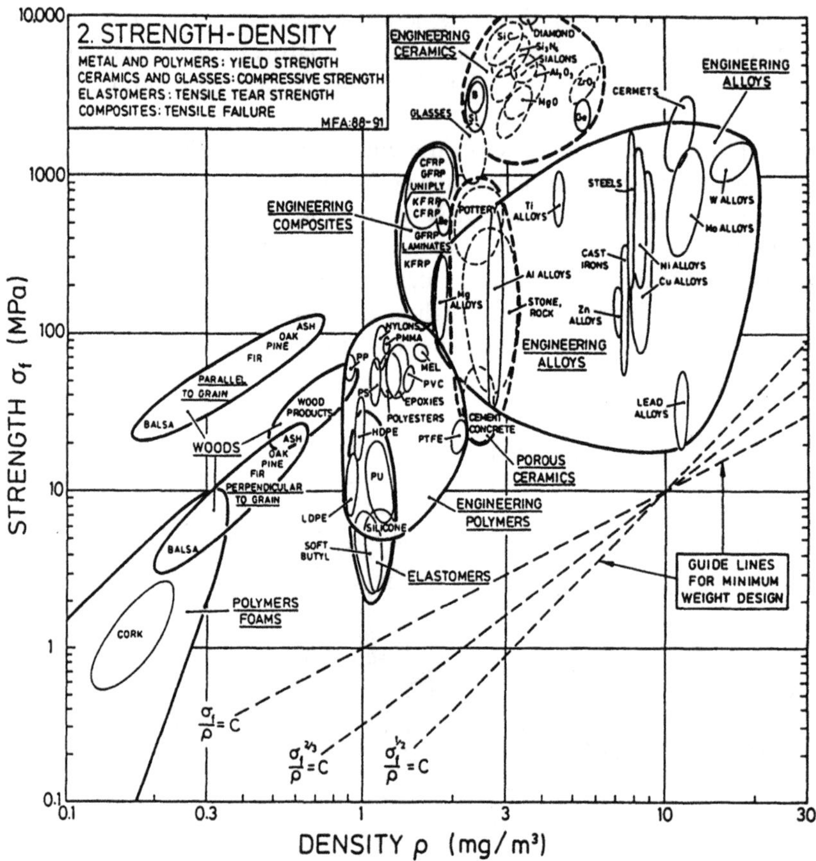

FIGURE 9.4 Materials selection charts showing attractive combinations of specific strength that can be obtained from engineering composites.

Similarly, the materials with the highest specific strengths, σ_f/ρ, are the materials at the top left-hand corner of the strength materials selection chart shown in Fig. 9.4. In both charts (Figs 9.3 and 9.4), polymer matrix composites such as carbon fiber-reinforced plastics (CFRPs), glass fiber-reinforced plastics (GFRPs), and kevlar fiber-reinforced plastics (KFRPs) emerge clearly as the materials of choice. For this reason, polymer matrix composites are often attractive in the design of strong and stiff lightweight structures.

A very wide range of synthetic and natural composite materials are possible. Conventional reinforcement morphologies include: particles (Fig. 9.5), fibers [Fig. 9.6(a)], whiskers [Figs 9.6(b) and 9.6(c)], and layers, Fig. 9.6(d). However, instead of abrupt interfaces which may cause stress concentrations, graded interfaces may be used in the design of coatings and interfaces in which the properties of the system are varied continuously from 100% A to 100% B, as shown schematically in Fig. 9.7. Such graded transitions in composition may be used to avoid abrupt changes in stress states that can occur at nongraded interfaces.

Furthermore, composite architectures can be tailored to support loads in different directions. Unidirectional fiber-reinforced architectures [Fig. 9.6(a)] are, therefore, only suitable for structural applications in which the loading is applied primarily in one direction. Of course, the composite fiber may be oriented to support axial loads in such cases. Similarly, bidirectional composite systems (with two orientations of fibers) can be oriented to support loads in two directions.

The fibers may also be discontinuous in nature [Figs 9.6(b) and 9.6(c)], in which case they are known as whiskers. Whiskers generally have high strengths due to their low defect densities. They may be aligned [Fig. 9.6(b)],

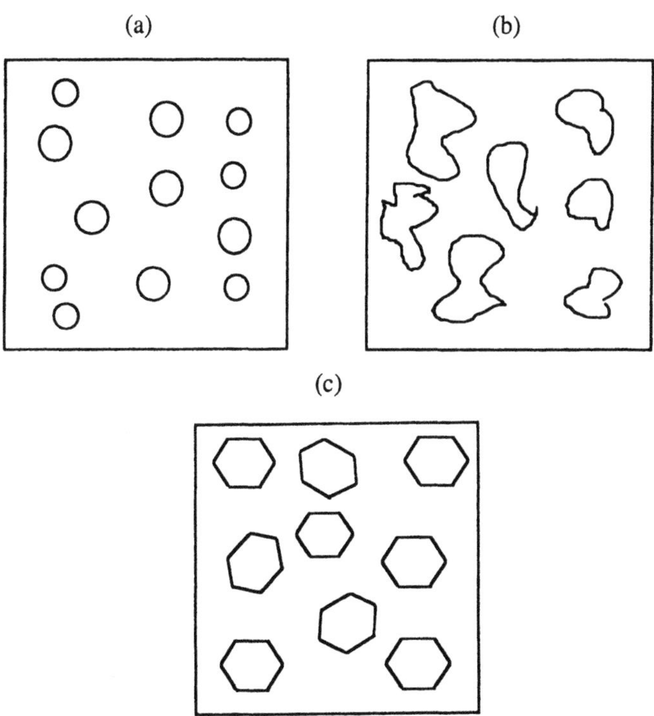

Figure 9.5 Schematic illustration of particulate reinforcement morphologies: (a) spherical; (b) irregular; (c) faceted.

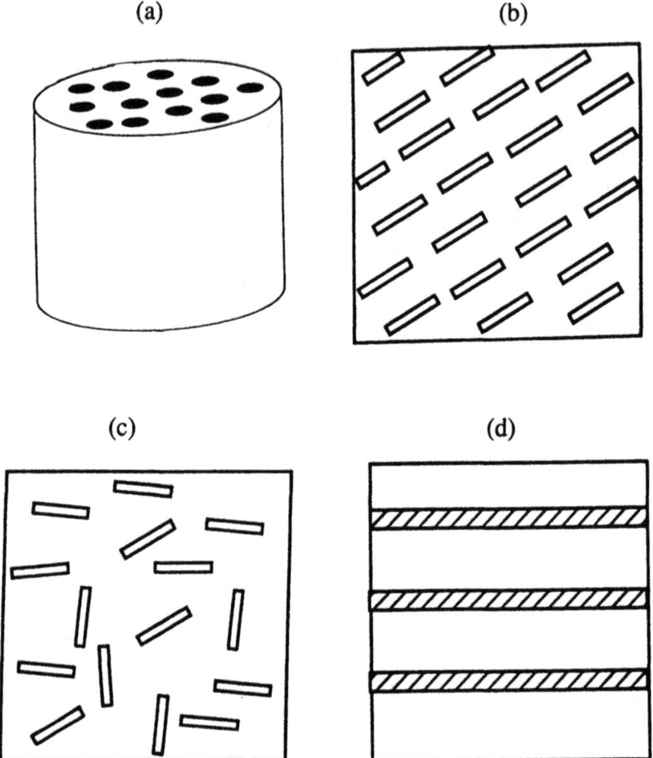

Figure 9.6 Examples of possible composite architectures: (a) unidirectional fiber reinforcement; (b) aligned whisker reinforcement; (c) randomly oriented whisker reinforcement; (d) continuous layers.

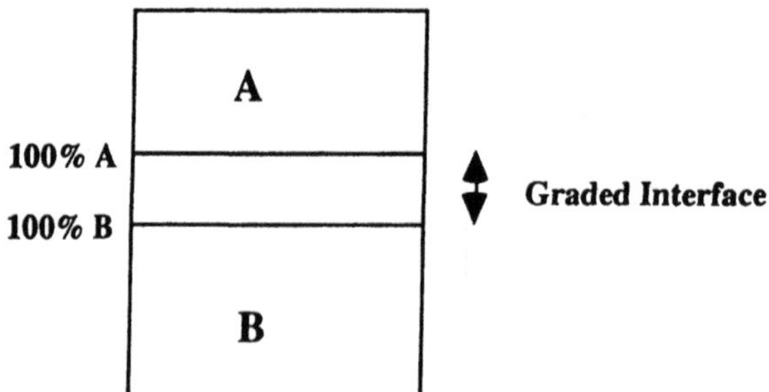

Figure 9.7 Schematic illustration of graded reinforcements.

or randomly oriented, Fig. 9.6(c). The reader may recognize intuitively that aligned orientations of whiskers or fibers will give rise to increased strength in the direction of alignment, but overall, to anisotropic properties, i.e., properties that vary significantly with changes in direction. However, random orientations of whiskers will tend to result in lower average strengths in any given direction, but also to relatively isotropic properties, i.e., properties that do not vary as much in any given direction.

In addition to the synthetic composites discussed above, several composite systems have been observed in nature. In fact, most materials in nature are composite materials. Some examples of natural composites include wood and bone. As discussed earlier, wood is a natural composite that consists of a lignin matrix and spiral hemicellulose fibers. Bone, on the other hand, is a composite that consists of organic fibers, inorganic crystals, water, and fats. About 35% of bone consist of organic collagen protein fibers with small rod-like (5 nm × 5 nm × 50 nm) hydroxyapatite crystals. Long cortical/cancellous bones typically have low fat content and compact structures that consist of a network of beams and sheets that are known as trabeculae.

It should be clear from the above discussion that an almost infinite array of synthetic and artificial composite systems are possible. However, the optimization of composite performance requires some knowledge of basic composite mechanics and materials concepts. These will be introduced in this chapter. More advanced topics such as composite ply theory and shear lag theory will be presented in Chap. 10.

9.3 RULE-OF-MIXTURE THEORY

The properties of composites may be estimated by the application of simple rule-of-mixture theories (Voigt, 1889). These rules can be used to estimate average composite mechanical and physical properties along different directions. They may also be used to estimate the bounds in mechanical/physical properties. They are, therefore, extremely useful in assessment of the combinations of basic mechanical/physical properties that can be engineered via composite reinforcement. This section will present constant-strain and constant-stress rules of mixture.

9.3.1 Constant-Strain and Constant-Stress Rules of Mixtures

An understanding of constant-strain and constant-stress rules of mixtures may be gained by a careful study of Fig. 9.8. This shows schematics of the same composite system with loads applied parallel [Fig. 9.8(a)] or perpendicular [Fig. 9.8(b)] to the reinforcement layers. In the case where the loads are applied parallel to the reinforcement direction [Fig. 9.8(a)], the strains in the matrix and reinforcement layers must be equal, to avoid relative sliding between these layers.

In contrast, the strains in the individual matrix and reinforcement layers are different when the loads are applied in a direction that is perpendicular to the fiber orientation, Fig. 9.8(b). Since the same load is applied to the same cross-sectional area in the reinforcement and matrix layers, the stresses in these layers must be constant and equal for a given load. The loading configuration shown in Fig. 9.8(b), therefore, corresponds to a constant stress condition.

Let us now return to the constant strain condition shown schematically in Fig. 9.8(a). If the initial length of each of the layers, L, and applied load, P, is partitioned between the load in the reinforcement, P_r, and the

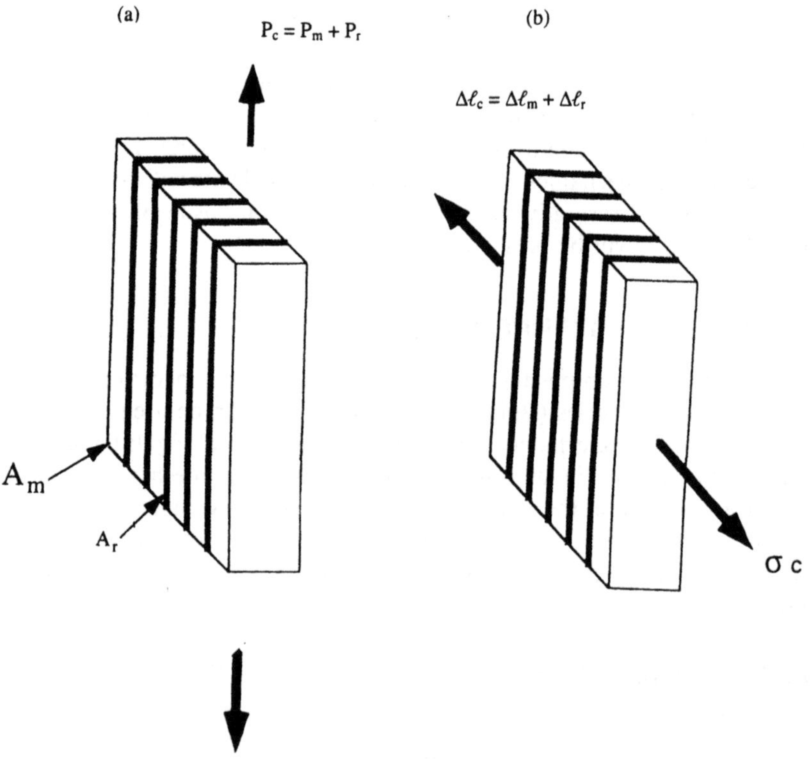

Figure 9.8 Schematic illustration of loading configurations for (a) constant-strain rule of mixtures and (b) constant-stress rule of mixtures.

load in the matrix, P_m, then simple force balance gives

$$P_c = P_m + P_r \tag{9.1}$$

However, from our basic definitions of stress, σ, we know that

$$\sigma = \frac{P}{A} \tag{9.2}$$

where P is the load and A is the cross-sectional area. Also, for uniaxial elastic deformation, Hooke's law gives

$$\sigma = E\varepsilon \tag{9.3}$$

where E is Young's modulus and ε is the uniaxial strain.

Substituting Eqs (9.2) and (9.3) into Eq. (9.1), and using subscripts c, m, and r to denote the composite, matrix, and reinforcement, respectively, gives

$$P_c = \sigma_c A_c = \sigma_m A_m + a\sigma_r \tag{9.4a}$$

and

$$P_c = E_c \varepsilon_c A_c = E_m \varepsilon_m A_m + E_r \varepsilon_r A_r \tag{9.4b}$$

where A_c is the area of composite, A_m is the area of matrix, and A_r is the area of the reinforcement. Noting that the strains in the composite, matrix, and reinforcement are equal, i.e., $\varepsilon_c = \varepsilon_m = \varepsilon_r$, we may simplify Eq. (9.4b) to obtain:

$$E_c = \left(\frac{A_m}{A_c}\right) E_m + \left(\frac{A_r}{A_c}\right) E_r \tag{9.5}$$

However, the ratio (A_m/A_c) corresponds to the area or volume fraction of matrix, V_m, while the area fraction (A_r/A_c) corresponds to the area or volume fraction of reinforcement, V_r. Equation (9.5) may, therefore, be simplified to give:

$$E_c = V_m E_m + V_r E_r \tag{9.6}$$

Similarly, substituting V_m and V_m into Eq. (9.4a) gives the strength of the composite, σ_c, as

$$\sigma_c = V_m \sigma_c + V_r \sigma_r \tag{9.7}$$

Equations (9.6) and (9.7) are the respective constant-strain rule-of-mixture expressions for *composite modulus* and *composite strength*. They represent the upper bound values for composite modulus and strength for the composite system shown schematically in Fig. 9.8. Furthermore, the constant-strain rule-of-mixture equations indicate that upper-bound composite properties are averaged according to the volume fraction of the composite constituents.

The fraction of the load supported by each of the constituents also depends on the ratio of the in moduli to the composite moduli. Hence, for most reinforcements, which typically have higher moduli than those of matrix materials (Tables 9.1 and 9.2) most of the load is supported by the fibers, since:

$$\frac{P_r}{P_c} = V_r \frac{E_r}{E_c} \tag{9.8}$$

Substituting typical numbers for engineering composites (mostly polymer matrix composites reinforced with ceramic fibers), $E_r/E_c \sim 10$ and $V_r \sim 0.55$, then P_r/P_c to 0.92. Hence, a very large fraction of the applied load is supported by the fibers due to their higher moduli.

Let us now consider the constant-stress rule-of-mixtures condition shown schematically in Fig. 9.8(b). In this case, the stresses are equal in the composite, matrix and reinforcement, i.e., $\sigma_c = \sigma_m = \sigma_r$. However, the composite displacement, $\Delta\ell_c$, is now given by the sum of the displacement in the matrix, $\Delta\ell_m$, and the displacement in the reinforcement, $\Delta\ell_r$. Hence, the composite displacement, $\Delta\ell_c$, may be expressed as

$$\Delta\ell_c = \Delta\ell_m + \Delta\ell_r \tag{9.9}$$

Noting that the engineering strain, ϵ, is defined as the ratio of length extension, $\Delta\ell$, to original length, ℓ, we may write:

$$\varepsilon_c \ell_c = \ell_m \varepsilon_m + \ell_r \varepsilon_r \tag{9.10}$$

Since the area or volume fractions now correspond to the length fractions of matrix and reinforcement, we may write

$$V_m = \frac{\ell_m}{\ell_c} \tag{9.11a}$$

and

$$V_r = \frac{\ell_r}{\ell_c} \tag{9.11b}$$

Dividing both the left- and right-hand sides of Eq. (9.10) by ℓ_c, and noting that $V_m = \ell_m/\ell_c$ and $V_r = \ell_r/\ell_c$ [from Eqs (9.11a) and (9.11b)] gives

$$\varepsilon_c = \left(\frac{\ell_m}{\ell_c}\right)\varepsilon_m + \left(\frac{\ell_r}{\ell_c}\right)\varepsilon_r \tag{9.12a}$$

or

$$\varepsilon_c = V_m \varepsilon_m + V_r \varepsilon_r \tag{9.12b}$$

The *composite strain* is, therefore, *averaged* between the *matrix* and *reinforcement* for the *constant stress* condition. The composite modulus for the constant stress condition may be obtained by substituting $\varepsilon = \sigma/E$ into Eq. (9.12b):

$$\varepsilon_c = \frac{\sigma_c}{E_c} V_m \frac{\sigma_m}{E_m} + V_r \frac{\sigma_r}{E_r} \tag{9.13}$$

However, since $\sigma_c = \sigma_m = \sigma_r$, Eq. (9.13) reduces to

$$\frac{1}{E_c} = \frac{V_m}{E_m} + \frac{V_r}{E_r} \tag{9.14a}$$

or inverting Eq. (9.14a) gives

$$E_c = \frac{E_m E_r}{V_m E_r + V_r E_m} \tag{9.14b}$$

Equation (9.14b) represents the loci of lower bound composite moduli for possible reinforcement volume fractions between 0 and 1 (Fig. 9.9). Note that the constant-strain rule-of-mixtures estimates represent the upper bound values in Fig. 9.9. Furthermore, the actual composite moduli for most systems are in between the upper and lower bound values shown schematically in Fig. 9.9. For example, particulate composites will tend to have composite moduli that are closer to the lower bound values, as shown in Fig. 9.9.

The expressions for modulus may be generalized for the wide range of possible composite materials between the constant strain (iso-strain) and constant stress (iso-stress) conditions (Fig. 9.9). This may be accomplished by the use of an expression of the form:

$$(X_c)^n = V_m(X_m)^n + V_r(X_r)^n \tag{9.15}$$

where X is a property such as modulus, n is a number between +1 and −1, and subscripts c, m, and r denote composite, matrix, and reinforcement, respectively. Equation (9.15) reduces to the constant strain and constant stress expressions at the limits of $n = +1$ and $n = -1$.

Also, the wide range of possible composite properties may be estimated for actual composites for which values of n are between −1 and +1. Furthermore, the values of n for many composites are close to zero. However, there are no solutions for $n = 0$, for which Eq. (9.15) gives the trivial solution $1 = 1$. Iterative methods are, therefore, required to obtain the solutions for $n = 0$.

Although the discussion so far has focused largely on iso-stress and iso-strain conditions, the above rule-of-mixture approach can be applied generally to the estimation of physical properties such as density, thermal conductivity, and diffusivity. The rule-of-mixture expressions can, therefore,

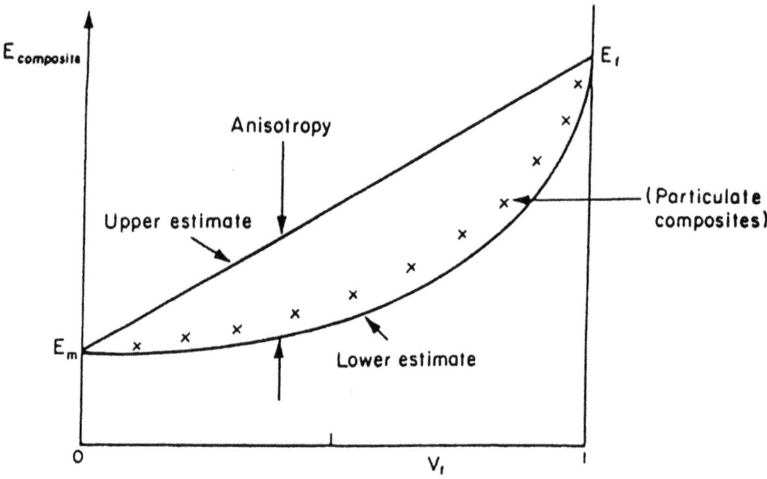

Figure 9.9 Schematic illustration of upper and lower bound moduli given by constant-strain and constant-stress mixture rules. Note that particulate-reinforced composites have moduli that are closer to lower bound values. (From Ashby and Jones, 1984. Reprinted with permission from Pergamon Press.)

be used to estimate the effects of reinforcements on many important physical properties. They may also be used to estimate the bounds in a wide range of physical properties of composite materials. Such rule-of-mixture calculations are particularly valuable because they can be used in simple "back-of-the-envelope" estimates to guide materials selection and design.

Finally, in this section, it is important to note that the simple averaging schemes derived above for two-phase composite systems can be extended to a more general case of any n-component system (where $n \geq 2$). This gives

$$(X_c)^n = \sum_{i=1}^{m} V_i (X_i)^n \tag{9.16}$$

where X_i may correspond to the physical/mechanical properties of matrix, reinforcement, or interfacial phases (Figs 9.1 and 9.2).

9.4 DEFORMATION BEHAVIOR OF UNIDIRECTIONAL COMPOSITES

Let us start this section by considering the uniaxial deformation of an arbitrary unidirectional composite reinforced with stiff elastic fibers.

During the initial stages of deformation, both the matrix and fibers deform elastically (Fig. 9.10). Furthermore, since the axial strains in the matrix and fiber are the same (iso-strain condition), then the stresses for a given strain, $\sigma(\varepsilon)$, are given simply by Hooke's law to be

$$\sigma_c(\varepsilon) = E_c\varepsilon = V_m E_m \varepsilon + V_f E_f \varepsilon \tag{9.17a}$$

$$\sigma_m(\varepsilon) = V_m E_m \varepsilon \tag{9.17b}$$

$$\sigma_f(\varepsilon) = V_f E_f \varepsilon \tag{9.17c}$$

where $\sigma(\varepsilon)$ denotes the stress corresponding to a given strain, ε, and subscripts c, m, and f correspond, respectively, to the composite, matrix, and fiber. The composite modulus in the elastic regime may also be estimated from Eq. (9.17a) by dividing by the uniaxial strain, ε. Also, each of the constituents in the composite will deform elastically until a critical (yielding or fracture) condition is reached in the matrix, fiber, or interface.

If we now consider the specific case of a ductile matrix composite reinforced with strong brittle fibers, matrix yielding is most likely to precede fiber fracture. In this case, the onset of composite yielding will correspond to the matrix yield strain, ε_{my}, as shown schematically in Fig. 9.11. Also, the composite yield stress, and the stresses in the matrix and fiber are given, respectively, by

$$\sigma_c(\varepsilon_{my}) = V_m \sigma_m(\varepsilon_{my}) + V_f \sigma_f(\varepsilon_{my}) \tag{9.18a}$$

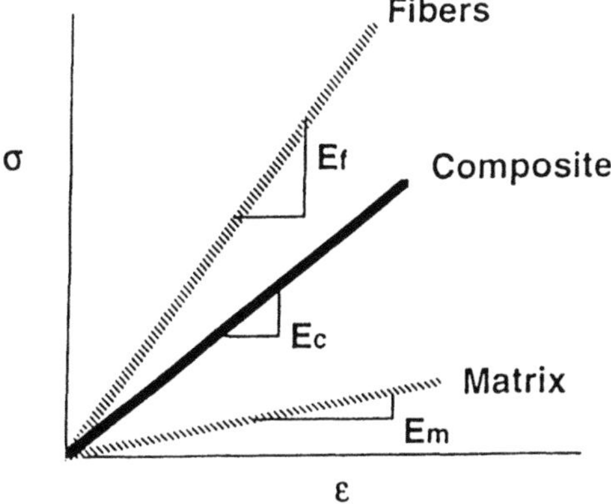

Figure 9.10 Stress–strain curves associated with uniaxial deformation of stiff elastic composite.

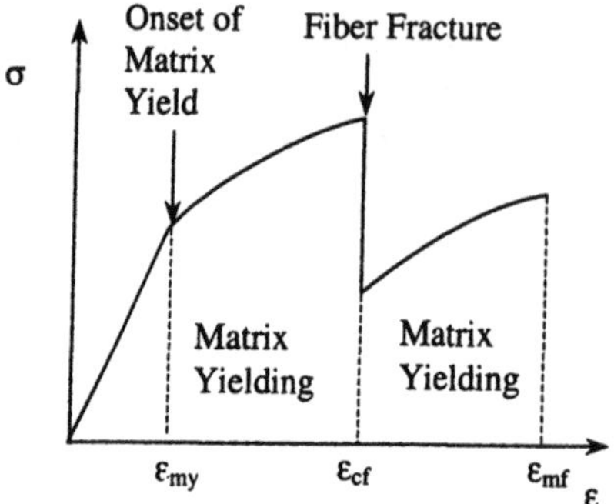

FIGURE 9.11 Schematic illustration of stages of deformation in an elastic–plastic composite.

$$\sigma_m(\varepsilon_{my}) = V_m\sigma_m(\varepsilon_{my}) \tag{9.18b}$$

$$\sigma_f(\varepsilon_{my}) = V_f\sigma_f(\varepsilon_{my}) \tag{9.18c}$$

where the subscript "my" corresponds to the matrix yield stress and subscripts c, m, and f denote composite, matrix, and fiber, respectively. Following the onset of matrix yielding, codeformation of the matrix and fibers continues until the fiber fracture strain, ε_{cf}, is reached. The stresses corresponding to this strain are again given by constant-strain rule-of-mixtures to be

$$\sigma_c(\varepsilon_{cf}) = V_m\sigma_m(\varepsilon_{cf}) + V_f\sigma_f(\varepsilon_{cf}) \tag{9.19a}$$

$$\sigma_m(\varepsilon_{cf}) = V_m\sigma_m(\varepsilon_{cf}) \tag{9.19b}$$

$$\sigma_f(\varepsilon_{cf}) = V_f\sigma_f(\varepsilon_{cf}) = V_f E_f \varepsilon_{cf} \tag{9.19c}$$

where all the above variables have their usual meaning. It is important to note that the onset of fiber fracture in many composite systems often coincides with the onset of catastrophic failure in the composite, Fig. 9.12(a). However, in other composites, matrix deformation may continue after fiber fracture. Such extended matrix deformation may continue until final failure occurs by matrix fracture. The resulting stress–strain behavior is shown schematically in Fig. 9.12(b).

The behavior shown schematically in Figs 9.11 and 9.12 may occur in either a ductile metal/polymer or matrix composites reinforced with strong/stiff elastic fibers. However, in the case of brittle matrix composites, such as a ceramic matrix or thermoset polymer matrix composites reinforced with strong/stiff fibers, the composite deformation is often restricted to the elastic

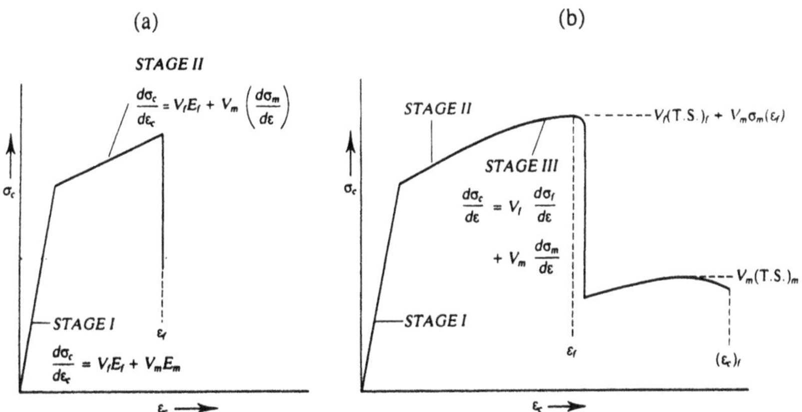

FIGURE 9.12 Schematic illustration of possible stress–strain curves in elastic–plastic composites: (a) composite failure coincides with fiber failure; (b) matrix deformation continues after fiber fracture. (From Courtney, 1990. Reprinted with permission from McGraw-Hill.)

regime, i.e., elastic deformation is truncated by composite failure. The above summary therefore provides a general framework for an appreciation of composite deformation in the different types of unidirectional composites that can be obtained by reinforcement with strong, brittle fibers.

9.5 MATRIX VERSUS COMPOSITE FAILURE MODES IN UNIDIRECTIONAL COMPOSITES

Depending on the volume fraction of matrix and fiber, and the ductilities of the matrix and fiber materials, different failure modes may occur in unidirectional composite materials. The composite "strength" will also depend on the sequence of matrix and fiber fracture. If fiber fracture occurs before matrix fracture, then the composite strength, σ_c, is given simply by the matrix contribution to be

$$\sigma_c = V_m \sigma_m(\varepsilon_{mf}) \tag{9.20}$$

where $\sigma_m(\varepsilon_{mf})$ is the stress in the matrix at the matrix fracture strain, ε_{mf}. Conversely, if matrix fracture somehow occurs before fiber failure, then the composite strength is given by the remaining fiber bundle strength:

$$\sigma_c = V_f \sigma_f(\varepsilon_{cf}) \tag{9.21}$$

Finally, if matrix and fiber failure occur simultaneously at the same strain, ε_c, then the composite fracture strength is given simply by constant-strain rule of mixtures to be

$$\sigma_c = V_m \sigma_m(\varepsilon_c) + V_f \sigma_f(\varepsilon_c) \tag{9.22}$$

The above summary provides a simple picture of the application of constant-strain rule of mixtures to the analysis of possible failure modes in unidirectional composites. However, it neglects the effects of stress concentrations (due to reinforcement geometries) and geometric constraints that are inherent in composite deformation. Nevertheless, such simple understanding of composite deformation and failure phenomena is an essential prerequisite to the development of an intuitive understanding of composite behavior.

In the case of fiber-reinforced composites, the failure modes are strongly affected by the volume fraction of fiber. This is illustrated in Fig. 9.13 in which the stress levels corresponding to matrix-dominated failure (composite failure at the matrix fracture strain) and composite-dominated failure are plotted. The composite failure stress increases linearly with increasing fiber volume fraction, since the fiber strength is generally greater than the matrix strength. However, the stress for matrix-dominated failure decreases continuously with increasing fiber volume fracture, since the matrix volume fraction undergoes a corresponding decrease as the fiber volume fraction increases ($V_m = 1 - V_f$).

Fig. 9.13 shows that composite strength is determined by the matrix-dominated failure locus for small fiber volume fractions. This is because the stresses required for matrix failure in this regime exceed those required for composite failure at low fiber volume fractions. The fiber volume fraction above which the composite failure stress exceeds the matrix-dominated failure stress is denoted by V_{min} in Fig. 9.13. This is obtained by equating the equation for composite failure Eq. (9.22) to that for matrix-dominated failure, Eq. (9.20):

$$V_c \sigma_f(\varepsilon_{cf}) + (1 - V_c)\sigma_m(\varepsilon_{cf}) = (1 - V_c)\sigma_m(\varepsilon_{mf}) \tag{9.23a}$$

or

$$V_c = \frac{\sigma_m(\varepsilon_{mf}) - \sigma_m(\varepsilon_{cf})}{\sigma_f(\varepsilon_{cf}) + \sigma_m(\varepsilon_{mf}) - \sigma_m(\varepsilon_{cf})} \tag{9.23b}$$

Similarly, we may obtain a condition for the minimum fiber volume fraction, V_{min}, at which the composite fracture strength, $\sigma_c(\varepsilon_{mf})$, exceeds the matrix fracture strength, $\sigma_m(\varepsilon_{mf})$:

$$\sigma_m(\varepsilon_{mf}) = (1 - V_{min})\sigma_m(\varepsilon_{cf}) + V_{min}\sigma_f(\varepsilon_{cf}) \tag{9.24a}$$

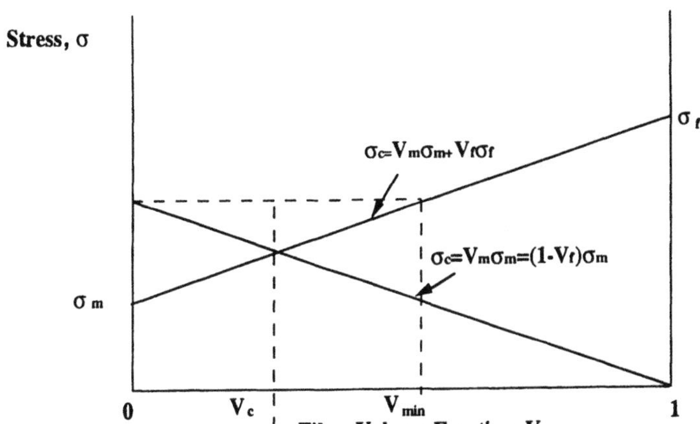

FIGURE 9.13 Loci of stress levels corresponding to matrix-dominated and composite failure modes.

or

$$V_{\text{min}} = \frac{\sigma_{\text{m}}(\varepsilon_{\text{mf}}) - \sigma_{\text{m}}(\varepsilon_{\text{cf}})}{\sigma_{\text{f}}(\varepsilon_{\text{cf}}) - \sigma_{\text{m}}(\varepsilon_{\text{cf}})} \tag{9.24b}$$

Typical values for V_c and V_{min} range between 0.02 and 0.10. Hence, relatively small volume fractions of fiber reinforcement are needed to improve the strengths of unidirectional fiber-reinforced composites, compared to those of the matrix. Also, composite failure modes are likely to occur in composites with volume fractions greater than V_c (Fig. 9.13).

9.6 FAILURE OF OFF-AXIS COMPOSITES

So far, we have focused primarily on the deformation behavior of unidirectional fiber-reinforced composites. However, it is common in several applications of composite materials to utilize fiber architectures that are inclined at an angle to the loading axis. Such off-axis composites may give rise to different deformation and failure modes, depending on the orientation of the fibers with respect to the loading axis.

To appreciate the possible failure modes, let us start by considering the loading of the arbitrary off-axis composite shown schematically in Fig. 9.14. The uniaxial force vector, F, may be resolved into two components: $F \cos \phi$ and F $\sin \phi$. The component $F \cos \phi$ results in loading of the fibers along the fiber axis. Since the area normal to the fibers is $A_0/\cos\phi$, then the force acting on the plane parallel to the fiber direction is given by

$$\sigma_\phi = \frac{F\cos\phi}{(A_0/\cos\phi)} = \frac{F}{A_0}\cos^2\phi = \sigma_0\cos^2\phi \tag{9.25}$$

where σ_ϕ is the stress along the fiber direction, is the angle of inclination of the fibers, σ_ϕ is the applied axial stress ($\sigma_0 = F/A_0$), and the other variables have their usual meaning. The resolved component of the applied stress along the fiber axis is given by Eq. (9.25). This results in the deformation of planes in the fiber and matrix that are perpendicular to the fiber direction.

Since the fibers are brittle, fiber failure will eventually occur when σ_ϕ is equal to the tensile strength of a composite inclined along the 0° orientation, i.e., at a stress level corresponding to the strength of a unidirectional fiber-reinforced composite. If we assume that this tensile strength is given by $(\text{T.S.})_0$, then the tensile strength corresponding to fiber failure in a composite reinforced with fibers inclined at an angle, ϕ, is given by Eq. (9.25) to be

$$(\text{T.S.})_\phi = \frac{(\text{T.S.})_0}{\cos^2\phi} \tag{9.26}$$

Since cos ϕ is less than 1 for $\phi > 0$, $(\text{T.S.})_\phi$ will increase with increasing ϕ between 0 and 90°. Hence, the composite strength will increase initially with increasing ϕ, for small values of ϕ, as shown in Fig. 9.15.

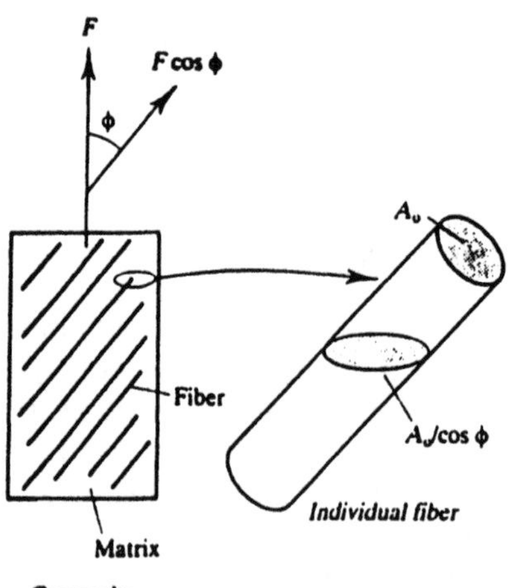

FIGURE 9.14 Arbitrary off-axis composite configuration showing deformed areas.

In addition to the axial loading on the fibers, the force component, $F\cos\phi$, will induce shear stresses on the planes that are parallel to the fiber directions. The matrix shear force, $F\cos\phi$, will act on cross-sectional areas of magnitude $A_0/\sin\phi$. Hence, the matrix shear stress, τ_ϕ, is given by

$$\tau_\phi = \frac{F\cos\phi}{(A_0/\sin\phi)} = \frac{F}{A_0}\sin\phi\cos\phi = \sigma_0\sin\phi\cos\phi \qquad (9.27)$$

In ductile matrix composites, the applied shear stress may cause matrix yielding to occur when the matrix shear yield strength, τ_{my}, is reached. For many materials, τ_{my} is approximately equal to half of the uniaxial matrix yield strength, σ_{my}. More precisely, the shear yield stress is given by the Von Mises yield criterion to be $\sigma_y/\sqrt{3}$, where σ_y is the uniaxial yield stress. In any case, the applied stress required to cause failure by matrix yielding is given by

$$\sigma_0 = \frac{\tau_{my}}{\sin\phi\cos\phi} \qquad (9.28)$$

Variations in σ_0 for matrix yielding by shear are illustrated in Fig. 9.15. Note that σ_0 for matrix shear failure is initially greater than the corresponding values for fiber failure for small values of ϕ. However, as ϕ increases, a critical condition is reached at which the stresses required for

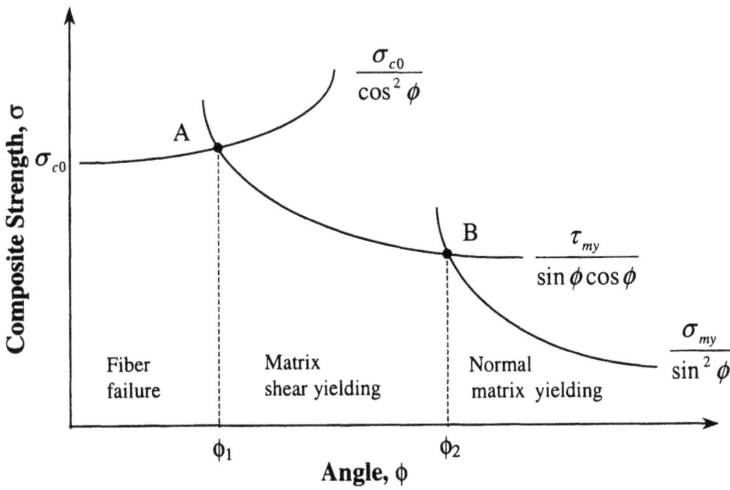

Figure 9.15 Dependence of composite strength and failure modes on fiber orientation.

fiber failure and matrix yielding are equal. This critical condition (for the transition from fiber-dominated failure to matrix shear yielding) is given by

$$\frac{\sigma_0}{\cos^2\phi} = \frac{\tau_{my}}{\sin\phi\cos\phi} \tag{9.29a}$$

or

$$\phi_{c1} = \tan^{-1}\left(\frac{\tau_{my}}{\sigma_0}\right) \tag{9.29b}$$

where ϕ_{c1} is the angle at which the transition occurs, σ_0 is the strength of the composite in the zero-degree orientation, i.e., $\sigma_0 = (\text{T.S.})_0 = V_m\sigma_m(\varepsilon_f) + V_f\sigma_f(\varepsilon_f)$. Typical values for ϕ_{c1} (Fig. 9.15) for most composites are below 10°. It is important to note here that fiber fracture dominates for $\phi > \phi_{c1}$, while matrix shear yielding occurs for $\phi > \phi_{c1}$, as shown in Fig. 9.15.

At even higher fiber angles ($\phi > \phi_{c1}$), a second transition can occur from matrix shear yielding to matrix failure in a direction normal/perpendicular to the fiber direction. The matrix normal stresses can be found by dividing the load component, $F\sin\phi$, by the area $A_0/\sin\phi$ (Fig. 9.14). The normal matrix stress, σ_n, is thus given by (Fig. 9.14) to be

$$\sigma_n = \frac{F\sin\phi}{(A_0/\sin\phi)} = \frac{F}{A_0}\sin^2\phi \tag{9.30}$$

For ductile matrix composites, matrix failure normal to the fiber direction occurs when σ_n is equal to the uniaxial matrix yield strength, σ_{my}. Hence, rearranging Eq. (9.30), and substituting σ_{my} with σ_n gives

$$\sigma_0 = \frac{\sigma_{my}}{\sin^2\phi} \tag{9.31}$$

where σ_0 is the applied stress required for failure to occur by normal matrix yielding. This stress becomes lower than that required for matrix shear yielding at higher values of σ (Fig. 9.15). The transition from matrix shear to normal matrix yielding occurs when σ_0 for matrix shear and normal matrix yielding are equal. From Eqs (9.28) and (9.31), this is given by

$$\frac{\sigma_{my}}{\sin^2\phi} = \frac{\tau_{my}}{\sin\phi\cos\phi} \tag{9.32a}$$

or

$$\phi_{c2} = \cot^{-1}\left(\frac{\tau_{my}}{\sigma_{my}}\right) \tag{9.32b}$$

Since $\tau_{my} \sim \sigma_{my}/2$, typical values for ϕ_{c2} at the transition from matrix shear to normal matrix yielding are $\sim \cot^{-1}(1/2) \sim 63.4°$. In summary, the composite strength depends strongly on fiber orientation angle, σ, as shown schematically in Fig. 9.15. Also, the strength dependence on ϕ will be determined by the underlying mechanisms of failure, and the critical conditions required for the transition from one mechanism to another.

9.7 EFFECTS OF WHISKER/FIBER LENGTH ON COMPOSITE STRENGTH AND MODULUS

So far, our discussion has focused on the behavior of long fiber-reinforced composites. However, such composites are often too expensive for practical applications, in spite of the attractive combinations of strength and stiffness that can be engineered by the judicious selection of appropriate fiber and matrix materials. In cases where moderate strength/stiffness and low/moderate cost are of the essence, it may be desirable to use composites that are reinforced with short fibers, which are also referred to as whiskers.

An example of a titanium boride (TiB) whisker-reinforced titanium matrix composite is presented in Fig 9.16. This shows aligned TiB whiskers in a titanium alloy matrix. The whiskers have been aligned by an extrusion process (Dubey et al., 1997). The resulting composite properties are intermediate between those of the titanium matrix alloy and an equivalent fiber-reinforced composite. However, the whisker-reinforced composite is much cheaper than possible fiber-reinforced composite alternatives.

Let us now return to answer the original question of why the whisker-reinforced composite has lower strength and modulus. We will begin by extracting a representative volume element or unit cell from the whisker-reinforced composite structure. This is illustrated in Figs 9.17(a) and (b). The representative volume element or unit cell is a microstructural configuration that captures the volume fraction and spatial geometry of the composite. Once it is obtained, we may proceed to do a force balance analysis to determine the stresses along the fibers.

Consider the conditions required for horizontal force equilibrium in the free body diagram shown in Figs 9.17(c) and (d). For force equilibrium, the axial force in the whisker must be balanced by the shear force in the matrix or interface, Fig. 9.17(d):

$$\frac{\pi d^2}{4}(\sigma + d\sigma - \sigma) - \tau(\pi d)dx = 0 \tag{9.33}$$

Simplifying Eq. (9.32) and integrating between appropriate limits give

$$\int_0^x d\sigma = \int_0^x \frac{4\tau}{d} dx \tag{9.34}$$

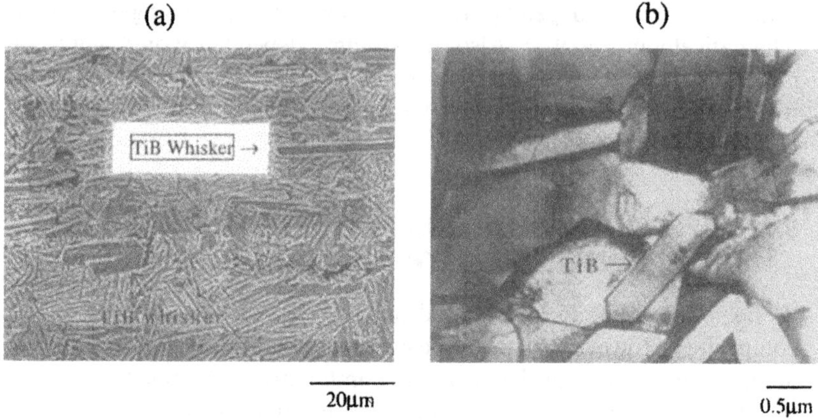

Figure 9.16 (a) SEM micrograph of Ti–6Al–4V–0.5B (704°C/1 h/AC) showing $\alpha + \beta$ Widmanstatten structure and second-phase TiB whiskers aligned in the extrusion direction; (b) TEM micrograph of the undeformed Ti–6Al–4V–0.5B showing α grains in a β matrix as well as TiB whiskers. Note that AC corresponds to air cool and the Widmanstatten structure consists of aligned colonies of x platelets with prior β grains. (From Dubey et al., 1997. Reprinted with permission from Elsevier.)

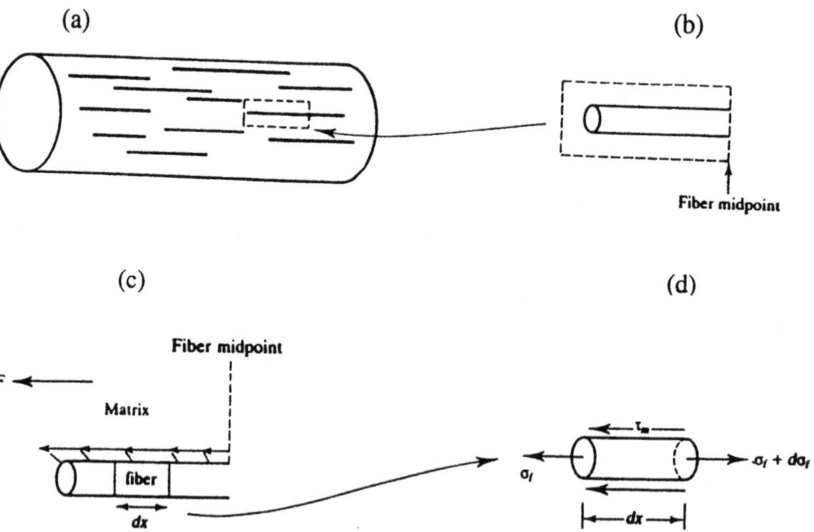

Figure 9.17 Sections of (a) whisker-reinforced composite, (b) half of a representative whisker, (c) loading, and (d) free body diagram.

For regions in the middle of the fiber, τ can be assumed to be constant. However, close to the fiber ends, τ may vary significantly in the matrix. Similarly, the stresses at the ends of the whiskers/fibers are equal to zero. Hence, neglecting the variations in τ near the fiber ends, Eq. (9.34) may be integrated to obtain the following approximate solution for the stress in the fiber, σ_f:

$$\sigma_f = \left(\frac{4\tau}{d}\right)x \tag{9.35}$$

Equation (9.35) suggests a linear increase in stress with increasing distance, x, from by the fiber ends. By symmetry, a similar linear increase in σ_f will be expected from the other end of the short fiber shown in Fig. 9.17. Hence, for small fiber/whisker lengths ($\ell < \ell_c$), the stress distribution along the whisker will have a triangular profile, as shown in Fig. 9.18. The peak fiber stress for short fibers/whiskers will, therefore, occur at the center of the whisker/short fiber.

As the whisker length increases, the peak stress at the center of the whisker increases, as shown in Fig. 9.18. This continues until the peak stress reaches the value that would be expected in a long fiber under constant strain conditions, $\sigma_f(\varepsilon_f)$. The fiber length corresponding to this critical condition corresponds to ℓ_c in Fig. 9.18. For values of $\ell > \ell_c$, the linear increase in stress occurs from both ends of the whisker/short fiber until $\sigma_f(\varepsilon_f)$ is reached at $x = \ell_c/2$. A constant fiber stress is then maintained for values of $x > \ell_c/2$. This is illustrated in Fig. 9.18 with the trapezoidal profiles for which $\ell > \ell_c$.

It should be clear from Fig. 9.18 that the average fiber stress, $\bar{\sigma}_f$, depends on whether the fiber length is less than or greater than ℓ_c. For fibers with lengths greater than ℓ_c, the average fiber stress is given by Fig. 9.18a to be:

$$\sigma_f = \sigma_f(\varepsilon_c)\left[1 - \left(\frac{\ell_c}{\ell}\right)\right] + \frac{1}{2}\sigma_f(\varepsilon_c)\left[\left(\frac{\ell_c}{2\ell}\right)\right] \tag{9.36}$$

Hence, applying the rule of mixtures to the composite with $\ell > \ell_c$ gives the composite strength to be

$$\sigma_c(\varepsilon_c) = V_f\bar{\sigma}_f + V_m\sigma_m(\varepsilon_c) = V_f\sigma_f(\varepsilon_c)\left[1 - \left(\frac{\ell_c}{2\ell}\right)\right] + V_m\sigma_m(\varepsilon_c) \tag{9.37}$$

where $\sigma_c(\varepsilon_c)$ is the stress in the composite at the critical condition at which long-fiber conditions are reached, and $\sigma_f(\varepsilon_c)$ are the corresponding stresses in the fiber and matrix, respectively. The expression $[\ell - (\ell_c/2\ell)]$ can be regarded as a fiber efficiency factor, η_f. Hence, substituting η_f into Eq. (9.37) gives

$$\sigma_c(\varepsilon_c) = \eta_f V_f\sigma_f(\varepsilon_c) + V_m\sigma_m(\varepsilon_c) \tag{9.38}$$

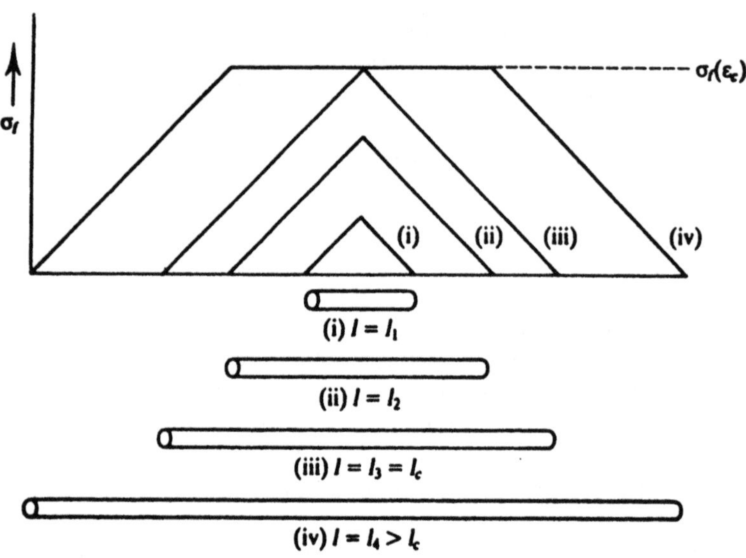

FIGURE 9.18 Stress variations along the length of a whisker/short fiber. (From Matthews and Rawlings, 1994. Reprinted with permission from Chapman and Hall.)

Similarly, we may consider the case of a subcritical short fiber/whisker-reinforced composite with $\ell \leqslant \ell_c$ (Fig. 9.18). In this case, the average fiber stress for the triangular profile is given by

$$\bar{\sigma}_f = \frac{1}{2}\sigma_f(\varepsilon)\frac{\ell}{\ell_c} \tag{9.39}$$

As before, the composite stress is given by simple rule of mixtures to be

$$\sigma_c(\varepsilon) = V_m\sigma_m(\varepsilon) + V_f\bar{\sigma}_f(\varepsilon) = V_m\sigma_m(\varepsilon) + \frac{1}{2}V_f\sigma_f(\varepsilon)\frac{\ell}{\ell_c} \tag{9.40}$$

Also, $\ell/(2\ell_c)$ may be considered to be a fiber efficiency factor, η_f, for whiskers of length, $\ell \leqslant \ell_c$. Hence, Eq. (9.40) may now be expressed as

$$\sigma_c(\varepsilon) = V_m\sigma_c(\varepsilon) + \eta_f V_f\sigma_f(\varepsilon) \tag{9.41}$$

The strengths of composites with whisker geometries may, therefore, be expressed in terms of constant-strain rule of mixtures and fiber efficiencies. Furthermore, expressions for composite moduli may be derived by noting that $\sigma = E\varepsilon$ for the composite, matrix, and fibers:

$$E_c = V_m E_m + \eta_f V_f E_f \tag{9.42}$$

Equation (9.42) applies to all lengths of whiskers ($\ell \leqslant \ell_c$ and $\ell > \ell_c$) provided that the appropriate expressions are used in the estimation of η_f.

So far, the equations presented in this section have been derived for aligned whiskers, as shown in Figs 9.17 and 9.18. However, in cases where the whiskers are randomly oriented [Fig. 9.6(c)], an orientation efficiency factor, η_0, must be introduced (Matthews and Rawlings, 1994). Detailed derivations of η_0 are beyond the scope of this text. It is simply sufficient to note here that the whisker orientation factors account for the average decrease in composite strength (in any given direction) due to the random orientations of the fibers. When this is taken into account, the modified rule-of-mixture expressions for composite strength and modulus are given by

$$\sigma_c(\varepsilon) = V_m \sigma_m(\varepsilon) + \eta_0 \eta_f V_f \sigma_f(\varepsilon) \tag{9.43}$$

and

$$E_c = V_m E_m + \eta_0 \eta_f V_f E_f \tag{9.44}$$

Typical values of η_0 are 0.375 for random in-plane two-dimensional arrays and 0.2 for three-dimensional random arrays. Also, η_0 is 1 for aligned longitudinal whiskers, and values of η_f are between 0 and 1.

9.8 CONSTITUENT AND COMPOSITE PROPERTIES

The rule-of-mixture expressions presented in the preceding sections can be used to facilitate our understanding of the effects of the constituents on composite strength and modulus. In most engineering composites, polymer or metal matrices are reinforced with strong/brittle fibers. However, our discussion in this section will be more general in nature. We will examine the properties of composite constituents, and how the constituent properties contribute to composite behavior.

9.8.1 Fibers and Matrix Materials

In Chap. 6, we showed that the theoretical strength of a solid is $\sim G/2\pi$, where G is the shear modulus. However, due to the existence of defects, the actual measured strengths of solids are generally a few orders of magnitude below the predicted theoretical strengths. In an effort to develop strengths that are closer to theoretical values, special processing techniques have been developed for the fabrication of composite fibers and whiskers with low defect content (small crack sizes).

The importance of defect size is illustrated in Fig. 9.19 in which composite strength is plotted against flaw size. Since the maximum possible crack size per unit volume increases with increasing fiber size, fiber strengths

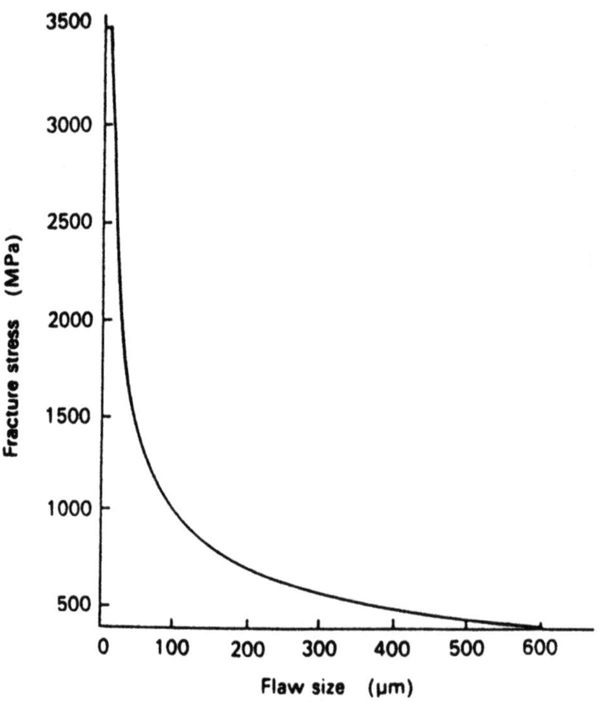

Figure 9.19 Variation in whisker/fiber strength as a function of flaw size.

will decrease with increasing fiber length. This is because failure is more likely to initiate from larger flaws, which are more likely to exist in longer fibers. The mechanical properties of fibers, therefore, exhibit statistical variations. These statistical variations are often well described by Weibull distributions (Weibull, 1951).

Typical values of the strengths and moduli for selected composite fibers are compared with those of their monolithic counterparts in Table 9.1. Note that the fiber strengths are approximately one order of magnitude greater than the strengths of the monolithic materials. Also, depending on the molecular orientation and structure of organic fibers (such as carbon and polyethylene fibers), the moduli of the fibers and their monolithic counterparts may be significantly different. This is because the long-chain polymer structures have stiffnesses that depend on how their covalent bonds and Van der Waals' forces are oriented with respect to the applied loads. Stiff polymer fibers can, therefore, be engineered by the careful orientation of strong covalent bonds along the fiber direction.

Table 9.3 Typical Mechanical and Physical Properties of Bulk Monolithic Materials

	Young's modulus (GPa)	Yield stress (MPa)	Tensile strength (MPa)	Ductility (%)
Al–Cu–Mg (2618A)	74	416	430	2.5
Al–Cu–Mg + 20%Al_2O_3	90		383	0.8
Al–Zn–Mg	~ 70		273	11.5
Al–Zn–Mg + 25%Al_2O_3	80 +		266	1.5
Titanium (wrought)	120	200	400	25
Titanium +35%SiC	213		1723	<1
Ti–Al–V (wrought)	115	830	1000	8
Ti–Al–V+35%SiC	190		1434	0.9

9.8.2 Composites

With the understanding of mixture rules developed in the earlier sections, it should be clear that the incorporation of stiff/strong fibers can be used to engineer composites with higher overall strengths and moduli than their monolithic counterparts. The strength and modulus of a composite may, therefore, be increased by increasing the volume fraction of strong/stiff fibers.

However, it is generally difficult to process uniform composite architectures with fiber volume fractions that are greater than 50–60%. This is because the viscosity of the composite "mix" (during composite processing) increases with increasing fiber volume fraction. This increase in viscosity makes it very difficult to achieve homogeneous mixing as the volume fraction of stiff/strong reinforcements is increased beyond about 50–60%. Furthermore, fibers are more likely to "swim" during processing, and thus come into contact after the fabrication of composites with reinforcement volume fractions greater than 50–60%. For these reasons, most fiber-reinforced composite systems are limited to maximum fiber volume fractions of ~ 50–60%.

9.8.2.1 Polymer Composites

Once the fibers are incorporated into the composite structure, matrix loads are transmitted to the fibers by shear. Since the fibers are stronger, they will support greater loads than the matrix can. This means that the load-carrying ability of most composites is provided by the fibers. This is always the case for polymer matrix composites in which the matrix strength and moduli are generally much less than those of the fibers (Table 9.4). The resulting composite properties are, therefore, dependent on the fiber properties, and the polymer matrix serves mostly as a "glue" that keeps the structure bonded, and the fibers separated from each other. The bonding between the matrix and the fiber materials also enables stresses to be transmitted from the matrix to the fiber via shear.

Since polymers have relatively open structures, they are typically less dense than ceramics and metallic materials (Fig. 9.4). Furthermore, the density, ρ, of a composite may be estimated from simple rule of mixtures:

$$\rho_c = V_m\rho_m + V_f\rho_f \tag{9.45}$$

where subscripts c, m, and f denote composite, matrix, and fiber, respectively. Composites reinforced with a significant fraction of less dense polymers will, therefore, have densities lower than those of most metal alloys and ceramic materials (Fig. 9.4).

For several applications of composites in which light weight is desirable, in addition to absolute strength of stiffness, it is useful to consider density normalized strength and modulus values. These are often referred to as specific strength and specific modulus, respectively. Due to their relatively low densities, polymer matrix composites have relatively high specific strengths. They are, therefore, used in the wings of military aircraft such as the Harrier jet.

The incorporation of stiff fibers into polymer matrices raises the composite moduli to levels that are sufficient to make engineering composites (mostly polymer matrix composites) very competitive with metallic and ceramic materials. This is shown clearly in Fig. 9.3. The most commonly used polymer matrix composites are epoxy matrix composites reinforced with stiff glass fibers (GFRPs) or carbon fibers (CFRPs).

It is of interest to note the dashed lines corresponding to the performance indicators (E/ρ, $E^{1/2}/\rho$, and $E^{1/3}/\rho$) in Fig. 9.3. These provide a normalized measure of how well a given system will perform under: tension without exceeding a design load (E/ρ); compression without buckling ($E^{1/2}/\rho$); and bending with minimum deflection ($E^{1/3}/\rho$). Further details on performance indices may be found in an excellent text by Ashby (1999).

9.8.2.2 Ceramic Matrix Composites

Due to their strong ionically and/or covalently bonded structures, ceramics tend to be relatively strong and stiff, compared to metals and polymers. However, ceramics are brittle, and are susceptible to failure by the propagation of pre-existing cracks. For this reason, there are relatively few applications of ceramic matrix composites (compared to those of polymer matrix

TABLE 9.4 Typical Mechanical Properties of Polymer Matrix Composites and Polymer Matrix Materials

	Density (mg/m^3)	Young's modulus (GPa)	Tensile strength (MPa)	Ductility (%)	Flexural strength (M Pa)	Specific modulus [(GPa)/(mg/m^3)]	Specific strength [(MPa)/(mg/m^3)]
Nylon 66 + 40% carbon fiber	1.34	22	246	1.7	413	16	184
Epoxide + 70% glass fibers							
unidirectional—longitudinal	190	42	750		1200	22	395
unidirectional—transverse	1.90	12	50			6	26
Epoxy +60% Aramid	1.40	77	1800			55	1286
Poly(ether imide) + 52%Kevlar		54	253				
Polyester + glass CSM	1.50	7.7	95		170	5	63
Polyester + 50%glass fiber							
undirectional–longitudinal	1.93	38	750	1.8		20	389
unidirectional—transverse	1.93	10	22	0.2		5	11

composites). Most of the applications of ceramic matrix composites take advantage of their excellent high-temperature properties. They include applications in nozzles, brakes, and heat shields such as the tiles of the space shuttle. Nevertheless, structural applications of ceramic composites have been difficult due to the problems associated with their inherent brittleness.

Typical strengths and moduli of selected ceramic matrix composites are presented in Table 9.5. Their moduli are relatively high due to the strong/stiff nature of the ionically or covalently bonded structures. However, the tensile strengths of these composites are moderate, due to the inherent susceptibility to internal/inherent populations of microcracks in the ceramic matrices.

9.8.2.3 Metal Matrix Composites

Unlike ceramics, metals are generally very ductile. Nevertheless, there are only a few applications of metal matrix composites (compared to numerous applications of polymer matrix composites) in engineering structures and components. The applications include connecting rods, struts, pistons, and valves in automobile engines (Saito et al., 1998). One example of a recent application of a titanium matrix composite is shown in Fig. 9.20. This shows an automotive valve fabricated from a low-cost in-situ titanium matrix composite reinforced with TiB whiskers (Fig. 9.16). The valve is currently being used in Toyota Alzetta motor vehicles in Japan. This selection of in-

Table 9.5 Typical Mechanical Properties of Ceramic Matrix Composites and Ceramic Materials

	Young's modulus (GPa)	Strength (MPa)	Toughness, K_{IC} (MPa $m^{1/2}$)
Alumina (99%purity)	340	300	4.5
Alumina + 25%SiC whiskers	390	900	8.0
Borosilicate glass (Pyrex)		70	0.7
Pyrex + 40%Al_2O_3 CF[a]		305	3.7
LAS[b] glass–ceramic	86	160	1.1
LAs + 50%SiC CF	135	640	17.0
Mullite		244	2.8
Mullite + 20%Sic whiskers		452	4.4

[a]CF = continuous fibers.
[b]LAS = lithium aluminosilicate.

situ titanium matrix composite valves was due to the improved performance and fuel savings that was achieved by the replacement of steel valves (with a density of ~ 7.8 g/cm^3) with the former valves (with a density of ~ 4.5 g/cm^3). Further details on in-situ titanium matrix composites may be found in papers by Saito (1994), Soboyejo et al. (1994), and Dubey et al. (1997).

In general, however, the applications of metal matrix composites have been limited by their cost and limited durability. Most of the applications have involved the use of lightweight aluminum matrix composites reinforced with SiC or Al_2O_3 particulates or whiskers (Fig. 1.16a). These take advantage of the low density ($\rho \sim 2.7$ g/cm^3) and low cost (generally less than $1 per pound) of aluminum and its alloys. Since aluminum alloys typically have relatively low strengths compared to structural alloys such as steels, the reinforcement of aluminum matrices with SiC or Al_2O_3 can significantly improve the strengths and moduli of aluminum matrix composites to levels where they are very competitive with other structural alloys/composites.

The strengths of aluminum/aluminum alloys and composites are compared with those of other structural metal/alloys and composites in Table

Figure 9.20 In-situ titanium matrix composite valve. (Courtesy of Tadahiko Furuta of Toyota Corporation, Japan.)

TABLE 9.6 Typical Mechanical Properties of Metal Matrix Composites

	Young's modulus (GPa)	Yield stress (MPa)	Tensile strength (MPa)	Ductility (%)
Al–Cu–Mg (2618A)	74	416	430	2.5
Al–Cu–Mg + 20%Al_2O_3	90		383	0.8
Al–Zn–Mg	~70		273	11.5
Al–Zn–Mg + 25%Al_2O_3	80 +		266	1.5
Titanium (wrought)	120	200	400	25
Titanium + 35%SiC	213		1723	<1
Ti–Al–V (wrought)	115	830	1000	8
Ti–Al–V + 35%SiC	190		1434	0.9

9.6. Also included in Table 9.6 are data for titanium matrix composites reinforced with SiC fibers. These have higher strengths and moduli than those of titanium alloys. However, they may undergo premature failure due to the initiation of damage from the brittle SiC fibers (Majumdar and Newaz, 1992; Soboyejo et al., 1997).

9.9 STATISTICAL VARIATIONS IN COMPOSITE STRENGTH

Due to the susceptibility of fiber strength to variations in crack populations, composite strengths may exhibit significant statistical variations. The sensitivity to defects (mostly cracks) is particularly strong in the case of brittle fibers. There is, therefore, a need to account for the variations in fiber/composite strength within a statistical framework. The statistical distributions that best describe the variations in strength depend to a large extent on ductility/brittleness. The variabilities in the strengths of most ductile phases are often well characterized by Gaussian distributions (Fig. 9.21). For such materials, $f(x)$, the frequency of failure at a given stress level, x, is given by [Fig. 9.21(a)]:

$$f(x) = \frac{1}{\sqrt{2\pi\mu}} \exp\left[-\frac{1}{2}\left(\frac{x-\overline{x}}{\mu}\right)^2\right] \tag{9.46}$$

where $\overline{x}$ is the mean stress, and μ is the standard deviation. The function $f(x)$ may be integrated to obtain an expression for the probability of failure, $F(x)$ or $P_f(x)$, at a given stress level, x:

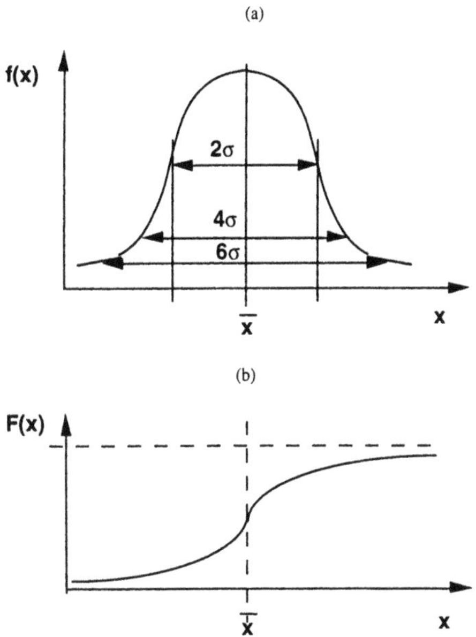

FIGURE 9.21 Normal or Gaussian probability density functions: (a) probability density function; (b) cumulative density function.

$$P_f(x) = \int_{-\infty}^{\sigma} f(x)\mathrm{d}x \tag{9.47}$$

The function $P_f(x)$, represents the cumulative density function, which is shown schematically in Fig. 9.21(b). The function $f(x)$ corresponds to the probability density function shown in Fig. 9.21(a). Conversely, the probability of survival, $P_s(x)$, may be determined from the following expression:

$$P_s(x) = 1 - P_f(x) = \int_{x}^{\infty} f(x)\mathrm{d}x \tag{9.48}$$

In the case of brittle materials, such as strong and stiff fibers, the statistical variations in strength do not often follow the Gaussian distribution. Instead, most brittle materials exhibit strength variations that are well characterized by a distribution function that was first proposed by Weibull (1951). The so-called Weibull distribution gives $P_s(V)$, the probability of survival of a stressed volume, V, that is subjected to a stress, x, as

$$P_s(V) = \exp\left\{-\frac{V}{V_0}\left(\frac{x - x_u}{x_0}\right)^m\right\} \tag{9.49a}$$

where V is the actual volume, V_0 is a reference volume, x is the applied stress, x_u is the stress corresponding to zero probability of failure, x_0 is the mean strength, and m is the Weibull modulus.

A schematic of the Weibull distribution is shown in Fig. 9.22(a); this shows the probability of survival plotted as a function of stress, x. Note that the probability of survival is 1 for $x = 0$, and 0 for $x = \infty$. Also, the median strength corresponds to $P_s(V_0) = 1/2$. Furthermore, for $V = V_0$ and $(x - x_u) = x_0$ in Eq. (9.49a), $P_s(V_0) = 1/e = 0.37$. Hence, at a stress of $x_0 + x_u$, 37% of the fibers will survive.

Equation (9.49a) may also be expressed in terms of the probability of failure, $P_f(V)$, of a stressed volume, V:

$$P_f(V) = 1 - P_s(V) = 1 - \exp\left\{-\frac{V}{V_0}\left(\frac{x - x_u}{x_0}\right)^m\right\} \tag{9.49b}$$

The Weibull modulus, m, is a key parameter in Eqs (9.49a) and (9.49b). It is a measure of the homogeneity of the strength data. Typical values of m are less than 10. Larger values of m are generally associated with less variable strength data, which would be expected from increased homogeneity.

Conversely, smaller values of m would be associated with increased variability or inhomogeneity. Typical values of m are between 1 and 10 for ceramics such as SiO_2, SiC, Al_2O_3 and Si_3N_4, and ~ 100 for most steels. In any case, taking natural logarithms twice on both sides of Eq. (9.49a) gives

$$\ell n\,\ell n\left(\frac{1}{P_s(V)}\right) = \ell n V - \ell n V_0 + m\ell n(x - x_u) - m\ell n x_0 \tag{9.50a}$$

or

$$\ell n\,\ell n\left(\frac{1}{P_s(V)}\right) = m\ell n X + C_1 \tag{9.50b}$$

where $C_1 = \ell n V - \ell n V_0 - m\ell n x_0$ and $X = x - x_u$. Also, C_1 is a constant for fixed values of V, V_0, x_u, and x_0. The value of m may be determined by plotting $\ell n\,\ell n(1/P_s(V))$ against $\ell n X$ on Weibull paper. The Weibull modulus may thus be determined from the negative of the slope of the Weibull plot. Typical Weibull plots are shown in Fig. 9.22(b).

It is important to note here that the median strength corresponds to $P_s(x) = P_f(x) = 0.5$. Similarly, depending on the defect content, two different stresses, x_1 and x_2, may be associated with the same failure or survival probabilities. Under such circumstances, substitution of x_1 and x_2 into Eq. (9.49a) gives

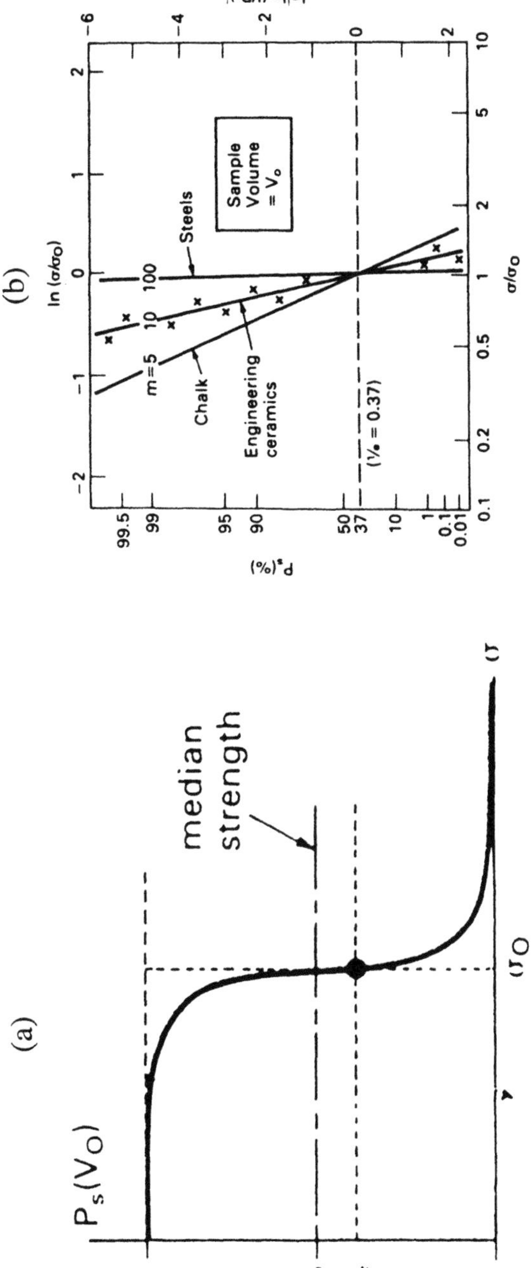

Figure 9.22 Weibull plots: (a) Weibull distribution function; (b) plot of survival probability on Weibull probability axes. (From Ashby and Jones, 1986. Reprinted with permission from Pergamon Press.

$$(x_2)^n V_1 = (x_2)^n V_2 \tag{9.51}$$

Equation (9.51) may be rearranged to obtain estimates of V or x. Appropriate statistical checks are needed to verify the applicability of any of the above distributions (Gaussian or Weibull) to the assessment of the variabilities in the strengths. These checks require the use of methods that are described in most introductory texts on statistics.

Finally in this section, it is interesting to comment on the behavior of fiber bundles, and the statistics of fiber-bundle fracture. Let us consider a fiber bundle that consists of N fibers that are subjected to an axial stress, σ. Before loading, the fiber bundle strength is σ_f. If the fibers are assumed to fail at the weakest link(s), with no load carrying capability in the individual broken fibers following the breaks, then we may obtain a first estimate of the fiber-bundle strength by considering the ratio of the remaining unbroken fibers, N, to the initial number of fibers, N_0. Hence, the fiber-bundle strength, σ_B, corresponding to N unbroken fibers is given by

$$\sigma_B = \frac{N}{N_0}\sigma_f \tag{9.52}$$

The probability of survival, S, at a stress of σ, is given by the ratio of N to N_0. Therefore, the probability of failure, $G(\sigma)$, may be obtained from

$$G(\sigma) = 1 - S(\sigma) = 1 - \left(\frac{N}{N_0}\right) \tag{9.53}$$

However, recalling from Eq. (9.52) that $\sigma_B = (N/N_0)\sigma_f$, Eq. (9.53) may now be rewritten as

$$\sigma_B = [1 - G(\sigma)]\sigma_f \tag{9.54}$$

The fiber-bundle strength, σ_B, is important because it characterizes the remaining strength of the fibers after fiber breaks. Since statistical fiber breaks are inherent to composite deformation, the estimation of composite properties is often based on fiber-bundle stress estimates that account for the effects of fiber breaks. Such estimates will generally result in lower composite strength levels than those predicted for undamaged fiber bundles. Hence, if the statistical function that describes the probability of failure, $G(\sigma)$, is known, then the fiber-bundle stress may be obtained from Eq. (9.54). The maximum fiber-bundle strength may also be obtained by differentiating Eq. (9.54) to obtain $d\sigma_B/d\sigma_f = 0$.

It is important to remember that the probability of failure is given by the expressions presented earlier for the different statistical distributions. For example, in the case of Gaussian distributions, the probability of failure, $G(\sigma)$, is given by Eq. (9.47). For strength variabilities that are well described by Weibull distributions, e.g., brittle fibers, $G(\sigma)$ is given by Eq. (9.49).

In general, however, the variabilities may not be well described by any of the well-known statistical functions (Gaussian, log normal, or Weibull). When this occurs, minimally biased entropy functions may be used to characterize the statistical variabilities, as proposed by Soboyejo (1973). More detailed descriptions of the statistics of fiber fracture may be found in papers by Sastry and Phoenix (1993, 1994), Curtin (1998) and Torquato (1991).

9.10 SUMMARY

This chapter presents a simple introduction to the deformation of composite materials. Following a brief review of the different types of composites, simple rule-of-mixture theories were introduced for the estimation of composite strength and moduli and the possible bounds in composite strength and moduli. A general framework was also described for the estimation of the physical properties of composites within a rule-of-mixture framework. Composite deformation behavior was then discussed before exploring the possible effects of fiber/whisker length and whisker orientation on composite strength and modulus. The failure modes in composites with off-axis fibers were examined before presenting an introduction to the statistical approaches that are used in the modeling of variabilities in individual fiber strengths and fiber-bundle strengths.

BIBLIOGRAPHY

Ashby, M.F. (1999) Materials Selection in Mechanical Design. Butterworth-Heineman, New York.

Ashby, M.F. and Jones, D.R.H. (1984) Engineering Materials 1. Pergamon Press, New York.

Ashby, M.J. and Jones, D.R.H. (1986) Engineering Materials 2. Pergamon Press, New York.

Courtney, T.H. (1990) Mechanical Behsvior of Materials. McGraw-Hill, New York.

Curtin, W.A. (1998) Adv Appl Mech. vol. 36, p 164.

Dubey, S., Lederich, R.J., and Soboyejo, W.O. (1997) Metall Mater Trans. vol 28A, pp 2037–2047.

Majumdar, B. and Newaz, G. (1992) Phil Mag A. vol. 66, pp 187–212.

Matthews, F.L. and Rawlings, R.D. (1994) Composite Materials: Engineering and Science. Chapman and Hall, New York.

Saito, T., Furuta, T., and Yamuguchi, T. (1994) Developments of low cost titanium alloy metal matrix composites. In: Recent Advances in Titanium Matrix Composites. TMS, Warrendale, PA, pp 33–40.

Sastry, A.M. and Phoenix, S.L. (1993) J Mater Sci Lett. vol. 12, pp 1596–1599.

Sastry, A.M. and Phoenix, S.L. (1994) SAMPE J. vol. 30, pp 60–61.
Shyue, J., Soboyejo, W.O., and Fraser, H.L. (1995) Scripta Metall Mater. vol 32, pp 1695–1700.
Soboyejo, A.B.O. (1973) Mater Sci Eng. vol. 12, pp 101–109.
Soboyejo, W.O., Lederich, R.J., and Sastry, S.M.L. (1994) Acta Mater. vol. 42, pp 2579–2591.
Soboyejo, W.O., Rabeeh, B.M., Li, Y., Chu, Y.C., Lavrentenyev, A., and Rokhlin, S.I. (1997) Metall Mater Trans. vol. 28A, pp 1667–1687.
Torquato, S. (1991) Appl Mech Rev. vol. 44, p 37.
Voigt, W. (1889) Wied Ann. vol. 38, pp 573–587.

10

Further Topics in Composites

10.1 INTRODUCTION

The introduction to composite deformation provided in the last chapter is adequate for the development of an intuitive understanding of composite deformation. However, the simple expressions presented in the last chapter cannot be used easily in the analysis of multi-ply composites with plies of arbitrary orientation. Furthermore, the models are only suitable for composites that contain simple reinforcement geometries, and the rule-of-mixture expressions provide only moderately accurate estimates of composite strength and modulus. More advanced composites concepts are, therefore, needed to complement the introductory initial framework presented in Chap. 9.

Following the brief introduction to the structure and deformation of composite materials in Chap. 9, this chapter presents further topics on composite deformation and design. The chapter is suitable for those that want to develop a more complete understanding of composites. It should probably be skipped in most undergraduate courses, and even in some graduate courses.

The chapter begins with systematic introduction to ply theory. This is done by first presenting a framework for the analysis of single composite plies, before explaining the assembly of global stiff matrices for multiply composites. Composite design concepts are then discussed briefly along with composite failure criteria. The shear lag theory is also described before briefly discussing the experimental methods that are used for the measurement of the interfacial strengths of fiber-reinforced composite materials.

10.2 UNIDIRECTIONAL LAMINATES

Let us begin by considering the elastic deformation of a unidirectional ply (Fig. 10.1). Typical plies in engineering composites (mostly polymer matrix composites) have thicknesses of ~ 0.125 mm and fiber volume fractions between 0.50 and 0.65. The plies are also transversely isotropic (orthothropic), which means that the transverse properties are symmetric about the longitudinal axis. However, the stiffnesses of the orthotropic composite plies are greater along the longitudinal/fiber directions than in any of the transverse directions.

Unlike isotropic materials, which require two independent elastic constants for the modeling of deformation, orthotropic materials require four elastic constants to describe their deformation. These include: the longitudinal Young' modulus, E_{11}, the transverse Young's modulus, E_{22}, the shear modulus, G_{12}, and the major Poisson's ratio, ν_{12}, or the minor Poisson's ratio, ν_{21}. Hooke's law for an orthotropic ply (Fig. 10.1) may thus be expressed as

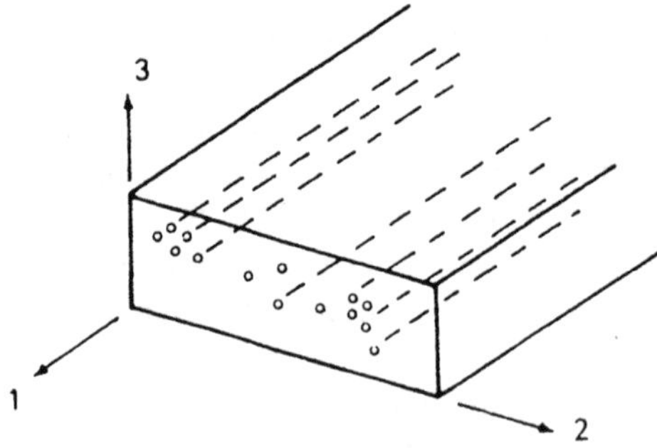

FIGURE 10.1 Deformation of a unidirectional ply. (From Matthews and Rawlings, 1994. Reprinted with permission from Chapman and Hall.)

$$\overset{\sigma_{12}}{\begin{bmatrix} \sigma_{11} \\ \\ \sigma_{22} \\ \\ \tau_{12} \end{bmatrix}} = \overset{Q}{\begin{bmatrix} \dfrac{E_{11}}{1-\nu_{12}\nu_{21}} & \dfrac{\nu_{21}E_{11}}{1-\nu_{12}\nu_{21}} & 0 \\ \\ \dfrac{\nu_{12}E_{22}}{1-\nu_{12}\nu_{21}} & \dfrac{E_{22}}{1-\nu_{12}\nu_{21}} & 0 \\ \\ 0 & 0 & G_{12} \end{bmatrix}} \overset{\varepsilon_{12}}{\begin{bmatrix} \varepsilon_{11} \\ \\ \varepsilon_{22} \\ \\ \gamma_{12} \end{bmatrix}} \tag{10.1a}$$

or, in short-hand notation:

$$\sigma_{12} = Q\varepsilon_{12} \tag{10.1b}$$

Alternatively, we may write strain as a function of stress by inverting the stiffness, Q, matrix:

$$\overset{\varepsilon_{12}}{\begin{bmatrix} \varepsilon_{11} \\ \\ \varepsilon_{22} \\ \\ \gamma_{12} \end{bmatrix}} = \overset{Q}{\begin{bmatrix} \dfrac{1}{E_{11}} & \dfrac{-\nu_{21}}{E_{22}} & 0 \\ \\ \dfrac{-\nu_{12}}{E_{11}} & 0 & 0 \\ \\ 0 & 0 & \dfrac{1}{G_{12}} \end{bmatrix}} \overset{\sigma_{12}}{\begin{bmatrix} \sigma_{11} \\ \\ \sigma_{22} \\ \\ \tau_{12} \end{bmatrix}} \tag{10.2a}$$

or

$$\varepsilon_{12} = S\sigma_{12} \tag{10.2b}$$

It is important to note here that the compliance matrix, S, is the inverse of the stiffness matrix, Q. The above expressions are, therefore, matrix versions of Hooke's law. For most composite plies, the elastic constants in Eqs (10.1a) and (10.2a) can be obtained from tables of materials properties. Also, the minor Poisson's ratio may be estimated from the following expression, if the other three elastic constants are known:

$$\nu_{21} = \frac{E_{22}}{E_{11}}\nu_{12} \tag{10.3}$$

Since $E_{22} < E_{11}$ (the transverse modulus is less than the longitudinal modulus), then the minor Poisson's ratio, ν_{21}, must be less than the major Poisson's ratio, ν_{12}. The stress–strain response of a unidirectional ply may, therefore, be modeled easily by the substitution of appropriate elastic constants into Eqs (10.1) and (10.2).

10.3 OFF-AXIS LAMINATES

Let us now consider the deformation of an off-axis ply reinforced with fibers inclined at an angle, θ, to the x–y axes (Fig. 10.2). The biaxial stresses σ_{xx}, σ_{yy}, and τ_{xy} are applied, as shown in Fig. 10.2. The components of stress that are parallel or perpendicular to the fibers are given by the resolved components σ_{11}, σ_{22}, and τ_{12}. Similarly, the components of the strain tensor may be expressed in terms of the 1–2 or x–y axes to be (ε_{11} ε_{22} $\gamma_{12}/2$) and ε_{xx} ε_{yy} $\gamma_{xy}/2$), respectively. The stress and strain components in the x–y and 1–2 co-ordinate systems are thus given by Rawlings and Matthews (1994) to be

$$\sigma_{12} = \{\sigma_{11}\ \ \sigma_{22}\ \ \tau_{12}\} \tag{10.4a}$$

$$\sigma_{xy} = \{\sigma_{xx}\ \ \sigma_{yy}\ \ \tau_{xy}\} \tag{10.4b}$$

$$\bar{\varepsilon}_{12} = \{\varepsilon_{11}\ \ \varepsilon_{22}\ \ \gamma_{12}/2\} \tag{10.4c}$$

$$\bar{\varepsilon}_{xy} = \{\varepsilon_{xx}\ \ \varepsilon_{yy}\ \ \gamma_{xy}/2\} \tag{10.4d}$$

We may transpose between strain or stress components in the x–y and 1–2 co-ordinate system by the simple use of a transposition matrix, T. This gives

$$\sigma_{12} = T\sigma_{xy} \tag{10.5a}$$

$$\bar{\varepsilon}_{12} = T\bar{\varepsilon}_{xy} \tag{10.5b}$$

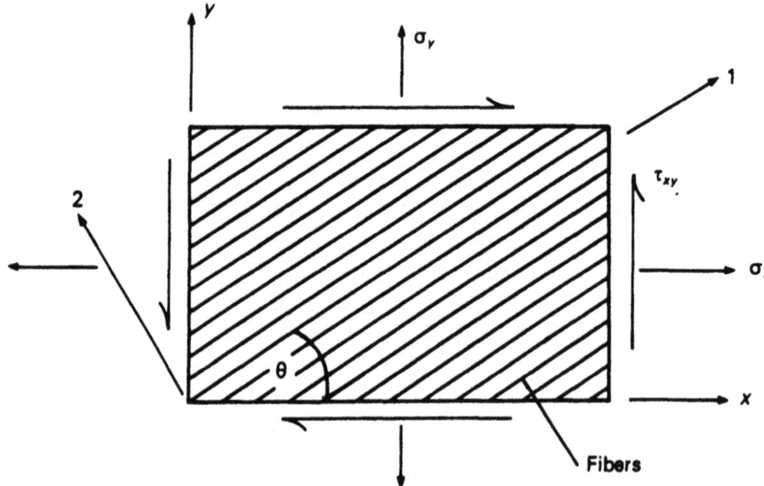

Figure 10.2 Deformation of an off-axis ply. (From Matthews and Rawlings, 1994. Reprinted with permission from Chapman and Hall.)

where the transposition matrix, T, is given by the following:

$$T = \begin{bmatrix} c^2 & s^2 & 2cs \\ s^2 & c^2 & -2cs \\ -cs & cs & (c^2 - s^2) \end{bmatrix} \tag{10.5c}$$

with $c = \cos\theta$ and $s = \sin\theta$. The strain and stress components in the x–y system may be obtained from the inverse transformation, T^{-1}. This gives:

$$\sigma_{xy} = T^{-1}\sigma_{12} \tag{10.6a}$$

$$\bar{\varepsilon}_{xy} = T^{-1}\bar{\varepsilon}_{12} \tag{10.6b}$$

where T^{-1} is given by

$$T^{-1} = \begin{bmatrix} c^2 & s^2 & -2cs \\ s^2 & c^2 & 2cs \\ cs & -cs & (c^2 - s^2) \end{bmatrix} \tag{10.6c}$$

Furthermore, we may describe the strain components $\varepsilon_{12} = \{\varepsilon_{11} \;\; \varepsilon_{22} \;\; \gamma_{12}\}$ and $\varepsilon_{xy} = \{\varepsilon_{xx} \;\; \varepsilon_{yy} \;\; \gamma_{xy}\}$. These may be obtained by multiplying the strain components of $\bar{\varepsilon}_{12}$ and $\bar{\varepsilon}_{xy}$ by a matrix, R, which gives

$$\overset{\varepsilon_{12}}{\begin{bmatrix} \varepsilon_{11} \\ \varepsilon_{22} \\ \gamma_{xy} \end{bmatrix}} = \overset{R}{\begin{bmatrix} 1 & 0 & 0 \\ 0 & 1 & 0 \\ 0 & 0 & 2 \end{bmatrix}} \overset{\bar{\varepsilon}_{12}}{\begin{bmatrix} \varepsilon_{11} \\ \varepsilon_{22} \\ \frac{\gamma_{12}}{2} \end{bmatrix}} \tag{10.7a}$$

or

$$\varepsilon_{12} = R\bar{\varepsilon}_{12} \tag{10.7b}$$

and

$$\overset{\varepsilon_{xy}}{\begin{bmatrix} \varepsilon_{xx} \\ \varepsilon_{yy} \\ \gamma_{xy} \end{bmatrix}} = \overset{R}{\begin{bmatrix} 1 & 0 & 0 \\ 0 & 1 & 0 \\ 0 & 0 & 2 \end{bmatrix}} \overset{\bar{\varepsilon}_{xy}}{\begin{bmatrix} \varepsilon_{xx} \\ \varepsilon_{yy} \\ \frac{\gamma_{xy}}{2} \end{bmatrix}} \tag{10.8a}$$

or

$$\varepsilon_{xy} = R\bar{\varepsilon}_{xy} \tag{10.8b}$$

Hence, we may write the following expressions for the strain components:

$$\varepsilon_{12} = R\bar{\varepsilon}_{12} \tag{10.9a}$$

$$\bar{\varepsilon}_{12} = R^{-1}\varepsilon_{12} \tag{10.9b}$$

$$\varepsilon_{xy} = R\bar{\varepsilon}_{xy} \tag{10.9c}$$

$$\bar{\varepsilon}_{xy} = R^{-1}\varepsilon_{xy} \tag{10.9d}$$

For linear elastic materials, the stress tensor is linearly related to the strain tensor via the transformed stiffness matrix, $\overline{Q}$. This gives:

$$\sigma_{xy} = \overline{Q}\varepsilon_{xy} \tag{10.10}$$

To find $\overline{Q}$, we must go through a series of matrix manipulations to transpose completely from the 1–2 co-ordinate system to the x–y co-ordinate system. From Eq. (10.6a), we may express σ_{xy} in terms of σ_{12}. This gives $\sigma_{xy} = T^{-1}\,\sigma_{12}$. Also, recalling from Eq. (10.1b) that $\sigma_{12} = Q\varepsilon_{12}$ and $\varepsilon_{12} = R\bar{\varepsilon}_{12}$ [Eq. (10.2b)], we may now rewrite Eq. (10.6a) as

$$\sigma_{xy} = T^{-1}\sigma_{12} = T^{-1}(Q\varepsilon_{12}) = T^{-1}Q(R\bar{\varepsilon}_{12}) \tag{10.11}$$

Furthermore, noting that $\bar{\varepsilon}_{12} = T\bar{\varepsilon}_{xy}$ [Eq. (10.6b)] and $\bar{\varepsilon}_{xy} = R^{-1}\,\varepsilon_{xy}$ [Eq. (10.9d)], we may simplify Eq. (10.11) to give

$$\sigma_{xy} = T^{-1}QR(\bar{\varepsilon}_{12}) = T^{-1}QR(T\varepsilon_{xy}) = (T^{-1}QRTR^{-1})\varepsilon_{xy} \tag{10.12}$$

or

$$\sigma_{xy} = \overline{Q}\varepsilon_{xy} = \overbrace{\begin{bmatrix} \overline{Q}_{11} & \overline{Q}_{12} & \overline{Q}_{13} \\ \overline{Q}_{21} & \overline{Q}_{22} & \overline{Q}_{23} \\ \overline{Q}_{31} & \overline{Q}_{32} & \overline{Q}_{33} \end{bmatrix}}^{\overline{Q}} \overbrace{\begin{bmatrix} \varepsilon_{xx} \\ \varepsilon_{yy} \\ \gamma_{xy} \end{bmatrix}}^{\varepsilon_{xy}} \tag{10.13}$$

where $\overline{Q}$ is the transformed stiffness matrix which is given by $\overline{Q} = T^{-1}QRTR^{-1}$. By substitution of the appropriate parameters into the T, Q, and R matrices, it is possible to show that the components of the $\overline{Q}$ matrix are

$$\overline{Q}_{11} = Q_{11}c^4 + 2(Q_{12} + 2Q_{33})s^2c^2 + Q_{22}s^4 \tag{10.14a}$$

$$\overline{Q}_{22} = Q_{11}s^4 + 2(Q_{12} + 2Q_{33})s^2c^2 + Q_{22}c^4 \tag{10.14b}$$

$$\overline{Q}_{12} = (Q_{11} + Q_{22} - 4Q_{33})s^2c^2 + Q_{12}(c^4 + s^4) \tag{10.14c}$$

$$\overline{Q}_{33} = (Q_{11} + Q_{22} - 2Q_{33})n^2c^2 + Q_{33}(c^4 + s^4) \tag{10.14d}$$

$$\overline{Q}_{13} = (Q_{11} - Q_{12} - 2Q_{33})sc^3 + (Q_{12} - Q_{22} + 2Q_{33})s^3c \tag{10.14e}$$

$$\overline{Q}_{23} = (Q_{11} - Q_{12} - 2Q_{33})s^3c + (Q_{12} - Q_{22} + 2Q_{33})sc^3 \tag{10.14f}$$

where the components of the stiffness matrix, Q_{ij}, are given by Eq. (10.1a). Alternatively, we may also express ε_{xy} as a function of σ_{xy}. In this case, $\varepsilon_{xy} = \overline{Q}^{-1}\sigma_{xy}$, where $\overline{Q}^{-1}$ corresponds to the compliance matrix, $\overline{S}$, which has the components:

$$\overline{S}_{11} = S_{11}c^4 + 2(S_{12} + S_{33})s^2c^2 + S_{22}s^4 \tag{10.15a}$$

$$\overline{S}_{22} = S_{11}s^4 + 2(S_{12} + S_{33})s^2c^2 + S_{22}c^4 \tag{10.15b}$$

$$\overline{S}_{12} = (S_{11} + S_{22} - S_{33})s^2c^2 + S_{12}(c^4 + s^4) \tag{10.15c}$$

$$\overline{S}_{33} = 2(2S_{11} + 2S_{22} - 4S_{12} - S_{33})s^2c^2 + S_{33}(c^4 + s^4) \tag{10.15d}$$

$$\overline{S}_{13} = (2S_{11} - 2S_{12} - S_{33})mn^3 + (2S_{12} - 2S_{22} + S_{33})cs^3 \tag{10.15e}$$

$$\overline{S}_{23} = (2S_{11} - 2S_{12} - S_{33})cs^3 + (2S_{12} - 2S_{22} + S_{33})c^3s \tag{10.15f}$$

As the reader can imagine, calculation of the components of the $\overline{S}$ and $\overline{Q}$ matrices can become rather tedious. For this reason, simple computer programs are often used to obtain the components of these matrices using the expressions presented above.

10.4 MULTIPLY LAMINATES

We are now in a position to consider the deformation of laminates that consist of multiple plies with different fiber orientations (Fig. 10.3). The overall stiffness of such a composite stack may be found by summing up the stiffness contributions from all of the individual plies. The overall strains, $\varepsilon_{xy}^{\text{tot}}$, in the composite may also be separated into axial, ε_{xy}^{0}, and bending, $\varepsilon_{xy}^{\text{b}}$, strain components (Fig. 10.4). Hence, applying the principle of linear superposition gives

$$\varepsilon_{xy}^{\text{tot}} = \varepsilon_{xy}^{0} + \varepsilon_{xy}^{\text{b}} = \varepsilon_{xy}^{0} + z\kappa \tag{10.16a}$$

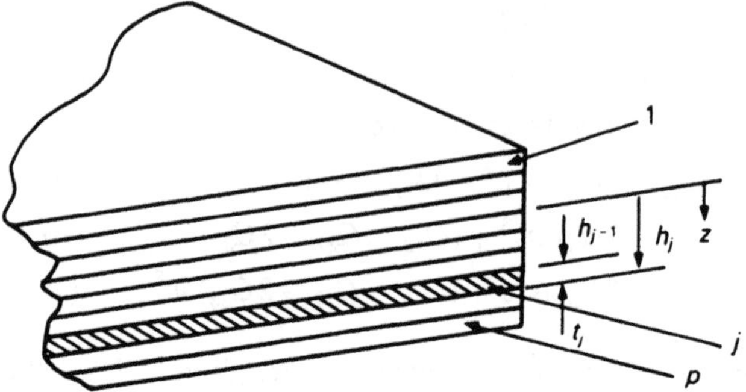

FIGURE 10.3 Schematic of a multiply laminate. (From Matthews and Rawlings, 1994. Reprinted with permission from Chapman and Hall.)

or

$$\begin{array}{ccc} \varepsilon_{xy} & \varepsilon^0_{xy} & \kappa \end{array}$$

$$\begin{bmatrix} \varepsilon_{xx} \\ \varepsilon_{yy} \\ \gamma_{xy} \end{bmatrix} = \begin{bmatrix} \varepsilon^0_{xy} \\ \varepsilon^0_{yy} \\ \gamma^0_{xy} \end{bmatrix} + z \begin{bmatrix} \kappa_x \\ \kappa_y \\ \kappa_{xy} \end{bmatrix} \tag{10.16b}$$

where κ represents the curvatures, z is the distance from the neutral axis, and the other parameters have their usual meaning.

For linear elastic deformation, Hooke's law applies. Hence, we may write:

$$\sigma_{xy} = \overline{Q}\varepsilon^{tot}_{xy} = \overline{Q}(\varepsilon^0_{xy} + \varepsilon^b_{xy}) = \overline{Q}(\varepsilon^0_{xy} + \kappa z) \tag{10.17a}$$

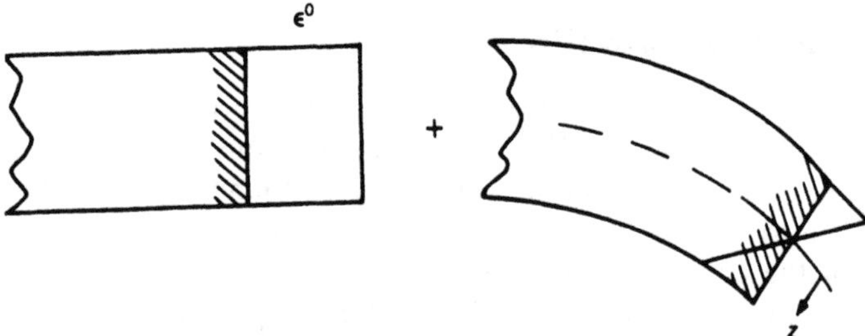

FIGURE 10.4 Superposition of axial and bending strain components. (From Matthews and Rawlings, 1994. Reprinted with permission from Chapman and Hall.)

$$\sigma_{xy} \qquad \varepsilon_{xy}^0 \qquad \kappa$$

$$\begin{bmatrix} \sigma_{xx} \\ \sigma_{yy} \\ \tau_{xy} \end{bmatrix} = \overline{Q} \begin{bmatrix} \varepsilon_{xx}^0 \\ \varepsilon_{yy}^0 \\ \gamma_{xy}^0 \end{bmatrix} + \overline{Q}_z \begin{bmatrix} \kappa_x \\ \kappa_y \\ \kappa_{xy} \end{bmatrix} \tag{10.17b}$$

where the above terms have their usual meaning, and the sign convention for the axial strain components and curvatures are given by Fig. 10.5. The force per unit length on each ply is obtained from the products of the stress components (σ_{xx} σ_{yy} τ_{xy}) and the layer thicknesses, t_i (Fig. 10.3). Hence, for a composite consisting of p plies, the overall force per unit length in any given direction is obtained from the following summation of the forces in each of the plies:

$$N_x = \sum_{i=1}^{p} \sigma_{xx} t_i \tag{10.18a}$$

$$N_y = \sum_{i=1}^{p} \sigma_{yy} t_i \tag{10.18b}$$

$$N_{xy} = \sum_{i=1}^{p} \tau_{xy} t_i \tag{10.18c}$$

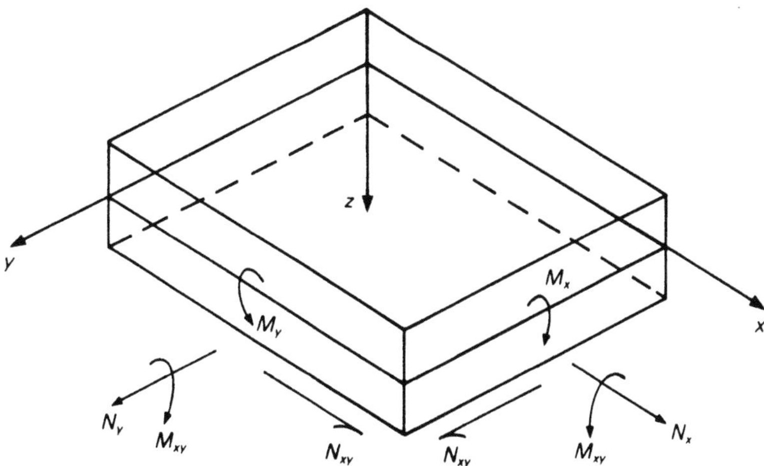

FIGURE 10.5 Sign convention for axial strain components, curvatures, and moments. (From Matthews and Rawlings, 1995. Reprinted with permission from Chapman and Hall.)

However, the thickness, t_i, may also be expressed in integral form since $t_i = \int_{z_{i-1}}^{z_i} dz$. Substituting this integral form into Eq. (10.18), and noting that $\sigma_{xy} = \overline{Q}\ \varepsilon_{xy}$, gives

$$N = \begin{bmatrix} N_x \\ N_y \\ N_{xy} \end{bmatrix} = \sum_{i=1}^{p} \varepsilon_{xy}^0 \int_{z_{i-1}}^{z_i} \overline{Q}\,dz + \sum_{i=1}^{p} \kappa \int_{z_{i-1}}^{z_i} Qz\,dz \tag{10.19}$$

Within each ply $\overline{Q}$, is independent of z. Also, $\varepsilon_{xy}^0$and κ are applied to the whole multiply composite and are independent of z or i. Hence, we may solve easily for the integrals in Eq. 10.19:

$$N = A\varepsilon_{xy}^0 + B\kappa \tag{10.20a}$$

where

$$A = \sum_{i=1}^{p} \overline{Q} \int_{zi-1}^{z_i} dz = \sum_{i=1}^{p} \overline{Q}_{rs,i}(z_i - z_{i-1}) \tag{10.20b}$$

and

$$B = \sum_{i=1}^{p} \mathrm{k}\overline{Q} \int_{z_{i-1}}^{z_i} z\,dz = \sum_{i=1}^{p} \overline{Q}_{rs,i}\frac{1}{2}(z_i^2 - z_{i-1}^2) \tag{10.20c}$$

Similarly, we may consider the moments per unit length associated with the stresses applied to the individual layers in the multiply laminate shown schematically in Fig. 10.3. For each layer, the bending moments (Fig. 10.5) are given by

$$M_i = \begin{bmatrix} M_{xi} \\ M_{yi} \\ M_{xyi} \end{bmatrix} = \sigma_{xy,i}t_iz_i = \int_{z_{i-1}}^{z_i} \sigma_{xy,i}z\,dz + \int_{z_{i-1}}^{z_i} \overline{Q}_iz(\varepsilon_{xy}^0 + \kappa z)dz \tag{10.21}$$

where z is the distance from the neutral axis, t is the thickness of the ply, and i corresponds to the number designation of the ply. As before, we may sum up the overall moments per unit length associated with the p plies in the composite:

$$M = \sum_{i=1}^{p} M_i \begin{bmatrix} M_x \\ M_y \\ M_{xy} \end{bmatrix} = \sum_{i=1}^{p} \varepsilon_{xy}^0 \int_{z_{i-1}}^{z_i} \overline{Q}_i \varepsilon_{xy}^0 z \, dz + \sum_{i=1}^{p} \int_{z_{i-1}}^{z_i} \overline{Q}_i \kappa z^2 dz \tag{10.22}$$

since Q_i is independent of z for a given ply, and and ε_{xy}^0 and κ are independent of z or i, we may solve the integrals in Eq. 10.22 to obtain:

$$\mathbf{M} = \begin{bmatrix} M_x \\ M_y \\ M_z \end{bmatrix} = \mathbf{B}\varepsilon_{xy}^0 + \mathbf{D}\kappa \tag{10.23}$$

where the **B** matrix may be obtained from Equation 10.20c and the **D** matrix is given by:

$$D = \sum_{i=1}^{p} \frac{\overline{Q}_i}{3} \kappa \left(z_i^3 - z_{i-1}^3 \right) \tag{10.24}$$

We may also combine Eqs (10.20a) and (10.23) to obtain the *plate constitutive equation* that describes the overall response of any multiply composite subjected to axial loads and moments (Figs 10.4 and 10.5):

$$\begin{bmatrix} N \\ M \end{bmatrix} = \begin{bmatrix} \mathbf{A} & \mathbf{B} \\ \mathbf{B} & \mathbf{D} \end{bmatrix} \begin{bmatrix} \varepsilon_{xy}^0 \\ \kappa \end{bmatrix} \tag{10.25}$$

where the *ABD* matrix is the global stiffness matrix. The terms A, B, and D in a global stiffness matrix each correspond to 3×3 matrices, with nine terms in each matrix. Equation (10.25) may be used to determine the axial strains and curvatures, ε_{xy}^0 and κ, associated with prescribed loads and displacements. This may be done by multiplying eq. (10.25) by the inverse of the *ABD* matrix, which gives

$$\begin{bmatrix} \varepsilon_{xy}^0 \\ \kappa \end{bmatrix} = \begin{bmatrix} A & B \\ B & D \end{bmatrix}^{-1} \begin{matrix} N \\ M \end{matrix} = \begin{bmatrix} A' & B' \\ B' & D' \end{bmatrix} \begin{bmatrix} N \\ M \end{bmatrix} \tag{10.26a}$$

where A', B', and D' are given by

$$A' = A^* + B^*[D^*]^{-1}[B^*]^t \tag{10.26b}$$

$$B' = B^* - [D^*]^{-1} \tag{10.26c}$$

$$D' = [D^*]^{-1} \tag{10.26d}$$

$$A^* = A^{-1} \tag{10.26e}$$

$$B^* = A^{-1}B \tag{10.26f}$$

$$D^* = D - BA^{-1}B \tag{10.26g}$$

Simple computer programs are often used to perform the matrix manipulations involved in solving for strains and curvatures or forces and moments. Furthermore, although the matrices presented in this section may look intimidating at first, most of the expressions may be viewed simply as *matrix expressions of Hooke's law*.

10.5 COMPOSITE PLY DESIGN

The design of multi-ply composite laminates for different structural functions can be accomplished to a large extent by the judicious selection of fiber orientations and ply stacking sequences. This section will briefly discuss the ways in which the individual plies within a multi-ply composite can be arranged to achieve different types of coupling between axial, bending, and twisting modes. The couplings between these different deformation modes are controlled by the components of the *ABD* matrix. These are shown completely in the following expanded version of Eq. (10.25):

$$\begin{bmatrix} N_x \\ N_y \\ N_{xy} \end{bmatrix} = \begin{bmatrix} A_{11} & A_{12} & A_{13} \\ A_{21} & A_{22} & A_{23} \\ A_{31} & A_{32} & A_{33} \end{bmatrix} \begin{bmatrix} \varepsilon_x^0 \\ \varepsilon_y^0 \\ \varepsilon_{xy}^0 \end{bmatrix} + \begin{bmatrix} B_{11} & B_{12} & B_{13} \\ B_{21} & B_{22} & B_{23} \\ B_{31} & B_{32} & B_{33} \end{bmatrix} \begin{bmatrix} \kappa_x \\ \kappa_y \\ \kappa_{xy} \end{bmatrix} \tag{10.27}$$

$$\begin{bmatrix} M_x \\ M_y \\ M_{xy} \end{bmatrix} = \begin{bmatrix} B_{11} & B_{12} & B_{13} \\ B_{21} & B_{22} & B_{23} \\ B_{31} & B_{32} & B_{33} \end{bmatrix} \begin{bmatrix} \varepsilon_x^0 \\ \varepsilon_y^0 \\ \varepsilon_{xy}^0 \end{bmatrix} + \begin{bmatrix} D_{11} & D_{12} & D_{13} \\ D_{21} & D_{22} & D_{23} \\ D_{31} & D_{32} & D_{33} \end{bmatrix} \begin{bmatrix} \kappa_x \\ \kappa_y \\ \kappa_{xy} \end{bmatrix}$$

The A, B, and D matrices are symmetric, with $A_{ij} = A_{ji}$, $B_{ij} = B_{ji}$, and $D_{ij} = D_{ji}$. Also, the axial, bending and shear forces are related to the components A_{ij}, B_{ij}, and D_{ij} of the ABD matrices. These give the following couplings:

1. A_{13} and A_{23} relate in-plane axial forces to in-plane shear strains, or in-plane shear forces to axial strains.

2. B_{13} and B_{23} relate in-plane direct forces to plate twisting, or torques to in-plane direct strains.
3. B_{11}, B_{12} and B_{13} relate in-plane bending moments to axial strains, or axial forces to in-plane curvatures.
4. B_{33} relates in-plane shear force to plate twisting, or torque to in-plane shear strain.
5. D_{13} and D_{23} relate bending moments to plate twisting, or torque to plate curvatures.

Since some of the couplings listed above may be undesirable in structural applications, it is common practice to select ply stacking sequences that result in zero values of the "undesirable" coupling parameters. For example, the coupling between in-plane strains and shear forces may be eliminated by choosing stacking sequences that result in $A_{23} = A_{13} = 0$. This may be achieved by choosing balanced composite lay-ups in which every $+\theta$ ply is balanced by a $-\theta$ ply. However, the stacking sequence in the composite does not need to be symmetric. Examples of balanced composite lay-ups include: $(+30°/-30°)$, $(0°/+45°/-45°)$, $(+60°/-60°)$, and $(0°/+75°/-75°)$.

It is also important to note that we may set $A_{23} = A_{13} = 0$ by choosing composites plies with 0° and/or 90° fiber orientations. Furthermore, we may eliminate bending membrane coupling by setting the B matrix to zero. This may be engineered by designing composite stacking sequences that are symmetric about the midplane. Examples of symmetric stacking sequences include: $(-30°/+30°/+30°/-30°)$, $(0°/+45°/+45°/0°)$, etc.

The coupling between bending and twisting is avoided by setting $D_{13} = D_{23} = 0$. This is achieved by the use of balanced antisymmetric lay-ups for which every $+\theta$ ply above the midplane is balanced by a $-\theta$ of identical thickness at the same distance below the midplane, and vice versa. Examples of such composite lay-ups include: $(+45°/-45°/0°/+45°/-45°)$ and $(90°/45°/0°/-45°/-90°)$.

The above discussion has focused on the stacking sequences required for the elimination of couplings. However, in many cases, composite designers may take advantage of couplings by using them to engineer coupled elastic responses that are not possible in relatively isotropic materials. For example, desirable aeroelastic responses may be engineered by the use of swept-wing profiles that are produced between twisting and bending. Judicious selection of coupling parameters D_{13} and D_{23} can be used to achieve aircraft wing profiles in which the wing twists downwards and bends upwards under aerodynamic loads. Such aeroelastic coupling may be used to achieve stable aerodynamic maneuvers in military aircraft during air combat. The tailoring of the composite ply lay-ups is, therefore, a key component of military aircraft design.

10.6 COMPOSITE FAILURE CRITERIA

Due to the relatively complex nature of composite structures (compared to the structure of relatively isotropic monolithic alloys), a wider range of failure modes is observed in such materials. As discussed in Sect. 9.5, composite failure may occur by fiber fracture, matrix shear, or matrix failure in tension or compression (perpendicular to fiber direction). The stress state is also inherently multiaxial within each of the composite plies. Hence, local failure criteria may vary widely within plies and across plies.

10.6.1 Critical Stress or Critical Strain Approaches

In this approach, we assume that failure occurs when critical conditions (such as failure stresses or strains) are reached locally within any of the regions in a composite. We neglect the possible changes in failure conditions due to multiaxial stress conditions and interactions between the failure modes. Failure is thus postulated to occur when local critical conditions are first exceeded in tension, compression, or shear.

If the local critical stress failure criteria in tension, compression, and shear are denoted by $\hat{\sigma}_{1T}$, $\hat{\sigma}_{2T}$, $\hat{\sigma}_{1c}$, $\hat{\sigma}_{2c}$, and $\hat{\tau}_{12}$, respectively, then we assume that local failure will occur when any of these strengths are first exceeded. Similarly, in cases where local failure is strain controlled, we assume that failure will occur when critical strains are exceeded in tension, compression or shear. These are denoted, respectively, by $\hat{\varepsilon}_{1T}$, $\hat{\varepsilon}_{2T}$, $\hat{\varepsilon}_{1c}$, $\hat{\varepsilon}_{2c}$, and $\hat{\gamma}_{12}$.

The discerning reader will realize that the failure of a single ply does not correspond to complete composite failure. Also, a ply may continue to have load-bearing capacity even after local failure criteria have been exceeded. It is, therefore, important to make some assumptions about the nature of the load-bearing capacity of individual plies after local failure criteria are exceeded.

In the most extreme cases, the failed plies are assumed to support low loads. This often leads to excessively conservative predictions of failure. Hence, it is more common to assume that the failed ply supports some fraction of the load that it carried before final "failure." Similar approaches are also applied to predictions in which interlaminar failure occurs between plies. However, these approaches often require iteration and good judgment in the determination of load-carrying capacity of the failed plies.

10.6.2 Interactive Failure Criteria

The second approach that is often used to predict failure involves the use of interactive failure criteria. These attempt to account for the possible interactions between stresses and failure criteria. The most commonly used failure criterion is the Tsai–Hill failure criterion, which was first proposed by Azzi and Tsai (1965). This empirical criterion predicts that failure of an anisotropic ply occurs when

$$\left(\frac{\sigma_{11}}{\hat{\sigma}_{11}}\right)^2 - \left(\frac{\sigma_{11}\sigma_{22}}{\hat{\sigma}_{11}^2}\right) + \left(\frac{\sigma_{22}}{\hat{\sigma}_{22}}\right)^2 + \left(\frac{\tau_{12}}{\hat{\tau}_{12}}\right)^2 \geq 1 \tag{10.28}$$

where the signs of the strength terms ($\hat{\sigma}_{11}, \hat{\sigma}_{22}$, and $\hat{\tau}_{12}$) correspond to the signs of the local stress components (σ_{11}, σ_{22}, and τ_{12}). The corresponding strength terms are, therefore, positive for tensile stresses, and negative for compressive stresses.

In cases where a uniaxial stress, σ_{xx}, is applied to off-axis plies (Fig. 10.2), the Tsai–Hill criterion becomes

$$\left(\frac{\sigma_{xx}\cos^2\theta}{\hat{\sigma}_{11}}\right)^2 - \frac{\sigma_{xx}^2\cos^2\theta\sin^2\theta}{\hat{\sigma}_{11}^2} + \left(\frac{\sigma_{xx}^2\sin^2\theta}{\hat{\sigma}_{22}}\right)^2 + \left(\frac{\sigma_{xx}\sin\theta\cos\theta}{\hat{\tau}_{12}}\right)^2 \geq 1 \tag{10.29}$$

Equation (10.29) can be solved to obtain a unique solution for σ_{xx} as a function of θ. This gives σ_{xx} as a continuous function of θ, as shown in Fig. 10.6. This is in contrast to the separate curves presented earlier in Chap. 9 for the composite matrix shear and matrix normal failure (Fig. 9.15).

It is also of interest to note that a reserve factor, R, can be described for cases in which the applied loads are insufficient to cause failure according to the Tsai–Hill criterion. The reserve factor corresponds to the factor by which the applied load can be scaled to induce failure according to the Tsai–Hill criterion. Hence, Eq (10.29) is now given by

$$\left(\frac{R\sigma_{11}}{\hat{\sigma}_{11}}\right)^2 - \left(\frac{R\sigma_{11}R\sigma_{22}}{\hat{\sigma}_{11}\hat{\sigma}''_{22}}\right) + \left(\frac{R\sigma_{22}}{\hat{\sigma}_{22}}\right)^2 + \left(\frac{R\tau_{12}}{\hat{\tau}_{12}}\right)^2 = 1 \tag{10.30}$$

or

$$\left(\frac{\sigma_{11}}{\hat{\sigma}_{11}}\right)^2 - \left(\frac{\sigma_{11}\sigma_{22}}{\hat{\sigma}_{11}^2}\right) + \left(\frac{\sigma_{22}}{\hat{\sigma}_{22}}\right)^2 + \left(\frac{\tau_{12}}{\hat{\tau}_{12}}\right)^2 = \frac{1}{R^2} \tag{10.31}$$

The reserve factor, R, is important because it provides a measure of how much the applied stress(es) can be increased before reaching the failure condition that is predicted by the Tsai–Hill criterion.

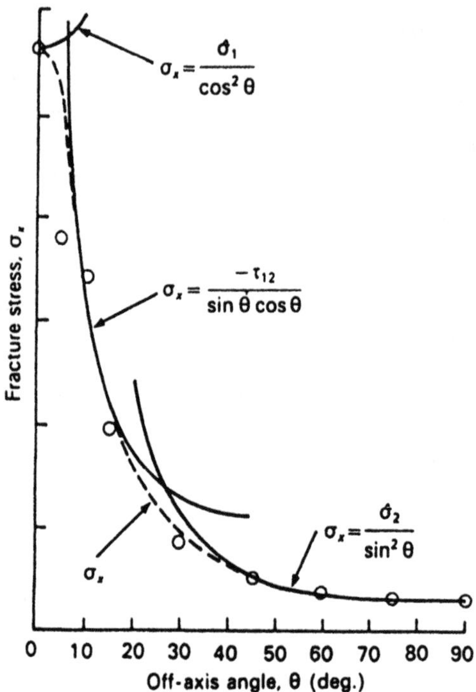

Figure 10.6 Dependence of σ_{xx} on θ.

10.7 SHEAR LAG THEORY

The shear lag model was first developed by Cox (1952) and later modified by Nardone and Prewo (1986). It is applicable to composites reinforced with whiskers or short fibers (Fig. 10.7). The model assumes that the applied load is transferred from the matrix to the whiskers/short fibers via shear. A representative volume element/unit cell of a whisker-reinforced composite is shown schematically in Fig. 10.7. This corresponds to a repeatable unit that can be used to model the behavior of a composite. Such unit cells are often used to obtain representative solutions to composite problems (Taya and Arsenault, 1989).

The Cox model assumes that the displacement gradient in the fiber, $d\sigma_f/dz$, is proportional to the difference between the displacement in the fiber and the matrix, i.e., u–v:

$$\frac{d\sigma_f}{dz} = h(u - v) \tag{10.32}$$

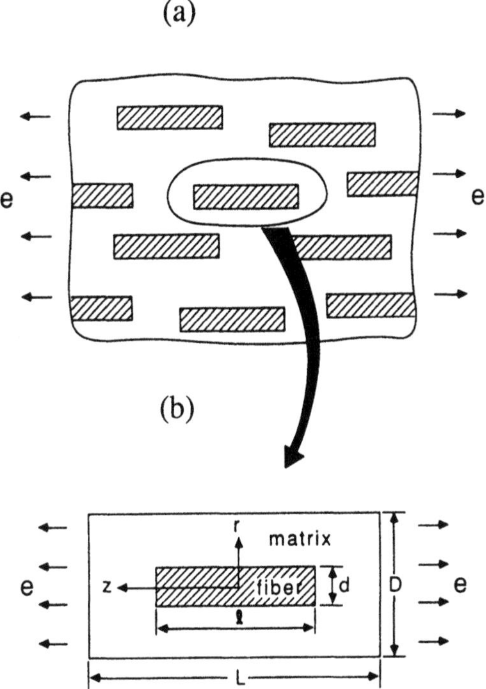

Figure 10.7 The shear lag model: (a) representative short fiber/whisker; (b) unit cell for shear lag analysis. (From Taya and Arsenault, 1989. Courtesy of Pergamon Press.)

where u is the axial fiber displacement, v is the axial matrix displacement, and h is a proportionality constant. Also, from the simple force balance in Fig. 10.7(b), we can show that

$$\sigma_f \frac{\pi d^2 \, dx}{4} + \tau_m \pi d . dx = 0 \tag{10.33a}$$

or

$$\sigma_f = -\frac{4}{d}\tau_m \tag{10.33b}$$

Substituting Eq. (10.33b) into Eq. (10.32) gives

$$\frac{d\sigma_f}{dz} = h(u - v) = -\frac{4\tau_m}{d} \tag{10.34a}$$

or

$$\sigma_f = \frac{4h}{d}(u - v) \tag{10.34b}$$

Note that, by definition, $du/dz = \varepsilon_f$, the strain in the fiber. Similarly, $dv/dz = \varepsilon_m = e$, the strain in the matrix. Also, the strain in the fiber, σ_f, is given by σ_f/E_f. Hence, if we now differentiate Eq. (10.34a) with respect to z, we obtain:

$$\frac{d^2\sigma_f}{dz^2} = h\left[\frac{du}{dz} - \frac{dv}{dz}\right] = h\left[\frac{\sigma_f}{E_f} - e\right] \tag{10.35}$$

Equation (10.35) can be solved to give

$$\sigma_f = E_f\varepsilon_f + C_1 \cosh\beta z + C_2 \sinh\beta z \tag{10.36}$$

where

$$\beta = \sqrt{\frac{h}{E_f}} \tag{10.37}$$

If we now apply the boundary conditions at the end of the fiber, we have $\sigma_f = \sigma_0$. Furthermore, in the middle of the fiber, $d\sigma_f/dz = 0$. We may then solve for C_1 and C_2 and show that

$$\sigma_f = E_f\varepsilon_m\left[1 + \frac{\left(\frac{\sigma_0}{E_f e} - 1\right)\cosh\beta z}{\cosh\left(\frac{\beta\ell}{2}\right)}\right] \tag{10.38}$$

Cox's original model (Cox, 1952) proposed that $\sigma_0 = 0$. However, Nardone and Prewo (1986) later recognized that $\sigma_0 \neq 0$ for strongly bonded fiber ends. Furthermore, the average fiber stress, $\bar{\sigma}_f$, may also be determined from:

$$\bar{\sigma}_f = \frac{2}{\ell}\int_0^{1/2} d\ell = E_f e\left[1 + \frac{\left(\frac{\sigma_0}{E_f e} - 1\right)\tanh\left(\frac{\beta\ell}{2}\right)}{\left(\frac{\beta\ell}{2}\right)}\right] \tag{10.39}$$

Now applying the rule of mixtures to obtain the average composite stress gives

$$\sigma_c = (1 - V_f)\bar{\sigma}_m + V_f\bar{\sigma}_f \tag{10.40}$$

where $\bar{\sigma}_m$ is the average matrix stress. Also, the stresses in the matrix and composite are, respectively,

$$\sigma_m = E_m e \tag{10.41}$$

and

$$\sigma_c = E_c e \tag{10.42}$$

Substituting Eqs (10.39), (10.41), and (10.42) into Eq. (10.40) now gives

$$E_c = (1 - V_f)E_m + V_f E_f \left[1 + \frac{\left(\dfrac{\sigma_0}{E_f e} - 1 \right) \tanh\left(\dfrac{\beta \ell}{2} \right)}{\left(\dfrac{\beta \ell}{2} \right)} \right] \tag{10.43}$$

For special cases when the fiber ends are stress free (Cox model), we have $\sigma_0 = 0$. This gives

$$E_c = (1 - V_f)E_m + V_f E_f \left[1 - \frac{\tanh\left(\dfrac{\beta \ell}{2} \right)}{\left(\dfrac{\beta \ell}{2} \right)} \right] \tag{10.44}$$

Also, when $\sigma_0 = \sigma_m = E_m e$, Eq. (10.44) may be expressed as

$$E_c = (1 - V_f)E_m + V_f \left[1 - \frac{\left(\dfrac{E_m}{E_f} - 1 \right) \tanh\left(\dfrac{\beta \ell}{2} \right)}{\left(\dfrac{\beta \ell}{2} \right)} \right] \tag{10.45}$$

The above expressions provide useful estimates of composite modulus. However, there are some inherent inconsistences in the use of the constant-strain condition in the shear lag formulation. For this reason, more rigorous tensor methods, such as those developed by Eshelby (1957, 1959), are needed when greater accuracy is required. However, these are beyond the scope of this book.

Finally, in this section, it is important to determine the constant, B, in Eqs (10.36)–(10.45). This can be found by analyzing the simple unit cell given in Fig. 10.7. Consider the relative displacement between a point at $r = D/2$ (where the displacement $= u$) and an arbitrary point in the matrix at $r = r$. From horizontal force balance we may write:

$$\tau_0(\pi d)dx = \tau(2\pi r)dx \tag{10.46}$$

Also, the shear strain, γ, is given by

$$\gamma = \frac{dw}{dr} = \frac{\tau}{G_m} = \frac{\tau_0 d}{2rG_m} \tag{10.47}$$

Integrating Eq. (10.47) between $r = d/2$ and $R = D/2$ gives

$$w = \frac{\tau_m D}{2G_m}\int_{d/2}^{D/2}\frac{1}{r} = \frac{\tau_m}{2G_m}\ln\left(\frac{D}{d}\right) \tag{10.48}$$

Hence, since $w = u - v$, we can combine Eqs (10.34a) and (10.48) to give

$$\frac{4\tau_m}{d} = h(u - v) = h\frac{\tau_m d}{2G_m}\ln\left(\frac{D}{d}\right) \tag{10.49}$$

or

$$h = \frac{UG_m}{d^2}\ln\left(\frac{D}{d}\right) \tag{10.50}$$

Therefore, since $\beta = \sqrt{h/E_f}$ [Equation (10.37)], we may thus substitute Eq. (10.50) into Eq. (10.37) to obtain:

$$\beta = \sqrt{\frac{8G_m}{d^2E_f}\ln\left(\frac{D}{d}\right)} = \frac{2\sqrt{2}}{d}\sqrt{\frac{G_m}{E_f}\ln\left(\frac{D}{d}\right)} \tag{10.51}$$

where G_m is the matrix shear modulus, E_f is the fiber Young's modulus, d is the fiber diameter, and D is the diameter of the unit cell shown schematically in Fig. 10.7.

10.8 THE ROLE OF INTERFACES

The above discussion has focused largely on the role of matrix and fiber materials in the deformation of composite materials. However, in several composite systems, the deformation characteristics and failure modes are strongly affected by the strong role of interfaces. In particular, weak or moderately strong interfaces tend to promote debonding or interfacial sliding between the matrix and the fiber. Since interfacial sliding and debonding may promote significant toughening in ceramic matrix and metal matrix composites (Evans and co-workers, 1990, 1991), considerable effort has been expended in the tailoring of interfaces with low/moderate shear strengths.

In the case of silicon carbide fibers that are used for the reinforcement of metal matrix composites, carbon coatings have been deposited on to the SiC fibers (Fig. 9.2) to obtain the desired interfacial sliding and debonding. However, the relatively high cost and limited durability of such composites have prevented potential structural applications in aeroengine structures and components (Soboyejo et al., 1997).

Since the interfacial properties of composite materials are of practical importance to the design of damage tolerant composites, significant efforts have been made to develop test techniques for the measurement of interfacial strength. The fiber pull-out test (Fig. 10.8) and the fiber push-out test (Fig. 10.9) have been the most commonly used techniques for interfacial strength measurement. These rely, respectively, on the pull-out or push-out of fibers from a thin slice of composite material that is polished and etched to reveal the fibers under an in-situ microscope. The applied load is measured with a load cell, while the displacement is often determined with a capacitance gauge. In this way, plots of load versus displacement can be obtained for subsequent analysis of fiber push-out behavior.

Typical plots of load versus displacement obtained from a fiber pull-out or push-out test and a fiber push-back test on a Ti-15-3/SiC composite are shown in Figs 10.10(a) and 10.10(b), respectively. During the fiber push-out test, the initial deformation involves elastic bending. This is associated with a linear increase in load until a critical load is reached at which debonding occurs between the matrix and the fiber. A significant load drop is observed at the onset of debonding, Fig. 10.10(a). This is followed by an increase in load that is associated with geometric decorrelation during the initially matching fiber and matrix geometries during fiber displacement (Mackin et al., 1992). Subsequent fiber push-out is accompanied by a load drop, as the length of fiber in sliding contact decreases during the fiber push-out stage, Fig. 10.10(a).

If the push-out fiber is now pushed back in, i.e., returned to its original location, a load drop is experienced as the fiber reseats into its original position, Fig. 10.10(b). This load drop is associated with the geometric "memory" of the debond surface between the fiber and the matrix.

The average shear stress, τ, experienced by the fiber may be expressed simply as

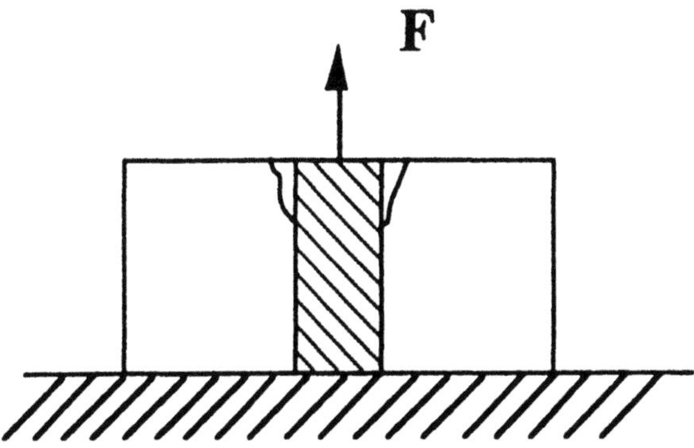

FIGURE 10.8 Schematic illustration of fiber pull-out test.

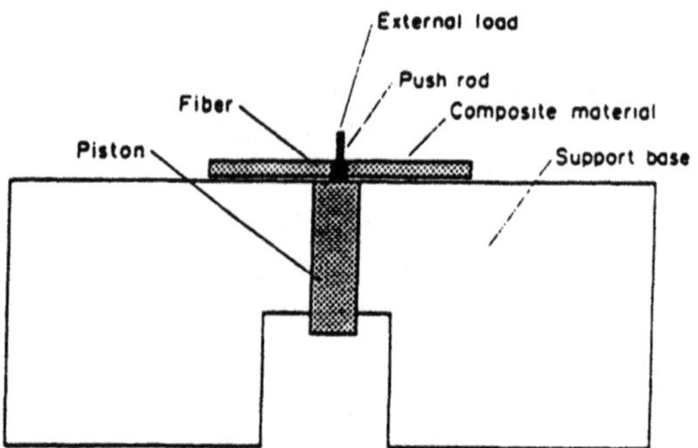

FIGURE 10.9 Schematic of fiber push-out test. (From Mackin et al., 1992. Reprinted with permission from Elsevier Scientific.)

$$\tau = \frac{P}{2\pi rt} \tag{10.52}$$

where P is the applied load, r is the fiber radius, and t is the thickness of the push-out or pull-out specimen (Figs 10.8 and 10.9). To avoid excessive bending during the fiber push-out test, t is usually selected to be ~ 2.5–3.0 times the fiber diameter. Hence, for SCS-6 fibers with diameters of $\sim 150\ \mu$m, t is $\sim 450\ \mu$m.

If we assume that the interfaces between the matrix and the fiber are smooth, fiber sliding may be analyzed using a generalized sliding friction law. This is given by Hutchinson and co-workers (1990, 1993) to be

$$\tau = \tau_0 - \mu\sigma_r \tag{10.53}$$

where τ_0 is the sliding stress, τ_0 is the contact sliding stress, μ is the friction coefficient, and σ_r is the radial clamping stress, which is due largely to the thermal expansion mismatch between the matrix and fibers. The fiber push-out stress has also be shown by Hutchinson and co-workers (1990, 1993) to be

$$\sigma_{\rm u} = \frac{\sigma_0[\exp(2\mu Bt/R) - 1](1-f)}{[f(E_{\rm f}/E_{\rm m})\exp(2\mu Bt/R) + 1 - f]} \tag{10.54}$$

where f is the volume fraction of fiber, R is the fiber radius, t is the embedded length of fiber in contact with the matrix, E is Young's modulus, subscripts m and f denote matrix and fiber, respectively, and

$$B = \nu E[E_{\rm f}(1+\nu) + E(1-\nu)]^{-1} \tag{10.55a}$$

(a)

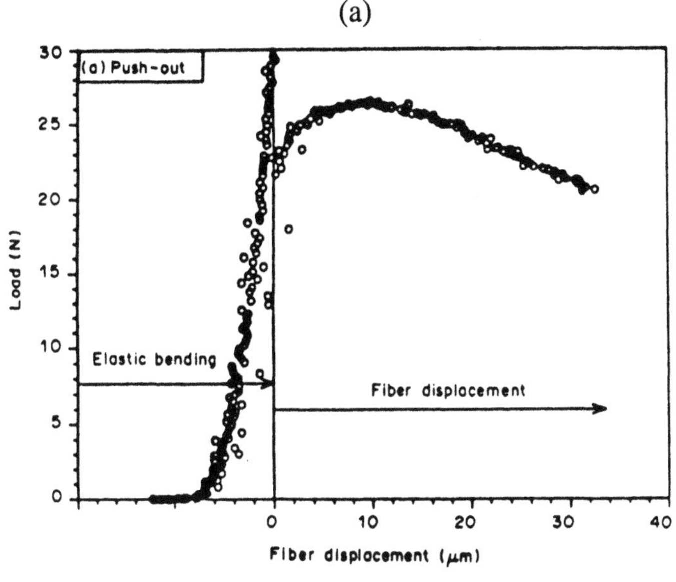

(b)

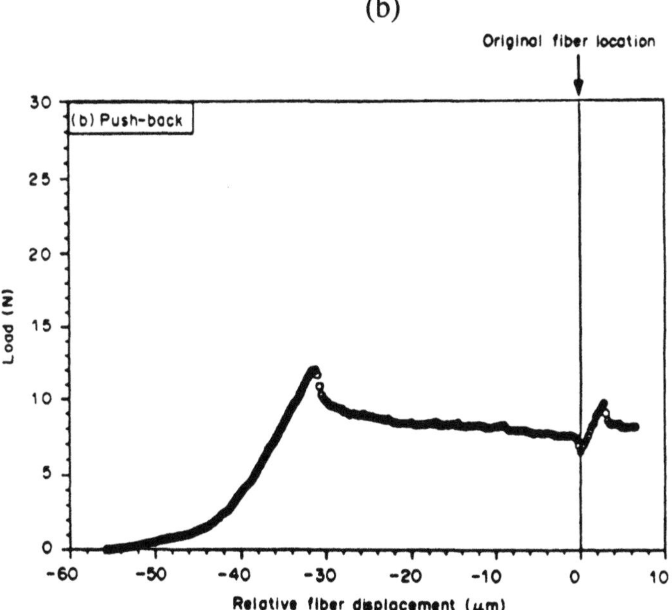

FIGURE 10.10 Load–displacement plot obtained from (a) fiber push-out test and (b) fiber push-back test on a Ti–15–3Cr–3Al–3Sn/SCS-6 composite. (From Mackin et al., 1992. Reprinted with permission from Elsevier Scientific.)

$$\sigma_0 = \frac{E_f \varepsilon_T}{\nu} + \left(\frac{\tau_0}{\mu}\right) \frac{E}{BE(1-f)} \tag{10.55b}$$

$$t = h - d \tag{10.55c}$$

where ν is Poisson's ratio, which is assumed to be the same in the fiber and the matrix, h is the specimen thickness and d is the sliding distance of the fiber.

However, as noted originally by Jero and co-workers (1990, 1991), the fiber roughness has a significant effect on interfacial sliding phenomena, in addition to the effects of interfacial clamping pressure. Detailed analyses by Carter et al. (1990) modeled the asperity contacts as Hertzian contacts that result in a sinusoidal variation of sliding stress. Work by Kerans and Parthasarathy (1991) has also considered the effects of asperity pressure on fiber debonding and sliding, as well as the effects of abrasion of fibers.

The interfacial roughness introduces a misfit, ε_T, that depends on the roughness along the interface:

$$\varepsilon_T = \varepsilon_a + \frac{\delta(z)}{R} \tag{10.56}$$

where ε_a is a misfit due to the thermal expansion mismatch between the matrix and the fiber (along the embedded fiber length), and $\delta(z)/R$ is the local misfit strain, which induces an additional pressure, p. If we assume that the local interfacial stress is predominantly Coulombic, then the interfacial stress is given by

$$\tau = \mu(\sigma_r + p) \tag{10.57}$$

If the asperity pressure is obtained from the integral of the asperity pressure over the fiber length, then the fiber push-out stress may be expressed as (Mackin et al., 1992):

$$\begin{aligned} \sigma_u(d) = {} & \frac{\varepsilon_a}{2B}[\exp(2\mu Bt/R(1 - d/t)) - 1] \\ & + (2E\mu/R^2)\exp(-2\mu Bd/R)\int_d^h [\exp(2_z/R)]\delta(z)dz \end{aligned} \tag{10.58}$$

where d is the push-out distance (Fig. 10.10) and the other constants have their usual meaning.

The above discussion has focused largely on the fiber push-out test. However, other tests may also be used to measure the interfacial shear strengths of composites. These include fiber pull-out tests (Fig. 10.8) and fiber fragmentation tests (Fig. 10.11) that relate the fiber length distributions of fractured segments (obtained by pulling individual fibers) to the inter-

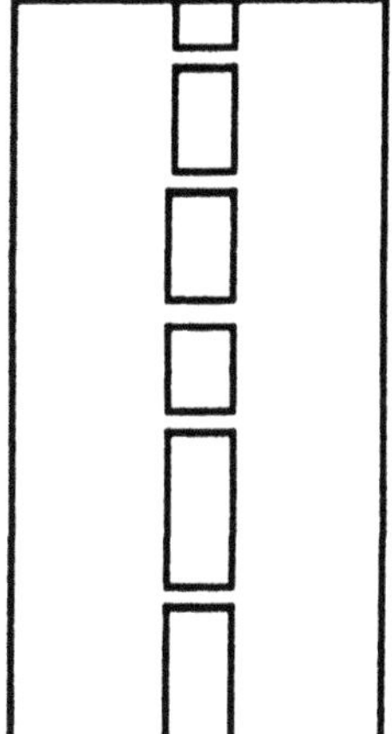

FIGURE 10.11 Schematic of fiber length distributions in the fiber fragmentation test

facial strengths between the matrix and fiber. Details of the fiber fragmentation test can be found in a paper by Ho and Drzal (1995).

10.9 SUMMARY

Further topics in composite deformation are presented in this chapter, following the introduction to composites in Chap. 9. The chapter begins with a systematic introduction to composite ply theory. This includes a detailed treatment of single- and multi-ply laminates. It is followed by a description of how the components of the *ABD* matrices can be used in composite design. Composite failure criteria are then presented before introducing the shear lag theory for whisker or short fiber-reinforced composites. Finally, the role of interfaces is considered in a section that includes considerations of debonding and fiber sliding during fiber push-out testing to determine interfacial properties.

BIBLIOGRAPHY

Azzi, V.D. and Tsai, S.W. (1965) Exp Mech. vol. 5, pp 283–288.
Carter, W.C., Butler, E.P., and Fuller, E.R. (1991) Scripta Metall Mater. vol. 25, p. 579.
Charalambides, P.G. and Evans, A.G. (1989) J Am Ceram Soc. vol. 72, pp 746–753.
Cox, H.L. (1952) J Appl Phys. vol. 3, p 72.
Eshelby, J.D. (1957) Proc Roy Soc. vol. 241, 1957, pp 376–396.

Eshelby, J.D. (1959) Proc Roy Soc. vol. A252, pp 561–569.
Evans, A.G. (1990) Perspective on the development of high toughness ceramics. J Am Ceram Soc. vol. 73, pp 187–206.
Evans, A.G. and Marshall, D.B. (1989) Acta Metall Mater. vol. 32, pp 2567–2583.
Evans, A.G., Zok, F.W., and Davis. J. (1991) Comp Sci Technol. vol. 42, pp 3–24.
Ho, H. and Drzal, L.T. (1995) Comp Eng. vol. 5, pp 1231–1244.
Hutchinson, J.W. and Jensen, H.M. (1990) Mech Mater. vol. 9, p 139.
Jero, P.D. and Kerans, R.J. (1990) Scripta Metall Mater. vol. 24, pp 2315–2318.
Jero, P.D., Kerans, R.J., and Parthasarathy, T.A. (1991) J Am Ceram Soc. vol. 74, pp 2793–2801.
Kerans, R.J. and Parthasarathy, T.A. (1991) J Am Ceram Soc. vol. 74, pp 1585–1590.
Liang, C. and Hutchinson, J.W. (1993) Mech Mater. vol. 14, pp 207–221.
Mackin, T.J., Warren, P.D., and Evans, A. G. (1992) Effects of fiber roughness on interface sliding in composites. Acta Metall. vol. 40, pp 1251–1257.
Matthews, F.L. and Rawlings, R.D. (1994) Composite Materials: Engineering and Science. Chapman and Hall, New York, pp 223–250.
Nardone, V.C. and Prewo, K. M. (1986) Scripta Metall. vol. 20, pp. 43–48.
Soboyejo, W.O., Rabeeh, B., Li, Y., Chu, W., Lavrentenyev, A., and Rokhlin, S. (1977) Metall Mater Trans. vol. 28A, pp 1667–1687.
Taya, M. and Arsenault, M. (1989) Metal Matrix Composites. Pergamon Press, New York, pp 25–28.

11

Fundamentals of Fracture Mechanics

11.1 INTRODUCTION

In the 17th century, the great scientist and painter, Leonardo da Vinci, performed some strength measurements on piano wire of different lengths. Somewhat surprisingly, he found that the strength of piano wire decreased with increasing length of wire. This length-scale dependence of strength was not understood until the 20th century when the serious study of fracture was revisited by a number of investigators. During the first quarter of the 20th century, Inglis (1913) showed that notches can act as stress concentrators. Griffith (1921) extended the work of Inglis by deriving an expression for the prediction of the brittle stress in glass. Using thermodynamic arguments, and the concept of notch concentration factors from Inglis (1913), he obtained a condition for unstable crack growth in brittle materials such as glass. However, Griffith's work neglected the potentially significant effects of plasticity, which were considered in subsequent work by Orowan (1950).

Although the work of Griffith (1921) and Orowan (1950) provided some insights into the role of cracks and plasticity in fracture, robust engineering tools for the prediction of fracture were only produced in the late 1950s and early 1960s after a number of well-publicized failures of ships and aircraft in the 1940s and early to mid-1950s. Some of the failures included the fracture of the so-called Liberty ships in World War II (Fig. 11.1) and

FIGURE 11.1 Fractured T-2 tanker, the S.S. Schenectady, which failed in1941. (Adapted from Parker, 1957—reprinted with permission from the National Academy of Sciences.)

the Comet aircraft disaster in the 1950s. These led to significant research and development efforts at the U.S. Navy and the major aircraft producers such as Boeing.

The research efforts at the U.S. Naval Research laboratory were led by George Irwin, who may be considered as the father of fracture mechanics. Using the concepts of linear elasticity, he developed a crack driving force parameter that he called the stress intensity factor (Irwin, 1957). At around the same time, Williams (1957) also developed mechanics solutions for the crack-tip fields under linear elastic fracture mechanics conditions. Work at the Boeing Aircraft Company was pioneered by a young graduate student, Paul Paris, who was to make important fundamental contributions to the subject of fracture mechanics and fatigue (Paris and coworkers, 1960, 1961, 1963) that will be discussed in Chap. 14.

Following the early work on linear elastic fracture mechanics, it was recognized that further work was needed to develop fracture mechanics approaches for elastic–plastic and fully plastic conditions. This led to the development of the crack-tip opening displacement (Wells, 1961) and the J integral (Rice, 1968) as a parameter for the characterization of the crack driving force under elastic–plastic fracture mechanics conditions. Three-parameter fracture mechanics approaches have also been proposed by McClintock et al. (1995) for the characterization of the crack driving force under fully plastic conditions.

The subject of fracture mechanics is introduced in this chapter. The chapter begins with a brief description of Griffith fracture theory, the Orowan plasticity correction, and the concept of the energy release rate. This is followed by a derivation of the stress intensity factor, K, and some illustrations of the applications and limitations of K in linear elastic fracture mechanics. Elastic–plastic fracture mechanics concepts are then introduced along with two-parameter fracture concepts for the assessment of constraint. Finally, the relative new subject of interfacial fracture mechanics is presented, along with the fundamentals of dynamic fracture mechanics.

11.2 FUNDAMENTALS OF FRACTURE MECHANICS

It is now generally accepted that all engineering structures and components contain three-dimensional defects that are known as cracks. However, as discussed in the introduction, our understanding of the significance of cracks has only been developed during the past few hundred years, with most of the basic understanding emerging during the last 50 years of the 20th century.

11.3 NOTCH CONCENTRATION FACTORS

Inglis (1913) modeled the stress concentrations around notches with radii of curvature, ρ, and notch length, a (Fig. 11.2). For elastic deformation, he was able to show that the notch stress concentration factor, K_t, is given by

$$K_t = \frac{\text{maximum stress around notch tip}}{\text{Remote stress away from notch}} = 1 + 2\sqrt{\frac{a}{\rho}} \tag{11.1}$$

Hence, for circular notches with $a = \rho$, he was able to show that $K_t \sim 3$. This rather large stress concentration factor indicates that an applied stress of σ is amplified by a factor of 3 at the notch tip. Failure is, therefore, likely to initiate from the notch tip, when the applied/remote stresses are significantly below the fracture strength of the un-notched material. Subsequent work by Neuber (1945) extended the work of Inglis to include the effects of notch plasticity on stress concentration factors. This has resulted in the publication of handbooks of notch concentration factors for various notch geometries.

Returning now to Eq. (11.1), it is easy to appreciate that the notch concentration factor will increase dramatically, as the notch-tip radius

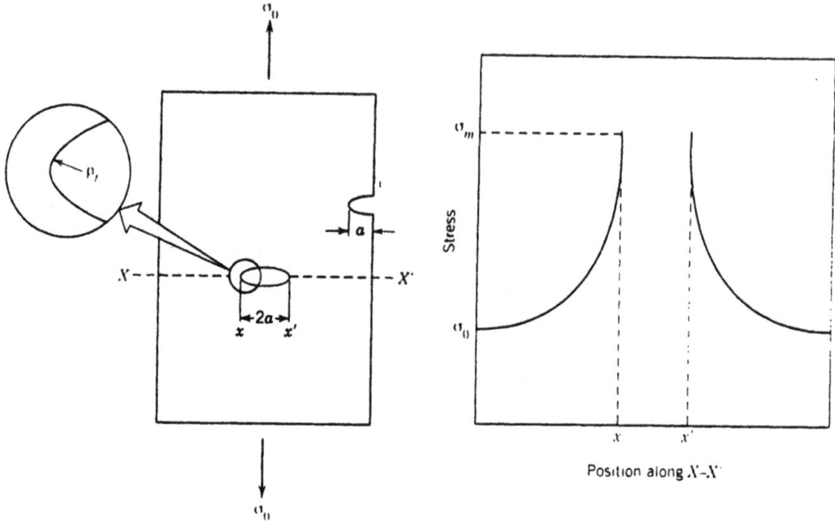

Figure 11.2 Stress concentration around a notch. (Adapted from Callister, 1999—reprinted with permission from John Wiley.)

approaches the limiting value corresponding to a single lattice spacing, b. Hence, for an atomistically sharp crack, the relatively high levels of stress concentration are likely to result in damage nucleation and propagation from the crack tip.

11.4 GRIFFITH FRACTURE ANALYSIS

The problem of crack growth from a sharp notch in a brittle solid was first modeled seriously by Griffith (1921). By considering the thermodynamic balance between the energy required to create fresh new crack faces, and the change in internal (strain) energy associated with the displacement of specimen boundaries (Fig. 11.3), he was able to obtain the following energy balance equation:

$$U_T = -\frac{\pi\sigma^2 a^2 B}{E'} + 4a\gamma_s B \tag{11.2}$$

where the first half on the right-hand side corresponds to the strain energy and the second half of the right-hand side is the surface energy due to the upper and lower faces of the crack, wihch have a total surface area of $4aB$. Also, σ is the applied stress, a is the crack length, B is the thickness of the specimen, $E' = E/(1-\nu^2)$ for plane strain, and $E' = E$ for plane stress,

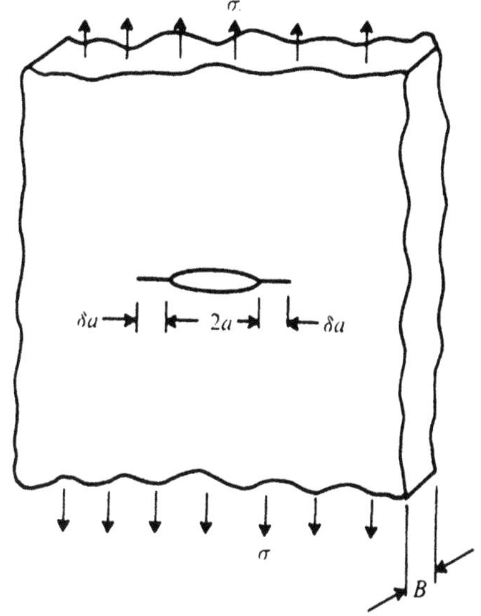

Figure 11.3 A center crack of length 2*a* in a large plate subjected to elastic deformations. (Adapted from Suresh, 1999—reprinted with permission from Cambridge University Press.)

where E is Young's modulus, ν is Poisson's ratio, and γ_s is the surface energy associated with the creation of the crack faces.

The critical condition at the onset of unstable equilibrium is determined by equating the first derivative of Eq. 11.2 to zero, i.e., $\mathrm{d}U_T/\mathrm{d}a = 0$. This gives

$$\frac{\mathrm{d}U_T}{\mathrm{d}a} = -\frac{2\pi\sigma^2 aB}{E'} + 4\gamma B = 0 \tag{11.3}$$

or

$$\sigma_c = \sqrt{\frac{2\gamma E'}{\pi a}} \tag{11.4}$$

where σ_c is the Griffith fracture stress obtained by rearranging Eq. (11.3), and the other terms have their usual meaning. Equation (11.4) does not account for the plastic work that is done during the fracture of most materials. It is, therefore, only applicable to very brittle materials in which no plastic work is done during crack extension.

Equation (11.4) was modified by Orowan (1950) to account for plastic work in materials that undergo plastic deformation prior to catastrophic failure. Orowan proposed the following expression for the critical fracture condition, σ_c :

$$\sigma_c = \sqrt{\frac{2(\gamma_s + \gamma_p)E'}{\pi a}} \tag{11.5}$$

where γ_p is a plastic energy term, which is generally difficult to measure independently.

Another important parameter is the strain energy release rate, G, which was first proposed by Irwin (1964). This is given by:

$$G = -\frac{1}{B}\frac{d(U_L + U_E)}{da} \tag{11.6}$$

where U_L is the potential energy of the loading system, U_E is the strain energy of the body, and B is the thickness of the body. Fracture should initiate when G reaches a critical value, G_c, which is given by

$$G_C = 2(\gamma_s + \gamma_p) \tag{11.7}$$

11.5 ENERGY RELEASE RATE AND COMPLIANCE

This section presents the derivation of energy release rates and compliance concepts for prescribed loading [Fig. 11.4(a)] and prescribed displacement [Fig. 11.4(b)] scenarios. The possible effects of machine compliance are considered at the end of the section.

11.5.1 Load Control or Deadweight Loading

Let us start by considering the basic mechanics behind the definition of the energy release rate of a crack subjected to remote load, F, Fig. 11.4(a). Also, u is the load point displacement through which load F is applied. The energy release rate, G, is defined as

$$G = -\frac{1}{B}\left(\frac{\partial \text{PE}}{\partial a}\right) \tag{11.8}$$

where B is the thickness of the specimen, PE is the potential energy, and a is the crack length. The potential energy for a system with prescribed load, F, is given by (Fig. 11.5):

$$\delta\phi = \text{PE} = \text{SE} - \text{WD} = \frac{1}{2}\text{F}\delta\text{u} - \text{F}\delta\text{u} = -\frac{1}{2}\text{F}\delta\text{u} \tag{11.9}$$

where SE is the strain energy and WD is the work done. By definition, the compliance, C, of the body is given simply by

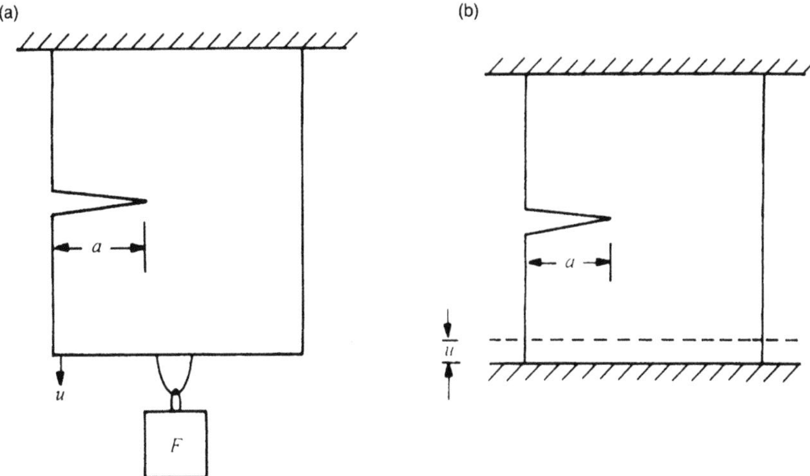

Figure 11.4 Schematic of notched specimens subject to (a) prescribed loading or (b) prescribed displacement. (Adapted from Suresh, 1999—reprinted with permission from Cambridge University Press.)

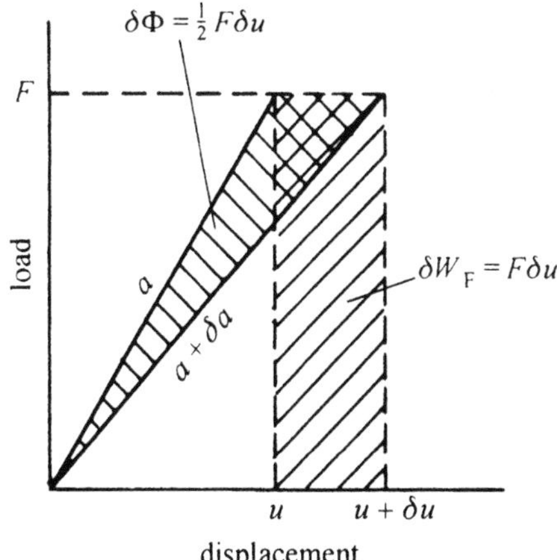

Figure 11.5 Schematic of load–displacement curve under prescribed load. (Adapted from Suresh, 1999—reprinted with permission from Cambridge University Press.)

$$C = \frac{u}{F} \tag{11.10}$$

where C depends on geometry and elastic constants E and ν. Hence, the potential energy is given by [Eqs (11.9) and (11.10)]:

$$\mathrm{PE} = -\frac{1}{2}\mathrm{Fu} = -\frac{1}{2}\mathrm{F(CF)} = -\frac{1}{2}\mathrm{F}^2\mathrm{C} \tag{11.11}$$

and the energy release rate is obtained by substituting Eq. (11.11) into Eq. (11.8) to give

$$G = \frac{1}{B}\frac{\partial \mathrm{PE}}{\partial a} = \frac{1}{2B}F^2\frac{dC}{da} \tag{11.12}$$

11.5.2 Displacement-Controlled Loading

Let us now consider the case in which the displacement is controlled [Fig. 11.4(a)], and the load, F, varies accordingly (Fig. 11.6). When the crack advances by an amount Δa under a fixed displacement, u, the work done is zero and hence the change in potential energy is equal to the strain energy.

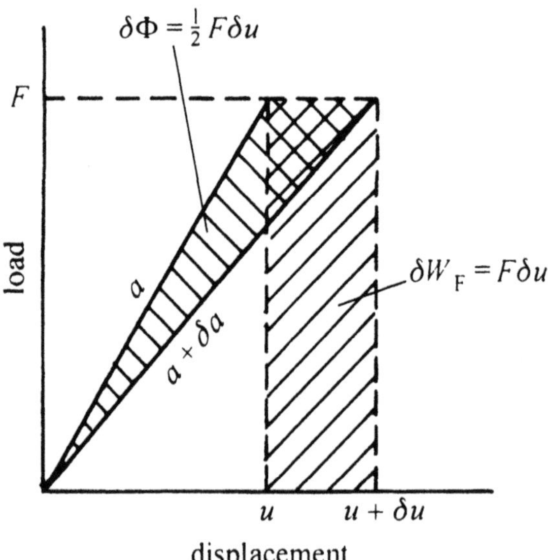

Figure 11.6 Schematic of load–displacement curve under prescribed displacement. (Adapted from Suresh, 1999—reprinted with permission from Cambridge University Press.)

$$\delta\phi\Big|_{\substack{u\\ \text{fixed}}} - \delta\text{PE}\Big|_{\substack{u\\ \text{fixed}}} = \text{SE} = -\text{G}\delta\text{A} = -\text{GB}\delta\text{a} \tag{11.13}$$

where δA is the change in crack area, dA = Bδa. Hence, rearranging Eq. (11.13) now gives

$$G = \frac{-1}{B}\left(\frac{\partial\text{PE}}{\partial a}\right)\Bigg|_{\substack{u\\ \text{fixed}}} \tag{11.14}$$

The strain energy, SE, is given simply by the shaded area in Figure 11.6:

$$\text{SE} = \frac{1}{2}\text{u}\delta\text{F} = -\frac{\text{F}}{2}\delta\text{F} \tag{11.15}$$

Since the compliance is u/F, then from Equation 11.15, we have:

$$G = \frac{-1}{B}\frac{\partial\text{PE}}{\partial a}\Bigg|_{\substack{u\\ \text{fixed}}} = \frac{-u}{2B}\frac{\partial F}{\partial a}\Bigg|_{\substack{u\\ \text{fixed}}} = \frac{F^2}{2B}\frac{\text{d}C}{\text{da}} \tag{11.16}$$

Hence, the expression for the energy release rate is the same for displacement control [Eq. (11.16)] and load control [Eq. (11.12)]. It is important to note the above equations for G are valid for both linear and non-linear elastic deformation. They are also independent of boundary conditions.

11.5.3 Influence of Machine Compliance

Let us now consider the influence of machine compliance, C_M, on the deformation of the cracked body shown in Fig. 11.7. The total displacement,

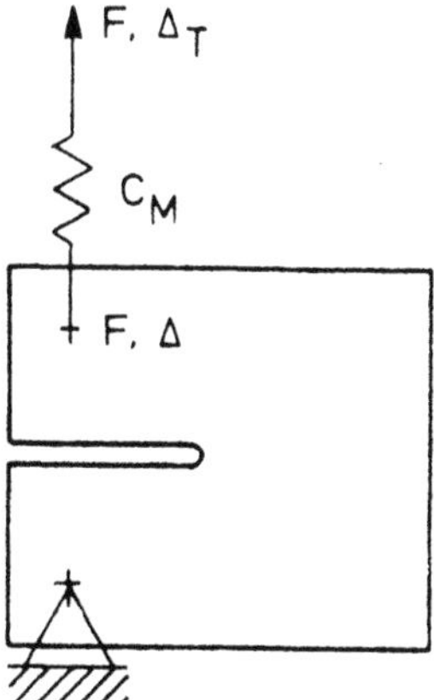

Figure 11.7 Schematic of deformation in a compliant test machine. (Adapted from Hutchinson, 1979—with permission from the Technical University of Denmark.)

Δ_T, is now the sum of the machine displacement, Δ_M, and the specimen displacement, Δ. If the total displacement is prescribed, then we have

$$\Delta_T = \Delta + \Delta_M = \Delta + FC_M \tag{11.17}$$

since $C = \Delta/F$, we may also write:

$$\Delta_T = \Delta + \frac{C_M}{C}\Delta \tag{11.18}$$

The potential energy is now given by

$$\mathrm{PE} = \mathrm{SE} + \frac{1}{2}C_M F^2 = -\frac{1}{2B}C^{-1}\Delta^2 + \frac{1}{2B}C_M^{-1}(\Delta_T - \Delta)^2 \tag{11.19}$$

and the energy release rate is

$$\begin{aligned} G = -\left(\frac{\partial \mathrm{PE}}{\partial a}\right)_{\Delta_T} &= -\left[\frac{C^{-1}\Delta - C_M^{-1}(\Delta_T - \Delta)}{B}\right]\left(\frac{\partial A}{\partial a}\right)_{\Delta_T} \\ &\quad + \frac{1}{2B}C^{-2}\Delta^2\frac{dC}{da} \\ &= \frac{1}{2B}C^{-2}\Delta^2\frac{dC}{da} = \frac{1}{2B}F^2\frac{dC}{da} \end{aligned} \tag{11.20}$$

Hence, as before, the energy release rate does not depend on the nature of the loading system. Also, the measured value of G does not depend on the compliance of the loading system. However, the experimental determination of G is frequently done with rigid loading systems that correspond to $C_M = 0$.

11.6 LINEAR ELASTIC FRACTURE MECHANICS

The fundamentals of linear elastic fracture mechanics (LEFM) are presented in this section. Following the derivation of the crack-tip fields, the physical basis for the crack driving force parameters is presented along with the conditions required for the application of LEFM. The equivalence of G and the LEFM cracking driving force (denoted by K) is also demonstrated.

11.6.1 Derivation of Crack-Tip Fields

Before presenting the derivation of the crack-tip fields, it is important to note here that there are three modes of crack growth. These are illustrated schematically in Fig. 11.8. Mode I [Fig. 11.8(a)] is generally referred to as the crack opening mode. It is often the most damaging of all the loading modes. Mode II [Fig. 11.8(b)] is the in-plane shear mode, while Mode III [Fig. 11.8(c)] corresponds to the out-of-plane shear mode. Each of the

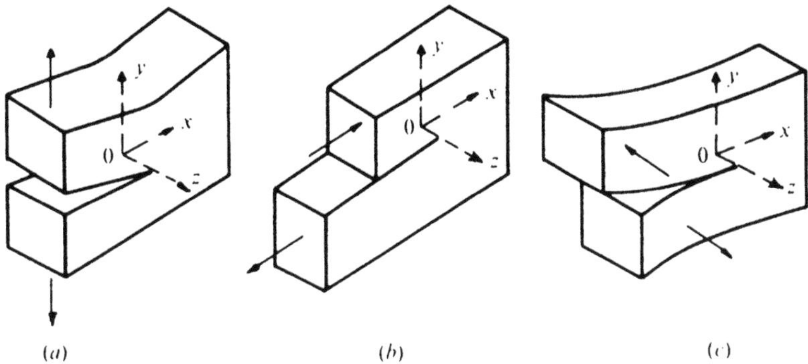

Figure 11.8 Modes of crack growth: (a) Mode I; (b) Mode II; (c) Mode III. (Adapted from Suresh, 1999—reprinted with permission from Cambridge University Press.)

modes may occur separately or simultaneously. However, for simplicity, we will derive the crack-tip fields for pure Mode I crack growth. We will then extend our attention to Modes II and III.

Now, let us begin by considering the equilibrium conditions for a plane element located at a radial distance, r, from the crack-tip (Fig. 11.9). For equilibrium in the polar co-ordinate system, the equilibrium equations are given by

$$\frac{\partial \sigma_{rr}}{\partial_r} + \frac{1}{r}\frac{\partial \sigma_{r\theta}}{\partial \theta} + \frac{\sigma_{rr} - \sigma_{\theta\theta}}{r} = 0 \tag{11.21a}$$

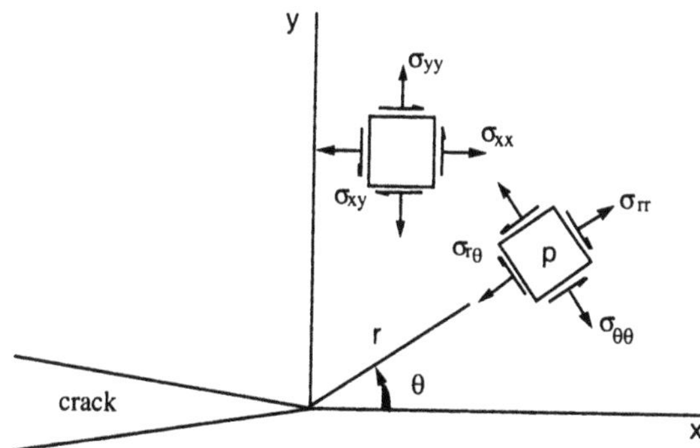

Figure 11.9 In-plane co-ordinate system and crack-tip stresses.

$$\frac{\partial \sigma_{r\theta}}{\partial r}+\frac{1}{r}\frac{\partial \sigma_{\theta\theta}}{\partial \theta}+\frac{2\sigma_{r\theta}}{r}=0 \tag{11.21b}$$

The in-plane strain components are given by

$$\varepsilon_{rr}=\frac{\partial u_r}{\partial r} \tag{11.21c}$$

$$\varepsilon_{\theta\theta}=\frac{u_r}{r}+\frac{1}{r}\frac{\partial u_\theta}{\partial \theta} \tag{11.21d}$$

$$\varepsilon_{r\theta}=\frac{1}{2}\left[\frac{1}{r}\frac{\partial u_r}{\partial \theta}+\frac{\partial u_\theta}{\partial r}-\frac{u_\theta}{r}\right] \tag{11.21e}$$

Finally, for strain compatibility we must satisfy

$$\frac{\partial^2 \varepsilon_{\theta\theta}}{\partial r^2}+\frac{2}{r}\frac{\partial \varepsilon_{\theta\theta}}{\partial r}-\frac{1}{r}\frac{\partial^2 \varepsilon_{r\theta}}{\partial r\partial \theta}-\frac{1}{r^2}\frac{\partial \varepsilon_{r\theta}}{\partial r\theta}+\frac{1}{r^2}\frac{\partial \varepsilon_{rr}^2}{\partial \theta^2}-\frac{1}{r}\frac{\partial \varepsilon_{rr}}{\partial r}=0 \tag{11.22}$$

For the in-plane problem (Modes I and II), the crack-tip strains are only functions of r and θ. Also, for the plane stress problem $\sigma_{zz}=0$. The relationship between stress and strain is given by Hooke's law:

$$E\varepsilon_{rr}=\sigma_{rr}-\nu\sigma_{\theta\theta} \tag{11.23a}$$

$$E\varepsilon_{\theta\theta}=\sigma_{\theta\theta}-\nu\sigma_{rr} \tag{11.23b}$$

$$2\mu\varepsilon_{r\theta}=\mu\gamma_{r\theta}=\sigma_{r\theta} \tag{11.23c}$$

As discussed in Chap. 4, the solutions to the above equations are satisfied by the Airy stress function via:

$$\sigma_{rr}=\frac{1}{r}\frac{\partial \chi}{\partial r}+\frac{1}{r^2}\frac{\partial^2 \chi}{\partial \theta^2} \tag{11.24a}$$

$$\sigma_{\theta\theta}=\frac{\partial^2 \chi}{\partial r^2} \tag{11.24b}$$

$$\sigma_{r\theta}=-\frac{\partial}{\partial r}\left(\frac{1}{r}\frac{\partial \chi}{\partial \theta}\right) \tag{11.24c}$$

When the compatibility condition is expressed in terms of the Airy stress function, we obtain a biharmonic equation of the form:

$$\nabla^2(\nabla^2 \chi)=0 \tag{11.25a}$$

where

$$\nabla^2 \equiv \frac{\partial}{\partial r^2} + \frac{1}{r}\frac{\partial}{\partial r} + \frac{1}{r^2}\frac{\partial^2}{\partial \theta^2} \tag{11.25b}$$

Since the crack faces are traction free, $\sigma_{\theta\theta} = \sigma_{r\theta} = 0$ at $\theta = \pm\pi$. Also, the Airy stress function (trial function) must be single valued and have the appropriate singularity at the crack-tip. One possible form of the airy stress function is

$$\chi = r^2 p(r, \theta) + q(r, \theta) \tag{11.26}$$

where $p(r, \theta)$ and $q(r, \theta)$ are both harmonic functions that satisfy Laplace's equations, i.e., $\nabla^2 p = 0$ and $\nabla^2 q = 0$. Following the approach of Williams (1957), let us consider expressions for $p(r, \theta)$ and $q(r, \theta)$ that are of separable form $\chi = R(r) \cdot \theta(\theta)$. From Williams (1957) we have

$$p(r, \theta) = A_1 r^\lambda \cos \lambda\theta + A_2 r^\lambda \sin \lambda\theta \tag{11.27a}$$

$$q(r, \theta) = B_1 r^{\lambda+2} \cos(\lambda + 2)\theta + B_2 r^{\lambda+2} \sin(\lambda + 2)\theta \tag{11.27b}$$

From Eq. (11.26) we may now write the following expression for χ:

$$\begin{aligned} \chi = {} & r^{\lambda+2}[A_1 \cos \lambda\theta + B_1 \cos(\lambda + 2)\theta] \\ & + r^{(\lambda+2)}[A_2 \sin \lambda\theta + B_2 \sin(\lambda + 2)\theta] \end{aligned} \tag{11.28}$$

The solution to the first expression in brackets (the symmetric part) corresponds to the Mode I solution for the crack-tip fields, while the solution to the second expression in brackets (the antisymmetric part) corresponds to the Mode II solution. For pure Mode I, we may obtain the crack-tip fields by substituting the terms in the brackets on the left-hand side into the Airy stress function expressions given by Eqs (11.24):

$$\sigma_{\theta\theta} = \frac{\partial^2 \chi}{\partial r^2} = (\lambda + 2)(\lambda + 1) r^\lambda [A_1 \cos \lambda\theta + B_1 \cos(\lambda + 2)\theta] \tag{11.29a}$$

$$\sigma_{r\theta} = [\frac{\partial}{\partial r}\left(\frac{1}{r}\frac{\partial \chi}{\partial \theta}\right) = (\lambda + 1) r^\lambda [\lambda A_1 \sin \lambda\theta + (\lambda + 2) B_1 \sin(\lambda + 2)\theta] \tag{11.29b}$$

Also, since the crack faces are traction free, $\sigma_{\theta\theta} = \sigma_{r\theta} = 0$ at $\theta = \pm\pi$. This gives

$$(A_1 + B)\cos(\lambda\pi) = 0 \tag{11.30a}$$

and

$$[\lambda A_1 + (\lambda + 2) B_1] \sin(\lambda\pi) = 0 \tag{11.30b}$$

Equations (11.30a) and (11.30b) are satisfied if

$$A_1 = -B \text{ and } \sin(\lambda\pi) = 0 \text{ (hence, } \lambda = Z, \text{ where } Z \text{ is an integer)} \tag{11.31}$$

or

$$A_1 = 0\frac{(\lambda+2)B_1}{\lambda} \text{ and } \cos(\lambda\pi) = 0 \left(\text{this gives A} = \frac{2Z+1}{2}\right) \tag{11.32}$$

since Equations (11.31) and (11.32) are linear, then any linear combination of solutions is admissible. Hence, $\lambda = Z/2$ is admissible. It is important to note that other linear combinations of solutions are also admissible. Furthermore, there is no basis to reject any value of λ. However, the solution can be chosen to reflect the lowest order singularity based on physical arguments. From Eq. (11.29), we note that $\sigma_{ij} \sim r^\lambda$ and $\varepsilon_{ij} \sim r^\lambda$. Hence, the strain energy density, ϕ, is given by

$$\phi = \frac{1}{2}\sigma_{ij}\varepsilon_{ij} \sim r^{2\lambda} \tag{11.33}$$

The total strain energy, Φ, in the annular region at the crack tip is given by the following integration:

$$\Phi = \int_0^{2\pi}\int_{r_0}^{R} \frac{1}{2}\sigma_{ij}\varepsilon_{ij} r\, dr\, d\theta = \int_0^{2\pi}\int_{r_0}^{2\pi} r^{2\lambda+1} dr\, d\theta \tag{11.34}$$

However, Φ must be finite ($\Phi < \infty$) and this requires that $\lambda > -1$. Also, since the displacements $u_I \sim r^{\lambda+1}$, finite displacement requires $\lambda > -1$. Hence, the admissible values of $\lambda = \frac{z}{2}$ are given by

$$\lambda = -\frac{1}{2}, 0, +\frac{1}{2}, 1, \frac{3}{2}, 2\ldots, \frac{Z}{2} \tag{11.35}$$

The most singular term is given by $\lambda = -1/2$, for which $B_1 = A_1/3$. Substituting these into Eq. (11.28) for χ now gives

$$\chi = r^{3/2}A_1\left[\cos\frac{\theta}{2} + \frac{1}{3}\cos\frac{3\theta}{2}\right] + O(r^2) + O(r^{5/2}) \tag{11.36a}$$

or

$$\sigma_{ij} = A_1 r^{-\frac{1}{2}}\tilde{\sigma}^{I}_{ij}(\theta) + \mathrm{O}_{ij}(Ir^0) + O_{ij}(r^{1/2}) \tag{11.36b}$$

The second term on the right-hand side (with an exponent of zero) is a nonsingular, nonvanishing term. The higher order terms, with exponents greater than zero, become insignificant as $r \to 0$. Hence, replacing A with $a/\sqrt{2\pi}$ now gives

$$\sigma_{ij} = \frac{\text{I}}{\sqrt{2\pi r}}\sigma_{ij}^{\text{I}}(\theta) + T\delta_{ix}\delta_{jx} + \text{ insignificant higher order terms} \tag{11.37}$$

where K_I is the Mode I stress intensity factor, $\sigma_{ij}^{\text{I}}(\theta)$ is a function of θ for Mode I conditions, T is the so-called T stress (Williams, 1957; Irwin, 1960; Larsson and Carlsson, 1973; Rice, 1974; Bilby et al., 1986), and δ_{ij} is the Kronecker delta defined in Chap. 4 (Section 4.6). The in-plane crack-tip stresses may thus be expressed as

$$\begin{bmatrix} \sigma_{xx} & \sigma_{xy} \\ \sigma_{xy} & \sigma_{yy} \end{bmatrix} = \frac{K_1}{\sqrt{2\pi r}} \begin{bmatrix} \tilde{\sigma}_{xx}^{1}(\theta) & \tilde{\sigma}_{xy}^{\text{I}}(\theta) \\ \tilde{\sigma}_{yx}^{\text{I}}(\theta) & \tilde{\sigma}_{yy}^{\text{I}}(\theta) \end{bmatrix} + \begin{bmatrix} T & 0 \\ 0 & 0 \end{bmatrix} \tag{11.38}$$

In general, the T term is only used to characterize the crack-tip fields when constraint (crack-tip triaxiality) is important. It is also used in the characterization of short fatigue cracks and mixed mode cracks (Suresh, 1999). In most cases, however, the T stress is not considered. For the Cartesian co-ordinate system, the resulting Mode I crack-tip fields are given by

$$\begin{bmatrix} \sigma_{xx} \\ \sigma_{yy} \\ \sigma_{xy} \end{bmatrix} = \frac{K_I}{\sqrt{2\pi r}}\cos\frac{\theta}{2} \begin{bmatrix} 1 - \sin\frac{\theta}{2}\sin\frac{3\theta}{2} \\ 1 + \sin\frac{\theta}{2}\sin\frac{3\theta}{2} \\ \sin\frac{\theta}{2}\cos\frac{3\theta}{2} \end{bmatrix} \tag{11.39a}$$

Using the cylindrical co-ordinate system, the Mode I stress fields are given by

$$\begin{bmatrix} \alpha_{rr} \\ \sigma_{\theta\theta} \\ \sigma_{r\theta} \end{bmatrix} = \frac{K_1}{\sqrt{2\pi r}}\cos\frac{\theta}{2} \begin{bmatrix} 1 + \sin^2\left(\frac{\theta}{2}\right) \\ \cos^2\left(\frac{\theta}{2}\right) \\ \sin\left(\frac{\theta}{2}\right)\cos\left(\frac{\theta}{2}\right) \end{bmatrix} \tag{11.39b}$$

$$\sigma_{zz} = \nu_1(\alpha_{rr}+\sigma_{\theta\theta}) \tag{11.40}$$

$$\sigma_{xz} = \sigma_{yz} = \sigma_{rz} = \sigma_{\theta z} = 0 \tag{11.41}$$

The corresponding displacements for Mode I are given by

$$\begin{bmatrix} u_x \\ \\ u_y \end{bmatrix} = \frac{K_1}{2E}\sqrt{\frac{r}{2\pi}}\begin{bmatrix} 1+\nu)\left[(2\kappa-1)\cos\frac{\theta}{2}-\cos\frac{3\theta}{2}\right] \\ (1+\nu)\left[(2\kappa-1)\sin\frac{\theta}{2}-\sin\frac{3\theta}{2}\right] \end{bmatrix} \tag{11.42a}$$

$$\begin{bmatrix} u_r \\ \\ u_\theta \end{bmatrix} = \frac{K_1}{2E}\sqrt{\frac{r}{2\pi}}\begin{bmatrix} 1+\nu)\left[(2\kappa-1)\cos\frac{\theta}{2}-\cos\frac{3\theta}{2}\right] \\ (1+\nu)\left[-(2\kappa-1)\sin\frac{\theta}{2}-\sin\frac{3\theta}{2}\right] \end{bmatrix} \tag{11.42b}$$

where

$$\kappa = \frac{3-\nu}{1+\nu},\ V_1 = V,\ V_2 = 0 \text{ (for plane stress)} \tag{11.43a}$$

and

$$\kappa = (3-4\nu),\ \nu_1 = \nu,\ \nu_2 = 0 \text{ (for plane strain)} \tag{11.43b}$$

The term K_I in the above expressions is the amplitude of the crack-tip field under Mode I conditions. It represents *the driving force for crack growth* under these conditions. It is also important to note here that K_I is independent of elastic constants. Similarly, the near-tip fields for Mode II can be derived by applying the boundary conditions to the antisymmetric part of Eq. 11.28. The resulting solutions are

$$\begin{bmatrix} \sigma_{rr} \\ \\ \sigma_{\theta\theta} \\ \\ \sigma_{r\theta} \end{bmatrix} = \frac{K_{II}}{2\pi r}\begin{bmatrix} -\sin\frac{\theta}{2}\left(1-3\sin^2\frac{\theta}{2}\right) \\ -3\sin\frac{\theta}{2}\cos^2\frac{\theta}{2} \\ \cos\frac{\theta}{2}\left(1-3\sin^2\frac{\theta}{2}\right) \end{bmatrix} \tag{11.44}$$

$$\sigma_{zz} = \nu_1(\sigma_{rr}+\sigma_{\theta\theta}) \tag{11.45}$$

$$\sigma_{rz} = \sigma_{\theta z} = 0 \tag{11.46}$$

The displacements in Mode II are given by

$$\begin{bmatrix} u_r \\ \\ u_\theta \end{bmatrix} = \frac{K_{II}}{2E}\sqrt{\frac{r}{2\pi}}\begin{bmatrix} (1+\nu)\left[-2\kappa-1)\sin\frac{\theta}{2}+\sin\frac{3\theta}{2}\right] \\ (1+\nu)\left[-(2\kappa+1)\cos\frac{\theta}{2}+3\cos\frac{3\theta}{2}\right] \end{bmatrix} \tag{11.47}$$

$$u_z = \frac{\nu_2 Z}{E}(\sigma_{xx} + \sigma_{yy}) = \frac{\nu_2 Z}{E}(\sigma_n + \sigma_{\theta\theta}) \tag{11.48}$$

For Mode III:

$$\begin{bmatrix} \sigma_{rz} \\ \sigma_{\theta z} \end{bmatrix} = \frac{K_{\text{III}}}{\sqrt{2\pi r}} \begin{bmatrix} \sin\frac{\theta}{2} \\ \cos\frac{\theta}{2} \end{bmatrix} \tag{11.49}$$

$$\sigma_{rr} = \sigma_{\theta\theta} = \sigma_{zz} = 0 \tag{11.50}$$

$$u_z = \frac{K_{\text{III}}}{2E}\sqrt{\frac{r}{2\pi}}\left[2(1+\nu)\sin\frac{\theta}{2}\right] \tag{11.51}$$

$$u_r = u_\theta = 0 \tag{11.52}$$

Hence, in general, the expressions for the crack-tip fields and displacements may be expressed, respectively, by equations of the form:

$$\tilde{\sigma}_{ij} = \frac{K_{\text{M}}}{\sqrt{2\pi r}\left[\tilde{\sigma}_{ij}^{\text{M}}(\theta)\right]} \tag{11.53}$$

$$u_i^{\text{M}} = \frac{K_{\text{M}}}{2E}\sqrt{\frac{r}{2\pi}}\,\tilde{u}_i^{\text{M}}\left[\tilde{\sigma}_{ij}^{\text{M}}(\theta)\right] \tag{11.54}$$

where K_{M} is the stress intensity factor for a given mode of failure (I, II, or III), and the other terms have their usual meeting. The stress intensity factor for each mode may also be defined as

$$K_1 = \lim_{r\to 0}\left[\sqrt{2\pi r}\,\sigma_{yy}|_{\theta=0}\right] \tag{11.55a}$$

$$K_{\text{II}} = \lim_{r\to 0}|\sqrt{2\pi r}\,\sigma_{xy}|_{\theta=0}] \tag{11.55b}$$

$$K_{\text{III}} = \lim_{r\to 0}\left[\sqrt{2\pi r}\,\sigma_{xz}|_{\theta=0}\right] \tag{11.55c}$$

Stress intensity factor solutions for selected geometries are presented in Fig. 11.10. More comprehensive solutions for a wide range of geometries can be found in the handbooks by Tada et al. (1999), Sih (1973), and Murakami (1987).

11.6.2 Crack Driving Force and Concept of Similitude

According to Eqs (11.55), the stress fields of all cracks are identical, except for the scaling factor, K. Therefore, K can be used to represent the amplitude of the crack-tip fields. More importantly, however, K represents the driving force for crack growth under LEFM conditions. The stress intensity factor, K, is generally given by expressions of the form (Fig. 11.10):

$$K = f(a/W)\sigma\sqrt{\pi a} \tag{11.56}$$

where σ is the applied stress, $f(a/W)$ is a function of the crack length, a, and W is the width of the specimen/component. As discussed earlier, geometry functions for different fracture mechanics geometries are presented in Fig. 11.10. More complete summaries of geometry functions can also be found in fracture mechanics handbooks such as the ones by Tada et al. (1999), Sih (1973), and Murakami (1987). Most of these solutions apply to two-dimensional cracks (mostly through-thickness cracks that are relatively easy to monitor in crack growth experiments). Such cracks are found commonly in standard fracture mechanics specimens that include single-edge notched bend (SENB), compact tension (C-T), and other specimen geometries (Fig. 11.10).

Type	Stress Intensity Formulation	Configuration
1. Compact specimen* C(T)	$K = \frac{P}{BW^{1/2}} f(a/W)$ where $f(a/W) = \frac{(2 + a/W)}{(1 - a/W)^{3/2}} [0.886 + 4.64a/W - 13.32(a/W)^2 + 14.72(a/W)^3 - 5.6(a/W)^4]$	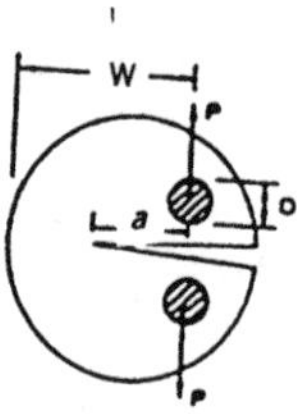 W, P, a, D, P B - THK
2. Disk-shaped compact specimen* DC(T)	$K = \frac{P}{BW^{1/2}} f(a/W)$ where $f(a/W) = \frac{(2 + a/W)}{(1 - a/W)^{3/2}} [0.76 + 4.8a/W - 11.58(a/W)^2 + 11.43(a/W)^3 - 4.08(a/W)^4]$	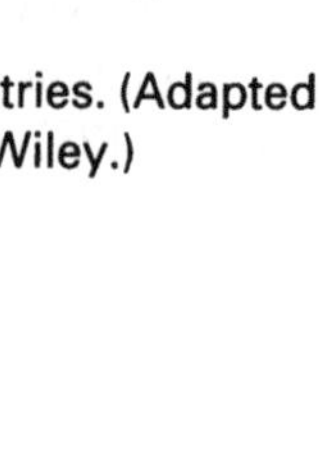 W, P, a, D, P B - THK

FIGURE 11.10a K solutions for common specimen geometries. (Adapted from Hertzberg, 1996—reprinted with permission from John Wiley.)

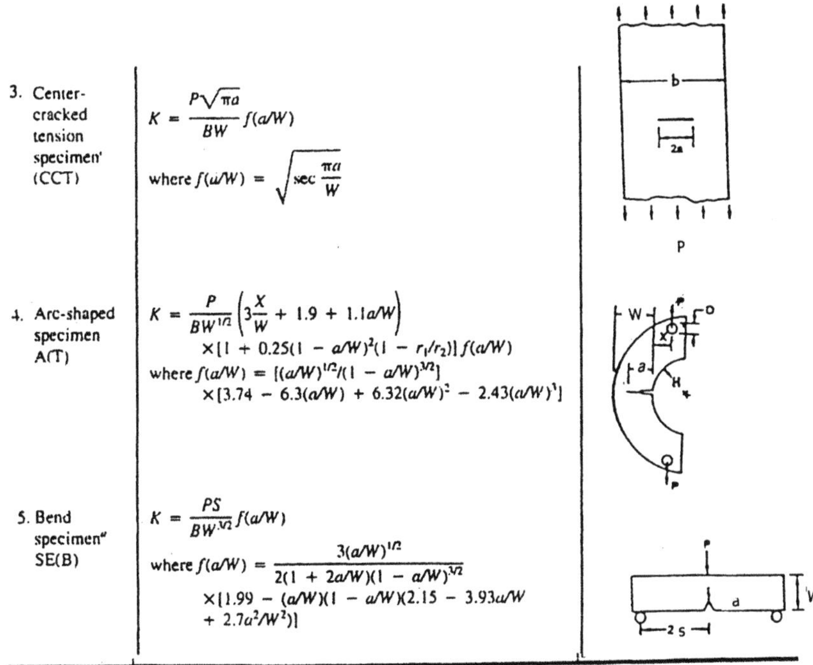

Specimen	Stress intensity factor
3. Center-cracked tension specimen[a] (CCT)	$K = \frac{P\sqrt{\pi a}}{BW} f(a/W)$ where $f(a/W) = \sqrt{\sec \frac{\pi a}{W}}$
4. Arc-shaped specimen A(T)	$K = \frac{P}{BW^{1/2}}\left(3\frac{X}{W} + 1.9 + 1.1a/W\right) \times [1 + 0.25(1 - a/W)^2(1 - r_1/r_2)] f(a/W)$ where $f(a/W) = [(a/W)^{1/2}/(1 - a/W)^{3/2}] \times [3.74 - 6.3(a/W) + 6.32(a/W)^2 - 2.43(a/W)^3]$
5. Bend specimen[a] SE(B)	$K = \frac{PS}{BW^{3/2}} f(a/W)$ where $f(a/W) = \frac{3(a/W)^{1/2}}{2(1 + 2a/W)(1 - a/W)^{3/2}} \times [1.99 - (a/W)(1 - a/W)(2.15 - 3.93a/W + 2.7a^2/W^2)]$

[a] ASTM Standard E 399-81, *Annual Book of ASTM Standards*, Part 10, 1981.

[b] C. E. Feddersen, ASTM *STP 410*, 1976, p. 77.

FIGURE 11.10 (continued).

However, most engineering structures and components contain surface cracks with semielliptical crack geometries (Fig. 11.11). The stress intensity factor solutions for such cracks have been obtained using detailed finite element solutions. The most comprehensive solutions for semielliptical cracks are by Newman and Raju (1982). These solutions show that the stress intensity factors vary with position around the semielliptical crack front. Also, depending on the aspect ratio (ratio of semimajor to semiminor axes) and the loading mode (tension or bending), the maximum stress intensity factor may occur either at the surface positions or at the deepest point along the crack front.

Finally in this section, it is important to note that the stress intensity factor has found widespread application due largely to the concept of *similitude*. This concept states simply that different crack geometries have the same crack driving force when the stress intensity factor at the crack tips are

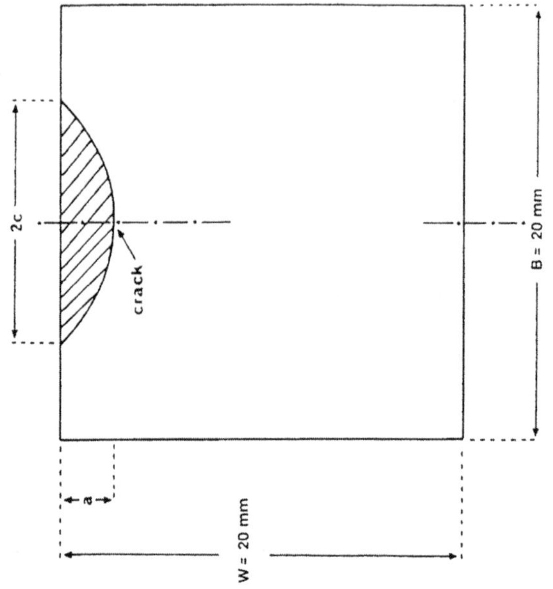

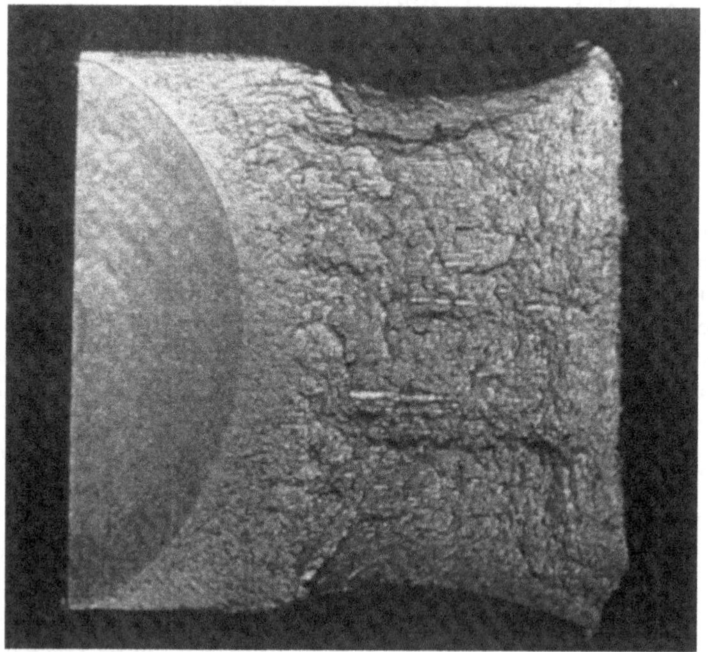

FIGURE 11.11 Semielliptical crack profiles in a Q1N pressure-vessel steel.

the same. Hence, the material response of a small laboratory specimen may be used to study the material response of a large structure to the same stress intensity factor.

11.6.3 Plastic Zone Size

11.6.3.1 Plastic Zone Size for Plane Stress and Plane Strain Conditions

The singularity of stress at the crack tip suggests that crack-tip stresses approach infinity. However, in reality, infinite stresses cannot be sustained. Hence, yielding occurs in an annular region around the crack-tip. The shape and size of this zone depends strongly on the stress state (plane stress versus plane strain) and the stress intensity factor, K, under small-scale yielding conditions (Fig. 11.12). In general, however, the size of the plastic zone may be estimated from the boundary of the region in which the tensile yield stress, σ_{ys}, is exceeded (Irwin, 1960). For the region ahead of the crack-tip in which $\theta = 0$, that gives

$$r_p = \frac{1}{3\pi}\left(\frac{K}{\sigma_{ys}}\right)^2 \qquad \text{(for plane strain)} \qquad (11.57a)$$

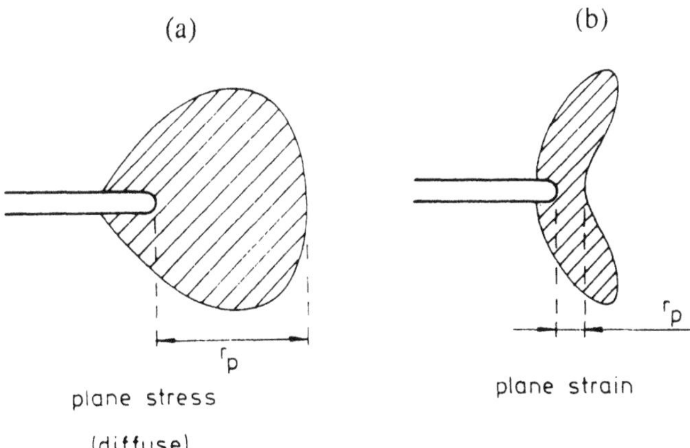

Figure 11.12 Schematics of plastic zones under (a) plane stress and (b) plane strain conditions. (Adapted from Hutchinson, 1979—reprinted with permission from the Technical University of Denmark.)

$$r_p = \frac{1}{\pi}\left(\frac{K}{\sigma_{ys}}\right)^2 \qquad \text{(for plane stress)} \tag{11.57b}$$

For plane strain conditions, it is important to note that the plastic zone size is less than the thickness, B, under small-scale yielding conditions in which $r_p \ll a$ ($r_p < a/50$). Also, the reader should note that the plastic zone size under plane strain conditions is less than that under plane stresses conditions.

11.6.3.2 Dugdale Model

Dugdale (1960) considered the problem of a Mode I crack in a thin plate of an elastic–perfectly plastic solid deformed under plane stress conditions. The Dugdale model proposes thin plastic strips of length, r_p, ahead of the two crack tips shown in Fig. 11.13. Tractions of magnitude $\sigma_{yy} = \sigma_{ys}$ are applied to these thin regions over the length, r_p. The tractions superimpose a negative stress intensity factor, K'', on the crack tips:

$$K'' = -\sigma_{ys}\sqrt{\pi(a + r_p)} + 2\sigma_{ys}\sqrt{\frac{a + r_p}{\pi}}\sin^{-1}\left(\frac{a}{a + r_p}\right) \tag{11.58}$$

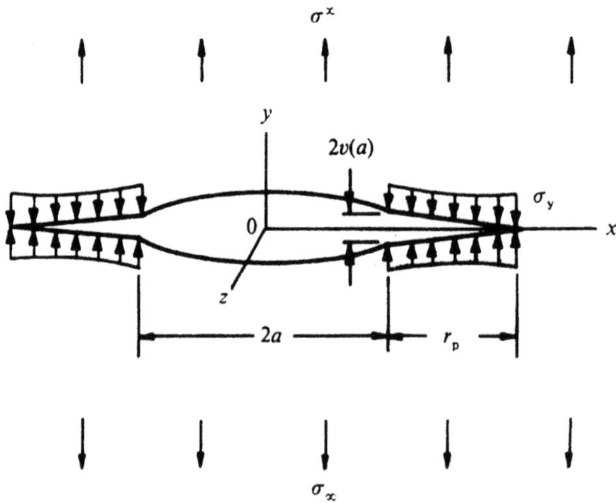

Figure 11.13 Schematic of the Dugdale plastic zone model. (Adapted from Suresh, 1999—reprinted with permission from Cambridge University Press.)

If the tractions were zero over the strips of length, r_p, the stress intensity factor, K, due to the remote applied stress, σ^{∞}, will be given by the solution for a center crack in an infinitely wide plate:

$$K = \sigma^{\infty}\sqrt{\pi(a + r_p)} \tag{11.59}$$

However, for the stresses to be bounded at the point $x = a + r_p$, $K + K'' = 0$. Solving for r_p now gives

$$\frac{r_p}{a} = \sec\left(\frac{\pi\sigma^{\infty}}{2\sigma_{ys}}\right) - 1 \tag{11.60}$$

Since the remote stress, σ^{∞}, is generally much less than the yield stress, σ_{ys}, the asymptotic series expansion of the above equation leads to the following approximate expression for the Dugdale plastic zone:

$$r_p = \frac{\pi}{8}\left(\frac{K}{\sigma_{ys}}\right)^2 \tag{11.61}$$

From the above equations, it is clear that the sizes of the Dugdale and plane stress plastic zones are comparable. However, the shapes of the plane stress and Dugdale plastic zones are somewhat different (Figs 11.12 and 11.13). Furthermore, it is important to note that the plastic zone size, r_p, only gives an approximate description of the plastic zone since the plastic zone boundary varies significantly ahead of the crack tip.

Before, closing, it is important to note that the Dugdale model can be used to estimate the crack-tip opening displacement, δ_t, at the points $x = \pm a$, and $y = 0$. The crack-tip opening displacement is given by

$$\delta_t = \frac{8\sigma_y a}{\pi E}\ln\left[\sec\left(\frac{\pi\sigma^{\infty}}{2\sigma_{ys}}\right)\right] \tag{11.62}$$

For most cases, however, $\sigma^{\infty} \ll \sigma_{ys}$, and hence the asymptotic form of the crack-tip opening displacement is given by

$$\delta_t = \frac{K^2}{\sigma_{ys}E} \quad \text{(for plane stress)} \tag{11.63}$$

Rice (1974) has shown that the crack-tip opening displacement under plane strain conditions may be expressed as

$$\delta_t = \frac{K^2}{2\sigma_{ys}E} \quad \text{(for plane strain)} \tag{11.64}$$

11.6.3.3 Barenblatt Model

Barenblatt (1962) has developed a model for brittle materials that is analogous to the Dugdale model (Fig. 11.12). However, in the Barenblatt model, the traction, σ_{yy}, is equal to the theoretical bond rupture strength of a brittle solid, $\sigma_{th} \sim E/10$ (Lawn, 1993). The critical condition for fracture may, therefore, be expressed in terms of a critical cohesive zone size, r_{co}, or in terms of the critical crack opening displacement, $\delta_c = 2\nu_c$ (Rice, 1968). The latter gives

$$G_c = 2\int_0^{\nu_c} \sigma_{yy} d\nu = \frac{8\sigma_{th}^2 r_{co}}{\pi E} = 2\gamma_s \tag{11.65}$$

where the terms in the above expression have their usual meaning. It is important to note here that $G_c \sim 2\gamma_s$ for a purely brittle solid. This was discussed in Sect. 11.4.

11.6.4 Conditions of K Dominance

The stress intensity factor, K is only valid within a small annular region at the crack tip where the asymptotic singular solutions (K and T terms) characterize the crack-tip fields to within 10%. Beyond the annular region, higher order terms must be included to characterize the crack-tip field. In general, however, the concept of K holds when the plastic zone, r_p, at the crack tip is small compared to the crack length ($r_p < a/50$). The concept of K also applies to blunt notches with small levels of notch-tip plasticity. Furthermore, it applies to scenarios where small-scale deformation occurs by mechanisms other than plasticity. These include stress-induced phase transformations and microcracking in brittle ceramics. Such mechanisms give rise to the formation of deformation process zones around the crack tip. The size of the region of K dominance is affected by the sizes of these process zones.

11.6.5 Equivalence of G and K

The relationship between G and K is derived in this section. Consider the generic crack-tip stress profile for Mode I that is shown in Fig. 11.14(a). The region of high stress concentration has strain energy stored over a distance ahead of the crack tip. This strain energy is released when a small amount of crack advance occurs over a distance, Δa. If G is the energy release rate, then energy released due to crack extension, GΔa, is related to the traction σ_{22} $(x, 0)$ via:

$$\mathrm{G}\Delta a = \frac{1}{2}\int \sigma_{22}(x, 0)[v(x, 0^{+} - v(x, 0^{-})] \tag{11.66}$$

where $v(x, 0^{+}) - v(x, 0^{-})$ is the crack face separation in the final position where the crack length is $a + \Delta a$. For small crack extension, Δa, this is given by

$$v(x, 0^{+} - v(x, 0^{-}) = K(a + \Delta a)\frac{1+\kappa}{G}\sqrt{\frac{\Delta a - x}{2\pi}} \tag{11.67}$$

where $K(a + \mathrm{d}a)$ is the value of K after crack extension. Hence, the energy balance becomes

$$G\Delta a = \frac{(1+\kappa)}{4\pi G}K(a)K(a + \ da)\int_{0}^{\Delta a}\left(\frac{\Delta a - x}{2\pi}\right) \tag{11.68a}$$

or

$$G\Delta a = \frac{(1+\kappa)}{8G}K(a)K(a + \mathrm{d}a)\Delta a \tag{11.68b}$$

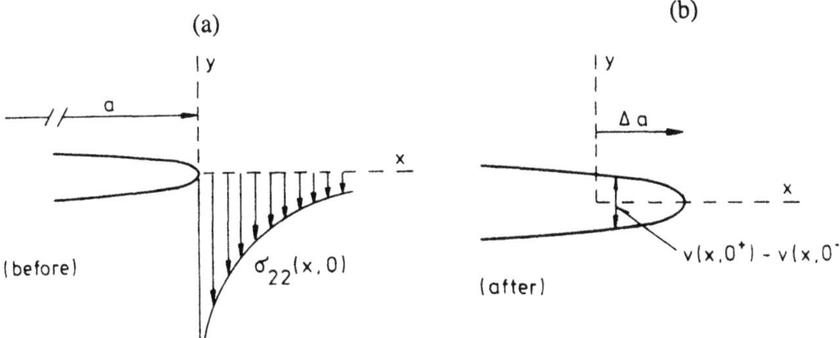

Figure 11.14 Schematic of crack-tip profile and stress state: (a) before incremental crack growth; (b) after incremental crack growth. (Adapted from Hutchinson, 1983—reprinted with permission from the Technical University of Denmark.)

Hence, from the definition of κ [Eq (11.43a) and (11.43b)], the following expressions are obtained for G under plane strain and plane stress conditions:

$$G = \frac{(1-\nu^2)}{E} K^2 \qquad \text{(for plane strain)} \tag{11.69a}$$

or

$$G = \frac{K^2}{E} \qquad \text{(for plane stress)} \tag{11.69b}$$

where the above terms have their usual meaning. Since energy is a scalar quantity, we may also add the energy release rate (for Modes I, II, and III) per unit length of crack edge. Hence, in a three-dimensional solid subjected to K_I, K_{II}, and K_{III}, we have

$$G = \frac{(1-\nu^2)}{E}(K_I^2 + K_{II}^2) + (1-\nu)K_{III}^2 \qquad \text{(for plane strain)} \tag{11.70a}$$

and

$$G = \frac{1}{E}(K_I^2 + K_{II}^2) \qquad \text{(for plane stress)} \tag{11.70b}$$

11.6.5.1 Worked Example—G and K for Double Cantilever Bend Specimen

Consider the example of the double cantilever bend specimen shown in Fig. 11.15:

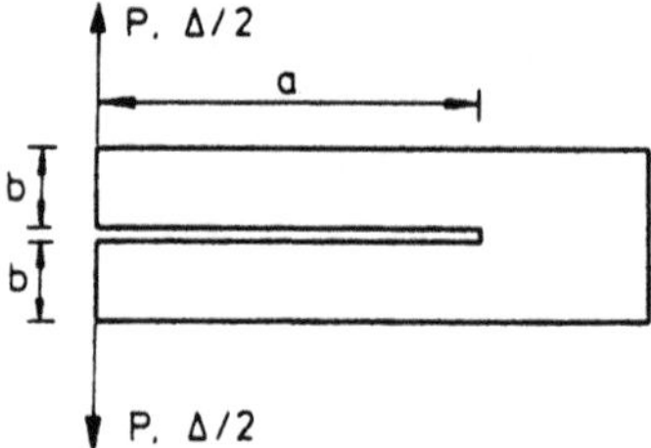

FIGURE 11.15 Schematic of a cantilever bend specimen. (Adapted from Hutchinson, 1979—reprinted with permission from the Technical University of Denmark.)

1. Using the approaches discussed in Sect., 11.5, derive an expression for the energy release rate, G.
2. Derive an expression for the stress intensity factor, K, from the result obtained in part 1.

Solution. 1. We may solve the problem by considering the in-plane deflection of two "cantilever" on either side of the crack. For crack length, a, and deflection, $\Delta/2$, on either side of the crack, we may write the following expression from beam theory:

$$\frac{\Delta}{2}=\frac{Pa^3}{3EI} \tag{11.71}$$

where I is the second moment of area. If the cantilever has a thickness, B, and a height, b, then the second moment of area, I, is given by

$$I=\frac{Bb^3}{12} \tag{11.72}$$

Hence, substituting Eq. (11.72) into Eq. (11.71) gives

$$\frac{\Delta}{2}=\frac{4Pa^3}{EBb^3} \tag{11.73}$$

By definition, the energy release rate [from Eq. (11.20)] is given by

$$G=\frac{P^2}{2B}\frac{dC}{da} \tag{11.74}$$

where C is the compliance which is given by

$$C=\frac{\Delta}{P}=\frac{8a^3}{Eb^3} \tag{11.75}$$

Hence, G is given by Eq. (11.74) to be

$$G=\frac{P^2}{2B}\frac{dC}{da}=\frac{12P^2a^2}{BEb^3} \tag{11.76}$$

2. For plane stress the stress intensity factor, K is given by

$$K=\sqrt{EG}=\left(\frac{12EP^2a^2}{BEb^3}\right)^{\frac{1}{2}}=\frac{2\sqrt{3}Pa}{Bb^{3/2}} \tag{11.77}$$

and for plane strain conditions:

$$K=\sqrt{\frac{EG}{(1-\nu^2)}}=\left(\frac{12EP^2a^2}{BE(1-\nu^2)b^3}\right)^{\frac{1}{2}}=\frac{2\sqrt{3}Pa}{B(1-\nu^2)^{1/2}b^{3/2}} \tag{11.78}$$

11.7 ELASTIC–PLASTIC FRACTURE MECHANICS

In many practical problems, the assumption of small-scale plasticity is not valid, especially when the plastic zone sizes are large compared to the crack size or specimen dimensions. Various approaches are presented in this section for the characterization of elastic–plastic fracture. The first approach involves the crack opening displacement (COD), which is a parameter that was first introduced by Wells (1961) and Cottrell (1961). The J integral is then described along with two-parameter approaches for the assessment of constraint/size effects.

11.7.1 Crack Opening Displacement

The COD is the amount of crack opening before crack extension (Figs 11.16 and 11.17). It was proposed independently by Cottrell (1961) and Wells (1961) as a parameter for characterizing the stress–strain field ahead of the crack tip. In plane stress, this is given by (Smith, 1962; Bilby et al., 1963; and Burdekin and Stone, 1966):

$$\delta = \left(\frac{8\sigma_{ys} a}{\pi E}\right) \ln \sec\left(\frac{\pi a}{2\sigma_{ys}}\right) \tag{11.79}$$

Expanding Eq. (11.79) and taking the first terms gives the COD under plane stress conditions as

$$\delta = \frac{K^2}{E\sigma_{ys}} \tag{11.80a}$$

Rice (1974) has also obtained the following expression for the COD under plane strain conditions, using finite element analyses:

$$\delta = \frac{K^2}{2E\sigma_{ys}} \tag{11.80b}$$

The COD is generally measured with a crack-mouth clip gauge. Using similar triangles (Fig. 11.16), accurate measurements with this gauge can be related easily to the crack-tip opening displacement (CTOD). The CTOD corresponds to the displacement between points on the crack face that are intersected by 45° lines from the center of the crack (Fig. 11.17). Values of the CTOD measured at the onset of fracture instability correspond to the fracture toughness of a material. Hence, the CTOD is often used to represent the fracture toughness of materials that exhibit significant plasticity prior to the onset of fracture instability. Guidelines for CTOD testing are given in the ASTM E-813 code.

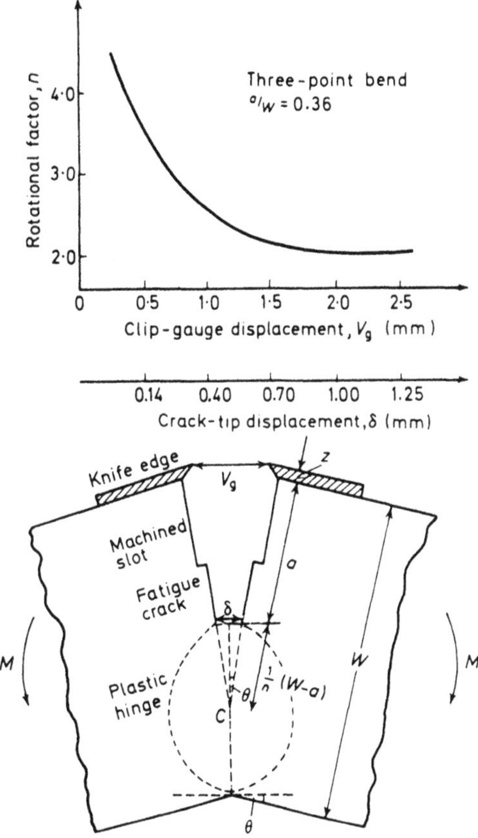

FIGURE 11.16 Diagram showing relationships between crack-tip displacement and knife-edge displacement for a rigid rotation about C. (Adapted from Knott, 1973—with permission from Butterworth.)

11.7.2 The J Integral

The J integral is a path-independent integral that relies on the determination of an energy term which expresses the change in potential energy when a crack is extended by an amount, da, in a manner analogous to the strain energy release rate, G, which is used for the linear elastic condition. The J integral was developed by Rice and Rosengren (1968) for a nonlinear elastic body, and is defined as

$$J = \int_{\Gamma} \left(W \mathrm{d}y - T \frac{\mathrm{d}u}{\mathrm{d}x} \mathrm{d}s \right) \tag{11.81}$$

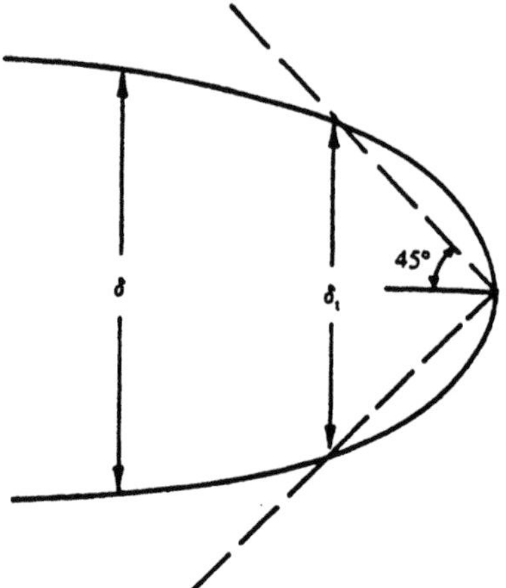

Figure 11.17 Measurement of the crack-tip opening displacement.

where W is the strain energy density, which is given by

$$W = W(x, y) = W(\varepsilon) = \int_0^{\varepsilon} \sigma_{ij} \mathrm{d}\varepsilon_{ij} \tag{11.82}$$

where Γ is an arbitrary closed contour followed counterclockwise in a stressed solid (Fig. 11.18), T is the traction perpendicular to Γ in an outward facing direction, u is the displacement in the x direction, and ds is an element of Γ (see Fig. 11.18).

For a nonlinear elastic solid, the J integral is defined as the change in potential energy, V, for a virtual crack extension, da (Rice, 1968):

$$J = -\frac{1}{B}\frac{\partial PE}{\partial a} \tag{11.83}$$

For a linear elastic material, $-1/B(\partial PE/\partial a) = G$, which means that $J = G$ for the linear elastic case. Also, under small scale yielding, J is uniquely related to K by

$$J = \frac{K^2}{E'} \tag{11.84}$$

where $E' = E$ for plane stress, $E' = E/(1 = \nu^2)$ for plane strain, and ν is Poisson's ratio.

Hutchinson (1968) and Rice and Rosengren (1968) have shown that the crack-tip fields during plane strain deformation of power law hardening materials under nonlinear elastic conditions are given by the so-called Hutchinson–Rice–Rosengren (HRR) fields. These are given by the following

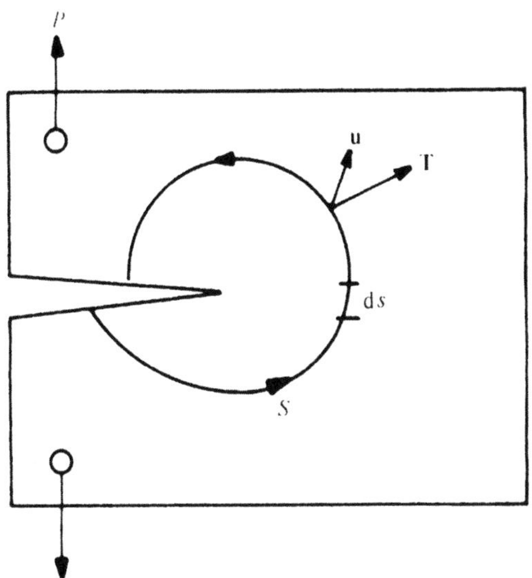

Figure 11.18 Schematic of the components of the path independent *J* integral. (Adapted from Suresh, 1999—reprinted with permission from the Cambridge University Press.)

expressions for stresses, strains, and displacements ahead of the near-tip, which exhibit, $r^{-1/(n+1)}$, $r^{-n/n(n+1)}$, and $r^{1/(n+1)}$ singularity, as shown in the following, respectively,

$$\sigma_{ij} = \sigma_Y \left(\frac{J}{\alpha \sigma_Y \varepsilon_Y I_n r} \right)^{1/(n+1)} \tilde{\sigma}_{ij}(\theta, n) \tag{11.85a}$$

$$\varepsilon_{ij} = \alpha \varepsilon_Y \left(\frac{J}{\alpha \sigma_Y \varepsilon_Y I_n r} \right)^{n/(n+1)} \tilde{\varepsilon}_{ij}(\theta, n) \tag{11.85b}$$

$$u_i = \alpha \varepsilon_Y \left(\frac{J}{\alpha \sigma_Y \varepsilon_Y I_n} \right)^{n/(n+1)} r^{1/(n+1)} \tilde{u}_i(\theta, n) \tag{11.85c}$$

where n is the hardening exponent in the Ramberg–Osgood equation:

$$\varepsilon/\varepsilon_Y = \sigma/\sigma_Y + \alpha\sigma/\sigma_Y)^n \tag{11.85d}$$

and σ_Y is the yield stress, ε_Y is the yield strain, r is the radial distance from the crack-tip, α is a dimensionless constant, I_n is an integration constant (Fig. 11.19), and $\tilde{\sigma}_{ij}(\theta, n)$, $\tilde{\varepsilon}_{ij}(\theta, n)$, and $\tilde{u}_i(\theta, n)$ can be found in most standard texts on fracture mechanics (Broek, 1978; Ewalds and Wanhill, 1984; Kanninen and Popelar, 1985). Plots of $\tilde{\sigma}_{ij}(\theta, n)$ are presented in Fig. 11.19.

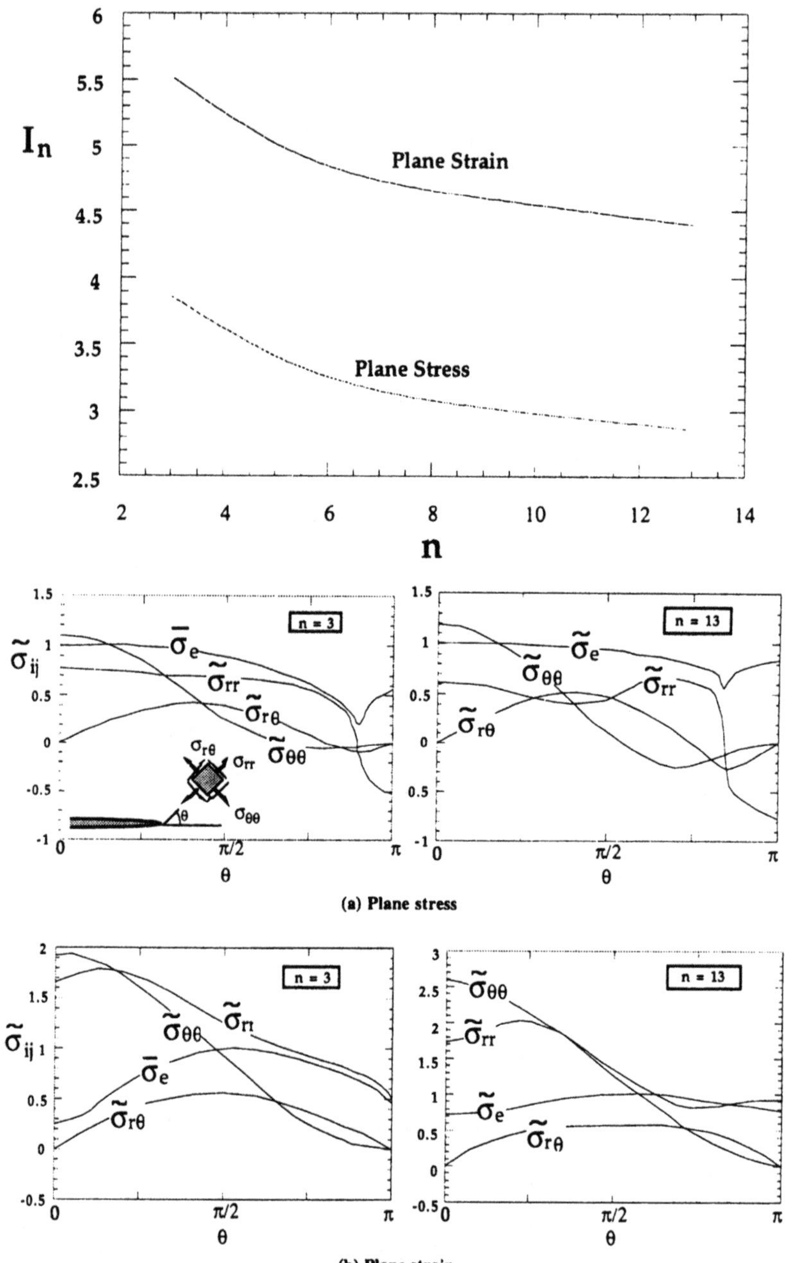

FIGURE 11.19 Dependence of I_n and $\tilde{\sigma}_{ij}$ on θ. (Adapted from Anderson, 1995—reprinted with permission from CRC Press.)

It is important to note here that J is the amplitude of the crack-tip field under elastic–plastic fracture mechanics conditions. However, the applicability of J to such scenarios was only established after careful work by Begley and Landes (1972a,b) and Landes and Begley (1977). Their work showed that J provides a measure of the driving force for crack growth under elastic–plastic fracture conditions.

11.7.3 Conditions of J Dominance

The J integral provides a unique measure of the amplitude of the crack-tip fields under nonlinear fracture conditions. However, it is only applicable when the conditions for J dominance are satisfied (Hutchinson, 1983). This requires that:

1. The J_2 deformation theory of plasticity must provide an adequate model of the small-strain behavior of real elastic–plastic materials under the monotonic loads being considered.
2. The regions in which finite strain effects are important and the region in which microscopic processes occur must be contained within the region of the small-strain solution dominated by the singularity fields.

The first condition is critical in any application of the deformation theory of plasticity. It is satisfied if proportional loading occurs everywhere, i.e., the stress components change in fixed proportion everywhere. Nevertheless, although the requirement for proportional loading is not fulfilled exactly in most cases under monotonic loading, the applications of uniaxial loads to stationary cracks do provide a good framework for the use of the deformation theory of plasticity.

The second condition is somewhat analogous to the so-called small-scale yielding condition for linear elastic fracture mechanics. It also provides a physical basis for the determination of the inner radius of the annular region of J dominance. This is illustrated schematically in Fig. 11.20, which is adapted from a review by Hutchinson (1983). The annular zone of J dominance corresponds to the region where the HRR field solutions are within $\sim 10\%$ of the full crack-tip field solutions obtained from finite element analyses (McMeeking, 1977; McMeeking and Parks, 1979). These analyses suggest that finite strain effects are significant over a distance of $\sim 3\delta_t$. This distance must be greater than the microstructural process zone for the J integral to be applied. The size of the microstructural process zone may correspond to the mean void spacing in the case of ductile dimpled fracture, or the grain size in the case of cleavage or intergranular fracture.

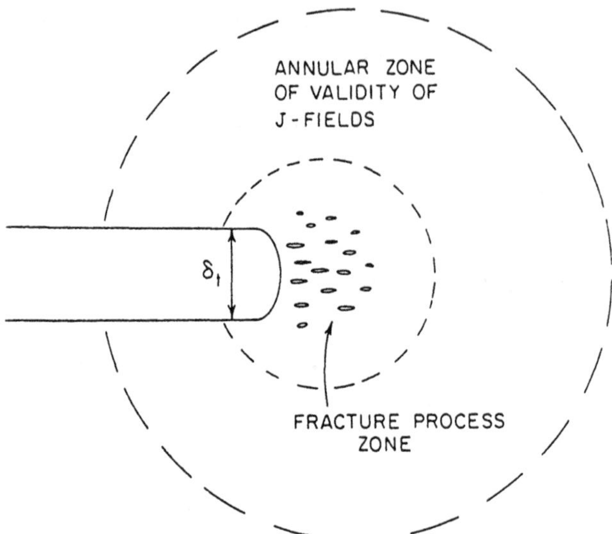

FIGURE 11.20 Schematic of annular region of J dominance. (From Hutchinson, 1983—reprinted with permission from the Technical University of Denmark.)

Under small-scale yielding conditions, the HRR field solutions have been shown to be applicable over a distance corresponding to ~ 20–25% of the plastic zone size. However, under large-scale yielding conditions, the region of J dominance is highly dependent on specimen configuration. In cases where the entire uncracked ligament is completely engulfed by the plastic zone, the size of the region of J dominance may be as low as 1% of the uncracked ligament for a center-cracked tension specimen or 7% of the uncracked ligament for a deeply cracked SENB or compact-tension (C-T) specimen (McMeeking and Parks, 1979).

Since the conditions for J dominance are so highly specimen dependent, standards have been developed to provide guidelines for the measurement of critical values of J. One of the most widely used codes is the ASTM E-813 code for J_{Ic} testing developed by the American Society for Testing and Materials, Conshohocken, PA. This code requires the use of deeply cracked SENB or C-T specimens with initial precrack length-to-specimen width ratios of 0.5. In general, such specimens provide a unique measure of the J_{Ic} when the remaining ligament, $W - a$, is

$$W - a \geq 25\left(\frac{J_{\mathrm{Ic}}}{\sigma_{\mathrm{ys}}}\right) \tag{11.86}$$

where the J_{Ic} for linear elastic conditions is given by

$$J_{\mathrm{Ic}} = \frac{K_{\mathrm{Ic}}^2}{E'} \tag{11.87}$$

where $E' = E$ for plane stress and $E' = E/(1 - \nu^2)$ for plane strain conditions. Furthermore, since the J integral does not model elastic unloading and the distinctly nonproportional loading that occurs near the crack tip when crack advance occurs, J-controlled crack growth requires that the regime of elastic unloading and nonproportional loading be confined within the zone of J dominance (Fig. 11.20). This gives

$$\Delta a \ll R \tag{11.88a}$$

and

$$\frac{\mathrm{d}J}{\mathrm{d}a} \gg \frac{J}{R} \tag{11.88b}$$

The above two conditions for J-controlled crack growth were first proposed by Hutchinson and Paris (1979). The first ensures that crack growth occurs within a zone that is uniquely characterized by J (Fig. 11.20). However, the second condition is less transparent. In any case, the HRR field solutions are for stationary cracks, and the growth of cracks causes deviations form these solutions. Work by Rice et al. (1980) has shown that the crack-tip fields for growing cracks are comparable to Prandtl slip line fields, except in regions behind the crack tip where differences of up to $\sim 10\%$ may occur. Growing cracks also exhibit logarithmic singularity that is somewhat weaker than the $1/r$ singularity for a growing crack.

In general, the initiation J integral, J_{i}, may be decomposed into an elastic component, J_{el}, and a plastic component, J_{pl}:

$$J_{\mathrm{i}} = J_{\mathrm{el}} + J_{\mathrm{pl}} \tag{11.89a}$$

For a plate deformed under three-point bending with a span-to-width ratio of 4, J_{i} is given by

$$J_{\mathrm{i}} = \frac{K_{\mathrm{i}}^2}{E'} + \frac{2A_{\mathrm{i}}}{B(W - a)} \tag{11.89b}$$

where B is the specimen thickness, A_{i} is the area under the load displacement curve, $W - a$ is the remaining ligament, and $E' = E$ for plane stress or $E' = E/(1 - \nu^2)$ for plane strain. For a compact tension specimen:

$$J_i = \frac{K_i^2}{E'} + A_i \frac{\left[2 + \frac{0.522(W-a)}{W}\right]}{B(W-a)} \tag{11.89c}$$

11.7.4 Two-Parameter J–Q

Under the condition of large-scale yielding, K–T theory cannot adequately define the stress field ahead of the crack tip. O'Dowd and Shih (1991) have proposed a family of crack-tip fields that are characterized by a triaxility parameter. The application of these fields has given rise to a two-parameter (J–Q) theory for the characterization of the effects of constraint on crack-tip fields. As before, the J integral characterizes the amplitude of the crack-tip field. However, the Q term now characterizes the contributions from hydrostatic stress. This has led to the $J - Q$ theory. In the region of the crack tip where $|\theta| < \pi/2$ and $J/\sigma_y < r < 5J/\sigma_{ys}$, the stress field is given by

$$\sigma_{ij} = (\sigma_{ij})_{HRR} + Q\sigma_{ys}\delta_{ij} \qquad |\theta| < \pi/2 \tag{11.90}$$

where Q is a measure of the crack-tip stress triaxiality, $(\sigma_{ij})_{HRR}$ is the HRR field, and σ_y is the yield stress. As an operational definition, Q is the difference, normalized by the yield strength, σ_{ys}, between the actual hoop stress at the crack tip and that given by the HRR singular field at a fixed distance $2J/\sigma_y$ directly ahead of the crack tip. the definitive equation for Q is

$$Q \equiv \frac{\sigma_{\theta\theta} - (\sigma_{\theta\theta})_{HRR}}{\sigma_y} \text{ at } \theta = 0, r = \frac{2J}{\sigma_y} \tag{11.91}$$

Figure 11.21 shows the variation of hoop stress, $\sigma_{\theta\theta}$, as a function of the normalized distance ahead of the crack tip, $r/(J/\sigma_y)$, for plane strain and for $E/\sigma_y = 500$ and $v = 0.3$. For comparison, Fig. 11.21(a) shows the HRR field solution for a small strain, while Fig. 11.21(b) shows the solution from the J–Q theory for a finite strain. It is clear from these figures that the location of the maximum hoop stress depends strongly on the triaxiality of stress (represented by Q in the finite-strain calculation) and the hardening exponent, n.

For small strain conditions the maximum hoop stress appears just at the tip of the crack. However, the maximum values of hoop stresses occur at certain values of $r/(J/\sigma)_y)$, which are related to the hardening exponent, n. Also, the distance from the crack tip correpsonding to the maximum hoop stress increases with increasing strain hardening exponent, n.

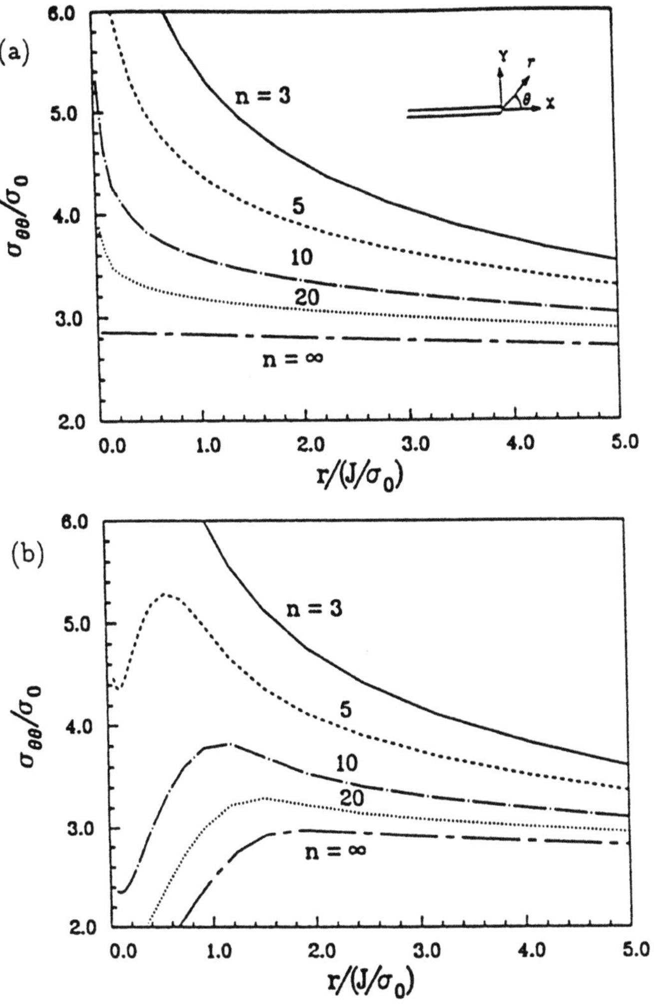

FIGURE 11.21 Plots of $\sigma_{\theta\theta}$ versus $r/(J/\sigma_{ys})$ for the *J–Q* field: (a) small strain; (b) large strain.

11.8 FRACTURE INITIATION AND RESISTANCE

Fracture initiation and/or resistance are typically studied using fracture mechanics specimens containing atomistically sharp fatigue precracks (Fig. 11.10). However, special specimen geometries may be selected to promote stable crack growth. For example, a finite crack in an infinite plate loaded by concentrated loads exhibits stress intensity factors that decrease with increasing crack length when the load is fixed. These are given by

$$K = \frac{Pa}{\sqrt{\pi a}} \tag{11.92}$$

Fracture specimens are typically loaded in incremental stages (under displacement or load control) until stable crack growth is observed to initiate. The crack driving force (K or J) at which this initiates is known as the initiation toughness, K_i or J_i. Perfectly brittle materials offer no resistance to crack growth [Fig. 11.21(a)] and hence crack growth continues as long as $K = K_c$, which is needed to propagate the crack. However, most materials exhibit some resistance-curve behavior, as shown in Fig. 11.21(b). Hence, higher crack driving forces, K or J, are needed to grow the crack with increasing crack growth. Furthermore, the resistance curve approaches a steady-state toughness if the specimen is large enough to permit sufficient stable crack growth.

The observed resistance-curve behavior depends strongly on the underlying energy dissipative (toughening) mechanisms that give rise to stable crack growth. These may involve plasticity at the crack tip, or crack-tip shielding concepts that will be discussed in detail in Chap. 13. In any case, the condition for continued crack advance is that the crack-driving force lies on the measured resistance curve, i.e., $K = K(\Delta a)$ or $J = J(\Delta a)$, where the terms on the left-hand side (J or K) correspond to the "applied" crack driving forces, and the terms on the right-hand side correspond to the material response. The condition for stable crack advance is now given by

$$\left(\frac{dK}{da}\right)_L < \frac{dK_R}{d\Delta a} \tag{11.93a}$$

or

$$\left(\frac{dJ}{da}\right)_L < \frac{dJ_R}{d\Delta a} \tag{11.93b}$$

Hence, from Eq. (11.93a) and (11.93b), crack growth instability occurs when

$$\left(\frac{dK}{da}\right)_L = \frac{dK_R}{d\Delta a} \tag{11.94a}$$

or

$$\left(\frac{dJ}{da}\right)_L = \frac{dJ_R}{d\Delta a} \tag{11.94b}$$

where the partial derivatives with respect to crack length, a, are taken with the prescribed loading conditions fixed. Hence, the transition from stable to unstable crack growth can be determined by the analysis of the resistance curve, as shown in Fig. 11.22. The dashed $K(L_i, a)$ curves depict K as a function of a at fixed loading conditions, L_i. These could correspond to prescribed loads or prescribed displacements. Also, $L_i < L_2 < L_3 < L_4$, and so on. The value of K at instability, K^*, in Fig. 11.22, depends on other R curves as well as the loading conditions. It is associated with L_4 in Fig. 11.22. Similar arguments may be proposed for $J(L_i, a)$ curves under elastic–plastic fracture conditions.

Alternatively, fracture toughness values may be obtained from simple tests in wihch the loads are increased monotonically until fracture occurs. The crack-mouth opening displacements are typically measured in such tests along with the corresponding applied loads. The fracture toughness is thus determined by the analysis of load–displacement plots to identify peak loads in accordance with ASTM E-399 criteria (Fig. 11.23). The most commonly used method for the determination of toughness is the 95% secant method. This is used to extract the peak load, as shown schematically in Fig. 11.23. The peak load is determined from the point of intersection of a radial line with a slope correpsonding to 95% of that of the original load–displacement line. Values of K corresponding to this load are known as the fracture toughness, K_Q.

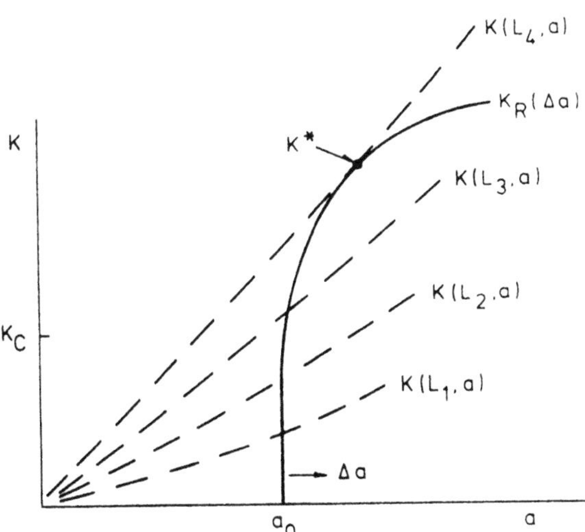

FIGURE 11.22 Schematic illustration of resistance-curve analysis. (Adapted from Hutchinson, 1983—reprinted with permission from the Technical University of Denmark.)

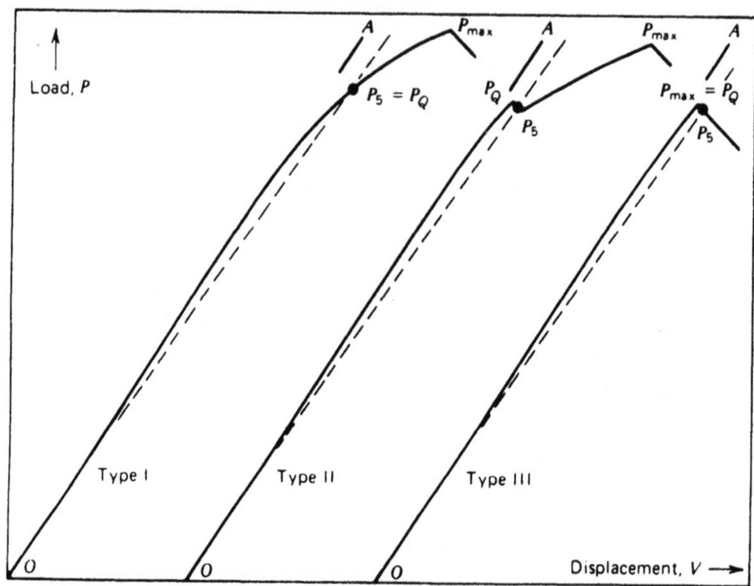

Figure 11.23 Schematics of load–displacement plots. (Taken from the ASTM E-399 Code.)

The fracture toughness values, therefore, correspond to points of instability on a rising *R* curve. A strict set of conditions must be satisfied before a critical stress intensity factor can be accepted as a material property called the fracture toughness. These conditions are specified in the ASTM E-399 code. The most critical criteria are associated with the need to maintain highly triaxial/plane strain conditions at the crack tip. This is generally achieved by using thick specimens in fracture toughness tests. In general, the tests must satisfy the following criteria:

$$B < 2.5\left(\frac{K_Q}{\sigma_y s}\right)^2 \tag{11.95a}$$

$$W < 2.5\left(\frac{K_Q}{\sigma_{ys}}\right)^2 \tag{11.95b}$$

$$(W - a) < 2.5\left(\frac{K_Q}{\sigma_{ys}}\right)^2 \tag{11.95c}$$

where *W*, *a*, and *B* correspond to the specimen dimensions (Fig. 11.10), and the other terms have their usual meaning. When the above conditions are

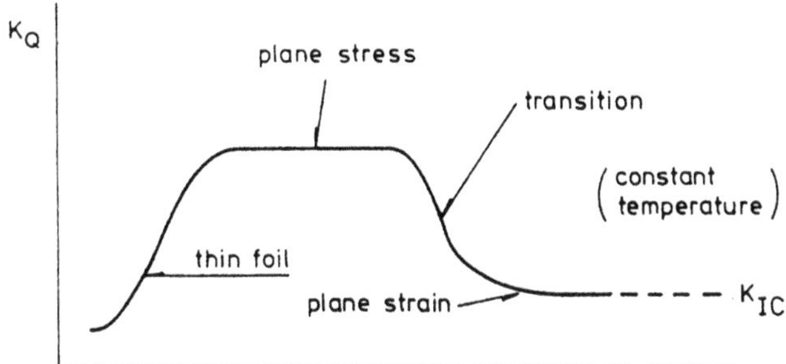

FIGURE 11.24 Dependence of fracture toughness on specimen thickness. (Adapted from Hutchinson, 1979—reprinted with permission from the Technical University of Denmark.)

satisfied [Eqs (11.97a–c)], the measured fracture toughness, K_Q, is independent of specimen thickness (Fig. 11.24). For Mode I loading conditions, the critical value of the stress intensity factor is generally referred to as the K_{Ic}. This is a material property just like yield strength and Young's modulus. Similarly, we may obtain fracture critical conditions, K_{IIc} and K_{IIIc} for Modes II and III, respectively.

Before closing, it is important to note that the thickness dependence of K_Q is illustrated in Fig. 11.24. This is associated largely with a transition from plane stress conditions (in thin specimens) to plane strain conditions (in thick specimens). Less triaxial or biaxial stress fields give rise to less intense damage, and hence the K_Q levels are greater under plane stress conditions. The lowest values of K_Q correspond to the conditions of high-stress triaxility that give rise to the most intense damage at the crack-tip.

Once plane strain conditions are attained at the crack tip, the K_Q value does not decrease with increasing thickness (Fig. 11.24). Hence, the measured values of K_Q for sufficiently thick specimens correspond to a material property that is independent of thickness. This property is known as the *fracture toughness*. Typical values of fracture toughness are presented in Fig. 11.25 for different materials. It is important to note here that fracture toughness values are dependent on composition, microstructure, and loading rates. These effects will be discussed in Chap. 12.

11.9 INTERFACIAL FRACTURE MECHANICS

In brittle solids, cracks extend at a specified toughness along a defined trajectory governed solely by the opening mode (Cottrell and Rice, 1980; Evans et al., 1999). Accordingly, the likelihood that a crack present in the

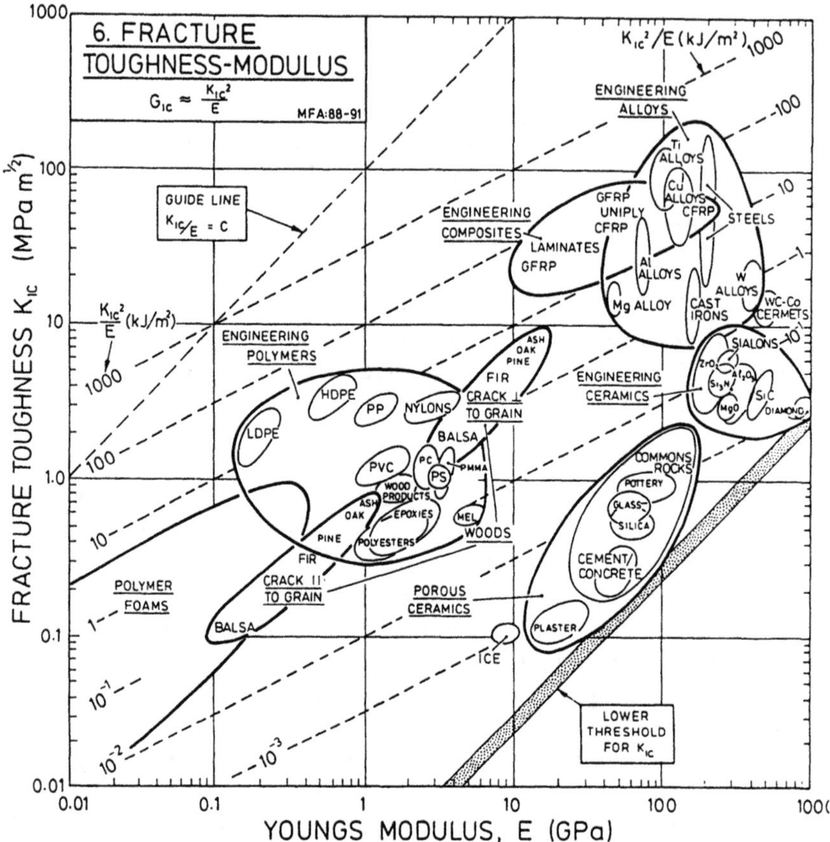

FIGURE 11.25 Fracture toughness values for different materials. (From Ashby, 1999—reprinted with permission from Butterworth Heinemann.)

material will extend upon loading, as well as its expected trajectory through the material, can be explicitly determined, once the opening mode toughness has been measured and provided that the stress intensity factors have been calculated from the applied loads using a numerical technique such as the finite element method (FEM). Fracture at interfaces is more complex, because cracks can remain at the interface in the presence of mode mixity, requiring that a mixity-dependent toughness be determined and used to analyze failure. The most basic relations governing interface crack growth are a follows. The amplitudes of the normal and shear stresses on an interface, distance x ahead of a plane strain interfacial crack, are characterized by two stress intensity factors, K_I and K_{II}.

$$\sigma_{22} = K_{\mathrm{I}}/\sqrt{2\pi x} \tag{11.96a}$$

and

$$\sigma_{12} = K_{\mathrm{II}}/\sqrt{2\pi x} \tag{11.96b}$$

These two quantities can be calculated for any interface crack using the finite element method. They can be combined to formulate two alternative parameters, G and ψ, found to constitute a more convenient practical measure of interface crack growth. One is designated the energy release rate, G:

$$G = \frac{1}{2}\left(\frac{1-\nu_1^2}{E_1} + \frac{1-\nu_2^2}{E_2}\right)(K_{\mathrm{I}}^2 + K_{\mathrm{II}}^2) \tag{11.97}$$

where the subscripts I and II denote Modes I and II, and the subscripts 1 and 2 denote the two dissimilar materials. The other is the mixity angle, ψ, defined as

$$\psi = \tan^{-1}\left(\frac{K_{\mathrm{II}}}{K_{\mathrm{I}}}\right) \tag{11.98}$$

The criterion for crack growth is that G atain a critical level, Γ_{i}: designated the interfacial toughness (Hutchinson and Suo, 1992). The integrity of an interface within a joint can be analyzed, through finite element calculations, once Γ_{i}^0 and ψ have been measured for that interface duplicated in laboratory specimens.

Conducting studies on interfaces is more challenging than the corresponding studies performed on their homogeneous counterparts. There are two main issues: (1) the geometric configurations encompassing interfaces of practical interest often constrain specimen design, and (2) large-scale inelastic deformations limit options, because of the different thermomechanical properties of the adjoining materials. When interfaces can be made in layered configurations by bonding procedures (such as sintering or adhesive attachment), a number of test geometries (Figure 11.26) are available. Most are limited to relatively narrow ranges of mode mixities, Figs 11.26(a) and 11.26(b). It is, therefore, common for serval specimen geometries to be used. The exceptions are the dissimilar mixed mode bend specimen (Soboyejo et al., 1999) and the Brazil specimen [shown in Figs 11.26(c) and 11.26(d)] (Hutchinson and Suo, 1992), which can be used to probe the full range of mixities.

There have been some studies of interfacial fracture between polymers and oxides (Cao and Evans, 1989; Ritter et al., 1999). These indicate mode mixity effects and reveal a strong-influence of moisture on the crack growth rates. Many more experimental results have been obtained for metal/oxide interfaces (Dalgleish et al., 1989; Reimanis et al., 1991; Bagchi and Evans,

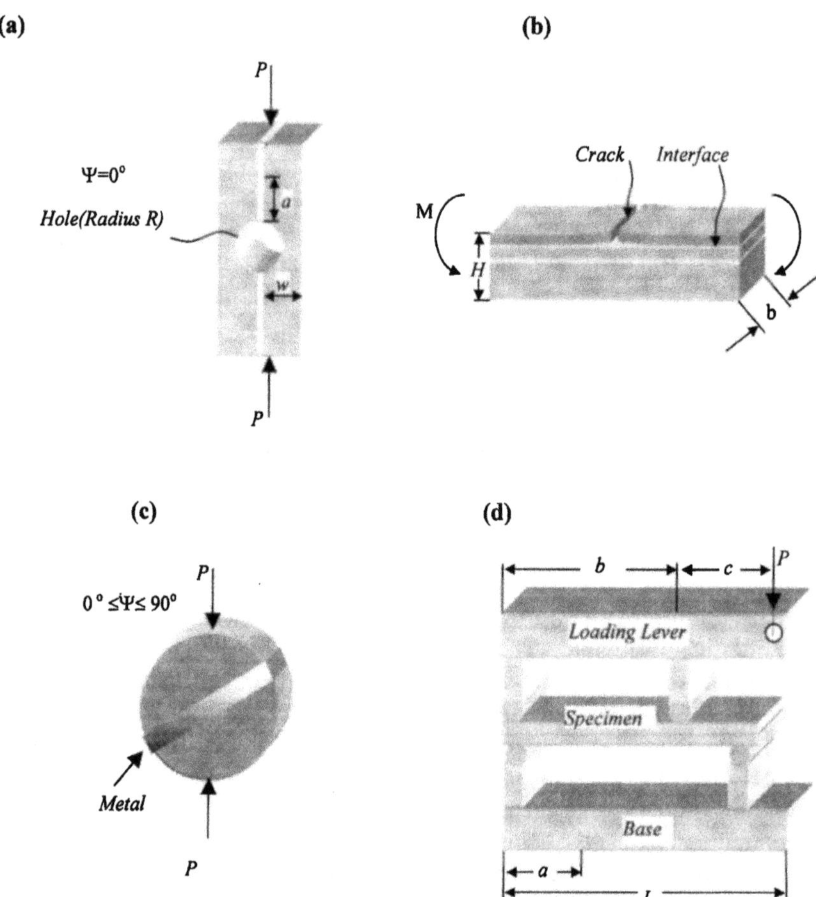

FIGURE 11.26 Types of interfacial fracture mechanics specimens: (a) double cleavage drilled compression ($\psi \sim 40°–45°$); (b) mixed-mode flexure specimen; (c) Brazil specimen ($0° \leq \phi \leq 90°$), (d) dissimilar mixed-mode bending specimen ($0 \leq \psi 90°$)

1996; McNaney et al., 1996; Turner and Evans, 1996; Gaudette et al., 1997; Lipkin et al., 1998; Evans et al., 1999). The systems that have been most extensively studied include Al_2O_3 bonded to Al (Dalgleish et al., 1989; McNaney et al., 1996), Au (Reimanis et al., 1991; Turner and Evans, 1996), and SiO_2 bonded to Cu (Bagchi and Evans, 1989).

The above systems exhibit resistance-curve behavor and fracture energies significantly greater than the thermodynamic work of adhesion. The high toughness has been attributed to: crack-tip shielding by ductile ligaments; plastic dissipation in the metallic constituent; and friction at asperities in the wake of the crack. These same phenomena give rise to a toughness that varies appreciably with mode mixity (Cao and Evans, 1989, Soboyejo et al., 1999).

Furthermore, crack-path selection criteria for cracks between dissimilar solids depend on the mode mixity and the ratio of interfacial toughness, Γ_i, to the toughness, Γ_s, of the adjoining material (Fig. 11.27) (Evans and Dalgleish, 1992). Note that the transition from interface cracking to material cracking is dependent not only on the ratio of Γ_i to Γ_s, but also on the mode mixity angle, ψ.

11.10 DYNAMIC FRACTURE MECHANICS

In 1951, Elizabeth Yoffe published a classical paper in which she analyzed the influence of crack velocity on the growth cracks. This pioneering work established the fundamental basis for future studies of dynamic fracture mechanics. Dynamic fracture is the study of crack initiation and growth under conditions in which the loading rates and crack velocities approach the Rayleigh wave speed, c_R. Such high crack velocities give rise to high local strain rates and local crack-tip phenomena such as dynamic shear

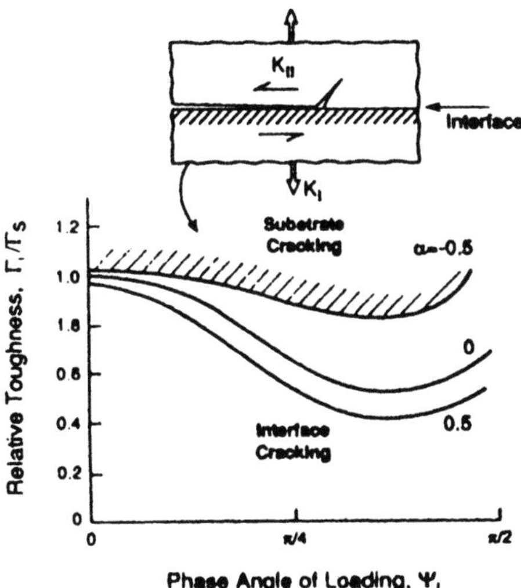

FIGURE 11.27 Diagram indicating region in which fracture deviates from interface. Using fracture energy/phase angle space. (From Evans and Dalgleish, 1992—reprinted with permission from Elsevier Science.)

localization or adiabatic shear bonding, and crack branching, which was predicted by the original analyses of Yoffe (1951). A common feature of all of these phenomena is the rapid loss of stress carrying ability at time scales such that inertial and/or material rate sensitivity effects are important (Rosakis and Ravichandran, 2000).

Dynamic fracture is associated with the rapid creation of new surfaces, while shear localization considers the deformation conditions that lead ultimately to the creation of high-velocity cracks. The resulting shear bands, whihc are produced under high strain rate or strain gradients, may lead ultimately to the initiation and growth of cracks under dynamic fracture conditions. Dynamic shear localization is also strongly influenced by material strain hardening, strain rate sensitivity, and thermal softening due to the intense heating associated with the dissipation of plastic work.

The subject of dynamic fracture mechanics may be divided into two major problem sreas: one associated with dynamically loaded cracks resulting in crack initiation times t, that are short compared to the transit time of a Rayleigh wave (wave speed, c_R), along a crack of length, $a(c_R t < a)$; and the other associated with cracks that are moving at speed that are greater than 20% of the Rayleigh wave speed.

Since the early work of Yoffe (1951), a large number of researchers have contributed significantly to the subject of dynamic fracture. They include: Broberg (1960); Atkinson and Eshelby (1968); Achenbach (1970, 1974); Kostrov and Nikitin (1970); Freund (1971); Willis (1975). A comprehensive text on dynamic fracture has also been published by Freund (1990).

For a running crack with length $a(t)$ at time, t, and an instantaneous crack speed, $\dot{a}(t)$ and load, $P(t)$, the dynamic stress intensity factor may be expressed as (Freund, 1990):

$$K_I^d(P(t), a(t), \dot{a}(t)) = k(\dot{a})K_I^d(P(t), a(t), 0) \tag{11.99}$$

where $K(\dot{a})$ is a universal function of crack-tip speed that decreases from 1 to 0, as the crack-tip speed increases from zero to the Rayleigh wave speed. In general, however, single crack fronts are not stable at high crack speeds (above $\sim$ 40–45% c_R). Hence, crack branching is generally observed to occur with increasing crack velocity (Kobayashi et al., 1974; Ravi-Chandar and Knauss, 1984; Gao, 1997; Suzuki et al., 1998). Crack branching occurs primarily because it reduces the overall energy required for subsequent crack extension. However, alternative explanations have also been proposed by researchers in the physics community. These are discussed in detail in the book by Freund (1990).

In the case of ductile materials subjected to dynamic fracture, the mechanics is less well developed due to the complexity of the associated phenomena. For growing two-dimensional cracks in idealized plastic solids, asymptotic analytical solutions have been reported by Achenbach et al. (1981), Freund and Douglas (1982), Lam and Freund (1985), and Deng and Rosakis (1994). However, although these results are of fundamental importance, they have not provided adequate measures of dynamic crack growth toughness or dynamic crack growth criteria in ductile materials. The few results are limited to small-scale yielding conditions in which $\dot{K}_{Ic}^{d}$ (K_{I}^{d}) is used as a measure of initiation toughness (Costin and Duffy, 1979; Nakamura et al., 1985; Owen et al., 1998). Under elastic–plastic conditions, Nakamura et al. (1988) and Guduru et al. (1997) have proposed the use of the J integral, J_{c}^{d})($\dot{J}^{d}$), as a measure of dynamic initiation toughness. Nishioka and Atluri (1983) have also proposed the use of the CTOD under elastic–plastic conditions.

In spite of the limited mechanistic understanding of dynamic fracture, dynamic fracture mechanics concepts have been applied to the analyses of a wide range of practical engineering problems. These include: the analyses of pipeline fracture (Kanninen et al., 1976; Kobayashi et al., 1988) and conventional pressure vessels (Kanninen and Popelar, 1985). Dynamic fracture concepts have also been applied to problems of armor penetration, high-speed machining, spacecraft/satellite shielding, and aircraft hardening.

11.11 SUMMARY

This chapter presents an introduction to the fundamentals of fracture mechanics. Following a brief review of Griffith fracture theory, the Orowan plasticity correction and the notion of the energy release rate, linear elastic, and fracture mechanics concepts are presented. These include: the basic ideas behind the derivation of the stress intensity factor parameter; the interpretation of the stress intensity factor parameter; the equivalence of the energy release rate and the stress intensity factor; and the applications of linear elastic fracture mechanics. Elastic–plastic fracture mechanics concepts are then examined within the framework of the J integral and the crack-tip opening displacement (CTOD). A two-parameter fracture J–Q approach is proposed for the assessment of constraint effects under elastic–plastic conditions before discussing the fundamentals of fracture testing. Finally, basic concepts in interfacial fracture mechanics and dynamic fracture are introduced.

BIBLIOGRAPHY

Achenbach, J.D. and Dunayevsky, V. (1981) Fields near a rapidly propagating crack tip in an elastic–perfectly plastic material. J Mech Phys Solids. vol. 29, pp 283–303.

Anderson, T.L. (1985) Fracture Mechanics: Fundamentals and Applications. 2nd ed. CRC Press, Boca Raton, FL.

Ashby, M.F. (1999) Materials Selection in Mechanical Design. Butterworth Heinemann, Boston, MA.

ASTM Standard E-813. (2001) Annual Book of ASTM Standards, 03-01, 74-8.

ASTM Standard E-399. (2001) Annual Book of ASTM Standards, 03-01.

Bagchi, A. and Evans, A.G. (1996) Interface Sci., vol. 3, p 169.

Barenblatt, G.E. (1962) Adv Aapl Mech. vol. 7, pp. 55–129.

Begley, J.A. and Landes, J.D. (1972a) ASTM STP 514, p 1.

Begley, J.A. and Landes, J.D. (1972b) ASTM STP 514, p 24.

Bilby, B.A. Cardhew, G.E., Goldthorpe, M.R., and Howard, I.C. (1986) A finite element investigation of the effect of specimen geometry on the fields of stress and strain at the tips of stationary cracks. In: Size Effects in Fracture. The Institution of Mechanical Engineers, London, pp 37–46.

Broek, D. (1978) Elementary Engineering Fracture Mechanics. Sitjhoff & Noordhoff, The Hague, The Netherlands.

Callister, W.D. (1999) Materials Science and Engineering: An Introduction. 5th ed. John Wiley, New York.

Cao, H.C. and Evans, A.G. (1989) Mech Mater, vol. 7, p 295.

Costin, L.S. and Duffy, J. (1979) J Eng Mater Technol. vol. 101, pp 258–264.

Cottrell, A.H. (1961) Iron and Steel Institute. Special Report. vol. 69, p 281.

Dalgleish, B.J., Trumble, K.P., and Evans, A.G. (1989) Acta Metall Mater. vol. 37, p 1923.

Deng, A. and Rosakis, A.J. (1994) Dynamic crack propagation in elastic–plastic solids under non-K-dominance conditions. Eur J Mech Solids. vol. 13, pp 327–350.

Dugdale, D.S. (1960) J Mech Phys Solids. vol. 8, p 100.

Ewalds, H.L. and Wanhill, R.J. H. (1984) Fracture Mechanics. Arnold, London.

Evans, A.G. and Dalgleish, B.J. (1992) Acta Metall Mater. vol. 40, p S295.

Evans, A.G., Hutchinson, J.W., and Wei, Y. (1999) Acta mater vol. 113, p 4093.

Freund, L.B. (1990) Dynamic Fracture Mechanics. Cambridge University Press, Cambridge, MA.

Freund, L.B. and Douglas, A.S. (1982) The influence of inertia on elastic plastic antiplane-shear crack-growth. J Mech Phys Solids. vol. 33, pp 169–191.

Gao, H. (1993) Surface roughening and branching instabilities in dynamic fracture. J Mech Phys Solids. vol. 1, pp 457–486.

Gaudette, F.A. Suresh, S., Evans, A.G., Dehm, G., and Ruhle, M. (1997) Acta Mater. vol. 45.

Griffith, A.A. (1921). The phenomena of rupture and flow in solids. Phil Trans Roy Soc, London, vol. A221, pp 163–197.

Guduru, P.R., Singh, R.P., Ravichandran, G., and Rosakis, A.J. (1997) Dynamic Crack Initiation In Ductile Steels. In: G. Ravichandran, A.J. Rosakis, M. Ortiz, Y.D.S. Rajapakse, K. Iyer, eds. Special Volume of the Special Issue J. Mech Phys Solids, devoted to Dynamic Failure Mechanics of Solids Honoring Professor Rodney Clifton. vol. 46(10), p 2016.

Hertzberg, R.W. (1996) Deformation and fracture mechanics of engineering materials. 4th ed. John Wiley, New York.
Hutchinson, J.W. (1968) J Mech Phys Solids. vol. 16, pp 13–31.
Hutchinson, J.W. (1979) Nonlinear Fracture Mechanics. Technical University of Denmark, Lyngby, Denmark.
Huthcinson, J.W. (1983) J Appl Mech. vol. 50, pp 1042–1051.
Hutchinson, J.W. (1987) Crack tip shielding by micro-cracking in brittle solids. Acta Metall. vol. 35, vol. 1605–1619.
Hutchinson, J.W. and Paris, P.C. (1979) Stability analysis of J-controlled crack growth. In: Elastic–Plastic Fracture, Special Technical Publication 668. ASTM, Conshohocken, PA, pp 37–64.
Hutchinson, J.W. and Suo, Z. (1992) Adv Appl Mech. vol. 29, p 63.
Inglis, C.E. (1913) Trans Inst Naval Architects. vol. 55, pp 219–241.
Irwin, G.R. (1957) J Appl Mech. vol. 24, pp 361–364.
Irwin, G.R. (1960) Plastic zone near a crack and fracture toughness. Proceedings of the 7th Sagamore Conference, Syracuse Uniersity, Syracuse, NY, vol. IV, pp 63–78.
Irwin, G.R. (1964) Structural aspects of brittle fracture. Phil Trans Roy Soc. vol. A221, pp 163–197.
Irwin, G.R. (1975) Trans ASME, J Appl Mech. vol. 24, p 361.
Kanninen, M.F. and Popelar, C.H. (1985) Advanced Fracture Mechanics. Oxford University Press, Oxford, UK.
Kanninen, M.F., Sampath, S.G., and Popelar, C.H. (1976) J Pressure Vessel Technol. vol. 98, pp 56–64.
Knott, J.F. (1973) Fundamentals of Fracture Mechanics. Butterworth, London, UK.
Kobayashi, A.S., Wade, B.G., Bradley, W.B., and Chiu, S.T. (1974) Eng Fract Mech. vol. 6, pp 81–92.
Kobayashi, T., Yamamoto, H., and Matsuo, K. (1988) Eng Fract Mech. vol. 30, pp 397–407.
Lam, P.S. and Freund, L.B. (1985). J. of Mech. and Phys. of Solids, 33, 153–167, 1985.
Landes, J.D. and Begley, J.A. (1977) ASTM ST 632, p 57.
Larsson, S.G. and Carlsson, A.J. (1973) J Mech Phys Solids. vol 21, pp 263–277.
Lawn, B. (1993) Fracture of Brittle Solids. 2nd ed. Cambridge University Press, Cambridge, UK.
McClintock, F.A., Kim, Y.-J., and Parks, D.M. (1995) Tests and analyses for fully plastic fracture mechanics of plane strain Mode I crack growth. In: W.G. Reuter, J.H. Underwood, J.C. Newman, eds. Fracture Mechanics. vol. 26. Conshohocken, PA, STP 1256, pp 155–222. ASTM.
McMeeking, R.M. (1977) Finite deformation analysis of crack tip opening in elastic–plastic materials and implications for fracture initiation. J Mech Phys Solids. vol. 25, pp 357–381.
McMeeking, R.M. and Parks, D.M. (1979) On the criteria for J dominance of crack tip fields in large-scale yielding. In: Elastic–Plastic Fracture. Special Technical Publication 667, ASTM, Conshohocken, PA, pp 175–94.

McNaney, J.M., Cannon, R.M., and Ritchie, R.O. (1996) Acta Mater. vol. 44, p 4713.

Murakami, Y., ed. (1987) Stress Intensity Factors Handbook. Pergamon Press, Oxford, UK.

Nakamura, T., Shih, C.F., and Freund, L.B. (1985) Eng Fract Mech. vol. 22, pp 437–452.

Nakamura, T., Shih, C.F., and Freund, L.B. (1988) Three-dimensional transient analysis of a dynamically loaded three-point-bend ductile fracture specimen. In: Nonlinear Fracture Mechanics. Special Technical Publication 995. ASTM, Conshohocken, PA, pp 217–141.

Neuber, H. (1946) Theory of Notch Stresses: Principle for Exact Stress Calculations. Edwards. Ann Arbor, MI.

Nishioka, T. and Atluri, S.N. (1983) Eng Fract Mech. vol. 18, pp 1–22.

O'Dowd, N.P. and Shih, C.F. (1991) Mech Phys Solids. vol. 39, pp 989–1015.

O'Dowd, N.P. and Shih, C.F. (1992) J Mech Phys Solids. vol. 40, pp 939–963.

Orowan, E. (1950) Fatigue and Fracture of Metals. MIT Press, Cambridge, MA, p 139.

Owen, D.M., Zhuang, S., Rosakis, A.J., and Ravichandran, G. (1998) Int J Fract vol. 90, pp 153–174.

Paris, P.C. (1960) The growth of cracks due to variations in loads. PhD thesis, Lehigh University, Bethlehem, PA.

Paris, P.C. and Erdogan, F. (1963) A critical analysis of crack propagation laws. J Basic Eng. vol. 85, pp 528–534.

Paris, P.C., Gomez, M.P., and Anderson, W.P. (1961) Trend Eng. vol. 13, pp 9–14.

Ravi-Chandar, K. and Knauss, W.G. (1984) Int J Fract. vol. 26, pp 141–154.

Reimanis, I., Dalgleish, B.J., and Evans, A.G. (1991) Acta Metall Mater. vol. 39, p 989.

Rice, J.R. (1968) J Appl Mech. vol. 35, pp 379–386.

Rice, J.R. (1974) J Mech Phys Solids. vol. 22, pp 17–26.

Rice, J.R. and Rosengren, G.F. (1968). J Mech Phys Solids. vol. 16, pp 1–12.

Rice, J.R., Drugan, W.J., and Sham, T.-L. (1980) Elastic–plastic analysis of growing cracks. In: Fracture mechanics, 12th Conference. Special Technical Publication 700, ASTM, Conshocken, PA, pp 189–221.

Ritter, J.E., Fox, J.R., Hutko, D.J., and Lardner, T.J. (1999) J Mater Sci. vol. 33, p 4581.

Rosakis, A.J. and Ravichandran, G. (2000) Int J Solids Struct. vol. 37, pp 331–348.

Sih, G.C.M. (1973) Handbook of Stress Intensity Factors. Lehigh university, Bethlehem, PA.

Soboyejo, W.O., Lu, G-Y., Chengalva, S., Zhang, J., and Kenner, V. (1999) Fat Fract Eng Mater Struct. vol. 22, p 799.

Suresh, S. (1999) Fatigue of Materials. 2nd ed. Cambridge University Press, Cambridge, UK.

Suzuki, S., Inayama, A., and Arai, N. (1998) Reflection type high-speed holographic microscopy to photograph crack bifurcation. Proceedings of the 11th International Conference on Experimental Mechanics, Oxford, UK, 1998.

Tada, H., Paris, P.C., and Irwin, G.R. (1999) The Stress Analysis of Cracks Handbook. 3rd ed., ASME, New York.

Turner, M. and Evans, A.G. (1996) Acta Metall Mater. vol. 44, p 863.

Wells, A.A. (1965) Brit Weld J. vol. 13, p 2.

Wells A.A. (1961) Proc. of Crack Propagation Symposium, Cranfield College of Astronautics, 1, p. 210.

Wells, A.A. (1968) Brit Weld J. vol. 15, p 221.

Wells, C.H. (1979) In: M. Meshii, ed. High-temperature fatigue in fatigue and microstructure. American Society for Metals, Metals Park, OH, pp 307–31.

Williams, M.L. (1957) On the stress distribution at the base of a stationary crack. J Appl Mech. vol. 24, pp 109–14.

Yoffe, E.H. (1951). The moving Griffith crack. Phil Mag. vol. 42, pp 739–750.

12

Mechanisms of Fracture

12.1 INTRODUCTION

Fracture occurs by the separation of bonds. However, it is often preceded by plastic deformation. Hence, it is generally difficult to understand the physical basis of fracture without a careful consideration of the deformation phenomena that precede it. Nevertheless, at one extreme, one may consider the case of brittle fracture with limited or no plasticity. This represents an important industrial problem that can be overcome largely by the understanding of the factors that contribute to brittle fracture. However, *ab initio* models for the prediction of brittle fracture are yet to emerge.

On the other hand, ductile fracture mechanisms represent another class of important fracture modes in engineering structures and components. They are somewhat more complex to analyze due to the nonlinear nature of the underying plasticity phenomena. However, a significant amount of scientific understanding of ductile fracture processes has facilitated the safe use of metals and their alloys in a large number of structural applications.

Most recently, there have been significant efforts to develop novel composite materials and engineered materials with improved fracture resistance. These efforts have led to an improved understanding of how to tailor the microstructure/architecture of a material for improved fracture toughness. The research that has been performed in the past 25–35 years has also led to identification of toughening mechanisms that can be used to engineer improved fracture toughness in all classes of materials. These will be discussed in detail in Chap. 13.

This chapter presents an introduction to the micromechanisms of fracture in different classes of materials. Following an initial review of brittle and ductile fracture mechanisms, the mechanisms of fracture in different classes of materials are discussed along with mechanics models that provide some additional insights into the mechanisms of fracture. Quantitative and qualitative approaches are also presented for the characterization of fracture modes before concluding with a section on the thermal shock response of materials.

12.2 FRACTOGRAPHIC ANALYSIS

To most people, there is a natural tendency to assume that macroscopic ductility is a clear indication of ductile fracture, e.g., during tensile fracture of smooth "dog-bone" specimens. However, although this may be true for many solids, evidence of macroscopic ductility is generally insufficient in fracture analysis. Instead, we are usually compelled to perform detailed analyses of the fracture surface(s) using scanning or transmission electron microscopy techniques. These provide the local evidence of microscopically ductile or brittle fractures.

In the case of scanning electron microscopy (Fig. 12.1), electrons are accelerated from an electron gun (cathode). The electron beam is collimated by a series of lenses and coils until it hits the specimen surface (fracture surface). The electrons are then reflected from the specimen surface after interacting with a small volume of material around the surface. The two types of electrons that are reflected back from the surface are secondary electrons and back-scattered electron. These are detected by detectors that are rastered to form a TV image. The second electron images usually provide good depth of field and clear images of surface topography, while the back-scattered electron images have the advantage of providing atomic number contrast that can be used to identify different phases (due to differences in chemical composition).

The fracture surfaces of conducting materials (mostly metals/intermetallics) can generally be viewed directly with little or no surface preparation prior to scanning electron microscopy. However, the fracture surfaces of nonconducting materials are generally coated with a thin (a few nanometers) layer of conducting material, e.g, gold, to facilitate fractographic examination in a scanning electron microscope (SEM). The SEM can be used to obtain images over a wide range of magnifications (100–100,000×).

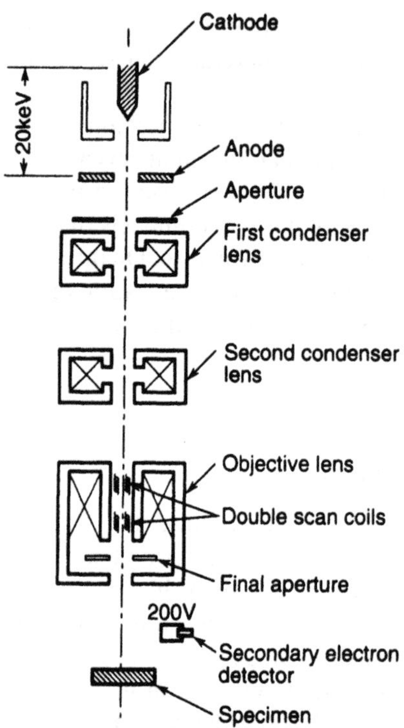

FIGURE 12.1 Schematic illustration of the operation of a scanning electron microscope. (From Reed-Hill and Abbaschian, 1991—reprinted with permission from PWS Publishing Co.)

Most elaborate fracture surface preparation is needed for the examination of fracture surfaces in the transmission electron microscope. These involve the preparation of replicas of the fracture surface. The centers of the replicas must also be thinned to facilitate the transmission of electrons. In the case of transmission electron microscopy, the collimated electron beams are transmitted through thinned specimens (Fig. 12.2). The transmitted electron beams may then be viewed in the diffraction mode [Fig. 12.29(a)], or in the imaging mode [Fig. 12.2(b)]. Some of the early studies of fracture were done using transmission electron microscopy analyses of the replicas of fracture surfaces in the 1950s and 1960s. However, with the advent of the SEM, it has become increasingly easier to perform fractographic analyses. Most of the images of fracture surfaces presented in this chapter will, therefore, be images obtained from SEMs. These have good depth of focus, and can produce images with resolutions of ~ 5–10 nm.

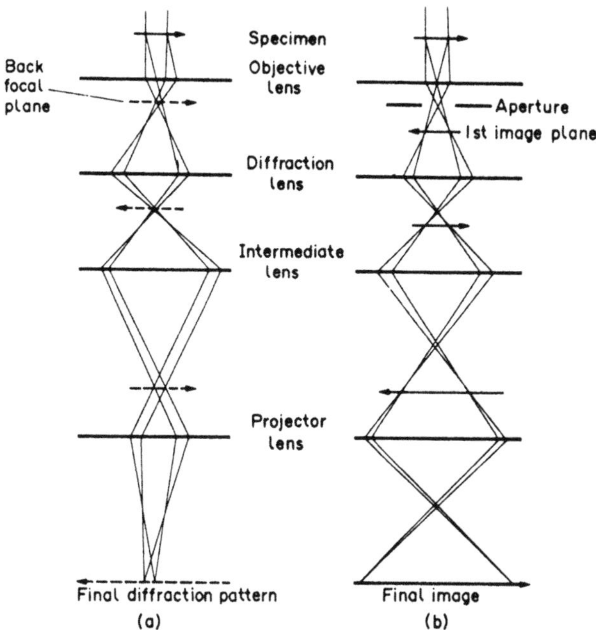

FIGURE 12.2 Schematic ray diagrams for (a) the diffraction mode and (b) the imaging mode of a transmission electron microscope. Most microscopes have more lenses than those shown here. (From Hull and Bacon, 1984—adapted from Loretto and Smallman, 1975—reprinted with permission from Pergamon Press.)

12.3 TOUGHNESS AND FRACTURE PROCESS ZONES

Fracture experiments are usually peformed on smooth or notched specimens. The experiments on smooth specimens generally involve the measurement of stress–strain curves, as discussed in Chap. 3. The smooth specimens are loaded continuously to failure at controlled strain rates, in accordance with various testing codes, e.g., the ASTM E-8 specification.

The area under the generic stress–strain curve is representative of the energy per unit volume required for the fracture. This is often described as the toughness of the material. Hence, in the representative stress–strain curves shown in Fig. 12.3, material B is the toughest, while materials A and C are not as tough. However, material A is strong and brittle, while material C is weak and ductile.

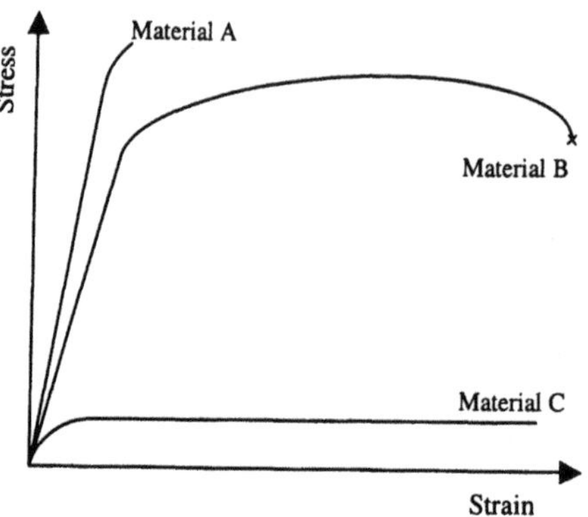

FIGURE 12.3 Illustration of toughness as the area under the stress–strain curves for materials A, B, and C.

The toughness or energy per unit volume, W, is given by

$$W = \int_0^{\varepsilon_f} \sigma \, d\varepsilon \tag{12.1}$$

where ε_f is the fracture strain, and σ is the applied stress, expressed as a function of strain, ε. The toughness must be distinguished from the fracture toughness, which may be considered as a measure of the resistance of a material to crack growth.

Materials with high toughness or fracture toughness generally require a significant amount of plastic work prior to failure. In contrast, the fracture toughness of purely brittle materials is controlled largely by the surface energy, γ_s, which is a measure of the energy per unit area required for the creating of new surfaces ahead of the crack tip. However, there is a strong coupling between the surface energy, γ_s, and plastic energy term, γ_p. This coupling is such that small changes in γ_s can result in large changes in γ_p, and the overall toughness or fracture toughness.

Finally in this section, it is important to note that the deformation associated with the fracture of tough materials generally results in the creation of a deformation process zone around the dominant crack. This is illustrated schematically in Fig. 12.4 The surface energy term is associated with the rupture of bonds at the crack tip, while the plastic work term is used partly in the creation of the deformation process zone. Details of the phenomena that occur in the deformation process zones are presented in the next few sections on fracture in the different cases of materials.

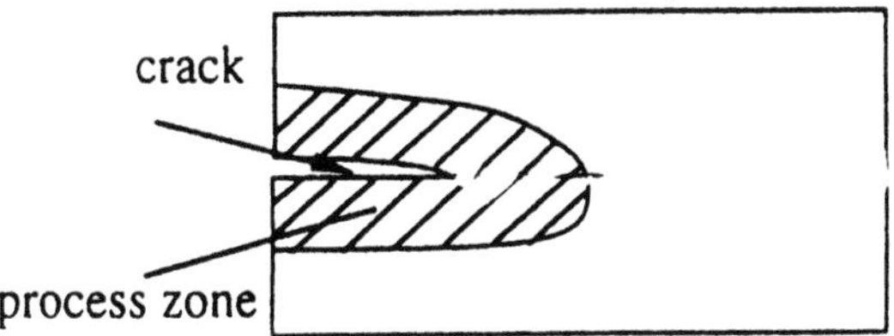

FIGURE 12.4 Schematic of the fracture process zone.

12.4 MECHANISMS OF FRACTURE IN METALS AND THEIR ALLOYS

12.4.1 Introduction

Fracture in metals and their alloys occurs by nominally brittle or ductile fracture processes. In cases where brittle fracture occurs without local plasticity along low index crystallographic planes, the failure is described as a cleavage fracture. Cleavage fracture usually occurs by bond rupture across gains. It is, therefore, often referred to as transgranular cleavage. However, bond rupture may also occur between grains, giving rise to a form of fracture that is known as intergranular failure.

In the case of ductile failure, fracture is usually preceded by local plasticity and debonding of the matrix from rigid inclusions. This debonding, which occurs as a result of the local plastic flow of the ductile matrix, is followed by localized necking between voids, and the subsequent coalescence of voids to form dominant cracks. It results in ductile dimpled fracture modes that are characteristic of ductile failure in crystalline metals and their alloys.

In contrast, the fracture of amorphous metals typically occurs by the propagation of shear bands, and the propagation of microcracks ahead of dominant cracks. These different fracture mechanisms are discussed briefly in this section for metals and their alloys.

12.4.2 Cleavage Fracture

As discussed earlier, cleavage fracture occurs by bond rupture along low index crystallographic planes. It is usually characterized by the presence of "river lines" (Fig. 12.5) that are formed as a result of the linkage of ledges produced by cracking along different crystallographic planes (Tipper, 1949). The river lines often resemble river lines on a map, and are relatively easy to identify. However, the presence of river lines alone may not be sufficient evidence to infer the occurrence of "pure" cleavage fracture in cases where

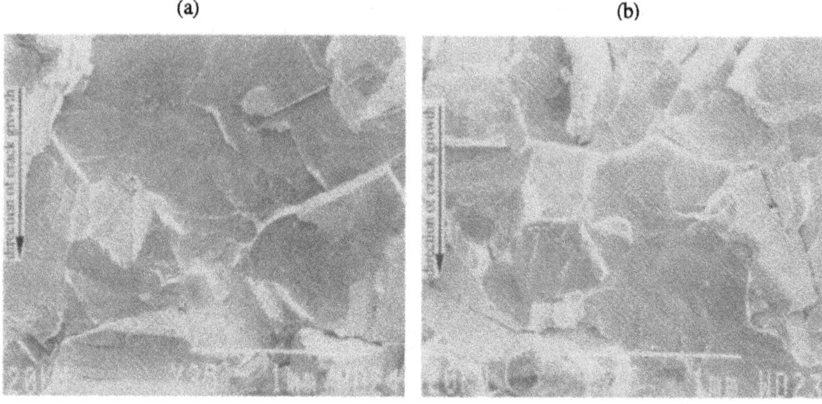

Figure 12.5 Cleavage fracture in niobium aluminide intermetallics: (a) Nb–15Al–10Ti, (b) Nb–15Al-25Ti. (From Ye et al., 1999.)

fracture is preceded by some local plasticity. Under such conditions, fracture is preceded by some local plasticity and the mirror halves of the two fracture surfaces do not match. This gives rise to a form of brittle fracture that is known as "quasi-cleavage" (Thompson, 1993).

Cleavage fracture is often obseved in metals at lower temperatures. Furthermore, a transition from brittle to ductile fracture is generally observed to occur with increasing temperature in body-centered cubic (b.c.c.) metals and their alloys, e.g., steels. This transition has been studied extensively, but is still not fully understood.

The first explanation of the so-called brittle-to-ductile transition (BDT) was offered by Orowan (1945) who considered the variations in the temperature dependence of the stresses required for yielding and cleavage (Fig. 12.6). He showed that the cleavage fracture stress exhibits a weak dependence on temperature, while the yield stress generally increases significantly with decreasing temperature. This is illustrated schematically in Fig. 12.6. The stresses required for yielding are, therefore, lower than those required for cleavage fracture at higher temperatures. Hence, failure above the BDT regime should occur by ductile fracture. In contrast, since the cleavage fracture stresses are less than the yield stresses below the BDT regime, cleavage fracture would be expected to occur below this regime.

Following the work of Orowan (1945), other researchers recognized the critical role that defects play in the nucleation of cleavage fracture. Stroh suggested that cleavage fracture occurs in a polycrystal when a critical value of tensile stress, $\sigma_{\theta\theta}$ is reached in an unyielded grain.

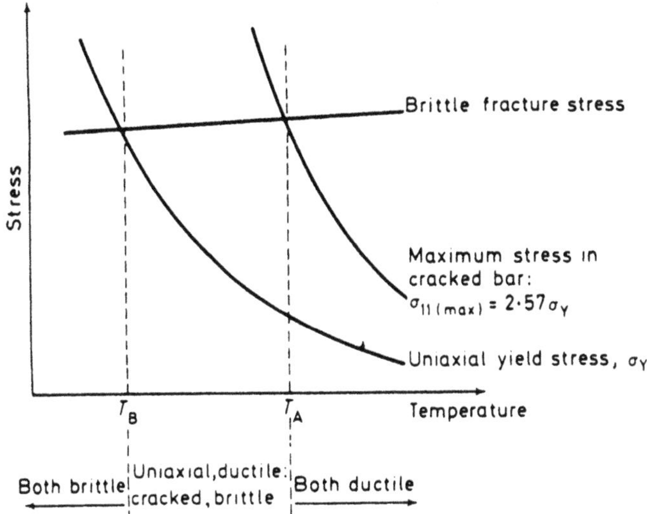

Figure 12.6 Schematic of Orowan ductile-to-brittle transition. (From Knott, 1973—reprinted with permission from Butterworth-Heinemann.

Using similar arguments to those employed in the Hall–Petch model (Chap. 8), Stroh (1954, 1957) derived the following relationship between the cleavage fracture stress, σ_c, and the grain size, d:

$$\sigma_c = \sigma_i + k_f d^{-\frac{1}{2}} \tag{12.2}$$

where k_f is the local tensile stress required to induce fracture in an adjacent grain under nucleation-controlled conditions. This theory correctly predicts the inverse dependence of the cleavage fracture stress on grain size, but it suggests a constant value of k_y that is not true for finer grain sizes.

Subsequent work by Cottrell (1958) showed that if the tensile stress is the key parameter, as suggested by experimental results, then cleavage fracture must be growth controlled. Cottrell suggested that cleavage fracture in iron occurs by the intersection of $\frac{a}{2}\langle\bar{1}\bar{1}1\rangle$ dislocations gliding on $\{101\}$ slip planes. This results in the following dislocation reaction (Fig. 13.7):

$$\frac{a}{2}\langle\bar{1}\bar{1}1\rangle_{(101)} + \frac{1}{2}\langle 111\rangle_{(\bar{1}01)} \rightarrow a[001] \tag{12.3}$$

Since the resulting $a[001]$ dislocation is sessile, this provides the first stage of crack nucleation that occurs due to the relative motion of material above and below the slip plane. Furthermore, the pumping of n pairs of dislocations into the wedge results in a displacement nb of length c. The total

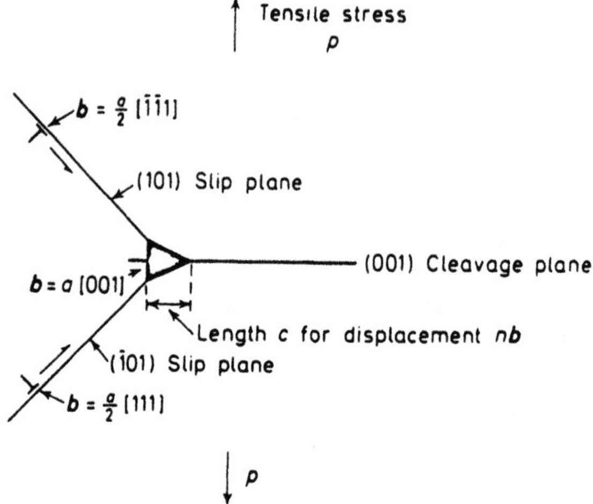

Figure 12.7 Cottrell's model of brittle fracture. (From Knott 1973—reprinted with permission from Butterworth-Heinemann.)

energy per unit thickness now consists of the following four components (Knott, 1973).

1. The Griffith energy of a crack of length, c, under tensile stress, p:

$$U_1 = \frac{-p^2(1-\nu^2)}{E}\left(\frac{c}{2}\right)^2 \tag{12.4a}$$

(for a crack of length $2a = c$)

2. The work done by the stress in forming the nucleus:

$$U_2 = -\frac{1}{2}pnbc \tag{12.4b}$$

3. The surface energy:

$$U_3 = 2\gamma c \tag{12.4c}$$

4. The strain energy of the cracked edge dislocation of Burgers vector nb:

$$U_v = \frac{\mu(nb)^2}{4\pi(1-\nu)}\ln\left(\frac{2R}{c}\right) \tag{12.4d}$$

where R is the distance over which the strain field is significant, μ is the shear modulus, and $c/2$ is the radius of the dislocation score of the cracked dislocation. The equilibrium crack lengths are found from $\partial/\partial c$

$(U_1 + U_2 + U_3 + U_4) = 0$. This gives a quadratic function with two possible solutions for the crack lengths. Alternatively, there may be no real roots, in which case the total energy decreases spontaneously. The transition point is thus given by

$$pnb = 2\gamma \tag{12.5a}$$

or for $b = 2[001]$:

$$pna = 2\gamma \tag{12.5b}$$

The tensile stress that is needed to propagate a nucleus is thus given by

$$p \geq \left(\frac{2\mu\gamma}{k_y}\right) d^{-\frac{1}{2}} \tag{12.5c}$$

Cottrell (1958) used the above expression to explain the results of Low (1954) for mild steel that was tested at 77 K (Fig. 12.8). By substituting $p = \sigma_y = 2\tau_y$ into the above expression, he obtained the following expressions for the stress to propagate an existing nucleus:

$$\tau_y \geq \left(\frac{\mu\gamma}{k_y}\right) d^{-\frac{1}{2}} \tag{12.5d}$$

or

$$\sigma_y \geq \left(\frac{2\mu\gamma}{k_y}\right) d^{-\frac{1}{2}} \tag{12.5e}$$

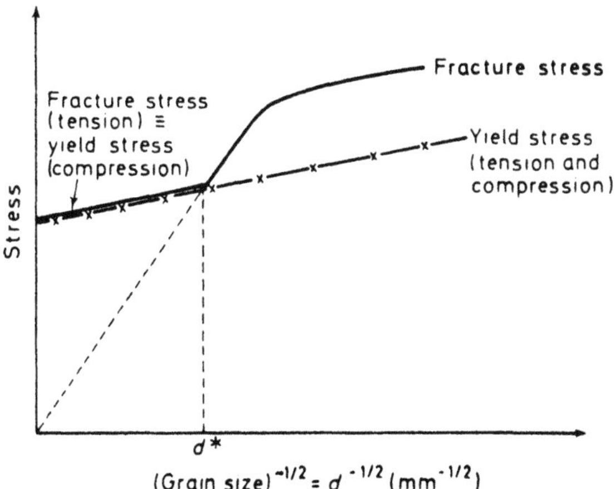

Figure 12.8 Yield and fracture stress in mild steel as functions of grain size. (From Low, 1954—reprinted with permission from ASM International.)

However, for grains coarser than d^* in Fig. 12.8, the fracture is not growth controlled. This is because it is necessary to first form the dislocations before a nucleus can be produced. The fracture stress is thus simply equal to the yield stress. However, for grains larger than d^*, yielding occurs prior to fracture, and the nucleus will spread when

$$\sigma_c \geq \left(\frac{2\mu\gamma}{k_y}\right) d^{-\frac{1}{2}} \tag{12.6}$$

The Cottrell model suggests that the cleavage fracture stress is controlled by the grain size ($d^{-1/2}$) and yielding parameters such as k_y. However, it does not account for the effects of carbides that can also induce cleavage fracture is steels. This was first recognized by McMahon and Cohen (1965) in their experiments on steels containing fine and coarse carbides. Materials containing coarse carbides were shown to be susceptible to cleavage, while those containing fine carbides were more ductile. The effects of carbides were later modeled by Smith (1966) who showed that the condition for growth-controlled fracture is

$$\left(\frac{d_0}{d}\right)\sigma_c^2 + \tau_{\text{eff}}^2\left[1 + \frac{4}{\pi}\left(\frac{C_0}{d}\right)^{\frac{1}{2}}\frac{\tau_i}{\tau_{\text{eff}}}\right]^2 \geq \frac{4E\gamma_p}{\pi(1-\nu^2)d} \tag{12.7}$$

where $\tau_{\text{eff}} = \tau_y - \tau_i$, τ_i is the intrinsic yield strength in the absence of grain boundaries, τ_y is the yield stress of a polycrystal, and the other terms have their usual meaning, as depicted in Fig. 12.9.

Work by Hull (1960) also recognized that cleavage fracture can be nucleated by interactions between dislocations twins. Similar results were later reported by Knott and Cottrell (1963) for fracture in polycrystalline mild steel. In the case of b.c.c. metals, twinning occurs by the movement of $a/6\langle\bar{1}11\rangle$ dislocations on $\{211\}$ planes. The twinning shear is 0.707. This can produce significant displacements normal to the cleavage plane. Estimates of γ associated with these displacements are ~ 20 J/m^2. These are approximately one order of magnitude greater than the surface energies.

Since the early work on cleavage fracture, subsequent research has shown that cleavage fracture is most likely to occur under conditions of high-stress triaxiality. In most cases, the so-called triaxiality factor, T.F., is expressed as a ratio of the hydrostatic stress to the Von Mises stress:

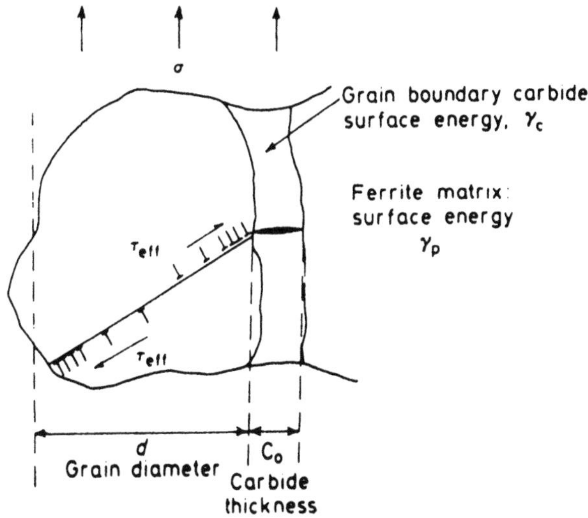

FIGURE 12.9 Schematic of cleavage fracture nucleation from carbides. (From Knott, 1973—reprinted with permission from Butterworth-Heinemann.)

$$\text{T.F.} = \frac{\frac{1}{3}(\sigma_{11} + \sigma_{22} + \sigma_{33})}{\frac{1}{\sqrt{2}}\left\{(\sigma_{11} + \sigma_{22})^2 + (\sigma_{22} + \sigma_{33})^2 - (\sigma_{33} + \sigma_{+}11)^2\right\}^{1/2}}$$
$$= \frac{\frac{1}{3}(\sigma_{xx} + \sigma_{yy} + \sigma_{zz})}{\frac{1}{\sqrt{2}}\left\{(\sigma_{xx} - \sigma_{yy})^2 + (\sigma_{yy} - \sigma_{zz})^2 + (\sigma_{zz} - \sigma_{xx})^2 + 6(\tau_{xy}^2 + \tau_{yz}^2 + \tau_{zx}^2)^2\right\}^{\frac{1}{2}}} \tag{12.8}$$

Typical values of T.F. are between 2 and 3. Furthermore, stress states with higher triaxiality factors are more likely to result in brittle cleavage fracture modes. It is important to note here that the occurrence of cleavage fracture alone is not necessarily an indication of the lack of ductility. This is particularly true for b.c.c. refractory metals in which cleavage fracture modes are likely to occur at high strain rates, or in the presence of notches. In any case, most b.c.c. metals and their alloys are ductile at room temperature, even in cases where final failure occurs by cleavage-or quasi-cleavage fracture mechanisms.

The critical role of crack-tip/notch-tip stress distributions in the nucleation of brittle cleavage fracture was first examined in detail by Ritchie, Knott, and Rice (1973). By considering the results of finite element calculations of the notch-tip fields of Griffith and Owens (1972),

Ritchie et al. (1973) postulated that cleavage-fracture nucleation is most likely to occur at a distance of 2–3 CTODs ahead of the notch-tip, Fig. 12.10(a). The so-called Ritchie–Knott–Rice (RKR) theory recognized the need for local tensile stresses to exceed the local fracture stress over a microstructurally significant distance ahead of a notch/crack tip, as shown in Fig. 12.10(a).

Subsequent work by Lin et al. (1986) resulted in the development of a statistical model for the prediction of brittle fracture by transgranular cleavage. Using weakest link statistics to characterize the strength distributions of the inclusions ahead of the notch tip/crack tip, they showed that the failure probability associated with the element of material in the plastic zone is given by

$$\delta\phi = 1 - \exp[-bfN\varepsilon K_1^4 \sigma_0^{2(n-2)} S_0^{-m} (\sigma_0 - S_u)^m \sigma^{-(2n+3)} d\sigma] \tag{12.9}$$

where b is the characteristic dimension along the crack tip, f is the fraction of particles that participate in the fraction initiation, N is the number of particles per unit volume, K_1 is the Mode I stress intensity factor, σ_0 is the yield or flow stress, S_0 is the Weibull scale parameter, S_u is the lower bound strength (of the largest feasible cracked particle), m is the shape factor, n is the work hardening exponent ($1 < n < \infty$), σ is the local stress within plastic zone, and ε is a strain term that is given by Lin et al. 1986 to be:

$$\varepsilon = 2(n+1)\left[\frac{1-\nu^2}{I_n}\right]^2 \int_0^{\pi} \tilde{\sigma}^{2(n+1)} d\theta \tag{12.10}$$

The elemental failure probability expressed by Eq. (12.9) exhibits a maximum at a characteristic distance, r^*, from the crack tip, which is given by ($d\delta\Phi = 0$):

$$r^* = \left[\frac{1-\nu^2}{I_n}\right]\left[\frac{2n+3-m}{2n+3}\right]^{n+1}\left(\frac{K_1}{\sigma_0}\right)^2\left(\frac{\sigma_0}{S_u}\right)^{n+1}\tilde{\sigma}^{n+1} \tag{12.11a}$$

occurring at the stress:

$$\sigma^* = \left[\frac{2n+3}{2n+3-m}\right]S_u \tag{12.11b}$$

as illustrated in Fig. 12.10(b), which shows the dependence of elemental fracture probability, $\delta\phi$, and tensile stress distribution on the same schematic plot.

A number of other researchers have made significant contributions to the modeling of cleavage fracture within a statistical framework. These include Curry (1980), Evans (1983), Beremin (1983), Mudry (1987),

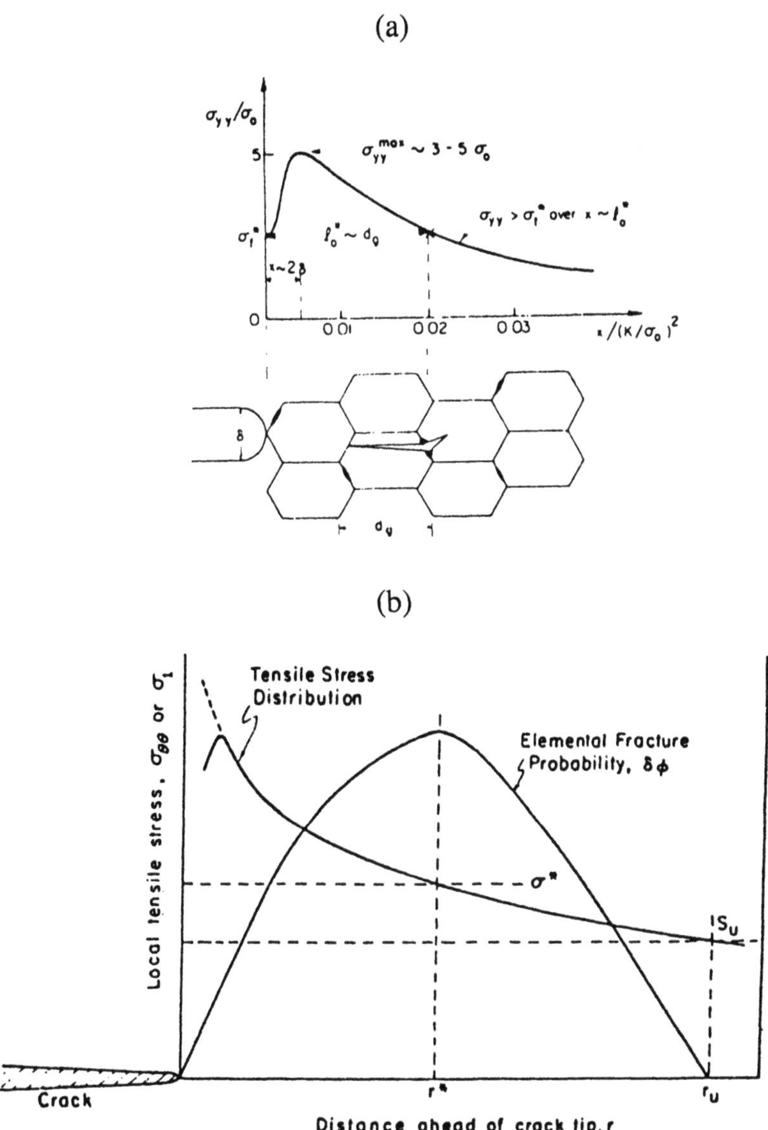

FIGURE 12.10 Stress distributions ahead of blunt notches: (a) Ritchie–Knott–Rice (RKR) model; (b) Lin–Evans–Ritchie model. (From Lin et al., 1986—reprinted with permission from Elsevier Science.)

Rousselier (1987), Fontaine et al. (1987), Rosenfield and Majumdar (1987), and Thompson and Knott (1993). All of these studies have suggested refinements to the approach of Lin et al. (1986). In general, however, the trends predicted by the different models are generally consistent. However, the size effect predicted by the Beremin (1983) and Evans (1983) theories are not the same.

12.4.3 Ductile Fracture

Ductile fracture in metals and their alloys is generally associated with the nucleation of voids around rigid inclusions. Since plastic flow of the matrix can occur around the inclusions, the matrix may become debonded from the rigid inclusions during plastic flow, Figs 12.11(a) and 12.11(b). Subsequent localized void growth [Fig. 12.11(c)] and deformation and necking [Figs 12.11(d) and 12.11(e)] may then occur prior to the coalescence of microvoids and final fracture, Fig. 12.11(f). This gives rise to the formation of larger dominant cracks that may propagate in a "stable" manner until catastrophic failure occurs. Not surprisingly, the fracture surfaces will contain dimples (Fig. 12.12) associated with the microvoid nucleation and propagation processes. The inclusions associated with nucleation and propagation may also be seen in some of the microvoids (Fig. 12.12).

Considerable experimental work has also been done to provide insights into the mechanisms of ductile fracture. The pioneering work in this area has been done by Knott and co-workers (1973, 1987). Other experimental researchers that have made significant contributions to the understanding of ductile fracture include Thompson and Knott (1993) and Ebrahimi and Seo (1996).

The studies by Ebrahimi and Seo (1996) explored the three-dimensional nature of crack initiation in ferritic and bainitic steels. They concluded that crack initiation occurs by the formation of disconnected cracks along the crack front. Inclusions and highly strained regions were found to be the sites for crack nucleation in ferritic–pearlitic steels, while geometrical inhomogenities associated with fatigue provided the sites for crack nucleation in bainitic steels.

In general, however, the dominant viewpoint is that voids nucleate at particles (Rice and Tracey, 1969). These include primary and secondary particles formed by phase transformations, and inclusions that are introduced largely during processing, e.g., during casting or powder processing.

The pioneering theoretical work on void growth was done by McClintock (1968), who analyzed the effects of stress rate on the growth of a long cylindrical void. Subsequent work by Rice and Tracey (1969)

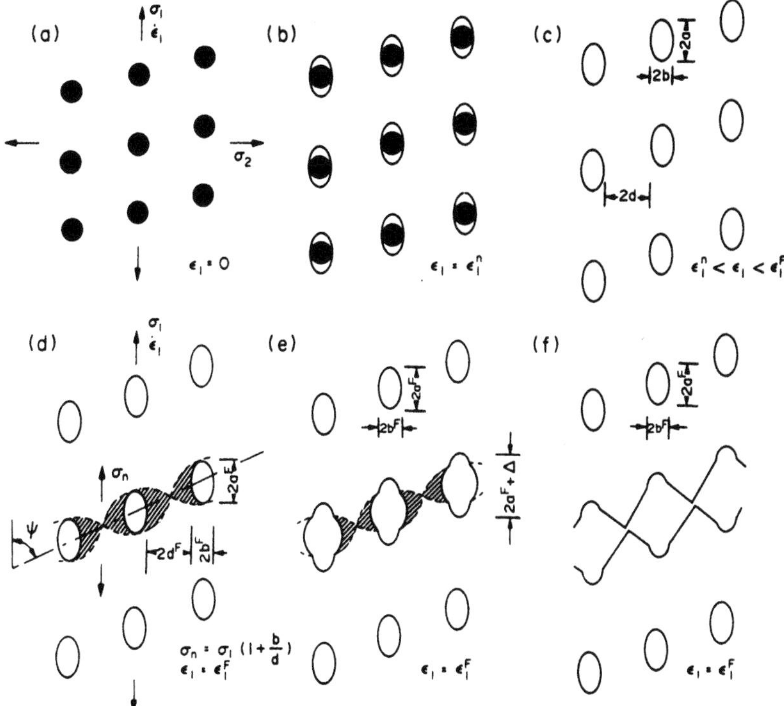

FIGURE 12.11 Schematics of ductile fracture processes in metals and their alloys: (a) onset of deformation; (b) microvoid nucleation; (c) void growth; (d) strain localization between voids; (e) necking between voids; (f) knife-edge separation. (From Thomason, 1990—reprinted with permission from Pergamon Press.)

proposed the well-known Rice–Tracey (RT) model for the characterization of the growth rate of an isolated spherical void (Fig. 12.13). The void growth rate predicted under continuum plasticity conditions in which $(\sqrt{3}\sigma_{app}\sigma_2 > 1)$ is

$$\ln\left(\frac{\overline{R}}{R_0}\right) = \int_0^{\varepsilon_q} 0.283\left(\frac{3}{2}\frac{\sigma_m}{\sigma_{ys}}\right)d\varepsilon_\nu^p \tag{12.12}$$

where ε_q is the equivalent Von Mises plastic strain, σ_{ys} is the yield stress (which may be replaced with the effective stress), σ_m is the mean stress, R_0 is the particle size, and $\overline{R}$ is given by the stress traxiality:

$$\overline{R} = \frac{R_1 + R_2 + R_3}{3} \tag{12.13}$$

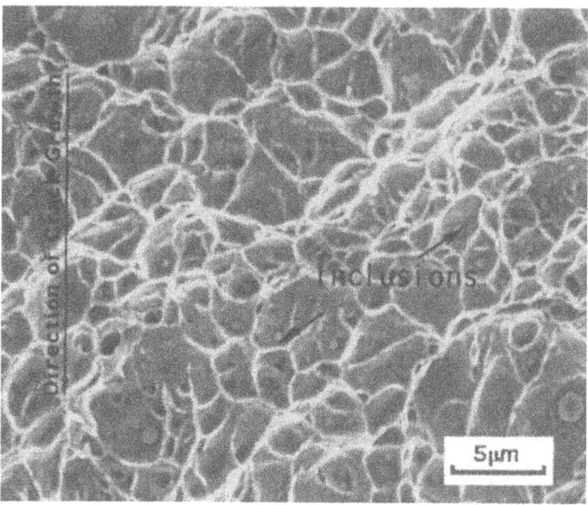

Figure 12.12 Ductile dimpled fracture in an A707 steel. (Courtesy of Jikou Zhou.)

However, in rear materials, more than one void nucleates almost at the same time, and there are interactions between the different growing voids. Also, although the RT model does not consider the interactions between voids, it is generally used to predict void growth behavior in ductile fracture processes.

Several modifications have been made to develop the RT model. By considering the nucleation and propagation of dislocation in the matrix between voids, Kameda (1989) deduced that the relationships between the void growth rate and the hydrostatic tensile stress and the void fracture is a function of the thermally activated shear stress and the activation volume for dislocation motiton in the matrix triaxility.

Idealizing the process of ductile fracture by confining void growth and coalescence to a material layer of initial thickness, D, ahead of the initial crack tip, Xia and Shih (1996) developed a mechanism-based cell model for the characterization of ductile tearing, and the transition from ductile-to-brittle fracture. The most important fracture process parameters are the initial void volume fracture in the cell, f_0, and the characteristic length of a cell, D, which should be interpreted as the mean spacing between the voids nucleated from the large inclusions. Microvoids nucleated from small inclusions assist with the process of hole link-up with the crack tip during the coalescence phase. The current void volume fraction, f, and the current flow stress of the matrix σ also change during the ductile fracture process.

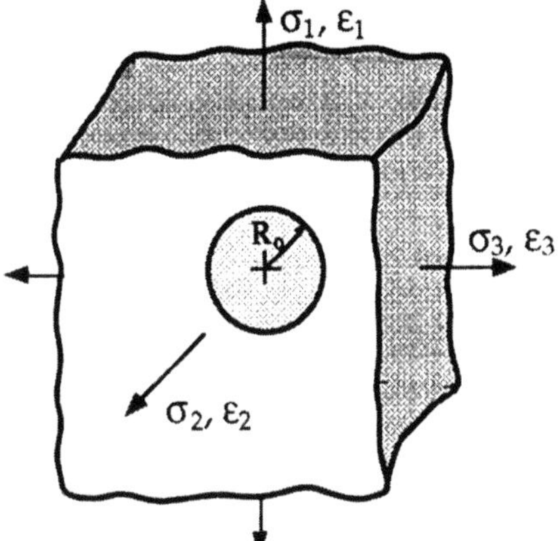

FIGURE 12.13 Schematic of spherical void growth in a solid subjected to a triaxial stress state. (From Anderson, 1994—reprinted with permission from CRC Press.)

The yield surface has been derived from approximate solutions for a single cell containing a centered spherical void. Gurson's yield condition (1977) has been used. This gives

$$\phi(\sigma_e, \sigma_m, \bar{\sigma}, f) \equiv \left(\frac{\sigma_e}{\bar{\sigma}}\right)^2 + 2q_1 f \cosh\left(\frac{3q_2\sigma_m}{2\bar{\sigma}}\right) = [1 + (q_1 f)^2] = 0 \tag{12.14}$$

where σ_e is the macroscopic effective Mises stress, σ_m if the macroscopic means stress, and q_1 and q_2 are the factors introduced by Tvergaard (1982) to improve the accuracy of the model.

Under small-scale yielding conditions, the energy balance of cell models is given by

$$\Gamma = \Gamma_0 + \Gamma_p + \Gamma_E \tag{12.15}$$

where Γ is the total work of fracture per unit area of crack advance, Γ_0 is work of the ductile fracture process, Γ_p is the plastic dissipation in the background material, and Γ_E is the additional contribution taking into account the work related to changes in the process zone size and the elastic energy variation within and just outside the plastic zone due to changes in the plastic zone size.

Tvergaard and Needleman (1984) have modified the original Gurson model by replacing f with an effective volume fraction, f^*. Thomason (1990) has also proposed a simple limit load model for internal necking between voids at a critical net section stress, $\sigma_n(c)$. This is illustrated schematically in Fig. 12.14 for a simple two-dimensional case. For in-plane void dimensions, $2a$ and $2b$, and spacing $2d$ between voids, the critical condition for necking is given by

$$\sigma_{n(c)} > \frac{d}{d+b} > \sigma_1 \tag{12.16a}$$

and the critical condition for growth is given by

$$\sigma_{n(c)} \frac{d}{d+b} = \sigma_1 \tag{12.16b}$$

where σ_1 is the remote principal stress.

Thomason (1990) has also applied the void growth model of Rice and Tracey (1969) to the prediction of void size and shape. The predicted failure strains were close to experimental observations, but an order of magnitude lower than those predicted by the Gurson model.

12.4.4 Intergranular Fracture

In cases where the grain boundary cohesion is reduced, crack growth may occur across the grain boundaries, giving rise to an intergranular fracture mode (Fig. 12.15). The grain boundary cohesion may be reduced by the segregation of atomic species such as sulfur or phosphorus to the grain

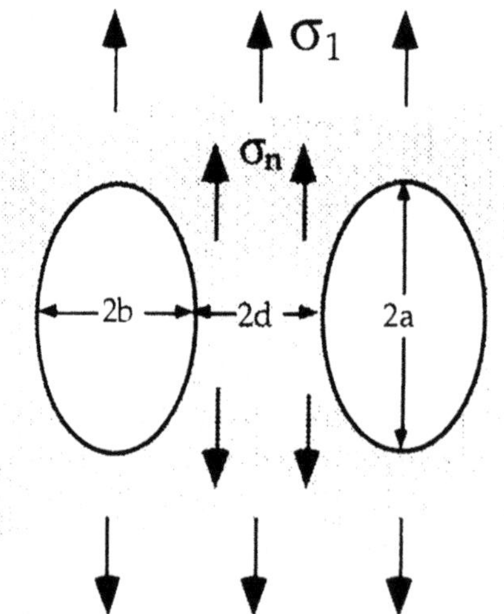

FIGURE 12.14 Mechanisms of ductile crack growth. (From Thomason, 1990—reprinted with permission from Pergamon Press.)

boundary regions (Briant, 1988). This is particularly true in steels and nickel base superalloys in which parts per million levels of sulfur and phosphorus have been shown to be sufficient to induce intergranular fracture. The resulting intergranular fracture modes are faceted, since the fracture paths follow the grain boundary facets, as shown in Fig. 12.15 Intergranular fracture may also occur as a result of the reduction in grain boundary cohesion by grain boundary precipitation and the stress-assisted diffusion of hydrogen. In some metals/alloys, intergranular fracture may also be induced at high strain rates or under cyclic loading. However, the fundamental causes of intergranular fracture in such scenarios are not well understood.

12.5 FRACTURE OF INTERMETALLICS

Intermetallics are compounds that form between metals and other metals. Examples include compounds between titanium and aluminum (titanium aluminides such as TiAl or Ti_3Al) or compounds between nickel and aluminum (nickel aluminides such as Ni_3al or NiAl). Due to their ordered or partially ordered structures, it is generally difficult to achieve the five independent slip systems required for homogeneous plastic deformation in intermetallic systems. Furthermore, even in cases where five independent slip systems are possible, reduced grain boundary cohesion of some intermetallics (such as Ni_3Al or Nial) may limit their potential for significant

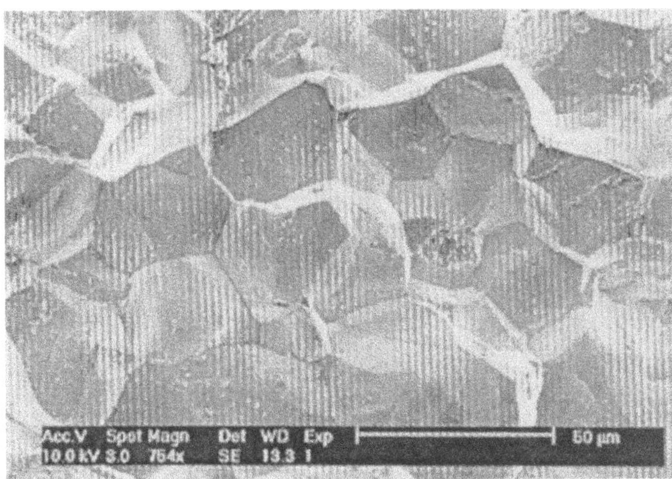

FIGURE 12.15 Intergranular fracture in a plain carbon steel. (Courtesy of Dr. Christopher Mercer.)

levels of deformation by slip (George and Liu, 1990; Liu et al., 1989). Room-temperature fracture, therefore, tends to occur by brittle cleavage or intergranular fracture modes in most intermetallic systems (Fig. 12.16).

(a)

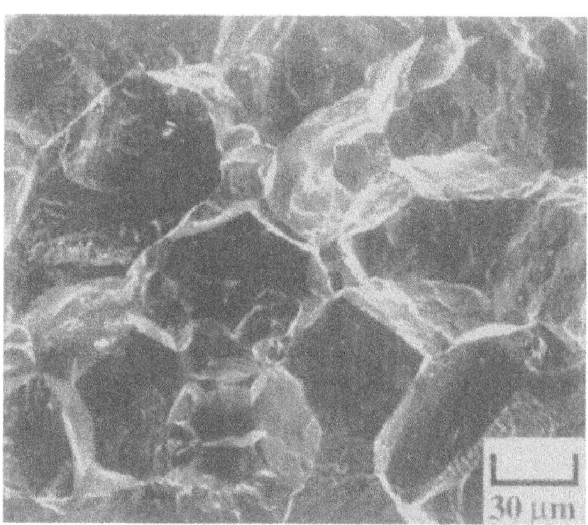

Figure 12.16 Room-temperature fracture modes in high-temperature intermetallics: (a) cleavage in Ti–48Al; (b) intergranular fracture in NiAl. (Courtesy of Dr. Padu Ramasundaram.)

However, since the grain boundary cohesion of some intermetallics (such as Ni_3Al and Fe_3Al) may be improved by alloying, it is possible to improve the room-temperature ductilities of such intermetallics by alloying. In the case of Ni_3Al, Liu et al. (1989) and George and Liu (1990) have shown that significant improvements in ductility can be achieved by alloying with boron, which is known to increase the grain boundary cohesion. Similarly, alloying with boron has been shown to improve the fracture resistance of Fe_3Al (Stoloff and Liu, 1994). Alloying with sufficient levels of boron also promotes a transition from intergranular fracture to ductile dimplied fracture.

Ductile dimpled fracture has also been reported to occur in partially ordered B2 niobium aluminide intermetallics such as Nb–15Al–40Ti (Ye et al., 1998, 1999). In such systems, the transitions from brittle to ductile fracture have been attributed to the effects of dislocation emission, which have been modeled using atomistic simulations (Farkas, 1997; Ye et al., 1999). Khantha et al. (1994) have also developed dislocation-based concepts that explain the ductile-to-brittle transition in a number of intermetallic systems.

The fracture modes in most intermetallic systems have been observed to exhibit strong temperature dependence. This is shown in Fig. 12.17 using images obtained from the fracture surfaces of Ti–48Al. In general, the transitions from brittle to ductile fracture modes have been attributed to the contributions from additional slip systems at elevated temperature (Soboyejo et al., 1992). However, thermally assisted grain boundary diffusion may result in intergranular fracture modes at intermediate temperatures (Fig. 12.17). A transition to a ductile transgranular fracture mode is also observed in TiAl-based alloys at even higher temperatures (between 815° and 982°C), as shown in Fig. 12.17.

Similar transitions in fracture modes have been observed in other intermetallic systems such as nickel and iron aluminides. However, the fracture mode transitions in these systems are also significantly influenced by environmental embrittlement processes (Stoloff and Liu, 1994).

12.6 FRACTURE OF CERAMICS

Ceramics are ionically and/or covalently bonded inorganic compounds. Due to their bonding and relatively large lattice spacings, slip is relatively difficult in cereamics. Hence, they tend to fail by brittle fracture modes (Lawn et al., 1993; Mecholsky and coworkers, 1989, 1991, 1994, 1997, 1999). The fracture surfaces of ceramics may be divided into four regions, as shown in Fig. 12.18. They include: the fracture source/initiation site; a smooth

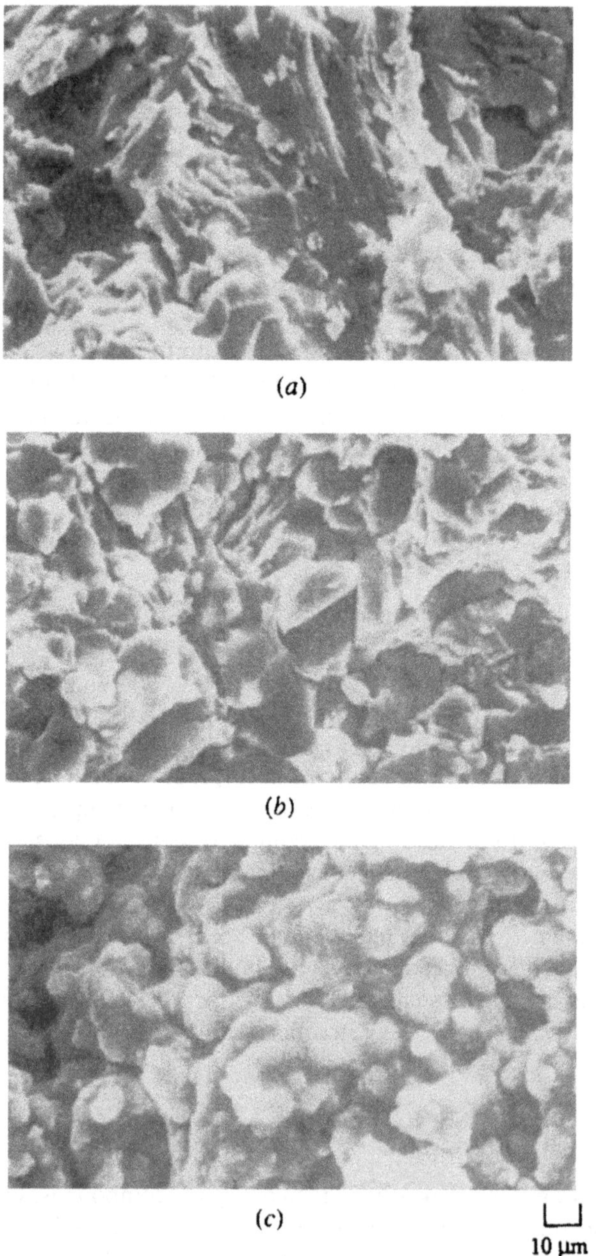

Figure 12.17 Effects of temperature on fracture modes in titanium aluminide alloys. (a) Cleavage and translamellar fracture at 25°C; (b) intergranular fracture at 700°C; and (c) transgranular fracture at 815°C.

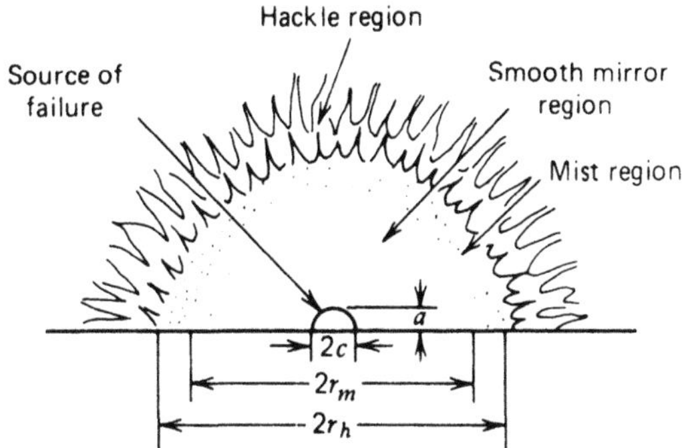

Figure 12.18 Schematic of four regions on the fracture surfaces of glassy ceramics. (From Mecholsky et al., 1978—reprinted with permission from Materials Research Society.)

mirror region with a highly reflective surface; a misty region that contains small radical ridges and microcrack distributions; and a hackle region that contains larger secondary cracks. The hackle region may also be bounded by branch cracks in some ceramic systems.

The dimensions of the mirror, mist, and hackle regions have been related to the fracture stress in work by Mecholsky et al. (1978). From their analysis of fracture modes in a wide range of ceramics, the following relationship has been proposed between the fracture stress, σ, and the radii of the mirror–mist, mist–hackle, and hackle–crack branching boundaries, $r_{\mathrm{m,h,cb}}$:

$$\zeta(r_{\mathrm{m,h,cb}})^{\frac{1}{2}} = M_{\mathrm{m,h,cb}} \tag{12.17}$$

where $M_{\mathrm{m,h,cb}}$ is a "mirror constant" corresponding to the mirror–mist, mist–hackle, and hackle–crack branching boundaries. Equation (12.17) has been shown to provide a good fit to experimental data obtained for a large number of ceramics. This is shown in Fig. 12.19 using data obtained for different ceramic materials by Mecholsky et al. (1976).

12.7 FRACTURE OF POLYMERS

A wide range of fracture modes is observed in polymers, depending on the underlying polymer structure and microstructure. Fracture in amorphous polymers tends to occur by craze formation due to the stretching of polymer microfibrils that give rise to gaps between polymer chains. These gaps are perceived as microcracks or crazes when viewed under a light microscope.

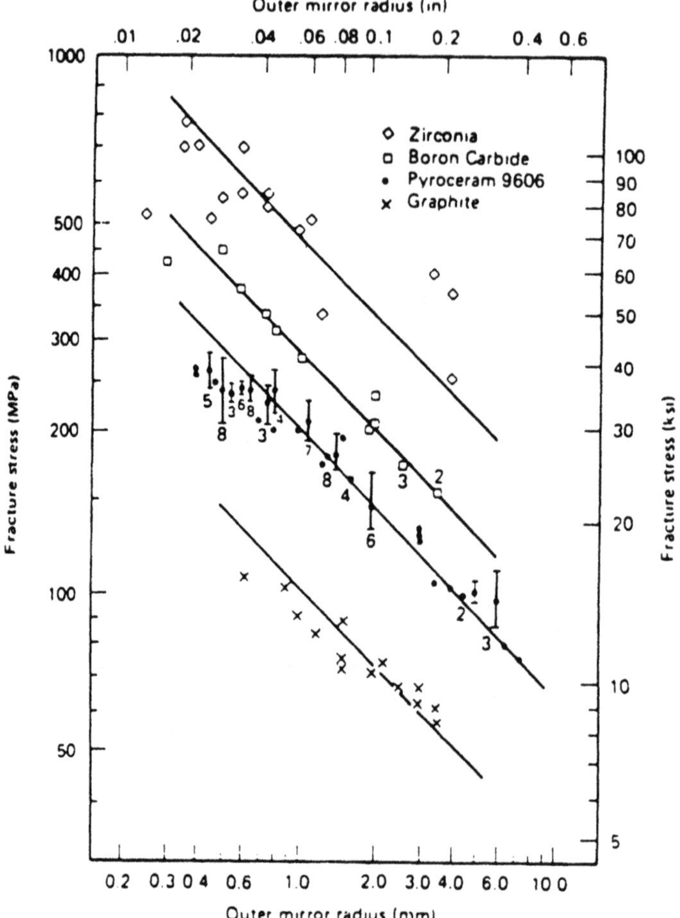

Figure 12.19 Dependence of fracture stress on mist–hackle radius. (From Mecholsky et al., 1976—reprinted with permission from Materials Research Society.)

Three stages of cracking are typically observed as a crack advances through an amorphous polymers. The first stage (Stage A in Fig. 12.20) involves crazing through the midplane. This results in the formation of a mirror area by the growth of voids along the craze. The second stage (Stage B in Fig. 12.20) involves crack growth between the craze/matrix interface. This results in so-called mackerel patterns. Finally, the third stage (Stage C in Fig. 12.20) involves cracking through craze bundles. This promotes the formation of hackle bands, as cracking occurs through bundles of crazes.

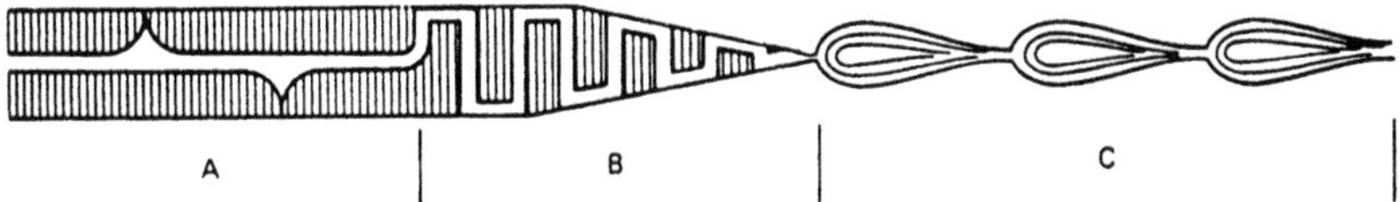

Figure 12.20 Schematic of stages of cracking in amorphous polymers. Region A: crack advance by void formation; region B: crack advance along alternate craze–matrix interfaces to form patch or mackerel patterns; region c: crack advance through craze bundle to form hackle bands. (Adapted from Hull, 1975—reprinted with permission from ASM.)

Hence, the fracture surfaces of amorphous polymers contain mirror, mist, and hackle regions (Fig. 12.21) that are somewhat analogous to those observed in ceramics. However, the underlying mechanisms associated with these different regions are very different in amorphous polymers. In any case, the mirror zones on the fracture surfaces of amorphous polymers often exhibit colorful patterns when viewed under a light microscope. A single color reflects the presence of a single craze of uniform thickness,

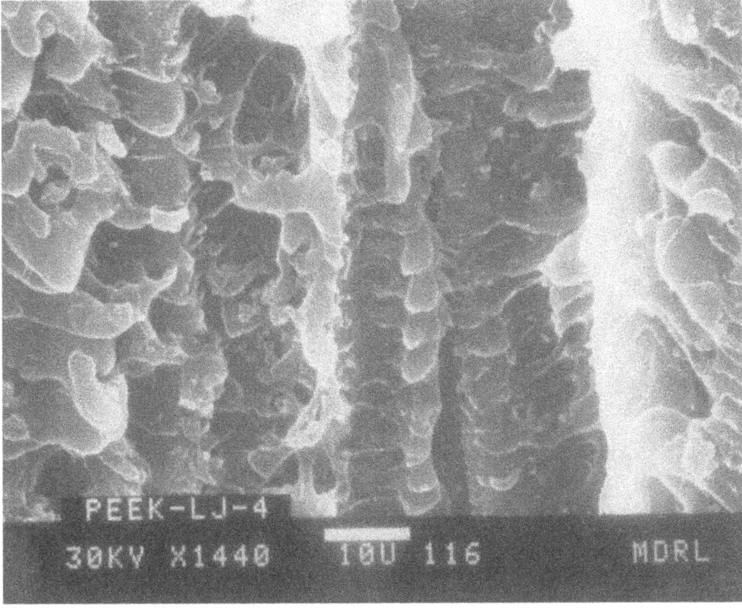

Figure 12.21 Band hackle markings in fast fracture region in poly(aryl ether ether ketone).

while multicolor fringes are formed by the reflection of light from multiple craze layers, or a single craze of variable thickness.

During the final stages of crack growth in amorphous polymers, crack advance outplaces the crack tip, and hackle bands form between the craze bundles. The resulting rough fracture surface has a misty appearance, and parabolic voids are observed on the fracture surfaces (Fig. 12.22). These voids are somewhat analogous to those observed on the fracture surfaces of ductile metals in the presence of shear.

In the case of semicrystalline polymers, the fracture modes depend very strongly on the interactions of the cracks with the underlying microstructure. The cracks may follow "interspherulitic" and trans-spherulitic" paths, as shown schematically in Fig. 12.23. The crack paths may also be significantly affected by crack velocity, with trans-spherulitic paths occurring at high velocity, and interspherulitic paths tending to occur at lower velocities. Relatively rough fracture modes are observed when extensive local plastic deformation precedes stable crack growth and catastrophic failure.

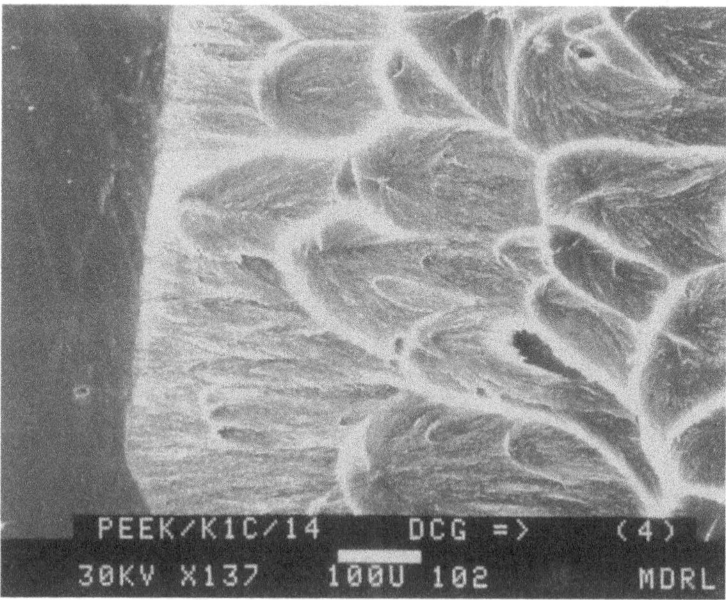

FIGURE 12.22 Tear dimples in poly(aryl ether ether ketone).

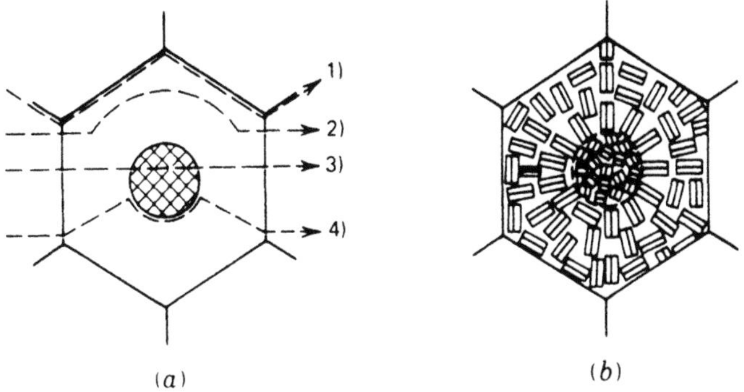

FIGURE 12.23 Schematics of (a) possible crack growth associated with spherulites in crystalline polymers and (b) orientations of crystal lamellae. (From Hertzberg, 1996—reprinted with permission from John Wiley.)

12.8 FRACTURE OF COMPOSITES

Since composite damage mechanisms may occur in the matrix, interface(s)/interphase(s), and reinforcement(s) in composite materials, a wide range of complex damage modes can occur in such materials under monotonic loading. Depending on the relative ductility of the matrix and reinforcement materials, and the interfacial strength levels, damage in composite materials may occur by:

1. Matrix or fiber cracking (single or multiple cracks).
2. Interfacial or interphase cracking or debonding.
3. Fiber pull-out or fiber cracking.
4. Delamination between piles.
5. Tunneling cracks between layers of different phases.

The complex sequence of damage phenomena is somewhat difficult to predict a priori. However, in the case of fiber-reinforced composites, the damage mechanisms that precede catastrophic failure have been studied by performing experiments in which composite damage mechanisms are observed in a microscopic at different stages of loading.

An example of the complex sequence of composite damage phenomena is presented in Fig. 12.24 for a titanium matrix composite reinforced with silicon carbide (SCS-6) fibers. The initial damage is concentrated in regions containing fibers that touch as a result of so-called "fiber swimming" processes during composite fabrication. Radial cracking and debonding is then observed in the predominantly TiC interphase between the

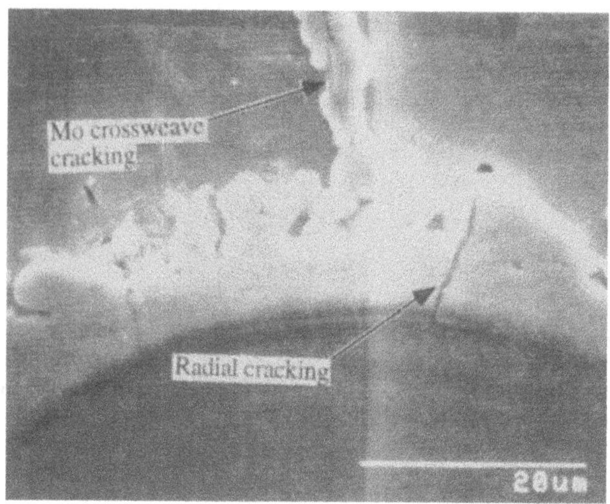

(a)

(b)

FIGURE 12.24 Damage mechanisms in $[\pm45°] - 2s$ and $[0/\pm45°]_{2s}$ Ti–15V–3Cr–3Al–3Sn composites reinforced with 35 vol % SiC (SCS-6) fibers: (a) debonding, microvoid nucleation, and radial cracking in $[\pm45°]_{2s}$ composite after loading to 0.1 of the ultimate tensile strength; (b) debonding and radical cracking in reaction layer in the $[0/\pm45°]_{2s}$ composite after loading to 0.1 of the ultimate tensile strength. (From Jin and Soboyejo, 1998.)

carbon coating and the titanium amtrix (Fig. 12.24). A summary of the observed damage modes is presented in Fig. 12.25.

At higher stresses, stress-induced precipitation occurs within the matrix before subgrain formation and circumferential fiber cracking. The matrix microvoids then coalesce, as radial cracking continues along the interphase, Fig. 12.23(e). Slip steps are then observed in the matrix along with complete debonding between the interphase and matrix. Final fracture occurs by ductile dimpled fracture in the matrix, and cleavage fracture of the silicon carbide fibers. There is also clear evidence of fiber pull-out prior to catastrophic failure.

As the reader can probably imagine, a different sequence of events would be expected from brittle polymer or ceramic matrix composite reinforced with carbon or glass fibers. Such composites tend to exihbit multiple matrix cracks with relatively uniform vertical spacings. These multiple matrix cracks evolve as a result of the stress states within the composites. There have been numerous studies of damage mechanisms in composite materials. It is, therefore, not possible to provide an adequate overview of composite damage mechanisms in this section. the interested reader is referred to texts on composites by Chawla (1987), Taya and Arsenault

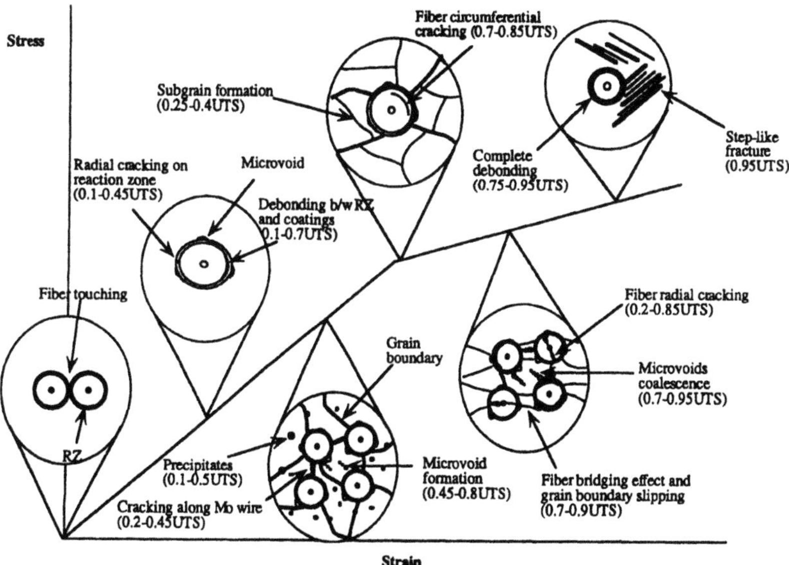

FIGURE 12.25 Summary of tensile damage mechanisms in $[\pm 45/-45]_{2s}$ Ti–15V–3Cr–3Al–3Sn reinforced with SiC (SCS-6) fibers. (From Jin and Soboyejo, 1998.)

(1989), Clyne and Withers (1993), and Matthews and Rawlings (1994) Hull and Clyne (1996).

12.9 QUANTITATIVE FRACTOGRAPHY

So far, our discussion of fracture modes has provided only qualitative descriptions of the failure mechanisms in the different types of materials. However, it is sometimes useful to obtain quantitative estimates of the features such as asperities on the fracture surfaces. Quantitative measurements of such features may be obtained by the use of stereo microscopes, profilometers (Talysurf measurement devices), and atomic force microscopes. Software packages have been developed to analyze the crack asperity profiles using roughness parameters that are described in this section. The fracture characteristics may also be described by fractal numbers that are presented at the end of this section.

The two types of roughness parameters that are used typically in stereology are the linear roughness parameter, R_L, and the surface roughness parameter, R_S. The linear roughness parameter is defined as the ratio of the true crack profile length, L_t, to the projected crack profile length, L_p:

$$R_L = \frac{L_t}{K_p} \tag{12.18}$$

Similarly, the surface roughness parameter may also be defined as the ratio of the true surface area, A_s, to the projected surface area, A_p:

$$R_s = \frac{H_t}{A_p} = \frac{4}{\pi}(R_L - 1) + 1 \tag{12.19}$$

The surface roughness parameter is generally difficult to measure directly. It is, therefore, useful to infer it from the linear roughness parameter, R_L, and the expression on the right-hand side of Eq. (12.19). In many cases, R_s is often greater than 2. A great deal of caution is, therefore, needed in the measurement of linear profiles/area parameters on fracture surfaces.

Finally, in this section, it is of interest to define a fractal framework that can be used to analyze fracture surface features. Fractals recognize that the length of an irregular profile depends on the size scale of the measurement. Also, the size of the measurement increases as the size of the measurement unit decreases. If the apparent length of the measurement unit is $L(\eta)$ and the measurement is η, then the size of the measurement is defined uniquely by

$$L(\eta) = L_0 \eta_0^{-(D-1)} \tag{12.20}$$

where L_0 is a constant with dimensions of length, and D is the fractal dimension. It is important to note that D is not a material constant since it may vary with microstructure and the measurement unit. However, the fractal parameter, D, may be related to fracture phenomena in different materials (Mecholsky and coworkers, 1976–1991, West et al., 1994; Hertzberg, 1996; Chen et al., 1997).

12.10 THERMAL SHOCK RESPONSE

Our discussion so far has focused on the growth of cracks due to monotonic (load increasing with time) mechanical loading. However, crack growth may also occur under "thermal" loading. When materials are subjected to rapid temperature changes that give rise to sudden changes in local stress/strain stages, crack nucleation and growth can occur due to cracking under down-quench (cold shock) or up-quench (hot shock) conditions. Such failure is often characterized as the thermal shock response of the material. It can occur in metals, polymers, or ceramics. However, it is most prevalent in ceramics due to their inherent brittleness. The phenomenon of thermal shock was first studied seriously by Kingery (1955) and Hasselman (1963).

12.10.1 Review of Thermal Shock

Kingery (1955) noted that under conditions where h, the heat transfer coefficient, is so large that the surface of the material is immediately changed to the surrounding temperature, the temperature difference to just initiate fracture is given by

$$\Delta T_c = S\frac{\sigma^*(1-\nu)}{E\alpha} = SR \tag{12.21}$$

and

$$R = \frac{\sigma^*(1-\nu)}{E\alpha} \tag{12.22}$$

where ΔT_c is the temperature difference to just initiate fracture, S is a shape factor, which is dependent on the geometry of the sample, σ^* is the fracture stress (shear or tension), ν is Poisson's ratio, E is the elastic modulus, and α is the coefficient of thermal expansion; R is known as the fracture resistance parameter, and increasing R will increase the resistance to fracture initiation due to thermal shock. For smaller, constant values of h, ΔT_c is given as:

$$\Delta T_c = S\frac{k\sigma^*(1-\nu)}{E\alpha}\frac{1}{h} = \frac{S}{hR'} \tag{12.23a}$$

and

$$R = \frac{k\sigma^*(1-\nu)}{E\alpha} \quad (12.23b)$$

where k is the thermal conductivity, h is the heat transfer coefficient, and R' is a second fracture resistance parameter. The above equations were developed for a homogeneous, isotropic body with physical properties that are independent of temperature. From this work, high values of thermal conductivity and strength, along with low values of the elastic modulus, Poisson's ratio, and thermal expansion, will lead to the best thermal shock resistance. Kingery (1955) also demonstrated that placing compressive surface stresses in spherical samples of zirconia leads to improved thermal stress reistance.

Hasselman (1963a) proposed that the temperature difference required to cause the fracture of a body of low initial temperature subjected to radiation heating is given by

$$\Delta T_c = \left[\frac{A}{\rho b}\right]^{\frac{1}{4}} \cdot \left[\frac{k\sigma^*(1-\nu)}{E\alpha}\right]^{\frac{1}{4}} \quad (12.24)$$

where A is a geometry constant, b is a size constant, ρ is the density, and k is the emissivity. Therefore, when a major portion of the heat transfer takes place by radiation, the emissivity becomes an important part of the thermal shock parameter. In addition to the parameters determined by Kingery (1955), a low value of emissivity will improve the resistance to fracture initiation due to thermal shock. Equation (12.24) does not apply to materials of high strength, or low values of α, k, or E.

Hasselman (1963b) later considered the factors that affect the propagation of cracks in systems subjected to thermal shock. It was originally proposed by Griffith that a crack will nucleate or propagate as long as the elastic energy released is equal to or greater than the effective surface energy. Hasselman noted that, for brittle, polycrystalline refractory materials, the effective surface energy is approximately equal to the thermodynamic surface energy. He determined the following two damage resistance parameters, R''' and r'''':

$$R''' = \frac{E}{(\sigma^*)^2(1-\nu)} \quad (12.25)$$

$$R'''' = \frac{E\gamma_{\text{eff}}}{(\sigma^*)^2(1-\nu)} \quad (12.26)$$

where γ_{eff} is the effective surface energy. Therefore, materials designed to have a high fracture resistance would have low damage resistance.

Hasselman conjectured that these parameters were only valid for the first thermal shock cycle.

Increasing the porosity has a negative effect on the fracture resistance parameter R', because increasing the porosity decreases the thermal conductivity. However, it increases the damage resistance because it decreases the elastic energy stored in the material. Increasing the surface energy increases R'''', and can be accomplished by including a second phase. This was demonstrated in some early work by Tinklepaugh (1960), who was one of the first to explore the use of ductile phase reinforcement in the toughening of cermets.

Nakayama and Ishizuka (1969) later tested five brands of commercial refractory firebricks. They were identically thermally shocked until a weight loss of 5% was reached. They concluded that the R'''' term correctly predicted the relative thermal shock resistance of the five types of bricks.

Hasselman (1969) presented a theory that unified the fracture initiation and crack propagation approaches. The conditions of this theory are as follows: the material is entirely brittle; it contains flaws in the form of circular, uniformly distributed Griffith microcracks; stress relaxation is absent; crack propagation occurs by simultaneous propagation of N cracks per unit volume; all cracks propagate radially; and neighboring cracks stress fields do not interact. The total energy (W_t) of the system per unit volume is the sum of the elastic energy plus the fracture energy of the cracks:

$$W_t = \frac{3(\alpha \cdot \Delta T)^2 E_0}{2(1-2\nu)}\left[1+\frac{16(1-\nu^2)Nl^3}{9(1-2\nu)}\right]^{-1} + 2\pi N l^2 G \tag{12.27}$$

where G is the surface fracture energy, E_0 is the elastic modulus of the uncracked matrix, and l is the radius of the cracks. Combining Eq. (12.27) with Griffith's approach (cracks are unstable between the limits for which):

$$\frac{dW_t}{dl} = 0 \tag{12.28}$$

leads to the critical temperature difference required for crack instability:

$$\Delta T_c = \left[\frac{\pi G(1-\nu^2)}{2E_0\alpha^2(1-\nu)}\right]^{1/2} \cdot \left[1+\frac{16(1-\nu^2)Nl^3}{9(1-2\nu)}\right]^{-1} \cdot [l]^{-1/2} \tag{12.29}$$

For initially short cracks, the term $[16(1-\nu^2)Nl^3/9(1-2\nu)]$ becomes small compared to unity. Thus, the ΔT_c for short cracks becomes:

$$\Delta T_c = \left[\frac{\pi G(1-2\nu)^2}{2E_0\alpha^2(1-\nu^2)}\right]^{1/2} \tag{12.30}$$

For long cracks, the term $[16(1-\nu^2)Nl^3/9(1-2\nu)]$ becomes negligible compared to unity. The ΔT_c for long cracks therefore becomes:

$$\Delta T_c = \left[\frac{128\pi G(1-\nu^2)N^2 l^5}{81E_0\alpha^2}\right]^{1/2} \tag{12.31}$$

Hasselman noted that Eq. 12.31 might not be valid because it ignores the interactions between the stress fields of neighboring cracks. Also, for initially short cracks, the rate of elastic energy released after initiation of fracture exceeds the surface fracture energy. The excess energy is converted into kinetic energy of the propagating crack. The crack still possesses kinetic energy when it reaches the lengths corresponding to the critical temperature difference in Eqs (12.30) and (12.31). Therefore, the crack continues to propagate until the potential energy released equals the total surface fracture energy. This process leads to the sudden drop in strength seen in many ceramics at ΔT_c, instead of a gradual decrease in strength, as would be expected, if the cracks propagated quasistatically. The final crack length can be expressed as

$$\frac{3(\alpha\cdot\Delta T_c)^2E_0}{2(1-2\nu)}\left\{\left[1+\frac{16(1-\nu^2)Nl_0^3}{9(1-2\nu)}\right]^{-1}-\left[1+\frac{16(1-\nu^2)Nl_0^3}{9(1-2\nu)}\right]^{-1}\right\} = 2\pi NG\left(l_f^2-l_0^2\right) \tag{12.32}$$

where l_0 and l_f are the initial and final crack lengths, respectively. When $l_f \gg l_0$ (initially short cracks), the expression for the final crack length simplifies to

$$l_f = \left[\frac{3(1-2\nu)}{8(1-\nu^2)l_0N}\right] \tag{12.33}$$

Equation (12.33) indicates that the final crack length of an initially short crack, except for Poisson's ratio, is independent of material properties. Longer cracks will not attain kinetic energy, and propagate in a quasistatic manner. This leads to a gradual decrease in strength after thermal shock, instead of the instantaneous decrease often seen in materials with small initial flaw sizes. Hasselman suggested crack propagation could be minimized by increasing the size of the Griffith flaw. This can be accomplished by increasing the grain size, or by introducing cracks large and dense enough that crack propagation does not occur kinetically.

Gupta (1972) tested the effect of grain size on the strength degradation of thermally shocked alumina. He used bars (0.3175 mm × 0.3175 mm ×2.8575 mm) of alumina with grain sizes of 10, 34, 40, and 85 μm. The bars were thermally shocked to varying degrees by heating the samples in a furnace and then dropping them into a bucket of 25°C water. The change in strength after thermal shock was measured by a four-point bend test. Figure 12.26 shows that the magnitude of the discontinuous decrease in strength at ΔT_c decreased as grain size increased. The 10 μm grain size samples exhibited at 75% drop in strength, while the strengths of the 34 and 40 μm samples dropped by 50 and 42.5%, respectively. The 85 μm grain size alumina did not exhibit a discontinuous drop in strength. This supports Hasselman's suggestion that increasing grain size cold minimize crack propagation by causing cracks to propagate in a quasistatic manner.

Evans and Charles (1977) studied conditions for the prevention of crack propagation in thermally shocked Al_2O_3 and ZrO_2 cylinders by using large precracks. Their criterion for the prevention of crack propagation is

$$\kappa < \kappa_C \equiv \frac{K_c(1-\nu)}{E\alpha\Delta T(r_0)^{1/2}} \tag{12.34}$$

where κ is the normalized crack tip stress intensity factor, ν is Poisson's ratio, E is the elastic modulus, α is the thermal expansion coefficient, ΔT is the temperature change of the quench, r_0 is the critical dimensional measurement, and K_c is the critical stress intensity factor. Crack propagation was prevented by using cracks of length 60 to 80% of the radius of the cylinder. Larger radii required larger relative crack lengths.

Faber et al. (1981) performed a different type of thermal shock test. Using disks that were 5 cm in diameter and 0.25 cm thick, they studied the thermal shock behavior of Al_2O_3. One surface was ground and polished, and precracked by a Knoop indenter with loads from 15 to 33 N. Fine grinding eliminated the residual stresses from precracking. The samples were individually tested by heating in a $MoSi_2$ furnace on a bed of fibrous insulation. The samples were thermally shocked by cooling with high-velocity air (100 m/s). If the crack did not extend, the experiment was repeated using a higher furnace temperature. The temperature difference required to extend the crack, ΔT_c, was recorded. Higher indentation loads led to a lower ΔT_c required for crack extension.

Bannister and Swain (1990) noted that the ratio of the work of fracture, γ_{WOF}, to the notch beam test work of fracture, γ_{NBT}, describes the

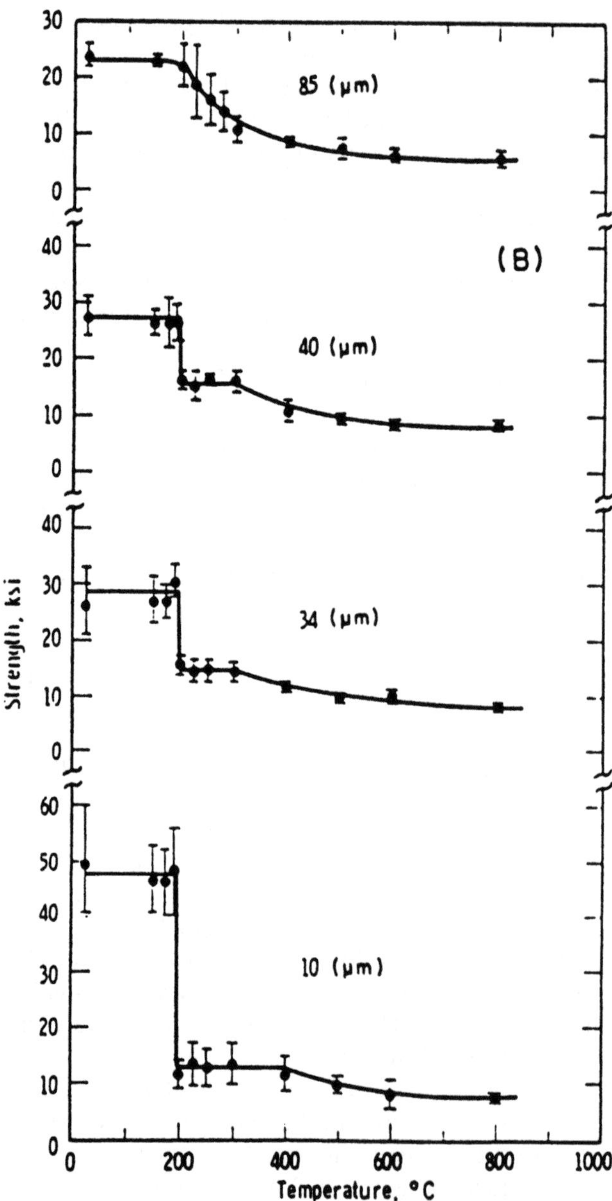

Figure 12.26 Thermal shock resistance of polycrystalline alumina as functions of quenching temperature and grain size. (From Gupta, 1972—reprinted with permission from American Ceramics Society.)

increasing crack resistance during crack extension, or the K^R curve. The K^R curve is not a material property, since it depends on the specimen dimensions, testing conditions, initial flaw, and evaluation method. The increasing resistance to crack propagation is due to crack deflection and/or crack branching, contact shielding processes, and/or stress-induced zone-shielding processes that will be described in the next chapter. Two required criteria for unstable crack growth are related to the K^R curve:

$$K_A \geq K_R(a) \tag{12.34a}$$

and

$$\frac{dK_A}{da} > \frac{dK_R(a)}{da} \tag{12.34b}$$

where K_A is the applied stress intensity factor due to thermal stresses, and $K_R(a)$ is the value of the K^R curve at the crack length, a. Crack arrest occurs when Eq. (12.34a) is not satisfied; K_A can be calculated from the through-thickness single-edged notched test method, which uses the equation:

$$K_A = Y\sigma_f\sqrt{a} \tag{12.35}$$

where Y is usually a polynomial of (a/W), W is the specimen width, and σ_f is the flexure stress.

The K^R curve (Bannister and Swain, 1990) and thermal shock behavior (Swain et al., 1991a,b) were determined for duplex ceramics containing spherical zones (second phase) homogeneously dispersed in a ceramic matrix of composition 3 vol% yttria-stabilized tetragonal zirconia polycrystals and 20 wt% Al_2O_3. The zones contained various fractions of monoclinic ZrO_2 particles, and expanded on cooling because of the ZrO_2 transformation. This produced radial compressive and tangential tensile hoop stresses arond the zones, and hydrostatic pressure within the zones. This microstructure retained the strength of the matrix, while increasing the crack growth resistance by encouraging crack deflection and branching.

Mignard et al. (1996) have tried to relate the flexural strength, toughness, and resistance to crack propagation from room temperature to 1000°C. The toughness and resistance to crack propagation were measured by single-edge notched beam techniques. The critical temperature difference, ΔT_c, was 682°C (from a 700°C furnace temperature to 18°C air temperature), as determined by a drop in retained strength, a first acoustic peak (measured in situ using acoustic emission signals), and the beginning of the decrease in elastic modulus. The first acoustic peak occurred 700 ms after cooling began. The temperature distribution with the sample was calculated for a temperature of 700°C after 700 ms of cooling. The largest surface stress was 115 MPa, and was located at the center of the largest face, which was at a temperature of 600°C. This corresponds to the bending strength measured at 600°C, which was 114 ± 16 MPa. The stress intensity factor at 700°C also reaches the K^R curve (at the initial flaw size) for the material at 600°C.

12.10.2 Materials Selection for Thermal Shock Resistance

Following the example of Ashby (1999), we may simplify the analysis of thermal shock phenomena by considering the simplest case in which sudden constrained thermal expansion leads to thermal stresses. Such stresses can give rise to relatively high stresses when they exceed the yield or fracture stress, σ_f. The condition at which this occurs is given by

$$\sigma_f \sim E\alpha\Delta T_c \quad \text{(for constrained expansion)} \tag{12.36a}$$

or

$$\sigma_f = \frac{E\alpha\Delta T_c}{C} \quad \text{(for unconstrained expansion)} \tag{12.36b}$$

where $C = 1$ for axial constraint, $(1 - \nu)$ for biaxial constraint or normal quenching, and $(1 - 2\nu)$ for triaxial constraint; ν is Poisson's ratio. Hence, the thermal shock resistance may be expressed as the critical temperature range that is obtained by rearranging Eqs (12.36a) and (12.36b).

However, this is only part of the story. This is because instant cooling requires an infinite heat transfer coefficient, h at the surface. Since this is never true in reality, the values of ΔT obtained at the surface can only be used to obtain approximate estimates of ΔT for the ranking of the thermal shock resistance of materials. Hence, in such cases, the thermal shock resistance is given by

$$B\Delta T = \sigma_f/\alpha E \tag{12.37}$$

where $B = C/A$ and typical values of A are given in table 12.1 for a 10 mm thick section. Also, an approximate value of A may be obtained from the following expression:

$$A = \frac{Hh/\lambda}{1 + Hh/\lambda} = \frac{\text{Bi}}{1 + \text{Bi}} \tag{12.38}$$

where Hh/λ is the so-called Biot number (Bi), H is a typical dimension, and λ is the thermal conductivity. Contours of $B\Delta T$, therefore, provide a measure of thermal shock resistance, as shown in Fig. 12.27. The materials with the highest thermal shock resistance have the highest values of $B\Delta T$. Since $B = C/A$ and A equals 1 for most materials, except high-conductivity metals (Table 12.1), the thermal shock resistance may be determined directly from the contours for most materials, with the appropriate correction for the

TABLE 12.1 Thermal Shock Parameters for 10 mm Thick Sections

Conditions	Foams	Polymers	Ceramics	Metals
Slow air flow ($h = 10$ W/m^{2}K)	0.75	0.5	3×10^{-2}	3×10^{-3}
Black body radiation (500°–0°C) ($h = 40$ W/m^{2}K)	0.93	0.6	0.12	1.3×10^{-2}
Fast air flow ($h = 10^2$ W/m^{2}K)	1	0.75	0.25	3×10^{-2}
Slow water quench ($h = 10^3$ W/m^{2}K)	1	1	0.75	0.23
Fast water quench ($h = 10^4$ W/m^{2}K)	1	1	1	0.1–0.9

Source: Ashby (1999)—reprinted with permission from Butterworth-Heinemann.

constraint factor, C. Also, for materials with $A \neq 1$, ΔT is larger by a factor of C/A.

If we now examine the possible differences between hot and cold shock conditions, we may obtain a range of expressions for different levels of heat transfer. For perfect heat transfer (Bi $= \infty$), the maximum sustainable heat transfer is given by

$$\Delta T = A_1 \frac{\sigma_f}{E\alpha} \tag{12.39}$$

where $A_1 \sim 1$ for cold shock and $A_1 \sim 3.2$ for hot shock. However, for poor heat transfer, the maximum sustainable temperature range is greater. In the limit, for small Biot number (Bi < 1), ΔT is given by

$$\Delta T = A_2 \frac{\sigma_f}{E\alpha} \frac{1}{\mathrm{Bi}} = A_2 \frac{\sigma_f}{E\alpha} \frac{k}{hH} \tag{12.40}$$

where k is the thermal conductivity, $A_2 \sim 3.2$ for cold shock, and $A_2 \sim 6.5$ for hot shock. In any case, for both cold and hot shock, the thermal shock resistance, ΔT, is maximized by selecting materials with high $\sigma_f/E\alpha$ for materials with high heat transfer (Bi $\to \infty$). However, for poor surface heat transfer (Bi < 1), the best materials have high $k\sigma_f/E\alpha$. The resulting materials selection map is shown in Fig. 12.28, which is taken from a paper by Lu and Fleck (1998). Materials with high thermal shock resistance lie to the top right of the diagram. These include glass ceramics and graphites. However, it should also be noted that the relative thermal shock resistance of the different materials depends significantly on the Biot number, Bi. Furthermore, when the Biot number if high, then the parameter of $\sigma_f/E\alpha$ becomes the relevant material selection index.

The above framework provides a strength-based approach to the selection of materials that are resistant to thermal shock. The resulting materials selection chart (Fig. 12.27) also shows that engineering ceramics and invar

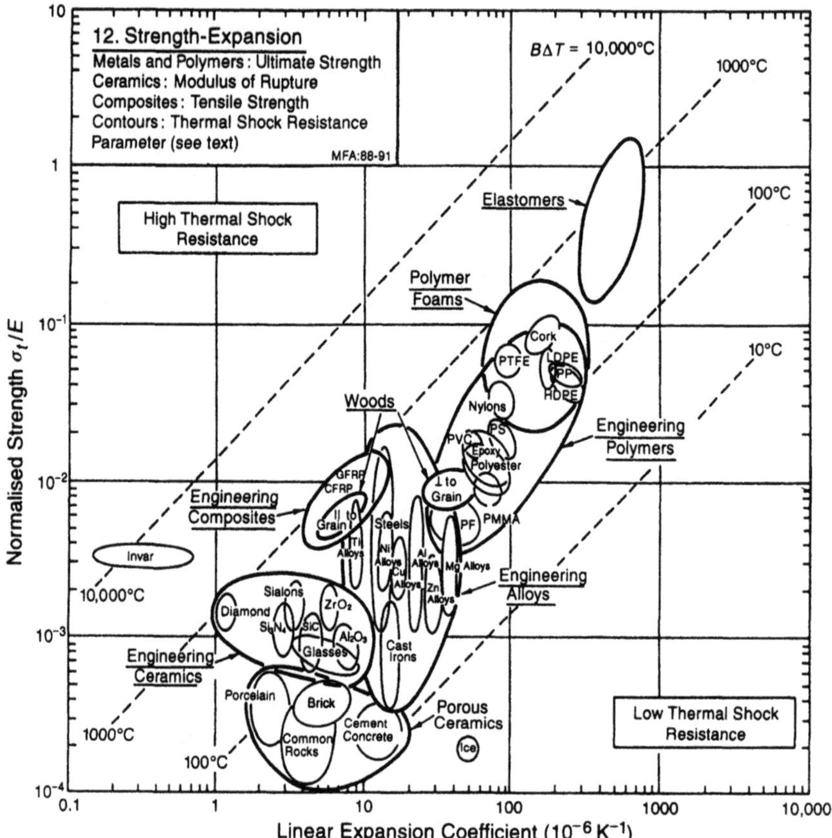

FIGURE 12.27 Materials selection chart for thermal shock resistance. Correction must be applied for constraint and to allow for the effect of thermal conductivity. (From Ashby, 1999—reprinted with permission from Butterworth-Heinemann.)

are some of the most thermal shock-resistant materials. However, engineering ceramics also have relatively low fracture toughness that can lead to crack nucleation and growth in systems containing pre-existing flows. There is, therefore, a need to use fracture mechanics approaches in the selection of materials that are resistant to thermal shock.

A fracture mechanics framework for the selection of thermal shock-resistant materials has been proposed by Lu and Fleck (1998). For cracking in an infinite plate subjected to cold shock, they consider the worst case in which a Mode I crack is induced from the edge (Fig. 12.29) where the transient stresses are greatest. In contrast, for hot shock conditions, crack-

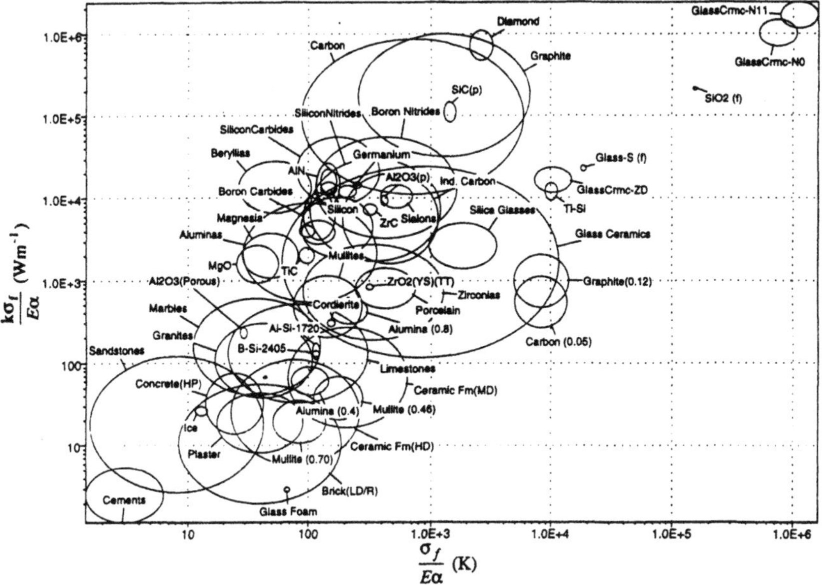

FIGURE 12.28 Materials selection chart for thermal shock resistance. (From Lu and Fleck, 1998—reprinted with permission from Elsevier.)

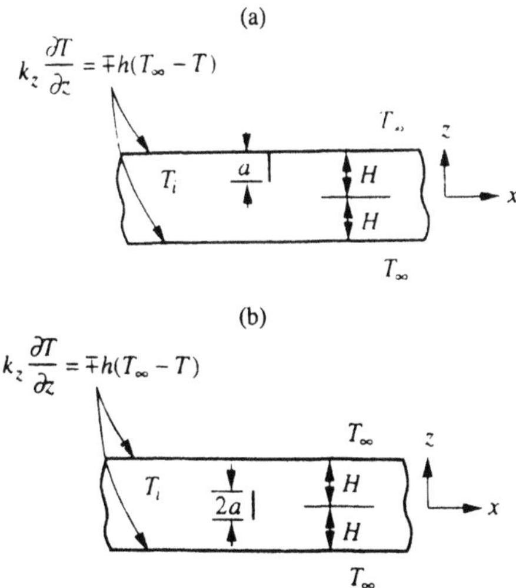

FIGURE 12.29 Cracks induced in a finite thickness plate exposed to convective medium of different temperatures: (a) cold shock; (b) hot shock. (From Lu and Fleck, 1998—reprinted with permission from Elsevier.)

ing is assumed to be induced at the center of the plate, Fig. 12.29(b). In any case, for fracture toughness-controlled cracking due to thermal shock, the maximum sustainable temperature range for perfect heat transfer (Bi $= \infty$) is given by Lu and Fleck (1998) to be

$$\Delta T = A_3 \frac{K_{Ic}}{E\alpha\sqrt{\pi H}} \tag{12.41}$$

where $A_3 \sim 4.5$ for cold shock, and ~ 5.6 for hot shock. Furthermore, since the sustainable temperature jump increases with decreasing Biot number (heat transfer), the limiting condition for small Biot number (Bi < 1) is of importance. This is given by

$$\Delta T = A_4 \frac{K_{Ic}}{E\alpha\sqrt{\pi H}} \frac{k}{hH} \tag{12.42}$$

where $A_4 \sim 9.5$ for cold shock, and ~ 12 for hot shock. Hence, from the above expressions, it is clear that the selection of thermal shock-resistant materials can be achieved by amximizing $K_{Ic}/E\alpha$ for perfect heat transfer (Bi $= \infty$), or by selecting high $kK_{Ic}/E\alpha$ for poor heat transfer (Bi < 1).

Materials selection charts for fracture mechanics-based selection are presented in Fig. 12.30. The relative positions of the different materials are similar to those presented in Fig. 12.28. However, glass ceramics and graphite have the best thermal shock resistance among ceramics. Also, for fracture toughness-controlled fracture, thermal shock resistance decreases with increasing plate thickness, $2H$, at all Biot numbers under hot shock and cold shock conditions. Hence, the actual material selection process requires a careful consideration of the section thickness, H, and strength or fracture toughness-based materials selection parameters.

Finally, it is important to note here that the most significant stresses are induced during the initial transient period associated with the thermal shock response of a material to hot shock (heating) or cold shock (cooling). This transient period, t^*, is controlled primarily by the Biot Number, Bi, which is a nondimensional heat transfer coefficient. For cold shock, the transient time, t^*, is given by Lu and Fleck (1998) to be

Figure 12.30 (a) Merit indices for strength-controlled failure $k\sigma_f/E\alpha$ at low Bi values vs. $\sigma_f/E\alpha$ at high Bi values; (b) merit indices for toughness-controlled failure $k_{IC}E\alpha$ vs. $K_{IC}/E\alpha$; (c) $K_{IC}/E\alpha$ vs. $\sigma_f/E\alpha$, with the guidelines H_f added to help in selecting materials according to both strength- and toughness-based fracture criteria. (From Lu and Fleck, 1998—reprinted with permission of Elsevier.)

(a)

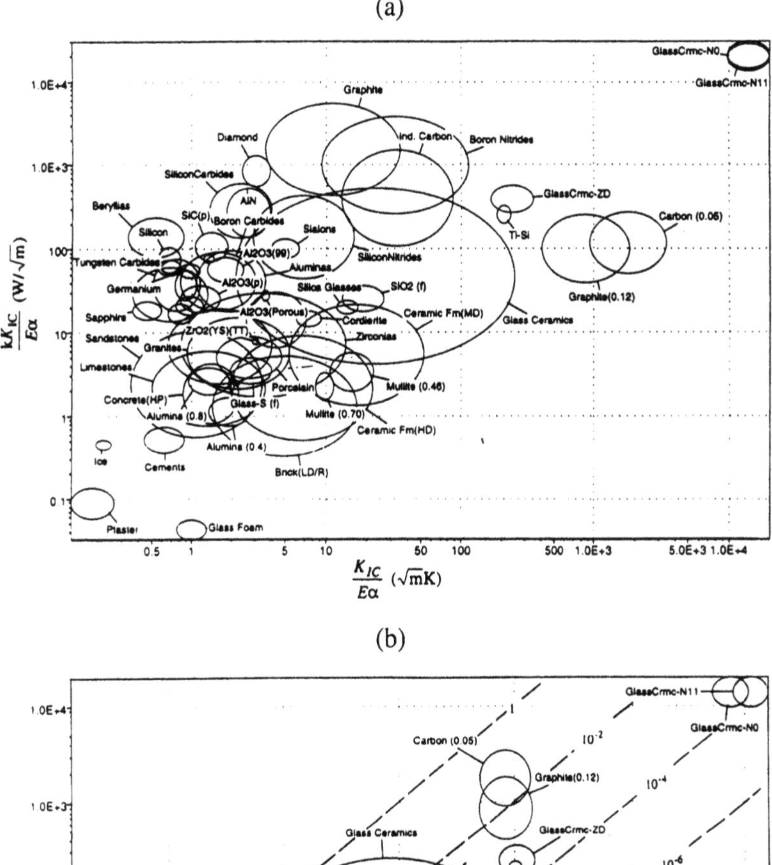
GlassCrmc-N0
GlassCrmc-N11
Graphite
Diamond
Ind. Carbon
Boron Nitrides
SiliconCarbides
AlN
Beryllias
SiC(p)
Boron Carbides
Sialons
Silicon
Al2O3(99)
SiliconNitrides
Tungsten Carbides
Aluminas
Germanium
Al2O3(p)
Silica Glasses
SiO2 (f)
Sapphire
Al2O3(Porous)
Cordierite
Ceramic Fm(MD)
Glass Ceramics
Sandstones
ZrO2(YS)(TT)
Zirconias
Granites
Limestones
Porcelain
Mullite (0.46)
Concrete(HP)
Glass-S (f)
Alumina (0.8)
Mullite (0.70)
Ceramic Fm(HD)
Alumina (0.4)
Ice
Cements
Brick(LD/R)
Plaster
Glass Foam
GlassCrmc-ZD
Ti-Si
Carbon (0.05)
Graphite(0.12)
$\frac{kK_{IC}}{E\alpha}$ (W/√m)
$\frac{K_{IC}}{E\alpha}$ (√mK)
(b)
GlassCrmc-N11
GlassCrmc-N0
Carbon (0.05)
Graphite(0.12)
GlassCrmc-ZD
Glass Ceramics
Boron Nitrides
Ti-Si
Ceramic Fm(MD)
Ind. Carbon
Graphite
Mullite (0.46)
Carbon
Silica
Ceramic Fm(HD)
Mullite (0.70)
Al2O3(Porous)
Mullites
Silica Glasses
Brick(LD/R)
Alumina (0.4)
Sialons
Aluminas
Cordierite
Porcelain
Diamond
Glass-S (f)
Sandstones
AlN
Marbles
Al-Si-1720
Ge: 100%
SiC(p)
Concrete(HP)
MgO
Al2O3(p)
Beryllias
Sapphire
ZrC
TiC
Limestones
TungstenCarbide
Glass Foam
Cements
Ice
Plaster
Magnesia
$\left(\frac{K_{IC}}{\sigma_f}\right)^2 = 10^{-8}$ m
$\frac{K_{IC}}{E\alpha}$ (√mK)
$\frac{\sigma_f}{E\alpha}$ (K)

$$t^* = \frac{0.48}{1 + 1.8\text{Bi}} \tag{12.43}$$

Similarly, for hot shock conditions, the transient period, t^*, is given by Lu and Fleck (1998) to be

$$t^* = \frac{0.115 + 0.45}{1 + 2.25\text{Bi}} \tag{12.44}$$

In any case, it should be clear that failure due to thermal shock can be designed by the careful selection of materials that are resistant to cracking under either hot shock or cold shock conditions. One way in which this can be achieved is by the use of reinforcements that reduce the overall crack driving force by bridging the crack faces. For example, this may be achieved by composite reinforcement with brittle or ductile phases. Alternatively, viscous bridging may be used to shield the crack tips from the applied thermal stress, as shown in Fig. 12.31 for a refractory ceramic. The concept of crack-tip shielding will be discussed in greater detail in Chap. 13.

12.11 SUMMARY

This chapter presents an overview of the micromechanisms of fracture in the different classes of materials. The chapter begins with a review of brittle and

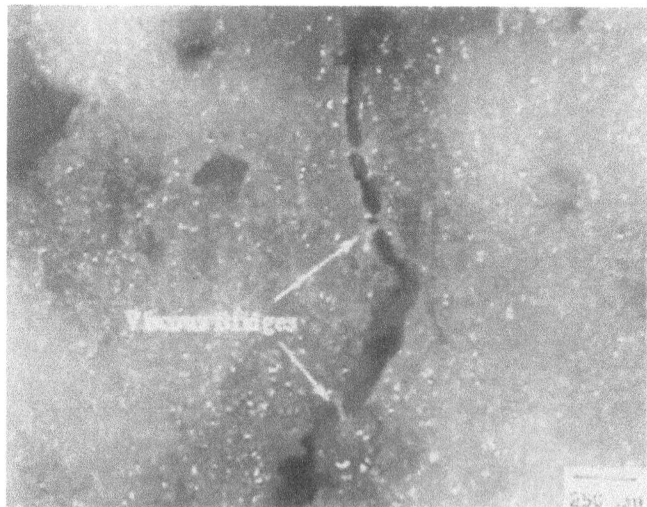

FIGURE 12.31 Crack bridging by viscous glassy phase in a refractory ceramic. (From Soboyejo et al., 2001.)

ductile fracture mechanisms in metals and their alloys. These include transgranular cleavage, intergranular fracture, and ductile dimpled fracture modes. The possible modes of failure are then elucidated for metallic, intermetallic, ceramic, and polymeric materials before exploring a selected range of possible failure modes in composite materials. The mechanisms of thermal shock are discussed after introducing simple concepts in quantitative stereology along with the notion of fractals.

BIBLIOGRAPHY

Anderson, T.L. (1994) Fracture Mechanics. Fundamentals and Applications. 2nd ed. CRC Press, Boca Raton, FL.

Ashby, M.F. (1999) Materials Selection in Mechanical Design. Butterworth-Heinemann, New York.

ASTME E-8 Code. (2001) American Society for Testing and Materials. vol. 3.01. West Conshohocken, PA.

Bannister, M.K. and Swain, M.V. (1990) Ceam Int. vol. 16, pp 77–83.

Batisse, R., Bethmont, M., Devesa, G., and Rousselier, G. (1987) Nucl Eng Des. vol. 106, p 113.

Beremin, F.M. (1983) Metall Trans. vol. 14A, pp 2277–2287.

Briant, C.L. (1985) Acta Metall. vol. 33, pp 1241–1246.

Briant, C.L. (1988) Metallurgical applications of Auger electron spectroscopy. Treatise on Materials Science and Technology. vol. 30. Academic Press, New York, p 111.

Chawla, K.K. (1987) Composite Materials. Springer Verlag, New York.

Chen, Z., Mecholsky, J.J., Joseph, T., and Beatty, C.L. (1997) J Mater Sci. vol. 32, p 6317.

Clyne, T.W. and Withers, P.J. (1993) An Introduction to Metal Matrix Composites. Cambridge University Press, Cambridge, UK.

Cottrell, A.H. (1958) Trans Am Inst Min Metall Petrol Eng. vol. 212, p 212, p 192.

Curry, D.A. (1980a) CEGB, Leatherhead, Surrey, UK. Rep. no. RD/L/N 154/80 (unpublished research).

Curry, D.A. (1980b) Met Sci. vol. 14, p 319.

Ebrahimi, F. and Seo, H.K. (1996) Acta Mater. vol. 44, p 831.

Englel, L., Klingele, H., Ehrenstein, G.W., and Schaper, H. (1981) An Atlas of Polymer Damage. Prentice-Hall, Englewood Cliffs, NJ.

Evans, A.G. (1983) Metall Trans. vol. 14A, p 1349.

Evans, A.G. and Charles, E.A. (1977) J Am Ceram Soc. vol. 60, pp 22–28.

Faber, K.T., Huang, M.C., and Evans, A.G. (1981) J Am Ceram Soc. vol. 64, pp 296–301.

Fontaine, A., Maas, E., and Tulou, J. (1987) J Nucl Mater Des. vol. 105, p 77.

George, E.P. and Liu, C.T. (1990) Mater Res. vol. 5, pp 754–762.

George, E., Liu, C.T., and Pope, D.P. (1996) Acta Metall. vol. 44, p 1757.

Gupta, T.K. (1972) J Am Ceram Soc. vol. 55, pp 249–253.

Gurson, A.L. (1977) J Eng Mater Technol. vol. 99, p 2.
Hasselman, D.P.H. (1963a) J Am Ceram Soc. vol. 46, pp 229–234.
Hasselman, D.P.H. (1963b) J Am Ceram Soc. vol. 46, pp 535–540.
Hasselman, D.P.H. (1969) J Am Ceram Soc. vol. 52, pp 600–604.
Hertzberg, R.W. (1996) Deformation and Fracture Mechanics of Engineering Materials. 4th ed. John Wiley, New York.
Hull, D. (1975) Polymeric Materials. ASM, Metals Park, OH, p 487.
Hull, D. and Bacon, D.J. (1984) Introduction to Dislocations. International Series on Materials Science and Technology. Vol. 37. Pergamon Press.
Hull, D. and Clyne, T.W. (1996) An Introduction to Composite Materials. 2nd ed. Cambridge University Press, Cambridge, UK.
Jin, O., Li, Y., and Soboyejo, W.O. (1998) Appl. Comp Mater. vol. 5, pp 25–47.
Kameda, J. (1989) Microscopic model for the void growth behavior. Acta Metall. vol. 37, p 2067.
Khantha, M., Vitek, V., and Pope, D.P. (1994). Phys Rev Lett. vol. 73, p 684.
Kingery, W.D. (1955) J Am Ceram Soc. vol. 38, pp 3–15.
Knott, J.F. (1973) Fundamentals of Fracture Mechanics. Butterworth, London, UK.
Lawn, B. (1993) Fracture of Brittle Materials. 2nd ed. Cambridge University Press, Cambridge, UK.
Lin, T., Evans, A.G. and Ritchie, R.O. (1986) J Mech Phys Solids. vol. 34, p 477.
Liu, C.T., Lee, E.H. and McKanney, C.G. (1989) Scripta Metall. vol. 23, p 875.
Loretto, M.H. and Smallman, R.E. (1975) Defect Analysis in Electron Microscopy. Chapman and Hall, New York.
Low, J.R. (1954) Symposium on Relation of Properties to Microstructure. vol. 163. ASM International, Materials Park, OH.
Lu, T.J. and Fleck, N. (1998) Acta Metall Mater. vol. 46, p 4744.
Mandelbrot, B.B. (1982) The Fractal Geometry of Nature. Will, Freeman, San Francisco, CA.
Matthews, F.L. and Rawlings, R.D. (1994). Composite Materials: Engineering and Science. Chapman and Hall, New York.
Mecholsky, J.J. and Freiman, S.W. (1991) Am Ceram Soc. vol. 74, 3136–3138.
Mecholsky, J.J. and Passoja, D.E. (1985) Fractal Aspects of Materials. In: R.B. Leibowitz, B.B. Mandelbrot and D.E. Passoja, eds. Materials Research Society, Pittsburgh, PA, p 117.
Mecholsky, J.J., Freiman, S.W., and Rice, R.W. (1976) J Mater Sci. vol. 11, p 1310.
Mecholsky, J.J., Freiman, S.W., and Rice, R. W. (1978) ATM ST 645, p 363.
Mecholsky, J.J., Passoja, D.E., and Feinberg-Ringel (1989) J Am Ceram Soc. vol. 72, pp 60–65.
Mignard, C., Olagnon, C., Fantozzi, G., and Saadaoui, M. (1996) J Mater Sci. vol. 31, pp 2437–2441.
Mudry, F. (1987) A local approach to cleavage fracture. Nucl Eng Des. vol. 105, p 65.
Nakayama, J. and Ishizuka, M. (1969) Cerami Bull. vol. 45, pp 666–669.
Orowan, E. (1945) Trans Inst Eng Shipbuilders, Scotland. vol. 89, p 1945.

Plangsangmas, L., Mecholsky, J.J., and Brannan, A.B. (1999) J Polym Sci. vol. 72, pp 257–268.
Ramasundaram, P., Ye, F., Bowman, R.R., and Soboyejo, W.O. (1998) Metall Mater Trans. vol. 29A, pp 493–505.
Reed-Hill, R.E. and Abbaschian, R. (1991) Physical Metallurgy Principles. 3rd ed. PWS Publishing Co. Boston, MA.
Rice, J.R. and Tracey, D.M. (1969) J Mech Phys Solids. vol. 17, p 201.
Rosenfield, A.R. and Majumdar, B.S. (1987) J Nucl Eng Des. vol. 105, p 51.
Rousselier, G. (1987) J Nucl Eng Des. vol. 105, p 97.
Saario, T., Wallin, K., and Törrönen, K. (1984) J Eng Mater Technol. vol. 106, pp 173–177.
Smith, E. (1966) Proceedings of Conference on Physical Basis of Yield and Fracture. Institute of Physics, Physical Society, Oxford, UK, pp 36–46.
Soboyejo, W.O Schwartz, D.S., and Sastry, S.M.L. (1992) Metall Trans. vol. 23A, pp 2039–2059.
Soboyejo, W.O., Mercer, C., Schymanski, J., and Van der Laan, S.R. (2001) An Investigation of thermal shock resistance in a high temperature refractory ceramic: a fracture mechanics approach. J Am Ceram Soc. vol. 84, pp 1309–1314.
Stroh, A.N. (1954) Proc Ro Soc. vol. A223, p 404.
Stroh, A.N. (1957) A theory of fracture of Metals. Adv Phys. vol. 6, pp 418–465.
Swain, M.V., Lutz, E.H., and Claussen, N. (1991a) J Am Ceram Soc. vol. 74, pp 11–18.
Swain, M.V., Lutz, E.H., and Claussen, N. (1991b) J Am Ceram Soc. vol. 60, pp 19–24.
Taya, M. and Arsenault, R.J. (1989) Metal Matrix Composites. Pergamon Press, New York.
Thomason, P.F. (1990) Ductile Fracture of Metals. Pegamon Press, Oxford, UK.
Thompson, A. (1993) Fracture Mechanisms in Titanium Aluminides. In: W.O. Soboyejo, T.S. Srivatsan, and D. L. Davidson, eds. Fatigue and Fracture of Ordered Intermetallics I. TMS, Warrendale, PA, pp 293–306.
Thompson, A.W. and Knott, J.F. (1993) Micromechanisms of brittle fracture. Metall Trans A. vol. 24A, p 523.
Tinklepaugh, J.R. (1960) Cermets. Reinhold, New York, pp 170–80.
Tipper, C.F. (1949) The fracture of metals. Metallurgia. vol. 39, p 133.
Tvergaard, V. and Needleman, A. (1984) Acta Metall. vol. 32, p 157.
Twead and Knott (1987) Acta Metall. vol. 35, p 140.
West, J.K., Mecholsky, J.J., and Hench, L.L. (1990) J Non-Crystall Solids. vol. 260, p 99.
West, J.K., Mecholsky, J.J., and Hence, L.L. (1999) J Non-Crystall Solids. vol. 260, pp 99–108.
Xia, L. and Shih, C.F. (1996) J Phys IV. vol. 6, p 363.
Ye, F., Farkas, D., and Soboyejo, W.O. (1998). Mat Sci Eng. vol. A264, pp 81–93.
Zhang, Y.W., Bower, A.F., Xia, L., and Shih, C.F. (1999) Three dimensional finite element analysis of the evolution of voids and thin films by strain and electromigration induced surface diffusion. JMPS. vol. 47, p 173.

13

Toughening Mechanisms

13.1 INTRODUCTION

The notion of designing tougher materials is not a new one. However, it is only in recent years that scientists and engineers have started to develop the fundamental understanding that is needed to guide the design of tougher materials. The key concept in this area is the notion of shielding the crack tip(s) from applied stress(es). When this is done, higher levels of remote stresses can be applied to a material before fracture-critical conditions are reached.

Various crack-tip shielding concepts have been identified by researchers over the past 30 years. These include:

1. Transformation toughening
2. Twin toughening
3. Crack bridging
4. Crack-tip blunting
5. Crack deflection
6. Crack trapping
7. Microcrack shielding/antishielding
8. Crazing

The above toughening concepts will be introduced in this chapter. However, it is important to note that toughening may also occur by some mechanisms that are not covered in this chapter (Fig. 13.1). In any case, the combined effects of multiple toughening mechanisms will also be discussed within a framework of linear superposition of possible synergistic interactions between individual toughening mechanisms.

13.1.1 Historical Perspective

Toughening concepts have been applied extensively to the design of composite materials. Hence, before presenting the basic concepts and associated equations, it is important to note here that even the simplest topological forms of composite materials are complex systems. In most cases, these incorporate interfaces with a wide range of internal residual stresses and

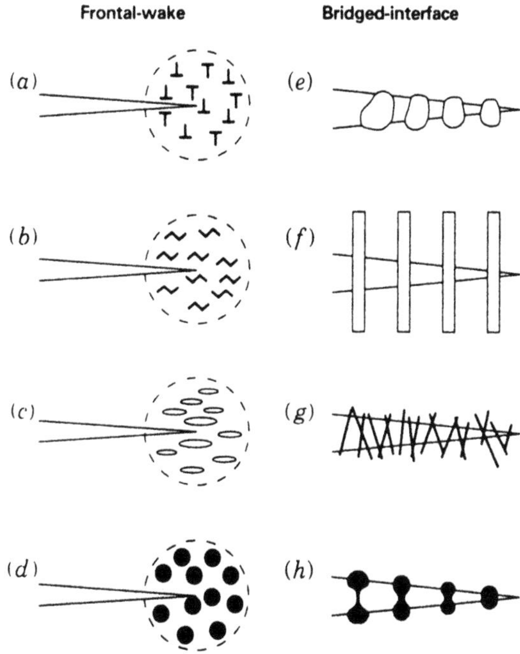

FIGURE 13.1 Crack-tip shielding mechanisms. Frontal zone: (a) dislocation cloud; (b) microcrack cloud; (c) phase transformation; (d) ductile second phase. Crack-wake bridging zone: (e) grain bridging; (f) continuous-fiber bridging; (g) short-whisker bridging; (h) ductile second phase bridging. From B. Lawn, reprinted with permission from Cambridge University Press.)

thermal expansion misfit. Also, most of the expressions presented in this chapter are, at best, scaling laws that capture the essential elements of complex behavior. In most cases, the expressions have been verified by comparing their predictions with the behavior of model materials under highly idealized conditions. However, due to the random features in the topologies of the constituent parts, the agreement between the models and experiments may be limited when the conditions are different from those captured by the models (Argon, 2000).

In any case, there are two types of toughening approaches. These are generally referred to as *intrinsic and extrinsic toughening*. In this chapter, intrinsic toughening is associated with mechanistic processes that are inherent to the normal crack tip and crack wake processes that are associated with crack growth. In contrast, extrinsic toughening is associated with additional crack tip or crack wake processes that are induced by the presence of reinforcements such as particulates, fibers, and layers. Available scaling laws will be presented for the modeling of intrinsic and extrinsic toughening mechanisms. Selected toughening mechanisms are summarized in Fig. 13.1.

13.2 TOUGHENING AND TENSILE STRENGTH

In most cases, toughening gives rise to resistance-curve behavior, as discussed in Chap. 11. In many cases, the associated material separation displacements are large. This often makes it difficult to apply traditional linear and nonlinear fracture concepts. Furthermore, notch-insensitive behavior is often observed in laboratory-scale specimens. Hence, it is common to obtain expressions for the local work of rupture, ΔW, and then relate these to a fracture toughness parameter based on a stress intensity factor, K, or a J-integral parameter.

If we now consider the most general case of material with an initiation toughness (energy release rate) of G_i and a toughening increment (due to crack tip or crack wake processes) of ΔG, then the overall energy release rate, G_c, may be expressed as:

$$G_c = G_i + \Delta G \tag{13.1}$$

Similar expressions may be obtained in terms of J or K. Also, for linear elastic solids, it is possible to convert between G and K using the following expressions:

$$G = K^2/E' \tag{13.2a}$$

or

$$K = \sqrt{E'G} \tag{13.2b}$$

where $E' = E$ for plane stress conditions, $E' = E/(1 - \nu^2)$ for plane strain conditions, E is Young's modulus, and ν is Poisson's ratio.

In scenarios where the material behaves linear elastically in a global manner, while local material separation occurs by nonlinear processes that give rise to long-range disengagement, it is helpful to relate the tensile strength and the work of fracture in specific traction/separation (T/S) laws. An example of a T/S law is shown in Fig. 13.2(a). These are mapped out in front of the crack, as is shown schematically in Fig. 13.2(b).

In the T/S law [Fig. 13.2(a)], the rising portion corresponds to the fracture processes that take the material from an initial state to a peak traction corresponding to the tensile strength, S. The declining portion SD corresponds to the fracture processes beyond the peak state, and the total area under the curve corresponds to the work of rupture of the material. It is also important to note here that the way in which the T/S laws affect the fracture processes ahead of an advancing crack can be very com-

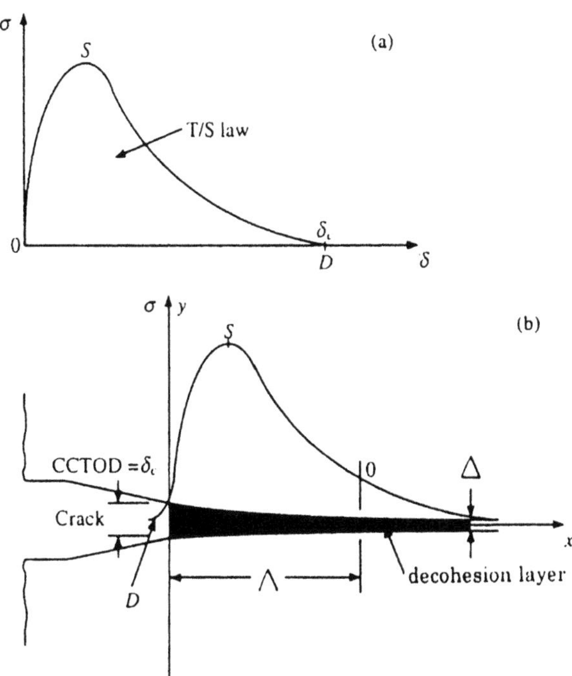

FIGURE 13.2 Schematic illustration of (a) traction/separation (T/S) across a plane and (b) T/S law mapped in front of a crack of limited ductility. (From Argon, 2000.)

plex. In any case, for a process zone of size, h, crack length, a, and width, W, the condition for small-scale yielding is given by Argon and Shack (1975) to be

$$\mathrm{CTOD_c} \ll h < a < W \tag{13.3}$$

Where $\mathrm{CTOD_c}$ is the critical crack tip opening displacement and the other variables have their usual meaning. For fiber-reinforced composites, the $\mathrm{CTOD_c}$ is $\sim$ 1–2 mm (Thouless and Evans, 1988; Budiansky and Amazigo, 1997), while in the case of fiber-reinforced cements, it is usually of the order of a few centimeters. Consequently, very large specimens are needed to obtain notch-sensitive behavior on a laboratory scale. Failure to use large enough specimens may, therefore, lead to erroneous conclusions on notch-insensitive behavior.

13.3 REVIEW OF COMPOSITE MATERIALS

An overview of composite materials has already been presented in Chaps 9 and 10. Nevertheless, since many of the crack-tip shielding mechanisms are known to occur in composite materials, it is important to distinguish between the two main types of composites that will be considered in this chapter. The first consists of brittle matrices with strong, stiff brittle reinforcements, while the second consists primarily of brittle matrices with ductile reinforcements. Very little attention will be focused on composites with ductile matrices such as metals and some polymers.

In the case of brittle matrix composites reinforced with aligned continuous fibers, the typical observed behavior is illustrated in Fig. 13.3 for tensile loading. In this case, the composite undergoes progressive parallel cracking, leaving the fibers mostly intact and debonded from the matrix. At the so-called first crack strength, σ_{mc}, the cracks span the entire cross-section, and the matrix contribution to the composite stiffness is substantially reduced. Eventually, the composite strength resembles the fiber bundle strength, and there is negligible load transfer between the matrix and the fibers. This leads to global load sharing, in which the load carried by the broken weak fibers is distributed to the unbroken fibers.

In the case of unrestrained fracture of all fibers, there would be only limited sliding/rubbing between broken fiber ends and the loosely attached matrix segments. This will result in the unloading behavior illustrated in Fig. 13.3(a). The associated area under the stress–strain curve would correspond to the work of stretching the intact fibers and the work of matrix cracking. However, this does not translate into fracture toughness improvement or crack growth resistance.

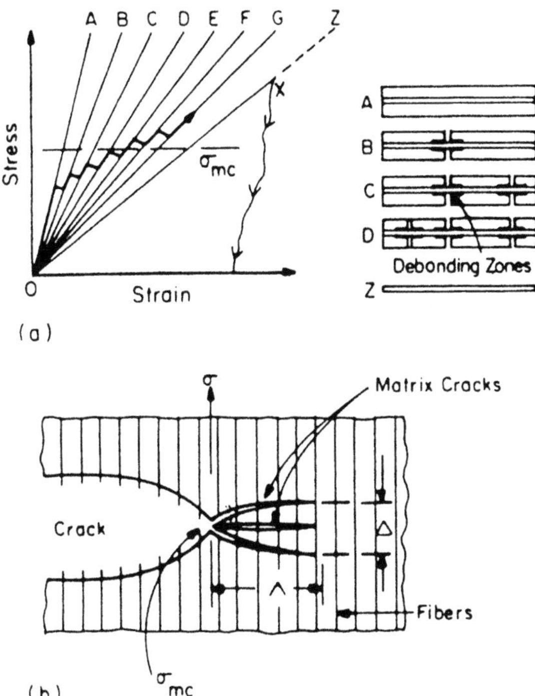

FIGURE 13.3 Schematic (a) progressive matrix cracking in a fiber-reinforced composite subjected to larger strain to fracture than the brittle matrix, leaving composite more compliant, and (b) macrocrack propagating across fibers at $\sigma < \sigma_{mc}$ with three matrix cracks in the process zone. (From Argon, 2000.)

For toughening or crack growth resistance to occur, the crack tip or crack wake processes must give rise to crack-tip shielding on an advancing crack that extends within a process zone in which the overall crack tip stresses are reduced. The mechanisms by which such reductions in crack tip stresses (crack tip shielding) can occur are described in the next few sections.

13.4 TRANSFORMATION TOUGHENING

In 1975, Garvie et al. (1975) discovered that the tetragonal (t) phase of zirconia can transform to the monoclinic (m) phase on the application of a critical stress. Subsequent work by a number of researchers (Porter et al., 1979; Evans and Heuer, 1980; Lange, 1982; Chen and Reyes-Morel, 1986;

Rose, 1986; Green et al., 1989; Soboyejo et al., 1994; Li and Soboyejo, 2000) showed that the measured levels of toughening can be explained largely by models that were developed in work by McMeeking and Evans (1982), Budiansky et al. (1983), Amazigo and Budiansky (1988), Stump and Budiansky (1989a, b), Hom and McMeeking (1990), Karihaloo (1990), and Stam (1994).

The increase in fracture toughness on crack growth was explained readily by considering the stress field at the crack tip, as well as the crack wake stresses behind the crack tip. The latter, in particular, are formed by prior crack-tip transformation events. They give rise to closure tractions that must be overcome by the application of higher remote stresses, Fig. 13.4(a). As the crack tip stresses are raised, particles ahead of the crack tip undergo stress-induced martensitic phase transformations, at speeds close to that of sound (Green et al., 1989). The unconstrained transformation yields a dilatational strain of $\sim 4\%$ and a shear strain of $\sim 16\%$, which are consistent with the lattice parameters of the tetragonal and monoclinic phases, Fig. 13.4(a) and Table 13.1.

The early models of transformation toughening were developed by McMeeking and Evans (1982) and Budiansky et al. (1983). These models did not account for the effects of transformation-induced shear strains, which were assumed to be small in comparison with those of dilatational strains. The effects of deformation-induced twinning were assumed to be small due to the symmetric nature of the twin variants which give rise to strain components that were thought to cancel each other out, Figs 13.4(c) and 13.4(d). However, subsequent work by Evans and Cannon (1986), Reyes-Morel and Chen (1988), Stam (1994), Simha and Truskinovsky (1994), and Li and Soboyejo (2000) showed that the shear components may also contribute to the overall measured levels of toughening.

For purely dilatant transformation, in which the transformations result in pure dilatation with no shear, the dependence of the mean stress, $\sum_{\mathrm{m}}$, on the dilational stress is illustrated in Fig. 13.5. In this figure, $\sum_{\mathrm{m}}^{\mathrm{c}}$ is the critical transformation mean stress, B is the bulk modulus and F is the volume fraction of transformed phase. For a purely isotropic solid, G is given by $G = E/[2(1+\nu)]$ and $B = E/[3(1-2\nu)]$.

Stress-induced phase transformations can occur when $\sum_{\mathrm{m}} > \sum_{\mathrm{m}}^{\mathrm{c}}$. They can also continue until all the particles are fully transformed. Furthermore, during transformation, three possible types of behavior may be represented by the slope $\overline{B}$ in Fig. 13.5. When $\overline{B} < -4G/3$, the transformation occurs spontaneously and immediately to completion. This behavior is termed supercritical. When $\overline{B} > -4G/3$, the behavior is subcritical, and the material can remain stable in a state in which only a part of the particle is transformed. This transformation also occurs gradually without any

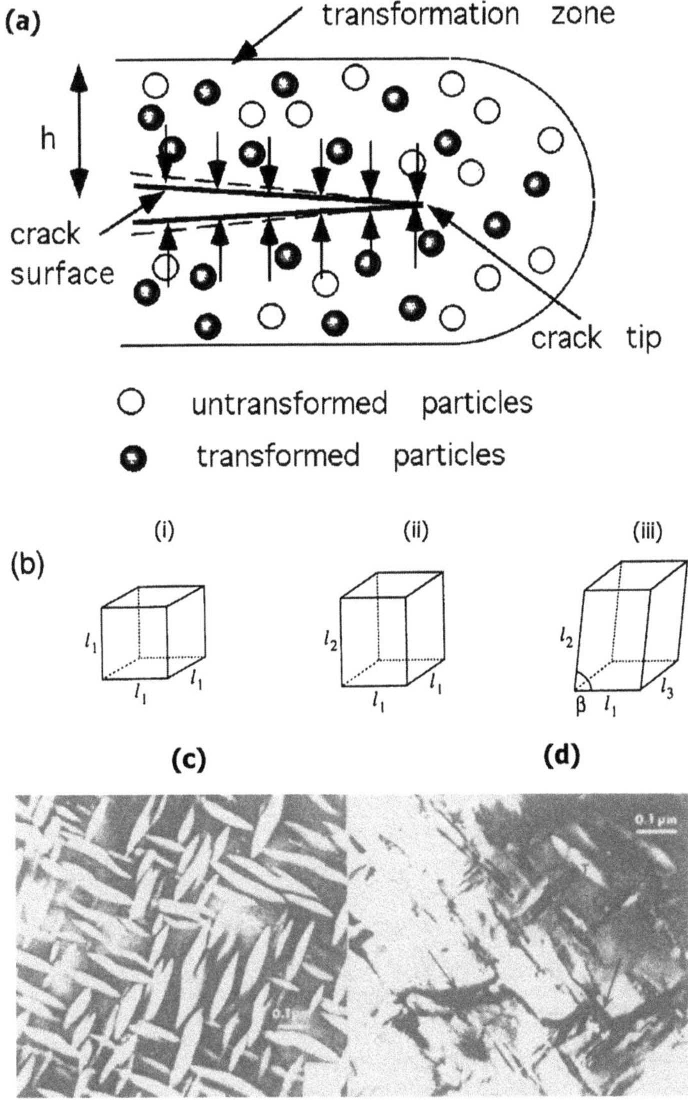

FIGURE 13.4 (a) Schematic illustration of transformation toughening; (b) the three crystal structures of zirconia; (c) TEM images of coherent tetragonal ZrO_2 particles in a cubic MgO–ZrO_2 matrix; (d) transformed ZrO_2 particles near crack plane—n contrast to untransformed ZrO_2 particles remote from crack plane. [(c) and (d) are from Porter and Heuer, 1977.]

TABLE 13.1 Lattice Parameters (in nanometers) Obtained for Different Phases of Zirconia at Room Temperature Using Thermal Expansion Data

	l_1	l_2	l_3	β
Cubic	0.507	0.507	0.507	90°
Tetragonal	0.507	0.507	0.516	90°
Monoclinic	0.515	0.521	0.531	~81°

Source: Porter et al. (1979).

jumps in the stress or strain states. Finally, when $\overline{B} = -4G/3$, the material is termed critical. This corresponds to a transition from subcritical to supercritical behavior.

Budiansky et al. (1983) were the first to recognize the need to use different mathematical equations to characterize the physical responses of subcritical, critical, and supercritical materials. The governing equations for subcritical behavior are elliptic, so that the associated stress and strain fields are smooth. Also, the supercritical transformations are well described by hyperbolic equations that allow for discontinuities in the stress and strain fields. The stress–strain relations are also given by Budiansky et al. (1983) to be

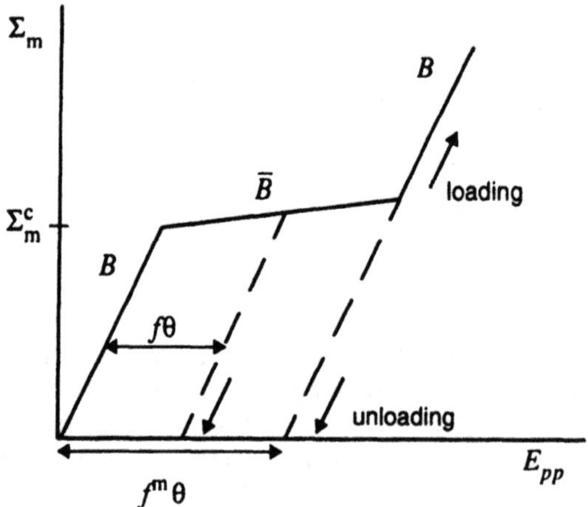

FIGURE 13.5 Schematic illustration of transformation toughening. (From Stam, 1994.)

$$E_{ij} = \frac{1}{2G}\dot{S}_{ij} + \frac{1}{3B}\dot{\sum}_{m}\delta_{ij} + \frac{1}{3}f\theta\delta_{ij} \tag{13.4}$$

or

$$\dot{\sum}_{ij} = 2G\left(\dot{E}_{ij} - \frac{1}{3}\dot{E}_{pp}\delta_{ij}\right) + B\left(\dot{E}_{pp} - \dot{f}\theta\right)\delta_{ij} \tag{13.5}$$

where $\dot{E}_{ij}$ are the stress rates of the continuum element, $\dot{S}_{ij} = \dot{\sum}_{ij} - \dot{\sum}_{m}\delta_{ij}$, $\dot{\sum}_{m} = \dot{\sum}_{pp}/3$, and $\dot{E}_{ij}$ represents the strain rates.

For transformations involving both shear and dilatant strains, Sun et al. (1991) assume a continuum element, consisting of a large number of transformable inclusions embedded coherently in an elastic matrix (referred to by index M). If we represent the microscopic quantities in the continuum element with lower case characters, the macroscopic quantities are obtained from the volume averages over the element. The relationship between macroscopic stresses ($\dot{E}_{ij}$) and microscopic stresses is, therefore, given by

$$\sum_{ij} = \langle\sigma_{ij}\rangle_{V} = \frac{1}{V}\int \sigma_{ij}\mathrm{d}V = f\langle\sigma_{ij}\rangle_{V_1} + (1-f)\langle\sigma_{ij}\rangle_{V_m} \tag{13.6}$$

where < > denotes the volume average of microscopic quantities, f is the volume fraction of transformed material. Note that f is less than f^{m}, the volume fraction of metastable tetragonal phase.

Furthermore, considerable effort has been expended in the development of a theoretical framework for the prediction of the toughening levels that can be achieved as a result of crack tip stress-induced transformations (Evans and coworkers, 1980, 1986; Lange, 1982; Budiansky and coworkers, 1983, 1993; Marshall and coworkers, 1983, 1990; Chen and coworkers, 1986, 1998). These transformations induce zone-shielding effects that are associated with the volume increase (~ 3–5% in many systems) that occurs due to stress-induced phase transformations from tetragonal to monoclinic phases in partially stabilized zirconia.

For simplicity, most of the micromechanics analyses have assumed spherical transforming particle shapes, and critical transformation conditions that are controlled purely by mean stresses, i.e., they have generally neglected the effects of shear stresses that may be important, especially when the transformations involve deformation-induced twinning phenomena (Evans and Cannon, 1986), although the possible effects of shear stresses are recognized (Chen and Reyes-Morel, 1986, 1987; Evans and Cannon, 1986; Stam, 1993; Simha and Truskinovsky, 1994; Li et al., 2000).

In general, the level of crack tip shielding due to stress-induced transformations is related to the transformation zone size and the volume fraction of particles that transform in the regions of high-stress triaxiality at the crack-tip. A transformation zone, akin to the plastic zone in ductile materials, is thus developed as a crack propagates through a composite reinforced with transforming particles. This is illustrated schematically in Fig. 13.4(a).

The size of the transformation zone associated with a Mode I crack under small-scale transformation conditions has been studied (McMeeking and Evans, 1982). Based on the assumption that the transformation occurs when the mean stress level at the crack tip exceeds a critical stress value (σ_C^T), McMeeking and Evans (1982) estimated the zone size for an idealized case in which all the particles within the transformation zone are transformed. Following a similar procedure, Budiansky et al. (1983) give the following equation for the estimation of the height of the transformation zone (Fig. 13.4(a):

$$h = \frac{\sqrt{3}(1+\nu)^2}{12\pi}\left(\frac{K}{\sigma_C^T}\right)^2 \tag{13.7}$$

where h is the half-height of the transformation wake, K is the far-field stress intensity factor, and ν is Poisson's ratio. For purely dilational transformation, the toughening due to the transformation can also be expressed as (Budiansky et al., 1983):

$$\Delta K_t = \frac{0.22 E_c f \varepsilon_C^T \sqrt{h}}{1-\nu} \tag{13.8}$$

where E_c is the elastic modulus of the composite, f is the volume fraction of transformed particles, and ε_C^T is the transformation volume strain. The above continuum model assumes the volume fraction of the transformed material to be constant with increasing distance, x, from the crack face. However, in reality, the actual volume fraction of transformed phase varies with increasing distance from the crack face. Equation (13.8) must, therefore, be expressed in an integral form to account for the variation in the degree of transformation with increasing distance from the crack face. This yields the following expression for the toughening due to stress-induced transformations (Marshall et al., 1990):

$$\Delta K_t = \frac{0.22 E_c \varepsilon_C^T}{1-\nu}\int_0^h \frac{f(x)}{2\sqrt{x}}\,dx \tag{13.9}$$

where $f(x)$ is a mathematical function that represents the fraction of transformed zirconia as a function of distance, x, from the crack. The critical transformation stress necessary to achieve the transformation can be expressed as a function of the total Gibb's free energy associated with the transformation from tetragonal to monoclinic phase. This may be estimated from (Becher, 1986):

$$\sigma_C^T = \frac{\Delta G}{\varepsilon_C^T} \tag{13.10}$$

where σ_C^T is the critical stress, and ΔG is the Gibb's free energy of the transformation. The above expression does not account for the effect of the enthalpy terms in the equivalent Kirchoff circuits for the transformation nor for the potential residual stresses that can be induced as a result of the thermal expansion mismatch between the constituents of composites reinforced with partially stabilized zirconia particles (Soboyejo et al., 1994). Depending on the thermal expansion coefficients, the zirconia particles may be subjected to either mean tension or compression. In general, however, if the mean stress is compressive, the far-field applied stress necessary for transformation will increase. On the other hand, the existence of tensile mean stress will trigger the transformation at a lower level of applied stress. As a result of this, the mean stress, σ_m, that is needed to induce the transformation of ZrO_2 particles is modified by the radial residual stress, σ_m. The modified critical condition for transformation is thus given by (Becher, 1986):

$$\sigma_m = (\sigma_c - \sigma_0) \tag{13.11}$$

For simplicity, the above discussion has focused on the toughening due to pure dilatational effects associated with stress-induced phase transformations. However, in reality, the shear stresses play a role that may sometimes cause significant differences between the experimental results and the theoretical predictions (Evans and Cannon, 1986). Since the shear strains associated with stress-induced phase transformations may be as high as $\sim 14\%$, it may be necessary to assess the effects of shear strains in the estimates of toughening. This has been estimated by Lambropoulus (1986) in an approximate analysis that gives

$$\Delta K_t = 055 \frac{E_c V_f \varepsilon_c^T \sqrt{h}}{(1 - \nu)} \tag{13.12}$$

where the transformations are induced by critical principal strains, and the transformation strains also develop in that direction. In this model, twinning is assumed to be induced by the shear stresses in the transforming particles. Also, the model assumes that there is no coupling between the crack tip fields and the development of the transformation zone. The initial work of Lambropoulus (1986) and Budiansky et al. (1983) has been followed by subsequent work by Stump (1991), Budiansky and Truskinovsky (1993), Simha and Truskinovsky (1994), Stam (1994), and Li et al. (2000).

13.5 CRACK BRIDGING

Crack bridging is illustrated schematically in Fig. 13.6. The bridging reinforcements restrict opening of cracks, and thus promote shielding of the crack tip. The effective stress intensity factor at the crack tip is, therefore, lower than the remote/applied stress intensity factor. In the case of stiff elastic fibers, interfacial sliding may occur during crack bridging. The tailoring of the interface to optimize frictional energy dissipation is, therefore, critical. This section will concentrate initially on bridging by ductile reinforcements. This will be followed by a focus on crack bridging by stiff elastic whiskers/fibers, as well as a section on debonding/fiber pull-out.

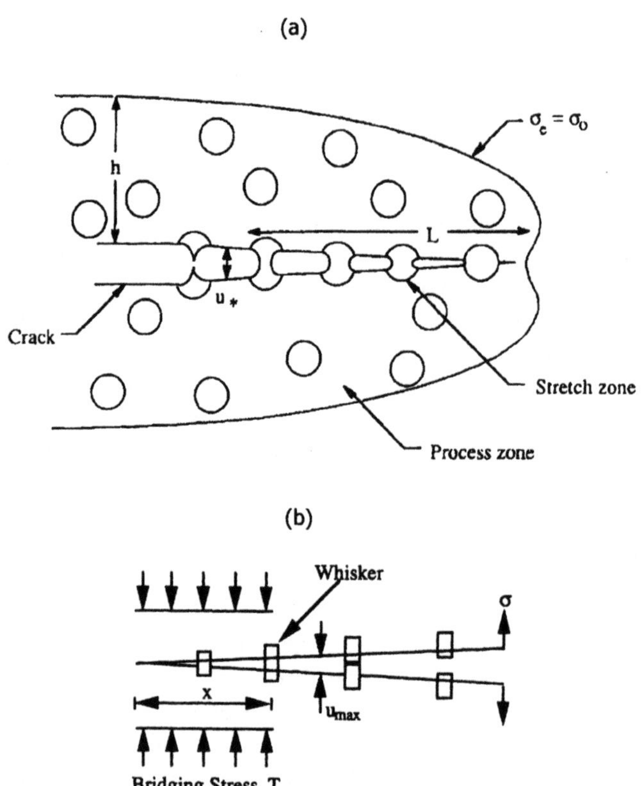

FIGURE 13.6 Schematic illustration of crack bridging by (a) ductile particles and (b) stiff elastic whiskers.

13.5.1 Bridging By Ductile Phase

An energy approach may be used to explain the toughening due to ductile phase reinforcement. Within this framework, ductile phase toughening by crack bridging may be attributed to the plastic work required for the plastic stretching of the constrained ductile spherical particles. For small-scale bridging in which the size of the bridging zone is much smaller than the crack length, the increase in strain energy, ΔG_{SS}, due to the plastic work required for the stretching of the ductile phase is given by (Soboyejo et al., 1996, Ashby et al., 1989; Cao et al., 1989; Shaw and Abbaschian, 1994; Kajuch et al., 1995; Bloyer et al., 1996, 1998; Lou and Soboyejo, 2001):

$$\Delta G_{SS} = V_f C \sigma_y \zeta \tag{13.13}$$

where V_f is the volume fraction of ductile phase that fails in a ductile manner (note that the actual reinforcement volume fraction is f), C is a constraint parameter which is typically between 1 and 6, σ_y is the uniaxial yield stress, and ζ is a plastic stretch parameter. The small-scale bridging limit may also be expressed in terms of the stress intensity factor, K:

$$K_{SS} = (K_0^2 + E' f \sigma_0 t \chi)^{\frac{1}{2}} \tag{13.14}$$

where K_{ss} is the steady-state (or plateau) toughness, K_0 is the crack-initiation toughness (typically equal to brittle matrix toughness), is $E' = E/(1 - \nu^2)$ for plane strain conditions, and $E' = E$ for plane stress conditions (where E is Young's modulus and ν is Poisson's ratio), f is the volume fraction, σ_0 is the flow stress of ductile reinforcement, t is equivalent to half of the layer thickness, and χ is the work of rupture, which is equal to the area under the load–displacement curve.

For small-scale bridging, the extent of ductile phase toughening may also be expressed in terms of the stress intensity factor. This gives the applied stress intensity factor in the composite, K_c, as the sum of the matrix stress intensity factor, K_m, and the toughening component due to crack bridging, ΔK_b. The fracture toughness of the ductile-reinforced composites may thus be estimated from (Budiansky et al., 1988; Tada et al., 1999):

$$K_c = K_m + \Delta K_b = K_m + \sqrt{\frac{2}{\pi}} \alpha V \int_0^L \frac{\sigma_y}{\sqrt{x}} d_x \tag{13.15a}$$

where K_m is the matrix fracture toughness, x is the distance behind the crack tip, and L is the bridge length (Fig. 13.6). The toughening ratio due to small-scale bridging (under monotonic loading) may thus be expressed as

$$\lambda_b = \frac{K_c}{K_m} = 1 + \frac{1}{K_m}\sqrt{\frac{2}{\pi}}\alpha V_t \int_0^L \frac{\sigma_y}{\sqrt{x}} dx \tag{13.15b}$$

Equations (13.15) can be used for the estimation of the toughening due to small-scale bridging under monotonic loading conditions (Li and Soboyejo, 2000; Lou and Soboyejo, 2001).

Alternatively, the toughening due to small-scale bridging by ductile phase reinforcements may be idealized using an elastic–plastic spring model (Fig. 13.7). This gives the toughening ratio, λ, as (Rose, 1987; Budiansky et al., Evans, 1988):

$$\lambda = \frac{K}{K_m} = \left[1 + \frac{\sigma_y^2}{kK_m^2}\left(1 + \frac{2u_p}{u_y}\right)\right]^{\frac{1}{2}} \tag{13.16a}$$

where K_m is the matrix toughness, σ_y is the uniaxial yield stress, u is the crack face displacement, k is a dimensionless spring-stiffness coefficient, u_y is the maximum elastic displacement, and u_p is the total plastic displacement to failure (Fig. 13.8). Equation (13.16a) can also be arranged to obtain the following expression for the toughening due to ductile phase bridging:

$$\Delta K_b = (\lambda - 1)K_m \tag{13.16b}$$

However, Eqs 13.16a and 13.16b do not include a bridging length scale. Hence, they cannot be readily applied to the prediction of toughening under large-scale bridging conditions in which the bridge lengths are comparable to the crack lengths.

In the case of large-scale bridging, where the bridging length is comparable to the specimen width, a somewhat different approach is needed for the estimation of K_c and λ_b. Mechanics models have been developed by Zok and Hom (1990), Cox (1992, 1996), and Odette et al. (1992), for the modeling of large-scale bridging. These researchers provide rigorous modeling of crack bridging by using self-consistent solutions of the crack opening pro-

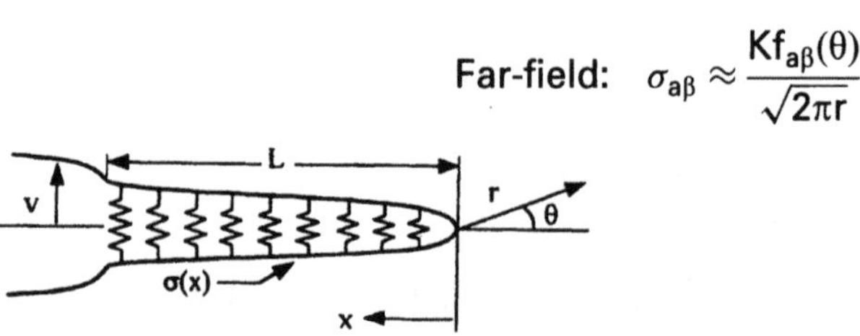

FIGURE 13.7 Schematic illustration of spring model of crack bridging.

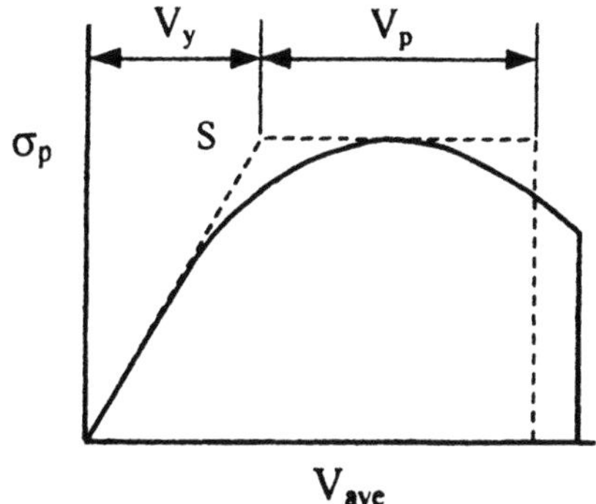

FIGURE 13.8 Schematic of elastic–plastic spring load–displacement function.

file, $u(x)$, and the crack face stress distribution, $\sigma(x)$, where x is the distance from the crack tip. Since these are not known precisely in most cases, they require experimental measurements of stress-displacement functions and iterative numerical schemes that often require significant computational effort.

In the case of the model by Odette et al. (1992), a trial $\sigma'(x)$ function is first used to calculate the reduction in the stress intensity factor (shielding) from the bridging zone. By applying Castigliano's theorem, the corresponding crack-face closure displacements, $u_b'(x)$, are also calculated. For common fracture mechanics specimen geometries, the solutions may be obtained from available handbooks such as that by Tada et al. (1999). The resulting integrals can also be solved numerically. However, for regions close to the crack tip, asymptotic solutions may be used.

The total trial applied stress intensity factor is taken as $K_p'(da) = K_t' + \Delta K_0$. A numerical integration scheme (again based on Castigliano's theorem) is then used to compute the trial crack opening displacement corresponding to the applied load, P'. The net crack opening is also computed as $u'(x) = u_p'(x) - u_b'(x)$. This is used to compute a trail $\sigma'(u)$ from the trial $\sigma'(x)$ and $u'(x)$. The difference between the specified and trial stress-displacement functions is then evaluated from $\varepsilon(u) = \sigma(u) - \sigma'(u)$. A solution is achieved when $\varepsilon(u)$ is less than the convergence criteria. Otherwise, the iterations are repeated until convergence is achieved.

Alternatively, the distributed tractions across the bridged crack faces may be estimated using weighting functions. These account for the weighted distribution of the bridging tractions along a bridge zone (Fett and Munz, 1994) (Fig. 13.9). The shielding due to large-scale bridging, ΔK_{lsb}, may thus be expressed as

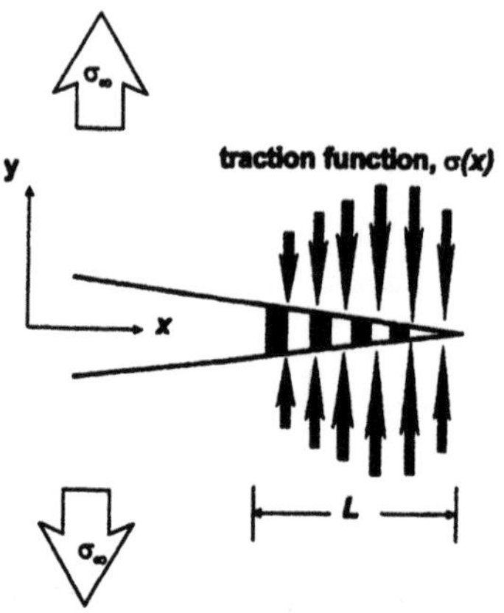

Figure 13.9 Schematic of weighted/distributed bridging tractions.

$$\Delta K_{\mathrm{Isb}} = \int_L \alpha\sigma(x)h(a,x)\mathrm{d}x \tag{13.17}$$

where L is the length of the bridge zone, α is a constraint/triaxiality factor, $\sigma(x)$ is a traction function along the bridge zone, and $h(a,x)$ is a weight function given by Fett and Munz (1994):

$$h(a,x) = \sqrt{\frac{2}{\pi a}}\frac{1}{\sqrt{1-\frac{x}{a}}}\left[1+\sum_{(\nu,\mu)}\frac{A_{\nu\mu}\left(\frac{a}{W}\right)}{\left(1-\frac{a}{W}\right)}\left(1-\frac{x}{a}\right)^{\nu+1}\right] \tag{13.18}$$

where a is the crack length and W is the specimen width. The Fett and Munz coefficients, $A_{\nu\mu}$, are given in Table 13.2 for a single-edge notched bend specimen. As before, the total stress intensity factor, K_c, may estimated

Table 13.2 Summary of Fett and Munz (1994) Parameters for Single-Edge Notched Bend Specimen Subjected to Weighted Crack Bridging Fractions

	μ				
ν	0	1	2	3	4
0	0.4980	2.4463	0.0700	1.3187	−3.067
1	0.5416	−5.0806	24.3447	−32.7208	18.1214
2	−0.19277	2.55863	−12.6415	19.7630	−10.986

from linear superposition ($K_c = K_m + \Delta K_b$) and the toughening ratio for large-scale bridging can be determined from the ratio of K_c to K_m. The simplified expressions presented in Eqs (13.17) and (13.18) have been successfully used to predict the resistance-curve behavior of ductile-layer toughened intermetallics composites (Bloyer et al., 1996, 1998; Li and Soboyejo, 2000).

13.5.2 Crack Bridging by Discontinuous Reinforcements

Bridging of the crack surfaces behind the crack tip by a strong discontinuous reinforcing phase which imposes a closure force on the crack is, at times, accompanied by pull-out of the reinforcement. The extent of pull-out (i.e., the pull-out length) of brittle, discontinuous reinforcing phases is generally quite limited due both to the short length of such phases, and the fact that bonding and clamping stresses often discourage pull-out. However, pull-out cannot be ignored, as even short pull-out lengths contribute to the toughness achieved.

Crack deflection by reinforcements has also been suggested to contribute to the fracture resistance. Often, out-of-plane (non–Mode I) crack deflections are limited in length and angle, and are probably best considered as means of debonding the reinforcement–matrix interface. Such interfacial debonding is important in achieving frictional bridging (bridging by elastic ligaments which are partially debonded from the matrix) and pull-out processes. Frictional bridging elastic ligaments can contribute significantly to the fracture toughness, as is described in the next section.

Here, we will concentrate on the toughening due to crack bridging by various brittle reinforcing phases, where the reinforcement simply bridges the crack surfaces and effectively reduces the crack driving force. This increases the resistance to crack extension. The bridging contribution to the toughness is given by

$$K_c = \sqrt{E_c J_m + E_c \Delta J_{cb}} \tag{13.19}$$

where K_c is the overall toughness of the composite, J_m is the matrix fracture energy, and the term ΔJ_{cb} corresponds to the energy change due to the bridging process.

The energy change associated with the bridging process is a function of the bridging stress/traction, T_u, and the crack opening displacement, u, and is defined as

$$\Delta J = \int_{0}^{u_{max}} T_u du \tag{13.20}$$

where u_{max} is the maximum displacement at the end of the zone.

One can equate the maximum crack opening displacement at the end of the bridging zone, u_{max}, to the tensile displacement in the bridging brittle ligament at the point of failure:

$$u_{max} = \varepsilon_f^l l_{db} \tag{13.21}$$

where ε_f^l represents the strain to failure of the whisker and l_{db} is the length of the f debonded matrix–whisker interface (Fig. 13.10). The strain to failure of the whisker can be defined as

$$\varepsilon_f^l = (\sigma_f^l / E^l) \tag{13.22}$$

where E^l is the Young's modulus of the reinforcing phase. The interfacial debond length depends on the fracture criteria for the reinforcing phase versus that of the interface and can be defined in terms of fracture stress or fracture energy:

$$l_{db} = (r\gamma^l / 6\gamma^i) \tag{13.23}$$

where γ^l/γ^i represents the ratio of the fracture energy of the bridging ligament to that of the reinforcement–matrix interface.

From Eq. (13.21), one quickly notices that the tensile strain displacement achieved in the bridging reinforcement, and hence the maximum crack opening displacement at the end of the bridging zone, increases as the debonded length/gauge length of the reinforcing ligament increases. Consideration of Eqs (13.22) and (13.23) also shows that increasing the reinforcing phase strength and/or enhancing interface debonding will con-

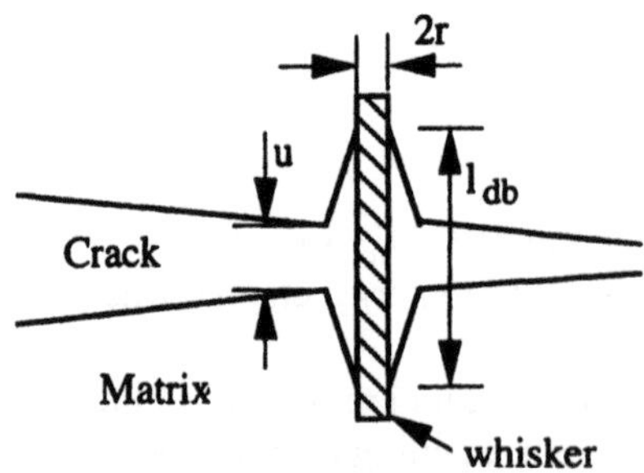

FIGURE 13.10 Schematic of debonding whisker.

tribute to greater tensile displacement within the reinforcing ligament. Increases in the crack opening displacement supported by the bridging zone will enhance the toughening achieved by such reinforcements. Therefore, debonding of the matrix–reinforcement interface can be a key factor in the attainment of increased fracture toughness in these elastic systems. In fact, in ceramics reinforced by strong ceramic whiskers, debonding is observed only in those systems which exhibit substantial toughening.

For the case of a bridging stress which increases linearly from zero (at the crack tip) to a maximum at the end of the bridging zone and immediately decreases to zero, Eq. (13.20) can be reduced to $T_{max}(u_{max})/2$. The maximum closure stress, T_{max}, imposed by the reinforcing ligaments in the crack tip wake is the product of the fracture strength of the ligaments, σ_f^l, and the real fraction of ligaments intercepting the crack plane, A^l:

$$T_{max} = \sigma_f^l A^l \approx \sigma_f^l V^l \tag{13.24}$$

where A^l is approximated by the volume fraction, V^l, for ligaments which have large aspect ratios (e.g., $l/r > 30$ for whiskers). Reinforcement by frictional bridging introduces a change in energy equal to

$$\Delta J^{flb} = \left[\sigma_f^l V^l \left(\sigma_f^l 1 E^l\right)\left(r\gamma^l l\gamma^i\right)\right]^{1/2} \tag{13.25}$$

From these results, the toughness contribution from frictional bridging by the reinforcing phase in the crack tip wake is

$$\Delta K^{flb} = \sigma_f^l \left[\left(r\gamma^l/12\right)\left(E^c/E^l\right)\left(\gamma^l/\gamma^i\right)\right]^{1/2} \tag{13.26}$$

13.5.3 Bridging by Stiff Fibers and Fiber Pull-Out

In the case of stiff fibers, the bridging provided by the fibers may also be expressed in the form of closure pressures (Marshall et al., 1985; Suo et al., 1993) (Fig. 13.11). For a bridged crack with a bridge length, c, much greater than the fiber spacing, the closure pressure $p(x)$ on the crack faces is given by

$$p(x) = T(x)V_f \tag{13.27}$$

where x is the distance from the crack tip to the position on the crack face, V_f is the fiber volume fraction, and T corresponds to the fiber fractions. For a uniform remote stress, σ_∞ acting along the crack faces, the net pressure is $[\sigma_\infty - p(x)]$. Also, the effective stress intensity factor for a straight crack in an infinite medium is given by Sih (1973):

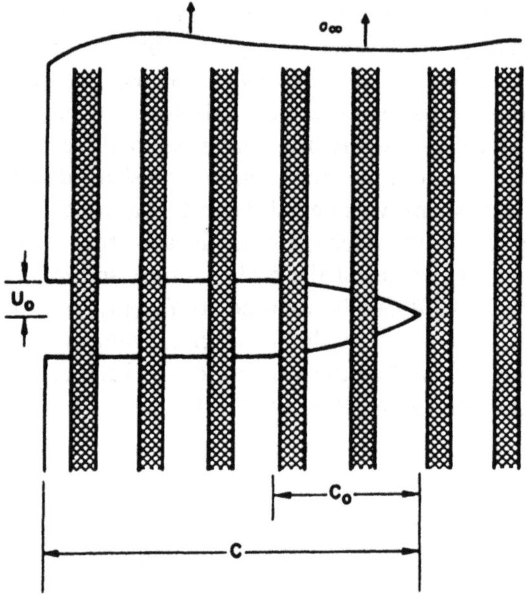

FIGURE 13.11 Crack bridging by stiff elastic fibers.

$$K^L = 2\left(\frac{c}{\pi}\right)^{\frac{1}{2}} \int_0^l \frac{[\sigma_\infty - p(X)]}{\sqrt{1 - X^{2l}}\mathrm{d}x} \tag{13.28a}$$

or

$$K^L = 2\left(\frac{c}{\pi}\right)^{\frac{1}{2}} \int_0^1 \frac{[\sigma_\infty - p(X)]}{\sqrt{1 - X^2}} X\mathrm{d}X \tag{13.28b}$$

for a penny-shaped crack.

In Eqs (13.28), $X = x/c$, c is the bridge length and x is the distance from the crack tip (Fig. 13.11).

If crack bridging is associated with fiber pull-out, then the closure pressure is related to the crack opening by (Marshall et al., 1985):

$$p = 2\left[u\tau V_f^2 E_f(1 + \eta)/R\right]^{\frac{1}{2}} \tag{13.29}$$

where $\eta = E_f V_v / E_m V_m$, R is the fiber radius, and τ is the sliding frictional stress at the interface. The crack opening displacement is then determined by the distribution of surface tractions (Sneddon and Lowengrub, 1969):

$$u(x) = \frac{4(1-\nu^2)c}{\pi E_c} \int_x^1 \frac{s}{\sqrt{s^2 - X^2}} \times \int_0^s \frac{[\sigma_\infty - p(t)]dt}{\sqrt{s^2} - t^2} ds \tag{13.30a}$$

for a straight crack, or

$$u(X) = \frac{4(1-\nu^2)c}{\pi E_c} \int_x^1 \frac{1}{\sqrt{s_x^2 - 2}} \times \int_0^s \frac{[\sigma_\infty - p(t)]dt}{\sqrt{s^2 - t^2}} ds \tag{13.30b}$$

where s and t are normalized position co-ordinates, and ν is the Poisson's ratio of the composite. Furthermore, for brittle matrix composites subjected to crack bridging, debonding, and pull-out, the steady-state pull-out toughening is given by

$$\Delta G_p^{SS} = 2f \int_0^\infty \sigma du \tag{13.31a}$$

where σ is the stress on the fiber between the crack surfaces, f is the fiber volume fraction, and u is the average crack opening. If the Mode II interfacial toughness is low enough for crack front and crack wake debonding to occur (Fig. 13.12), then the steady-state toughening associated with fiber pull-out is given by

$$\Delta G_{pp}^{SS} = -\frac{fE_f e_T a}{\mu \xi} \int_0^{F_p} \frac{F_p, dF_p}{(1 - \nu F_p)^2} \tag{13.31b}$$

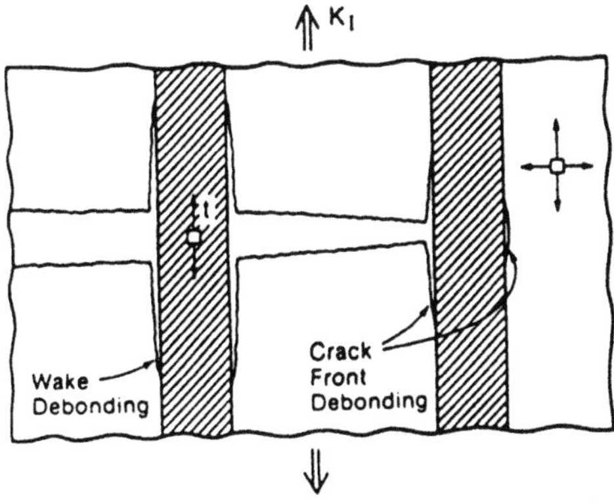

Figure 13.12 Schematic of crack front and crack wake debonding. (From Charalambides and Evans, 1989.)

where F_p is a nondimensional pull-out stress, e_T is a stress-free strain $\Delta\alpha\Delta T$), a is the fiber radius, μ is the fraction coefficient, ξ is $(1-f)/[X(1+f)+(1-f)(1-2\nu)]$, and ν is Poisson's ratio. In general, the above results indicate that bridging by stiff brittle fibers and fiber pull-out may be engineered by the design of weak or moderately strong interfaces in which the interfacial sliding conditions are controlled by coatings. The interested reader is referred to detailed discussion on the modeling of debonding by Charalambides and Evans (1989) and Hutchinson and Jenson (1990).

13.6 CRACK-TIP BLUNTING

Crack tips can be blunted when they move from a brittle phase into a ductile phase, Fig. 13.13(a). Crack-tip blunting may also occur by debonding along the interface of a composite, Fig. 13.13(b). Crack growth in the ductile phase can be considered in terms of a critical strain criterion, which assumes that

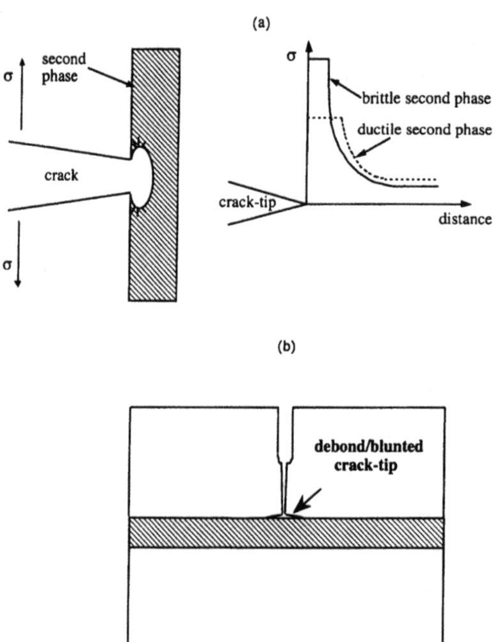

Figure 13.13 Schematic illustration of (a) crack-tip blunting by ductile phase and associated change in crack-tip stress fields, and (b) crack-tip blunting by debonding along two-phase boundary. (Courtesy of Dr. Fan Ye.)

fracture occurs when the strain at a characteristic distance from the crack tip exceeds a critical value. When the crack-tip is blunted by a ductile phase, the critical strain value is increased by the presence of a ductile phase, Fig. 13.13(b). Shielding due to crack tip blunting effects can be estimated using a micromechanics model developed by Chan (1992). In his model, he postulated that the near-tip effective strain distribution in the matrix/composite in a ductile-phase reinforced brittle matrix composite could both be described by the Hutchinson-Rice–Rosengren (HRR) field expressions given below:

$$\bar{\varepsilon}_m = \alpha_m \varepsilon_m^y \left[\frac{J_m}{\alpha_m \varepsilon_m^y \sigma_m^y I_{n_m} r} \right]^{n_m/n_m+1} \bar{\varepsilon}(\theta, n_m) \tag{13.32}$$

and

$$\bar{\varepsilon}_c = \alpha_c \varepsilon_c^y \left[\frac{J_c}{\alpha_c \varepsilon_c^y \sigma_c^y I_{n_c} r} \right]^{n_c/n_c+1} (\varepsilon(\theta, n_c) \tag{13.33}$$

where subscripts m and c represent matrix and composite, respectively, superscript y represents yield stress, J_m and J_c are path-independent parameters (J integrals) that vary with the applied load, crack length, and the geometry of the specimen, I_{n_m} and I_{n_c} are numerical constants that depend on the stress–strain relationship (constitutive equations) of the material, $\bar{\varepsilon}(\theta, n_m)$ and $\varepsilon(\theta, n_c)$ are also numerical constants related to the angle away from the crack plane at a particular n value, and α_m, α_c, n_m, and n_c are constants in the Ramberg–Osgood stress–strain relation, which is given by

$$\frac{\varepsilon}{\varepsilon_Y} = \frac{\sigma}{\sigma_y} + \alpha \left[\frac{\sigma}{\sigma Y} \right]^n \tag{13.34}$$

where σ_y and ε_y are yield stress and yield strain, respectively, n is the strain hardening exponent, and α is a dimensionless material constant. The second term on the right-hand side of Eq. (13.34) describes the plastic or nonlinear behavior. By assuming that the stress–strain behavior is the same in both the matrix and composite, i.e., $\alpha_m = \alpha_c = \alpha$ and $n_m = n_c = n$, the following expression is obtained by dividing Eq. (13.33) by Eq. (13.32):

$$\frac{\bar{\varepsilon}_c}{\bar{\varepsilon}_m} = \frac{\varepsilon_c^y}{\varepsilon_m^y} \left[\frac{J_c}{J_m} \right]^{n/n+1} \left[\frac{\varepsilon_m^y \sigma_m^y}{\varepsilon_c^y \sigma_c^y} \right]^{n/n+1} \tag{13.35}$$

By invoking $\varepsilon_c^y = \sigma_c^y / E_c$, $\varepsilon_m^y = \sigma_m^y / E_m$, $J_c = (1 - \nu_c^2) K_c^2 / E_c$, and $J_m = (1 - \nu_m^2) K_m^2 / E_m$ and rearranging both sides of the resulting equation, Eq. (13.35) reduces to

$$\frac{K_c}{K_m} = \left[\frac{\sigma_c^y}{\sigma_m^y}\right]^{n-1/2n}\left[\frac{\varepsilon_| c}{\bar{\varepsilon}_m}\right]^{n+1/2n}\left[\frac{E_c}{E_m}\right]^{n+1/2n} \tag{13.36}$$

The toughening ratio, which is defined as the ratio of the applied stress intensity factor to the stress intensity factor in the matrix, is thus given by

$$\lambda_{bl} = \frac{K_c}{K_m} = [1 + V_f(\Sigma - 1)]^{\frac{n-1}{2n}}[1 + V_f(\Gamma - 1)]^{\frac{n+1}{2n}}\left[\frac{E_c}{E_m}\right]^{\frac{n+1}{2n}} \tag{13.37}$$

or

$$\Delta K_{bl} = (\lambda_{bl} - 1)K_m \tag{13.38}$$

with

$$\Sigma = \frac{\sigma_y^d}{\sigma_y^m} \tag{13.39}$$

and

$$\Gamma = \frac{\varepsilon_f^d}{\varepsilon_f^m} \tag{13.40}$$

where K_∞ is the applied stress intensity factor, K_m is the matrix stress intensity factor, σ_y is the yield stress, ε_f is the fracture strain, n is the inverse of the strain hardening exponent, E_m is the matrix modulus, and superscripts m and d denote matrix and ductile phases, respectively.

In composites with brittle matrices, an elastic crack tip stress field is more likely to describe the stress distribution in the matrix (Soboyejo et al., 1997). In the plastic region, the strain field in the composite can be described by the HRR field expression given by Eq. (13.32):

$$\bar{\varepsilon}_c = \alpha_c\varepsilon_c^y\left[\frac{J_c}{\alpha_c\varepsilon_c^y\sigma_c^y I_{n_c} r}\right]^{n_c/n_c+1}\bar{\varepsilon}(\theta, n_c) \tag{13.41}$$

In the elastic region, the strain in the composite is

$$\bar{\varepsilon}_c = \frac{K_c}{E_c}\frac{1}{(2\pi r)^{1/2}}\bar{\varepsilon}(\theta) \tag{13.42}$$

where K_c and E_c are the stress intensity factor and Young's modulus of the composite, $\bar{\varepsilon}(\theta)$ is a function of orientation, and r is the distance from the crack tip. Equations (13.32) and (13.33) can be reformulated as

$$\left(\frac{\bar{\varepsilon}_c}{\alpha_c\varepsilon_c^y}\right)^{(n+1)/n} = \frac{J_c}{\alpha_c\varepsilon_c^y}\frac{1}{I}[\bar{\varepsilon}_c(\theta, n_c)]^{(n+1)/n} \tag{13.43}$$

and

$$(\bar{\varepsilon}_c)^2 = \left(\frac{K_c}{E_c}\right)^2 \frac{1}{2\pi r}\left[\bar{\varepsilon}_c(\theta)\right]^2 \quad (13.44)$$

At the junction between the elastic and plastic regions, $\bar{\varepsilon}_c = \varepsilon_c^y$. Equations (13.34) and (13.35) can be combined with

$$J_c = \frac{(1-\nu_c^2)K_c^2}{E_c} \quad (13.45)$$

leading to

$$\frac{[\bar{\varepsilon}_c(\theta, n_c)]^{(n+1)/n}}{[\bar{\varepsilon}_c(\theta)]^2}\frac{1}{I_n r}\frac{(1-\nu_c^2)}{\alpha_c(1/\alpha_c)^{(n+1)/n}} = \frac{1}{2\pi} \quad (13.46)$$

For the matrix, the near-tip strain field can be expressed as

$$(\bar{\varepsilon}_m)^2 = \left(\frac{K_m}{E_m}\right)^2 \frac{1}{2\pi r}[\bar{\varepsilon}_m(\theta)]^2 \quad (13.47)$$

Since $\bar{\varepsilon}_m(\theta) = \bar{\varepsilon}_c(\theta)$, combining Eqs (13.34–3.38) gives

$$\frac{K_c}{K_m} = \frac{E_c}{E_m}\frac{(\bar{\varepsilon}_c)^{(n+1)/2n}}{\bar{\varepsilon}_m}(\varepsilon_c^y)^{(n-1)/2n} \quad (13.48)$$

Hence, the modified blunting toughening ratio is given by

$$\lambda_b = \frac{K_\infty}{K_m} = \frac{K_c}{K_m} = \frac{E_c}{E_m}\frac{(\bar{\varepsilon}_c)^{(n+1)/2n}}{\bar{\varepsilon}_m}(\varepsilon_c^y(n-1)/2n \quad (13.49)$$

Toughening ratio estimates obtained from the modified crack tip blunting model are slightly higher than those predicted by the model of Chan (1992). This result is intuitively obvious since the HRR field expressions may overestimate the crack tip field in an elastic material. Conversely, the assumption of purely elastic behavior in the brittle matrix material is likely to underestimate the actual crack tip fields in nearly elastic materials. The most representative toughening ratios are, therefore, likely to be in between those predicted by Chan's model and the model by Soboyejo et al. (1997).

Finally, it is also interesting to note here that crack tip blunting may occur due to debonding along the interface between two phases. This is illustrated in Fig. 13.13(b). The stress redistribution associated with such debonding has been modeled by Chan (1993).

13.7 CRACK DEFLECTION

Second-phase particles located in the near-tip field of a propagating crack will perturb the crack path, as shown in Fig. 13.10, causing a reduction in the stress intensity. The role of in-plane tilting/crack deflection and out-of-plane twisting can be assessed using the approach of Bilby et al. (1977) and Cotterell and Rice (1980). The possible tilting and twisting modes are shown schematically in Figs 13.14(a)–13.14(c). An example of crack deflection by tilting and twisting around spherical niobium particles is also presented in Fig. 13.14(d).

The local Modes I, II, and III stress intensity factors K_1, K_2, and K_3 at the tip of the reflected crack can be expressed in terms of the far-field stress intensities for the equivalent linear crack (K_{I} and K_{II}). These give

$$K_1^1 = \alpha_{21}(\theta)K_1 + \alpha_{12}(\theta)K_{11} \tag{13.50}$$

$$K_2^1 = \alpha_{21}(\theta)K_1 + \alpha_{22}(\theta)K_{11} \tag{13.51}$$

where θ is the tilt angle, $\alpha_{ij}(\theta)$ are angular functions, and superscript t denotes the reduction in crack driving force due to tilting through an angle, θ. The corresponding expressions for K_1 and K_3, due to twisting through an angle θ, are

$$K_1^{\mathrm{T}} = b_{11}(\phi)K_1^t + b_{12}(\phi)K_2^1 \tag{13.52}$$

$$K_3^{\mathrm{T}} = b_{31}(\phi)K_1^{\mathrm{t}} + b_{32}(\phi)K_{32}^{\mathrm{t}} \tag{13.53}$$

where the angular functions $b_{ij}(\theta)$ are given. The effective crack driving force in terms of energy release rate G or K_{eff} is, therefore, given by

$$EG = K_1^2(1-\nu^2) + K_2^2(1-\nu^2) + K_3^2(1+\nu^2) \tag{13.54}$$

or

$$K_{\mathrm{eff}} = \sqrt{EG} \tag{13.55}$$

For continuous crack deflection at all possible angles around spherical reinforcements [Fig. 13.14(d)], the toughening from pure tilt-induced deflection is given by Faber and Evans (1983):

$$G_{\mathrm{c}} = G_{\mathrm{m}}(1 + 0.87V_{\mathrm{f}}) \tag{13.56}$$

Assuming linear elastic behavior of the composite, this yields:

$$K_{\mathrm{c}} = \sqrt{E_{\mathrm{c}}G_{\mathrm{c}}} \tag{13.57}$$

where K_{c} is the fracture toughness of the composite and E_{c} is Young's modulus of the composite.

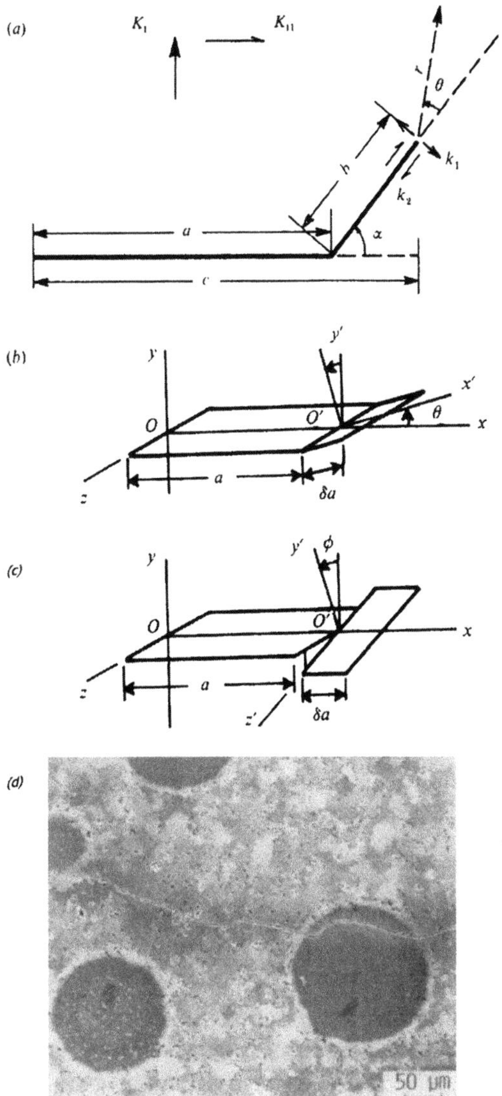

Figure 13.14 Crack deflection mechanisms: (a) schematic of putative kink; (b) deflection by tilting; (c) deflection by twisting; (d) tilting and twisting around spherical Nb particles in an $MoSi_2$/Nb composite. (From Venkateswara Rao et al., 1992.)

Furthermore, crack-tip shielding due to deflection by pure tilting can be estimated from the Modes I and II stress intensity factors, K_I and K_{II}, induced at the crack tip as a result of crack deflection through an angle ϕ (Suresh, 1985). For a crack with equal undeflected and deflected segments, this results in K_I and K_{II} values that are given by a simple geometrical model by Suresh (1985) to be

$$K_I = \cos^2(\phi/2)K_\infty \tag{13.58}$$

and

$$K_{II} = \sin(\phi/2)\cos^2(\phi/2)K_\infty \tag{13.59}$$

where K_∞ is the applied stress intensity factor, and ϕ is deflection angle.

Assuming that crack growth is driven purely by K_I, Eq. (13.58) can be rearranged to obtain the following expression for λ_d, the toughening ratio due to crack deflection:

$$\lambda_d = \frac{K_\infty}{K_m} = \frac{1}{\cos^2(\phi/2)} \tag{13.60}$$

Equation 13.58 is a useful expression to remember for simple estimation of the toughening due to crack deflection. It shows clearly that the toughening due to deflection by pure tilting is relatively small with the exception of the tilting cases when angles are large (> 45°). Experimental evidence of toughening due to crack deflection is presented in work by Suresh (1985), Claussen (1990), and Venkateswara Rao et al. (1992).

13.8 TWIN TOUGHENING

The presence of twin process zones around cracks in gamma titanium aluminide alloys exerts stress fields that will act in opposition to the stress intensity factors at the crack tips, causing a shielding effect. This phenomenon is known as twin toughening and may be associated with nonlinear deformation in the crack process zone. The concept of twin toughening is illustrated schematically in Fig. 13.15. The twin toughening ratio, λ_t, is defined as the stress intensity at the crack tip with the presence of a twin process zone, K_t, divided by the stress intensity in the absence of deformation-induced twinning, K_0. This is given by Mercer and Soboyejo (1997) to be

$$l_t = K_t/K_0 = (1 + V_t E \Delta G_{SS}/K_0^2)^{\frac{1}{2}} \tag{13.61}$$

Twin toughening appears to be very promising and Dève and Evans (1991), report two-fold toughness increases due to the presence of deformation-induced twinning around the crack tip. The original Dève and Evans

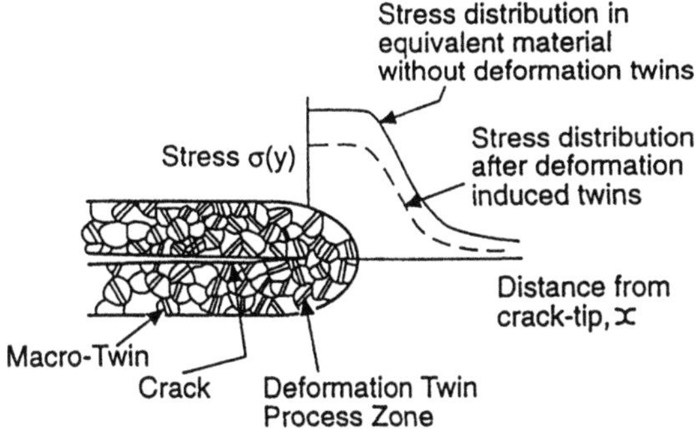

FIGURE 13.15 Schematic of twin toughening. (From Soboyejo et al., 1995a.)

model (1991) has also been extended in later work by Soboyejo et al. (1995a) and Mercer and Soboyejo (1997).

13.9 CRACK TRAPPING

The toughness enhancement due to crack-trapping (Fig. 13.16) has been considered theoretically by Rose (1975, 1987), Gao and Rice (1990), and Bower and Ortiz (1991). Briefly, the crack front is pinned at several points by particles/fibers that are tougher than the matrix. This results in a "bowed" crack front, with the stress intensity factor along the bowed segments being lower than the far-field stress intensity factor, while the stress intensity factor at the pinned points is correspondingly larger. Crack-growth occurs when the stress intensity factor at the pinned points exceeds the fracture toughness of the reinforcements.

Apart from the fracture toughness, the important variable that determines the shape of the crack front and the overall fracture toughness is the volume fraction of the reinforcement and the average distance between trapping points. Rose (1975) derived the following expression for the toughening due to crack trapping when the obstacles are penetrated by the crack, i.e., when $K_c^{par} < 3K_c^{mat}$:

$$\lambda_{tr} = \frac{K_\infty}{K_{Ic}^m} = \left\{1 + \frac{2R}{L}\left[\left(\frac{K_{Ic}^{par}}{K_{Ic}^m}\right)^2 - 1\right]\right\}^{1/2} \tag{13.62}$$

where K_∞ is the fracture toughness of the composite, K_{Ic}^m is the fracture toughness of the matrix, R is the radius of the reinforcement, L is the average distance between the centers of adjacent pinning points (typically the distance between the particles where the crack is trapped, but not neces-

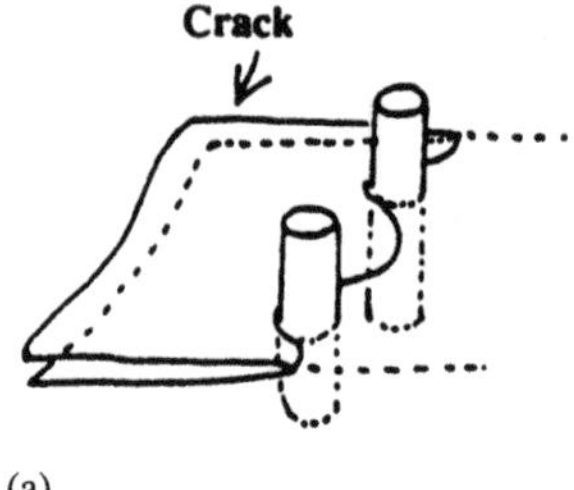

(a)

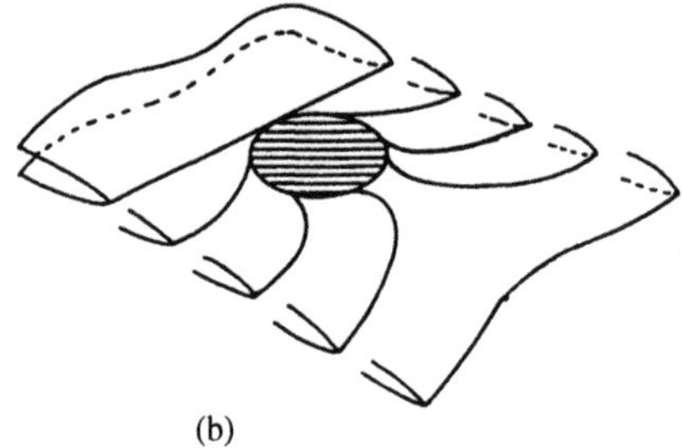

(b)

Figure 13.16 Schematic illustration of (a) crack trapping (From Argon et al., 1994) and (b) crack/reinforcement interactions that give rise to combination of crack bridging and crack trapping mechanisms.

sarily the particle spacing, since not all particles may trap the crack front if the crack front is not planar), and $K_{\mathrm{Ic}}^{\mathrm{par}}$ is the fracture toughness of the reinforcement.

In the case where the cracks do not penetrate the particles ($K_{\mathrm{c}}^{\mathrm{par}}/K_{\mathrm{c}}^{\mathrm{mat}} \gg 3$), both crack bridging and crack trapping occur (Fig. 13.16b) simultaneously. Under these conditions, Bower and Ortiz (1991) have shown that

$$\frac{K^{\infty}}{K_{\mathrm{c}}^{\mathrm{mat}}} = 3.09\frac{R}{L}\frac{K_{\mathrm{c}}^{\mathrm{par}}}{K_{\mathrm{c}}^{\mathrm{mat}}} \tag{13.63}$$

where R is the particle radius, L is the particle spacing, $K_{\mathrm{c}}^{\mathrm{par}}$ is the particle toughness, and $K_{\mathrm{c}}^{\mathrm{mat}}$ is the material toughness. Equation (13.62) applies to all ratios of R/L, while Eq. (13.63) only applies to $R/L < 0.25$. Experimental evidence of crack trapping has been reported by Argon et al. (1994) for toughening in transparent epoxy reinforced with polycarbonate rods. Evidence of crack trapping has also been reported by Ramasundaram et al. (1998) for NiAl composites reinforced with Mo particles, while Heredia et al. (1993) have reported trapping by Mo fibers in

NiAl/Mo composites. In these studies, the models were found to provide reasonable estimates of the measured toughening levels in model composites.

13.10 MICROCRACK SHIELDING/ANTISHIELDING

Depending on the spatial configurations of cracks, either shielding or antishielding may occur due to the effects of microcrack distributions on a dominant crack, Figs 13.1(b) and 13.17. The shielding or antishielding due to distributions of microcracks has been modeled by a number of investigators (Kachanov (1986); Rose, 1986; Hutchinson, 1987). In cases where a limited number of microcracks with relatively wide separations are observed ahead of a dominant crack, the shielding effects of the microcracks can be treated individually using Rose's analysis (1986). This predicts that the change in the stress intensity of the dominant crack, ΔK_{i}, is given by

$$\Delta K_{\mathrm{i}} = \Delta\sigma_{\mathrm{i}}^{\mathrm{P}}\sqrt{2\pi R_1} \quad F_{\mathrm{i}}(S, R_1, \theta_1, \alpha_1, \Delta\sigma_{\mathrm{i}}^{\mathrm{P}}) \tag{13.64}$$

where $2S$ is the length of the microcrack located at radial distance, R_1, and an angle of θ_1, from the crack tip, α_1, is the orientation of the microcrack with respect to the stress axis, and the values of $F_{\mathrm{i}}(S, R_1, \theta_1, \alpha_1, \Delta\sigma_{\mathrm{i}}^{\mathrm{P}})$ are given by mathematical expressions due to Rose (1986). The toughening ratio due to microcracking λ_{m} is given by

$$\lambda_{\mathrm{m}} = \frac{K_\infty}{K} = \frac{K_\infty}{\left[(K_\infty + \Delta K_1^{\mathrm{p}})^2 (\Delta K_1^{\mathrm{p}})^2\right]^{\frac{1}{2}}} \tag{13.65}$$

where in Eq. (13.65) is less than unity, then microcracks ahead of the crack tip result in antishielding, i.e., lower fracture toughness. Conversely, values of λ_m greater than unity result in toughening by the microcracks. The extent of antishielding or extrinsic toughening depends largely on the angle between the microcracks and the dominant crack. Experimental evidence of microcrack toughening has been reported by Rühle et al. (1987) and Bischoff and Rühle (1986).

13.11 LINEAR SUPERPOSITION CONCEPT

In cases where multiple toughening mechanisms operate, the total toughening ratio can be estimated by applying the principle of linear superposition This neglects possible interactions between the individual toughening mechanisms. Hence, for toughening by transformation toughening, crack bridging, crack-tip blunting, and crack deflection, the overall toughening is given by

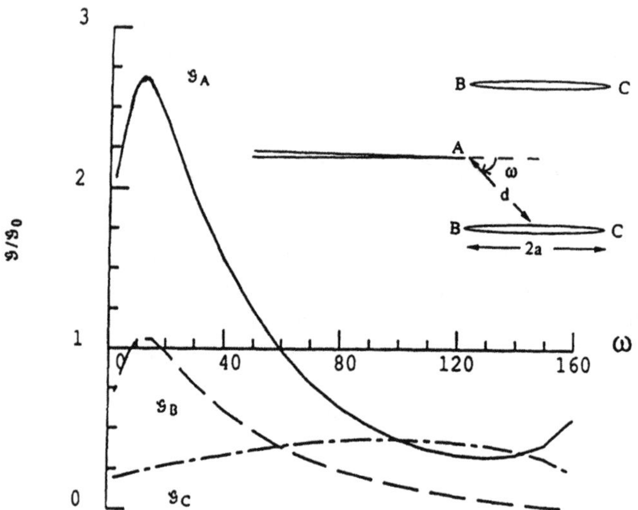

FIGURE 13.17 Schematic of microcrack zone around dominant crack. (From Shin and Hutchinson, 1989.)

$$\lambda_{tot} = \lambda_T + (\lambda_b - 1) + (\lambda_{bl} - 1) + (\lambda_{dl} - 1) \tag{13.66}$$

or

$$\Delta K_{tot} = \Delta K_T + \Delta K_b + \Delta K_{bl} + \Delta K_{dl} \tag{13.67}$$

13.12 SYNERGISTIC TOUGHENING CONCEPT

In cases where multiple toughening mechanisms operate, the total toughening can be estimated from the sum of the contributions due to each mechanism. Such linear superposition concepts neglect the possible interactions between individual mechanisms. However, synergistic interactions between individual toughening mechanisms may promote a greater degree of toughening than the simple sum of the toughening components (Amazigo and Budiansky, 1988).

Amazigo and Budiansky (1988) conducted an original theoretical study of the possible interactions between toughening effects of crack bridging and transformation toughening. They showed that it is possible to induce synergy, depending on the parametric ranges of bridging and transformation toughening. In cases where the interaction is synergistic, the overall increase in toughness is greater than the sum of the toughening due to crack bridging and transformation toughening alone. The interaction of crack bridging and transformation toughening is characterized in terms of the following key parameters:

1. The modified toughening ratio due to particulate crack bridging is given by

$$\Lambda_p = \frac{\lambda_p}{\sqrt{1-c}} = \frac{K_p}{K_m\sqrt{1-c}} \tag{13.68}$$

where λ_p is the toughening ratio due to crack bridging, c is the volume concentration of ductile particles, K_p is the increased toughness due to bridging by ductile particulate reinforcements, and K_m is the fracture toughness of the matrix.

2. The toughening ratio due to transformation toughening is given by

$$\lambda_T = \frac{K_T}{K_m} \tag{13.69}$$

where K_T is the increased toughness from transformation toughening.

3. The combined modified toughening ratio is now given by

$$\Lambda = \frac{K}{K_m\sqrt{1-c}} \tag{13.70}$$

where K is the total increased toughness of the composite system.

4. The degree of synergy may be assessed through a coupling parameter, ρ:

$$\rho = \frac{(1+\nu)cS}{\sigma_m^c} \tag{13.71}$$

where ν is Poisson's ratio, S is the strength of the ductile particle, and σ_m^c is the mean stress in transformation zone. This coupling parameter governs the interaction between particulate and transformation toughening when they occur simultaneously during steady crack growth. Representative numerical results are presented in Fig. 13.18 for Λ versus λ_T, for $\Lambda_p = 2$ and 4, respectively. The individual curves in each figure are for selected values of the coupling parameter in the range (0–∞). The limiting results for $\rho = \infty$ and 0 are of special interest:

$$\Lambda = \Lambda_p \lambda_T \quad \text{for } \rho \to \infty \tag{13.72}$$

and

$$\Lambda = \left[\Lambda_p^2 + \lambda_T^2\right]^{\frac{1}{2}} \text{ for } \rho \to 0 \tag{13.73}$$

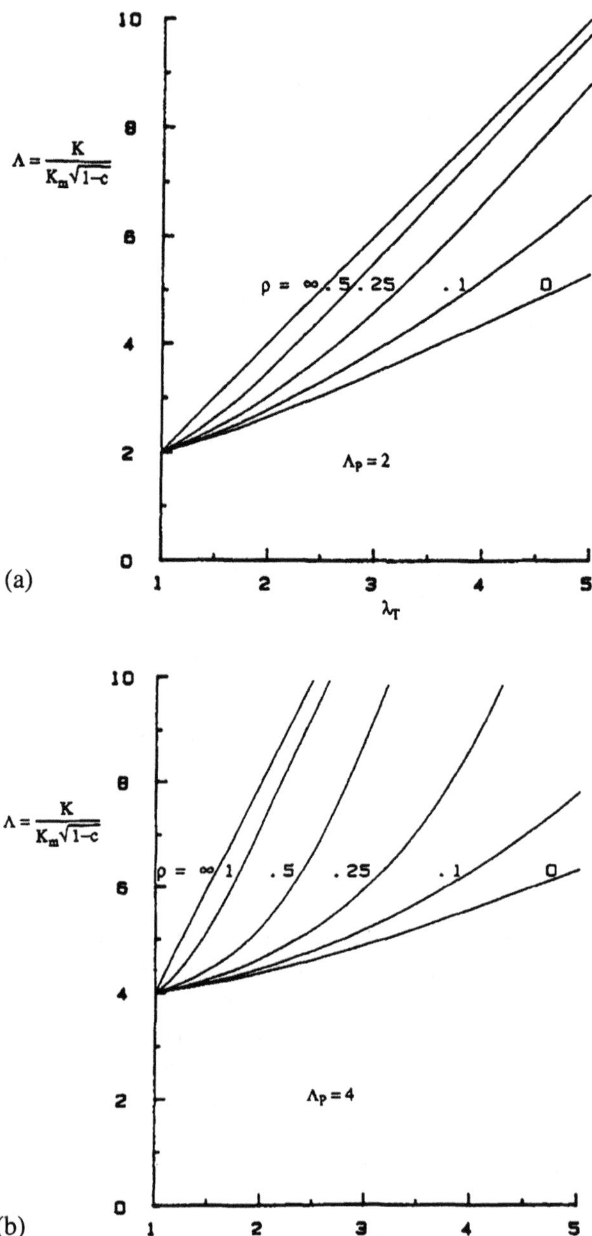

FIGURE 13.18 Coupling parameter, ρ, for: (a) $\Lambda_p = 2$, and (b) $\Lambda_p = 4$. (From Amazigo and Budiansky, 1988b.)

The anticipated product rule $\lambda = \lambda_p \lambda_T$ for the combined toughening ratio holds in the first limiting case. For sufficiently large finite values of ρ, bridging and transforming particles interact synergistically, producing a larger increase in the fracture toughness than the sum of the individual contributions from crack bridging and transformation toughening. On the other hand, for ρ near zero, the combined toughness can be substantially less than the cumulative.

Due to difficulties in obtaining sufficient information to calculate the above coupling parameter, ρ, a new coupling parameter, η, is defined as

$$\eta = \frac{H_T(1 - c)}{L_p} \tag{13.74}$$

where H_T is the transformation zone height and L_p is the bridging length for pure particulate toughening. These are process zone characteristics that are relatively easy to measure. Hence, η is relatively easy to calculate compared to ρ. An appropriate choice for can then be made on the basis of observations of separate toughening mechanisms. For quite small values of η, it is enough to provide results close to those for $\eta = \rho = \infty$, for which the synergistic product rule applies. It means that synergism is not precluded despite the fact that transformation toughening zone heights tend to be less than particulate bridging lengths.

Another study by Cui and Budiansky (1993) has also shown that transforming particles and aligned fibers may interact synergistically to increase the effective fracture toughness of a brittle matrix containing a long, initially unbridged crack. The results obtained from this analysis are qualitatively similar to those obtained from the model by Amazigo and Budiansky (1988). Recently, Ye et al. (1999) have shown that the synergy is associated with increased levels of stress-induced phase transformations induced by the layer tractions (Figs. 13.19 and 13.20). Li and Soboyejo (2000) have also confirmed that the Amazigo and Budiansky model provides good predictions of toughening in synergistically toughened NiAl composites reinforced with molybdenum particles and zirconia particles that undergo stress-induced phase transformations.

13.13 TOUGHENING OF POLYMERS

Before closing, it is important to make a few comments on the toughening mechanisms in polymers. Significant work in this area has been done by Argon and coworkers (1994, 1997, 2000). There has also been some recent work by Du et al. (1998). The work of Argon et al. (1994) has highlighted the importance of stress-induced crazing in cross-linked thermosetting poly-

(a)

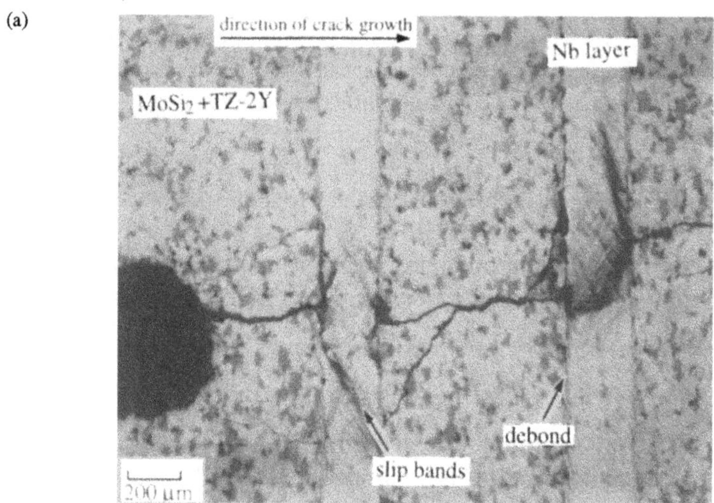

(b)

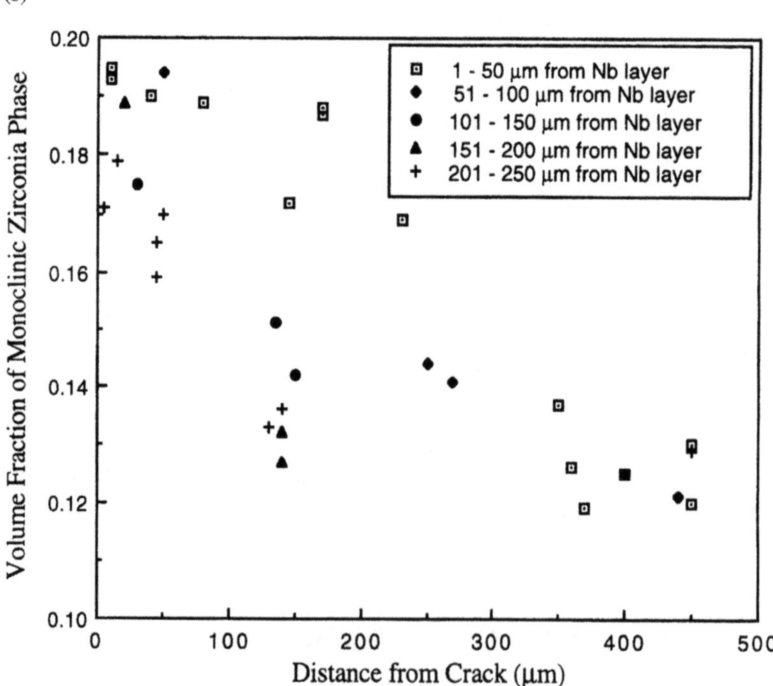

FIGURE 13.19 (a) Crack bridging in hybrid $MoSi_2$ + Nb layer (20 vol %) + 2 mol% Y_2O_3-stabilized ZrO_2 (20 vol %) composite; (b) transformation toughening volume fraction of transformed phase as a function of distance in the same composite. (From Ye et al., 1999.)

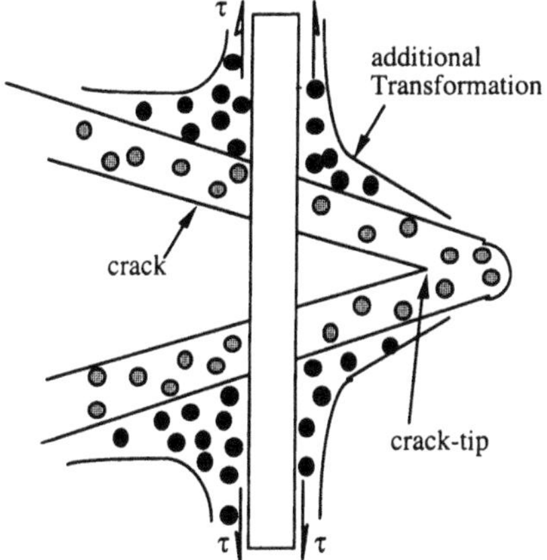

● Prior Crack-Tip Stress-Induced Transformation

● Additional Transformation Induced by Bridging Tractio

FIGURE 13.20 Schematic illustration of sources of synergism due to interactions between transformation toughening and crack bridging. (Courtesy of Dr. Fan Ye.)

mers (such as epoxies) and certain flexible-chain thermoelastic polymers. When these materials undergo crazing, they exhibit brittle behavior (tensile brittleness and relatively low fracture toughness) in tension and significant plastic deformation in compression.

Argon et al. (1994) have also shown that the crack trapping mechanism is particularly attractive for the toughening of epoxies. This can be achieved by the use of well-adhering particles that cause the cracks to bow around the fibers, Fig. 13.16(a). Also, crazable polymers can be toughened by decreasing the craze flow stress below the threshold strength defined by the size of the inclusions. The interested reader is referred to papers by Argon and coworkers (1994, 1997, 2000).

13.14 SUMMARY AND CONCLUDING REMARKS

This chapter has presented an overview of the current understanding of toughening mechanisms. Whenever appropriate, the underlying mechanics of crack tip shielding, or antishielding, has been presented to provide some basic expressions for scaling purposes. The expressions are intended to be approximate, and are obtained generally under idealized conditions. Hence, they must be used with this in mind. Nevertheless, the expressions presented in this chapter should serve as useful tools in the microstructural design of toughened materials.

After a brief review of basic composite concepts, the following toughening mechanisms were examined in this chapter: transformation toughening; crack bridging; crack-tip blunting; crack deflection; twin toughening; crack trapping; and microcrack shielding/antishielding. The linear superposition of multiple toughening mechanisms was then discussed before presenting an introduction to possible synergisms that can be engineered via interactions between multiple toughening mechanisms. Finally, a qualitative discussion of polymer toughening mechanisms was presented.

BIBLIOGRAPHY

Argon, A.S. (2000) In: A. Kelly and C. Zweben. Comprehensive Composite Materials. vol. 1 (ed. by T.-W. Chou). Pergamon Press, Oxford, UK, pp 763–802.

Argon, A.S. and Shack, W.J. (1975) Fiber-Reinforced Cement and Concrete. The Construction Press Ltd, Lancaster, UK, p 39.

Argon, A.S., Cohen, R.E., and Mower, T.M. (1994) Mater Sci Eng. vol. A176, pp 79–80.

Argon, A.S., Bartczak, Z., Cohen, R.E., and Muratoglu, O.K. (1997) Novem mechanisms for toughening in semi-crystalline polymers. Proceedings of the 18th Risø International Symposium on Materials Science: Polymeric Composites—Expanding the Limits. Risø National Laboratory, Roskilde, Denmark.

Ashby, M.J., Blunt, F.J., and Bannister, M. (1989) Acta Metall. vol. 37, pp 1847–1857.

Becher, P.F. (1986) Acta Metall. vol. 34, pp 1885–1891.

Bilby, B.A. Cardhew, G.E., and Howard, I.E. (1977) Stress intensity factors at the tips of kinked and forked cracks. In: Fracture 1977, vol. 3, ICF4 Waterloo, Canada. Pergamon Press, Oxford, p 197–200.

Bischoff, E. and Rühle, M. (1986) Microcrack and transformation toughening of zirconia-containg alumina. Science and Technology of Zirconia III. In: S. Somiya, N. Yamamoto and H. Hanagida, eds. A Advances in Ceramics, vol. 24 A&B. American Ceramic Society, Westerville, OH, p 635.

Bloyer, D.R., Venkateswara Rao, K.T., and Ritchie, R.O. (1996) Mater Sci Eng. vol. A216, pp 80–90.

Bloyer, D.R., Venkatewara Rao, K.T., and Ritchie, R.O. (1998) Metall Mater Trans. vol. 29A, pp 2483–2495.

Bower, A.F. and Ortiz, M. (1991) J Mech Phys Solids. vol. 39, pp 815–858.

Budiansky, B. and Amazigo, J.C. (1997). In: J.R. Willis, ed. Non-linear Analysis of Fracture, Kluwer, Dordrecht, The Netherlands.

Budiansky, B. and Truskinovsky, L. (1993) J Mech Phys Solids. vol. 41, pp 1445-1459.

Budiansky, B., Hutchinson, J.W., and Lambropoulos, J.c. (1983) Int J Solids Struct, vol. 19, pp 337–355.
Budiansky, B., Amazigo, J., and Evans, A.G. (1988) J Mech Phys Solids. vol. 36, p 167.
Cao, H.c., Dalgleish, B.J., Dève, H.E., Elliott, C., Evans, A.G., Mehrabian, R., and Odette, G.r. (1989) Acta Metall. vol. 37, pp 2969–2977.
Chan, K.S. (1992) Metall Trans. vol. 23A, pp 183–199.
Charalambides, P.G. and Evans, A.G. (1989) J Am Ceram Soc. vol. 72, pp 746–753.
Chen, I.W. and Reyes-Morel, P.E. (1986) J Am Ceram Soc vol. 69, pp 181–189.
Chen, I.W. and Reyes-Morel, P.E. (1987) Transformation plasticity and transformation toughening in Mg-PSZ and Ce-TZP. Proceedings of Materials Research Society Symposium, Proc. 78, Boston, MA, pp 75–78.
Claussen, N. (1990) Ceramic platelet composite. In: J.J. Bentzen et al., eds. Structural Ceramics Processing, Microstructure and Properties. Risø National Laboratory, Roskilde, Denmark, pp 1–12.
Cotterell, B. and Rice, J.R. (1980) Int J Fract. vol. 16, pp 155–169.
Cox, B.N. and Lo, C.S. (1992) Acta Metall Mater. vol. 40, pp 69–80.
Cox, B.N. and Rose, L.R.F. (1996) Mech Mater. vol. 22, pp 249–263.
Cui, Y.L. and Budiansky, B. (1993) J Mech Phys Solids. vol. 41, pp 615–630.
Dève, H. and Evans, A.G. (1991) Acta Metall Mater. vol. 39, p 1171.
Du, J., Thouless, M.D., and Yee, A.F. (1998) Int J Fract. vol. 92, pp 271–285.
Du, J., Thouless, M.D., and Yee, A.F. (2000) Acta Mater. vol. 48, pp 3581–3592.
Evans, A.G. (1990) J Am Ceram Soc. vol. 73, pp 187–206.
Evans, A.G. and Cannon, R.M. (1986) Acta Metall. vol. 34, pp 761–860.
Evans, A.G. and Heuer, A. (1980) J Am Ceram Soc. vol. 63, pp 241–248.
Faber, K.T. and Evans, A.G. (1983) Acta Metall. vol. 31, pp 565–576, 577–584.
Fett, T. and Munz, D. (1994) Stress Intensity Factors and Weight Functions for One Dimensional Cracks. Institut fur Materialforschung, Kernforschungzentrum, Karlsuhe, Germany.
Gao, H. and Rice, J.R. (1990) J of Appl Mech. vol. 56, pp 828–836.
Garvie, R.C., Hannink, R.H.J., and Pascoe, R.T. (1975) Nature. vol. 258, pp 703–704.
Heredia, F., He, M.Y., Lucas, G.E., Evans, A.G., Déve, H.E., and Konitzer, D. (1993) Acta Metall Mater. vol. 41, pp 505–511.
Hom, C.L. and McMeeking, R.M. (1990) Int J Solids Struct. vol. 26, pp 1211–1223.
Huang, X., Karihaloo, B.L., and Fraser, W.B. (1973) Int J Solids Struct. vol. 30, pp 151–160.
Hutchinson, J.W. (1987) Acta Metall. vol. 35, pp 1605–1619.
Hutchinson, J.W. and Jenson, H.M. (1990) Models for Fiber Debonding and Pullout in Brittle composites with Fraction. Harvard University, Cambridge, MA, Rep. no. MECH-157.
Kachanov, M. (1986) Int J Fract. vol. 30, p R65–R72.
Kajuch, J., Short, J., and Lewandowski, J.J. (1995) Acta Metall Mater. vol. 43, pp 1955–1967.

Karihaloo, B. (1990), Contribution to the t→m Phase Transformation of ZTA. In: J.J. Bentzen et al. eds, Structural Ceramics Processing. Microstructure and Properties. Risø, Denmark, pp 359–364.

Lambropoulos, J.C. (1986) Int J Solids Struct. vol. 22, pp 1083–1106.

Lange, F.F. (1982) J Mater Sci (Parts 1–5). vol. 17, pp 225–263.

Lawn, B. (1993) Fracture of Brittle Solids, 2nd ed. Cambridge University Press, Cambridge, UK.

Li, M. and Soboyejo, W.O. (1999) Mater Sci Eng. vol. A271, pp 491–495.

Li, M. and Soboyejo, W.O. (2000) Metall Mater Trans. vol. 31A, pp 1385–1399.

Li, M., Schaffer, H., and Soboyejo, W.O. (2000) J Mater Sci. vol. 35, pp 1339–1345.

Lou, J. and Soboyejo, W.O. (2001), Metall Mater Trans. vol. 32A, pp 325–337.

Marshall, D.B. and Evans, A.G. (1985) J Am Ceram Soc. vol. 68, p 225.

Marshall, D.B., Evans, A.G., and Drory, M. (1983) Transformation toughening in Ceramics. In: R.C. Bradt et al. eds. Fracture Mechanics of Ceramics. vol. 6. Plenum Press, New York, pp 289–307.

Marshall, D.B., Cox, B.N., and Evans, A.G. (1985) Acta Metall. vol. 11, pp 2013–2021.

Marshall, D.B., Shaw, M.C., Dauskardt, R.H., Ritchie, R.O., Readey, M.J., and Heuer, A.H. (1990) J Am Ceram Soc. vol. 73, pp 2659–266.

McMeeking, R.M. and Evans, A.G. (1982) J Am Ceram Soc. vol. 65, pp 242–246.

Mercer, C. and Soboyejo, W.O. (1997) Acta Metall Mater. vol. 45, pp 961–971.

Muju, S., Anderson, P.M., and Mendelsohn, D.A. (1998) Acta Mater. vol. 46, pp 5385–5397.

Odette, G.R., Chao, B.L., Sheckhard, J.W., and Lucas, G.E. (1992). Acta Metall Mater. vol. 40, p 2381.

Ramasundaram, P., Bowman, R., and Soboyejo, W.O. (1998) Mater Sci Eng. vol. A248, pp 132–146.

Reyes-Morel, P.E. and Chen, I.W. (1988) J Am Ceram Soc. vol. 71, pp 343–353.

Rose, L.R.F. (1975) Mech Mater. vol. 6, p 11.

Rose, L.R.F. (1986) Int J Fract. vol. 31, pp 233–242.

Rose, L.R.F. (1987) J Mech Phys Solids. vol. 35, p 383–403.

Rühle, M., Evans, A.G., McMeeking, R.M., and Charalambides, P.G. (1987) Acta Metall mater. vol. 35, pp 2701–2710.

Shaw, L. and Abbaschian, R. (1994) Acta Metall Mater. vol. 42, pp 213–223.

Shum, D.K.M. and Hutchinson, J.W. (1989) On Toughening by Microcracks. Rep no. Harvard University, Cambridge, MA, MECH-151.

Sigl, L.S. and Evans, A.G. (1988) Effects of Residual Stress and Frictional Sliding on Cracking and Pull-out in Brittle Matrix Composites. University of California, Santa Barbara, CA, Rep. M88–20.

Sih, S.C. (1973) Handbook of Stress Intensity Factors. Lehigh University, Bethlehem, PA.

Simha, N. and Truskinovsky, L. (1994) Acta Metall Mater. vol. 42, p 3827.

Sneddon, I.N. and Lowengrub, M. (1969) Crack Problems in the Classical Theory of Elasticity. John Wiley, New York.

Soboyejo, W.O., Rabeeh, B., Li, Y-L., Chu, W., Larentenyev, A., and Rokhlin, S. (1987). Metall Mater Trans. vol. 27A, pp 1667–1687.
Soboyejo, W.O., Aswath, P.B., and Mercer, C. (1995a) Scripta Metall Mater. vol. 33, pp 1169–176.
Soboyejo, W.O., Brooks, D., Chen, L.C., and Lederich, R.J. (1995b) Am Ceram Soc. vol. 78, pp 1481–1488.
Sobeyojo, W.O., Ye, F., Chen, L-c., Bahtishi, N., Schwartz, D.S., and Lederich, R.J. (1996) Acta Metall Mater. vol. 44, pp 2027–2041.
Soboyejo, W.O., Aswath, P.B. and Xu, L. (1997) J Mater Sci. vol. 32, pp 5833–5847.
Stam, G. (1994) A micromechanical approach to transformation toughening in ceramics, PhD thesis, Delft University. Delft, The Netherlands.
Stump, D.M. (1991) Phil Mag. vol. A64, pp 879–902.
Stump, D.M. and Budiansky, B. (1988a) Int J Solids Struct. vol. 28, pp 669–689.
Stump, D.M. and Budiansky, B. (1988b) Int J Solids Struct. vol. 25, pp 635–646.
Sun, Q.P., Hwang, K.C., and Yu, S.U. (1991) J Mech Phys Solids. vol. 39, pp 507–524.
Suo, Z., Ho, S., and Gong, X. (1993) J Eng Mater Technol. vol. 115, pp 319–326.
Suresh, S. (1985) Metall Trans. vol. 16A, pp 249–260.
Tada, H., Paris, P.C., and Irwin, G.R. (1999) Stress Analysis of Cracks Handbook. ASME, New York.
Thouless, M.D. and Evans, A.G. (1988) Acta Metall. vol. 36, pp 517–522.
Venkateswara Rao, K.T., Soboyejo, W.O., and Ritchie, R.O. (1992) Metall Mater Trans. vol. 23A, pp 2249–2257.
Ye, F., Li, M., and Soboyejo, W.O. (1999) J Am Ceram Soc. vol. 82, pp 2460–2464.
Zok, F. and Hom, C.L. (1990) Acta Metall Mater. vol. 38, pp 1895–1904.

14

Fatigue of Materials

14.1 INTRODUCTION

Fatigue is the response of a material to cyclic loading by the initiation and propagation of cracks. Fatigue has been estimated to account for up to 80–90% of mechanical failures in engineering structures and components (Illston et al., 1979). It is, therefore, not surprising that a considerable amount of research has been carried out to investigate the initiation and propagation of cracks by fatigue. A summary of prior work on fatigue can be found in a comprehensive text by Suresh (1999). This chapter will, therefore, present only a general overview of the subject.

The earliest work on fatigue was carried out in the middle of the 19th century, following the advent of the industrial revolution. Albert (1838) conducted a series of tests on mining cables, which were observed to fail after being subjected to loads that were below the design loads. However, Wöhler (1858–1871) was the first to carry out systematic investigations of fatigue. He showed that fatigue life was not determined by the maximum load, but by the load range. Wohler proposed the use of S–N curves of stress amplitude, S_a (Fig. 14.1), or stress range, ΔS (Fig. 14.1), versus the number of cycles to failure, N_f, for design against fatigue. Such data are still obtained from machines of the type shown in Fig. 14.2. He also identified a

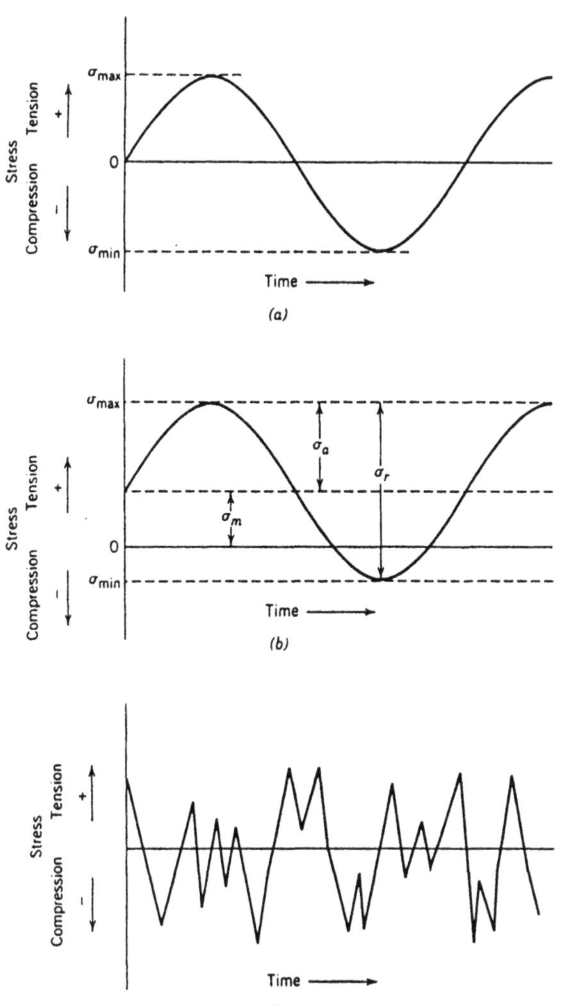

FIGURE 14.1 Basic definitions of stress parameters that are used in the characterization of fatigue cycles. (From Callister, 2000—reprinted with permission of John Wiley & Sons.)

"fatigue limit" below which smooth specimens appeared to have an infinite fatigue life.

Rankine (1843) of mechanical engineering fame (the Rankine cycle) noted the characteristic "brittle" appearance of material broken under repeated loading, and suggested that this type of failure was due to recrys-

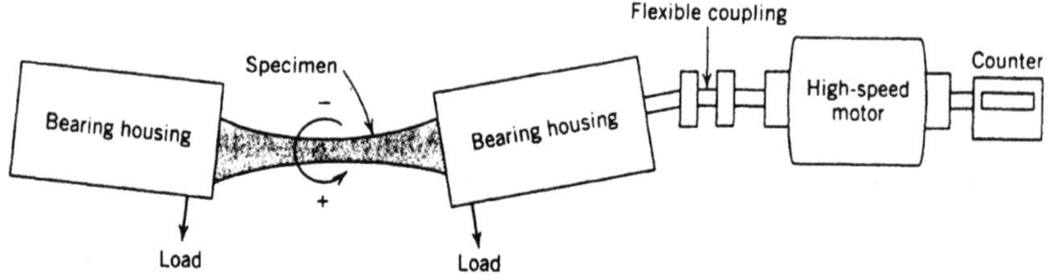

Figure 14.2 Schematic of rotating bending test machine. (From Keyser 1973—reprinted with permission of Prentice-Hall, Inc.)

tallization. The general opinion soon developed around this concept, and it was generally accepted that because these failures appeared to occur suddenly in parts that had functioned satisfactorily over a period of time, the material simply became "tired" of carrying repeated loads, and sudden fracture occurred due to recrystallization. Hence, the word fatigue was coined (from the latin word "fatigare" which means to tire) to describe such failures.

This misunderstanding of the nature of fatigue persisted until Ewing and Humfrey (1903) identified the stages of fatigue crack initiation and propagation by the formation of slip bands. These thicken to nucleate microcracks that can propagate under fatigue loading. However, Ewing and Humfrey did not have the modeling framework within which they could analyze fatigue crack initiation and propagation. Also, as a result of a number of well-publicized failures due to brittle fracture (Smith, 1984), the significance of the pre-existence of cracks in most engineering structures became widely recognized. This provided the impetus for further research into the causes of fatigue crack growth.

As discussed earlier in Chapter 11, Irwin (1957) proposed the use of the stress intensity factor (SIF) as a parameter for characterizing the stress and strain distributions at the crack tip. The SIF was obtained using a representation of the crack-tip stresses, proposed initially by Westergaard (1939) for stresses in the vicinity of the crack tip. It was developed for brittle fracture applications, and was motivated by the growing demands for developments in aerospace, pressure vessels, welded structures, and in particular from the U.S. Space Program. This led to the rapid development of fracture mechanics, which has since been applied to fatigue crack growth problems.

Paris et al. (1961) were the first to recognize the correlation between fatigue crack growth rate, da/dN, and the stress intensity range, ΔK. Although the work of Paris et al. (1961) was rejected initially by many of the leading researchers of the period, it was soon widely accepted by a global audience of scientists and engineers. Paris and Erdogan (1963) later showed that da/dN can be related to ΔK through a simple power law expression. This relationship is the most widely used expression for the modeling of fatigue crack growth.

In general, however, the relationshp between da/dN and ΔK is also affected by stress ratio, $R = K_{min}/K_{max}$. The effects of stress ratio are particularly apparent in the so-called near-threshold regime, and also at high SIF ranges. The differences in the near-threshold regime have been attributed largely to crack closure (Suresh and Ritchie, 1984a, 1984b), which was first discovered by Elber (1970) as a graduate student in Australia. The high crack growth rates at high ΔK values have also been shown to be due to the additional contributions from monotonic or "static" fracture modes (Ritchie and Knott, 1973).

Given the success of the application of the SIF to the correlation of the growth of essentially long cracks, it is not surprising that attempts have been made to apply it to short cracks, where the scale of local plasticity often violates the continuum assumptions of linear elastic fracture mechanics (LEFM) that were made in the derivation of K by Irwin (1957). In most cases, anomalous growth short cracks have been shown to occur below the so-called long-crack threshold. The anomalous behavior of short cracks has been reviewed extensively, e.g., by Miller (1987), and has been attributed largely to the combined effects of microstructure and microtexture localized plasticity (Ritchie and Lankford, 1986). Various parameters have been proposed to characterize the stress–strain fields associated with short cracks. These include the fatigue limit, Coffin–Manson type expressions for low cycle fatigue (Coffin, 1954; Manson, 1954), and elastic–plastic fracture mechanics criteria such as ΔJ and the crack opening displacement (Ritchie and Lankford, 1986).

Considerable progress has also been made in the understanding of fatigue crack initiation and propagation mechanisms. Although Ewing and Humfrey observed the separate stages of crack initiation by slip-band formation and crack propagation as early as in 1903, it was not until about 50 years later that Zappfe and Worden (1951) reported fractographs of striations associated with fatigue crack propagation. However, they did not recognize the one-to-one correspondence of striations with the number of cycles. This was first reported by Forsyth (1961), a year before Laird and Smith (1962) proposed the most widely accepted model of crack propagation. Since then, a great deal of research has been carried out to investigate various aspects of fatigue. A summary of the results obtained from well-established prior research on the fatigue of materials is presented in this chapter.

14.2 MICROMECHANISMS OF FATIGUE CRACK INITIATION

Microcracks tend to initiate in regions of high stress concentration such as those around notches and inclusions. They may also initiate in the central regions of grains, or in the grain boundaries, even when no macroscopic stress raisers are present. In general, however, microcracks initiate as a result of slip processes (Wood, 1958) due to stress or plastic strain cycling. Dislocations either emerge at the surface or pile up against obstacles such as grain boundaries, inclusions, and oxide films, to form slip bands, which were first observed by Ewing and Humfrey (1903). Thompson et al. (1956) later showed that if these slip bands are removed by electropolishing, they will reappear when fatiguing is recommenced, and so they referred to them as persistent slip bands (PSBs).

The resistance to the initiation of slip at the central portion decreases with increasing grain size, following the Hall–Petch relation (Hall, 1951; Petch, 1953). The resistance of the grain boundary regions can also be weakened in soft precipitate free zones (PFZs) (Mulvihill and Beevers, 1986) at the regions of intersection of grain boundaries, e.g., triple points (Miller, 1987), by embrittlement due to grain boundary segregation (Lewandowski et al., 1987), and also by stress corrosion effects (Cottis, 1986). Hence, cracking can occur within grains or at grain boundaries.

The initiation of microcracks may also be influenced by environment. Laird and Smith (1963) showed that initiation of fatigue cracks was slower in vacuum than in air, and they attributed this largely to the effects of the irreversibility of slip in air.

Four main stages of crack initiation have been identified. They involve:

1. Localized strain hardening or softening due to the accumulation of slip steps at the surface. This occurs at sufficiently high alternating plane strain amplitudes. A slip step of one Burgers vector is created when a dislocation emerges at the surface. Since dislocations emerge during both halves of each fatigue cycle, slip steps can accumulate in a local region, and this leads to severe roughening of the surface.
2. The formation of intrusions and extrusions (Fig. 14.3). Cottrell and Hull (1957) have postulated that these can be formed when sequential slip occurs on two intersecting slip planes, as illustrated in Fig. 14.4. Slip occurs in the first slip system and then in the second during the first half of the cycle, to give the indentation shown in Fig. 14.4(c). The slip systems may operate con-

secutively or simultaneously during the reverse cycle to give rise to pairs of intrusions and extrusions, as shown in Figs 14.4(d) and 14.4(e). It is also possible that intrusions and extrusions may form as a result of a dislocation avalanche along parallel neighboring slip planes containing dislocation pile-ups of opposite signs, as postulated by Fine and Ritchie (1979). This is illustrated in Fig. 14.5. Although it is unlikely that intrusions and extrusions form exactly by either of these mechanisms, they do illustrate the kind of slip processes that must be operative.

3. The formation of microcracks. This is often defined by the resolving power of the microscope or the resolution of the nondestructive inspection tool that is used. It is still not clear how intrusions and extrusions evolve into microcracks. These cracks often propagate initially along crystallographic planes of maximum shear stress by Mode II (Forsyth Stage II) shear mechanisms (Forsyth, 1961). Since the plasticity associated with the crack tips of these microcracks is often less than the controlling microstructural unit size, microstructural barriers, such as grain boundaries and dispersed precipitates may cause discontinuities in the crack growth.
4. The formation of macrocracks (usually larger than several grain sizes) as a result of microcrack coalescence or crack growth to a particular crack size where the crack begins to propagate by

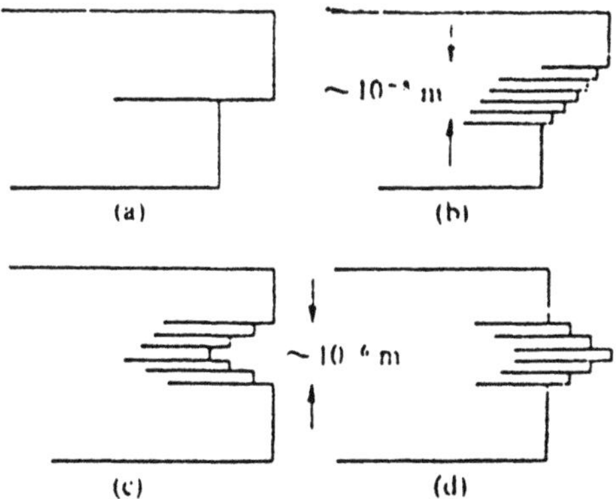

FIGURE 14.3 Formation of surface cracks by slip. Static slip forms unidirectional step: (a) optical microscope; (b) electron microscope. Fatigue slip by to-and-fro movements in slip band may form notch (c) or peak (d). (From Wood, 1958—reprinted with permission of Taylor & Francis Ltd.)

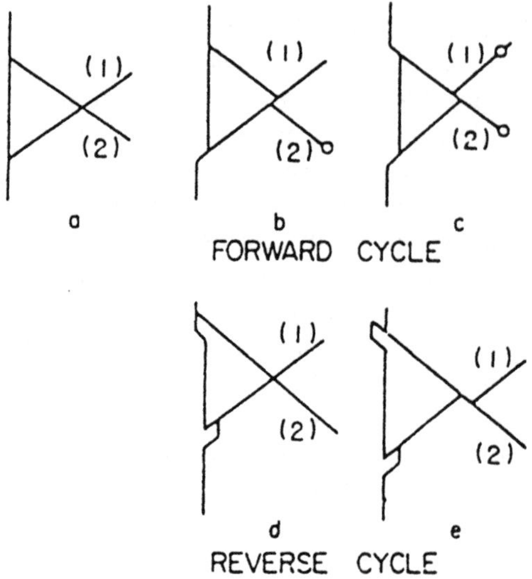

FIGURE 14.4 Cottrell–Hull model for the formation of intrusions and extrusions. (From Cottrell and Hull, 1957—reprinted with permission from the Royal Society.)

Mode I (Forsyth Stage II) mechanisms (Forsyth, 1961), with the direction of crack propagation being perpendicular to the direction of the principal axis. There is no universally accepted definition of the transition from microcrack to macrocrack behavior, although a fatigue macrocrack is usually taken to be one that is sufficiently long to be characterized by LEFM.

14.3 MICROMECHANISMS OF FATIGUE CRACK PROPAGATION

Various models of fatigue crack propagation have been proposed (Forsyth and Ryder, 1961; Laird and Smith, 1962; Tomkins, 1968; Neumann, 1969, 1974; Pelloux, 1969, 1970; Tomkins and Biggs, 1969; Kuo and Liu, 1976). However, none of these models has been universally accepted. It is also unlikely that any single model of fatigue crack growth can fully explain the range of crack extension mechanisms that are possible in different materials over the wide range of stress levels that are encountered in practice. Nevertheless, the above models to provide useful insights into the kinds of processes that can occur at the crack tips during crack propagation by

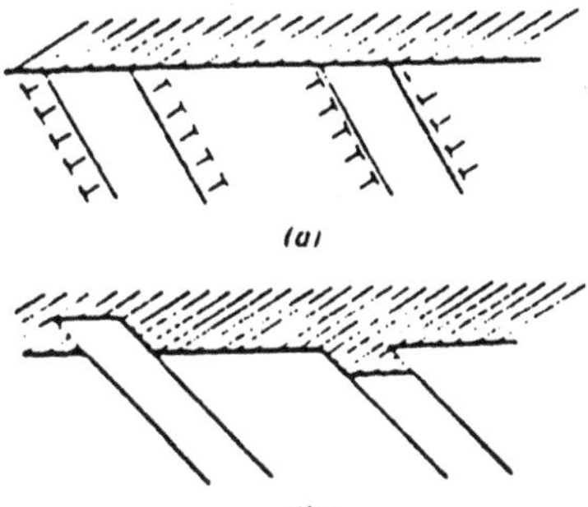

FIGURE 14.5 Paired dislocation pile-ups against obstacle on metal surface grow with cyclic straining until they reach a critical size at which an avalanche occurs to form intrusions and extrusions. (From Fine and Ritchie, 1979—reprinted with permission of ASM International.)

fatigue. Many of them are based on the alternating shear rupture mechanism which was first proposed by Orowan (1949), and most of them assume partial irreversibility of slip due to the tangling of dislocations and the chemisorption of environmental species on freshly exposed surfaces at the crack tip.

One of the earliest models was proposed by Forsyth and Ryder (1961). It was based on observations of fatigue crack growth in aluminum alloys. They suggested that fatigue crack extension occurs as a result of bursts of brittle and ductile fracture (Fig. 14.6) and that the proportion of brittle and ductile fracture in a situation depends on the ductility of the material. They also proposed that crack growth could occur in some cases by void linkage. These voids are formed during the forward cycle around particles that fracture during the previous reverse cycle. Cracking then occurs by the necking down of intervening material until the void links up with the crack, as shown in Figs. 14.7.

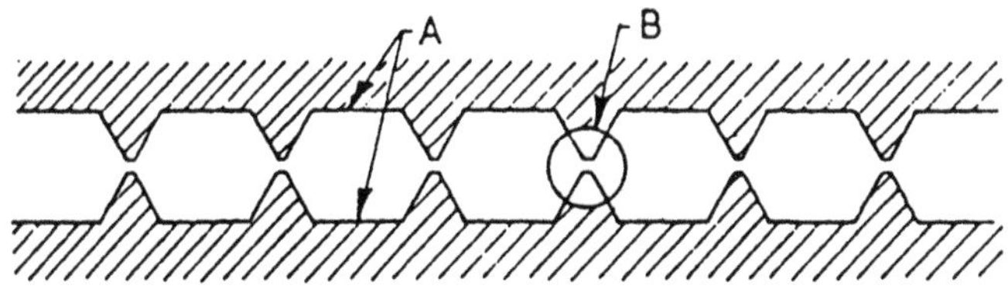

FIGURE 14.6 Bursts of brittle fracture (A) and ductile fracture (B) along striation profile. (From Forsyth and Ryder, 1961—reprinted with permission from Cranfield College of Engineers.)

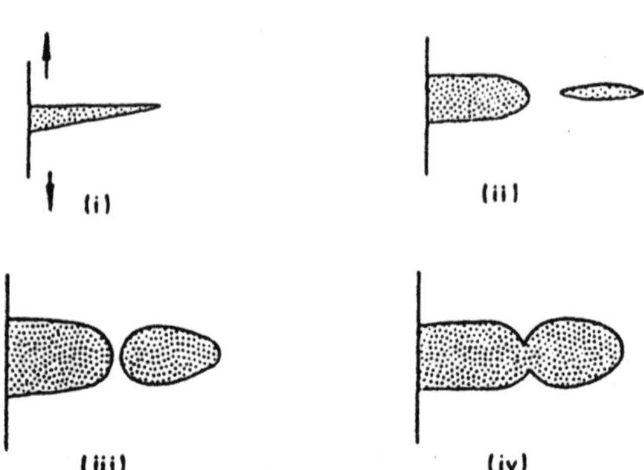

FIGURE 14.7 Forsyth and Ryder model of crack extrusion by void linkage. (From Forsyth and Ryder, 1961—reprinted with permission of Metallurgica.)

Laird and Smith (1962) and Laird (1967) proposed an alternative model based on the repetitive blunting and sharpening of the crack tip due to plastic flow. In this model, localized slip occurs on planes of maximum shear oriented at $\sim 70°$ to the crack tip (Irwin, 1957; Williams, 1957) on the application of a tensile load. As the crack opens during the forward cycle, the crack tip opens up, Figs 14.8(a) and 14.8(b). Further straining results in the formation of ears [Fig. 14.8(c)], which they observed clearly at the peak tensile strain, and the broadening of the slip bands, Fig. 14.8(c). The crack tip is also blunted progressively [Figs 14.8(b) and 14.8(c)] as a

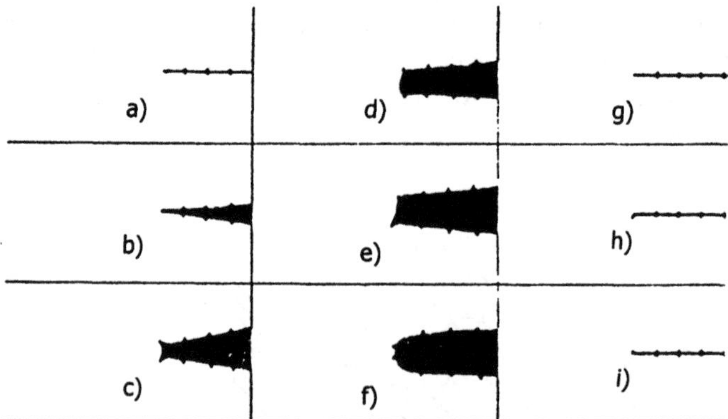

FIGURE 14.8 Schematic representation of fatigue crack advance by Laird and Smith's plastic blunting model. (From Laird and Smith, 1962—reprinted with permission of Taylor and Francis Ltd.

result of plastic flow, which is reversed on unloading, Fig. 14.8(d). The crack faces are brought together as the crack closes, but the adsorption of particles in the environment at the crack tip on to the freshly exposed surfaces prevents complete rewelding, and hence perfect reversibility of slip. Also, the newly created surfaces buckle as the crack extends by a fracture of the crack opening displacement, during the reverse half of the cycle, Figs 14.8(e) and 14.8(f). The corresponding crack-tip geometries obtained on compressing the specimen during the reverse cycle are shown in Figs 14.8(g–i).

Tomkins and Biggs (1969) and Tomkins (1968) have proposed a model that is similar to Laird and Smith's plastic bunting model. They suggest that new crack surfaces are formed by plastic decohesion on available shear planes, at the limit of tensile straining. This model also applies to Stage I growth where they hypothesize that slip will only occur on one of the two available slip planes. Crack extension by this model is illustrated in Fig. 14.9 for Stage II fatigue propagation.

Pelloux (1969, 1970) has formulated a different model based on alternating shear. The behavor of the crack tip is simulated using fully plastic specimens containing sharp notches (Fig. 14.10)—this can be justified when the plastic zone is several times the size of the striation spacing. Pelloux's model is illustrated in Fig. 14.11. Crack extension occurs on intersecting slip planes as a result of alternating slip, which takes place sequentially or

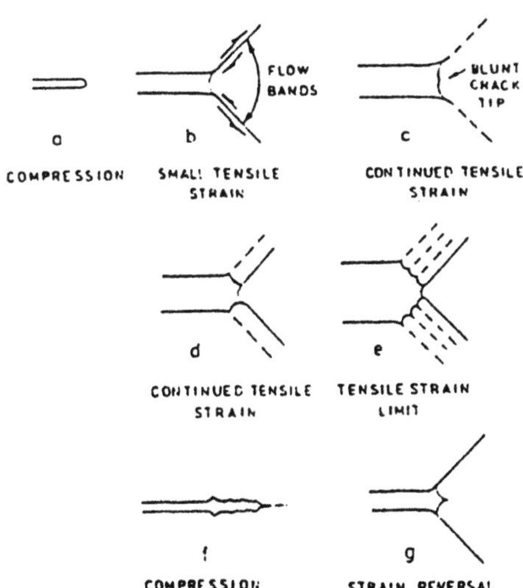

FIGURE 14.9 Plastic flow model of crack advance proposed by Tomkins and Biggs (1969). (Reprinted with permission of Taylor & Francis Ltd.)

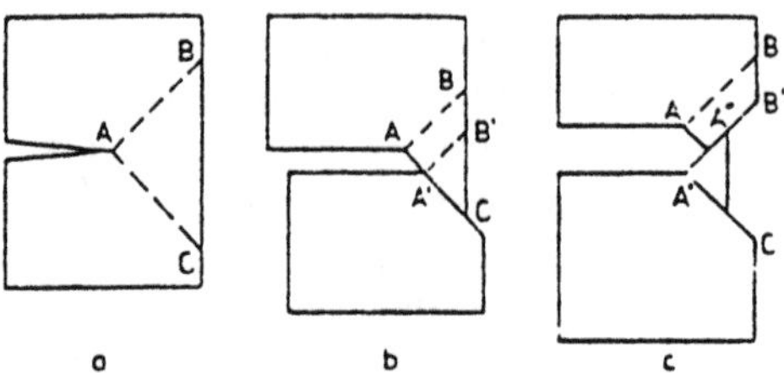

FIGURE 14.10 Pelloux's fully plastic specimen. (Pelloux, 1969, 1970—reprinted with permission of ASM International.)

simultaneously. Complete reversibility is prevented in active environments, e.g., laboratory air, by the formation of oxide layers on the fresly exposed surfaces during the reverse cycle. The slower cycle growth rates that are generally observed in vacuum can also be explained using Pelloux's alternating shear model, since reversed slip would be expected in a vacuum due to the absence of oxide layers. Pelloux (1969, 1970) has proposed a model for crack growth in vacuum which is illustrated in Fig. 14.11i–m.

Similar models based on alternating slip have been proposed by Neumann (1969, 1974) and Kuo and Liu (1976). Neumann's coarse slip model (Fig. 14.12) was proposed for high fatigue crack propagation rates where more than one pair of slip planes are activated per cycle. However,

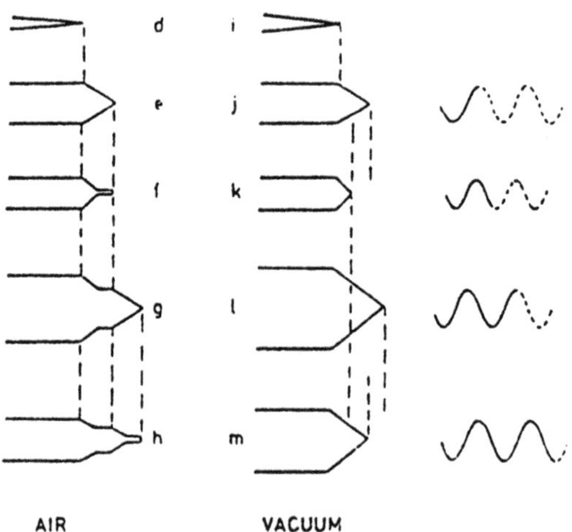

FIGURE 14.11 Crack extension by Pelloux's alternating shear model for laboratory air and vacuum. (From Pelloux, 1970—reprinted with permission of ASM International.)

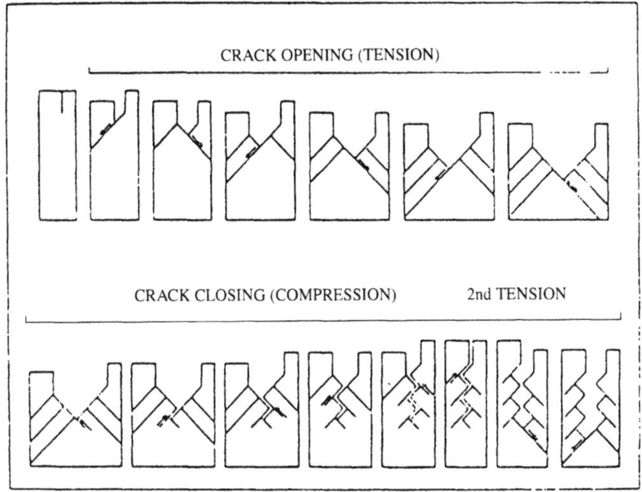

Figure 14.12 Neumann's coarse slip model of crack advance. (From Neumann, 1974—reprinted with permission of AGARD.)

although it requires crack extension to occur as a result of irreversibility of slip, it does not include crack blunting and sharpening stages. The "unzipping" model by Kuo and Liu (1976) is a simple variant of Pelloux's alternating slip model, with the added restriction that only shear at the crack tip will contribute to crack growth. They define a single point with an upper and lower part A^+ and A^- [Figs. 14.13(a–d)] and argue that crack extension will only occur for a sharp crack when these points are physically separated. They also suggest that, although plastic deformation may occur at the crack tip, it will not contribute to crack extension. Cracking by the model is only allowed by unzipping along slip line fields, as shown schematically in Fig. 14.13(e–j).

14.4 CONVENTIONAL APPROACH TO FATIGUE

14.4.1 Stress Amplitude or Stress Range Approach

Since the original work by Wöhler (1958–1871), the conventional approach to fatigue has relied on the use of *S–N* curves. These are curves usually derived from tests on smooth specimens by applying constant amplitude load ranges in tension–compression tests with zero mean load, or in rotating–bending tests, e.g., BS 3518 (1963) They show the

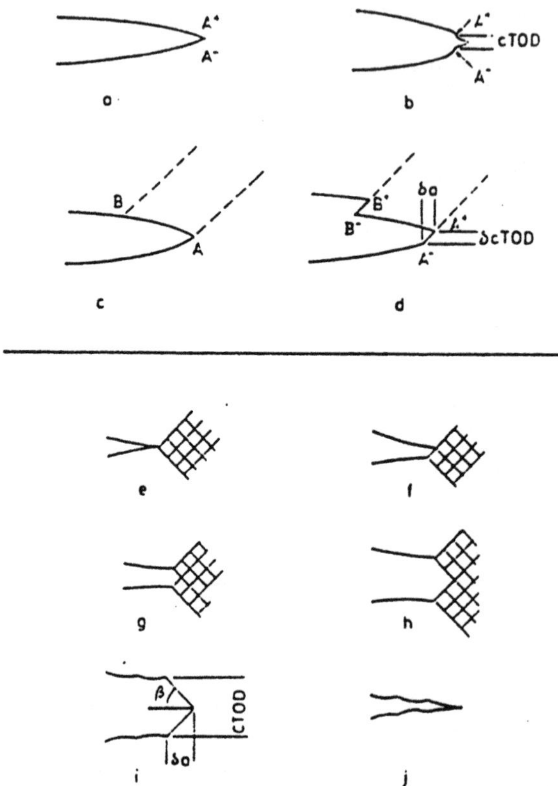

FIGURE 14.13 Kuo and Liu's "unzipping" model of crack growth. (From Kuo and Liu, 1976—reprinted with permission of Elsevier Science.)

dependence of a stress parameter (usually the stress amplitude, S_a or σ_a, or the stress range, ΔS or $\Delta\sigma$, not the number of cycles required to cause failures, N_f.

Strain aging materials, such as mild steel show a sharp "fatigue limit" below which no fatigue takes place, and the specimens appear to last indefinitely. Nonaging materials do not show a sharp fatigue limit, and it is conventional to specify an "endurance limit" for design purposes, which is usually defined as the alternating stress required to cause failure in 10^8 cycles. Typical *S–N* curves are presented in Fig. 14.14. It is important to note that, although the fatigue limits are less than the yield stress in mild steels (typically half the yield stress), they are generally greater than the yield stress and less than the ultimate tensile strength (UTS). In most cases, the fatigue limits of aging steels are typically $\sim$ (UTS)/2 in steels.

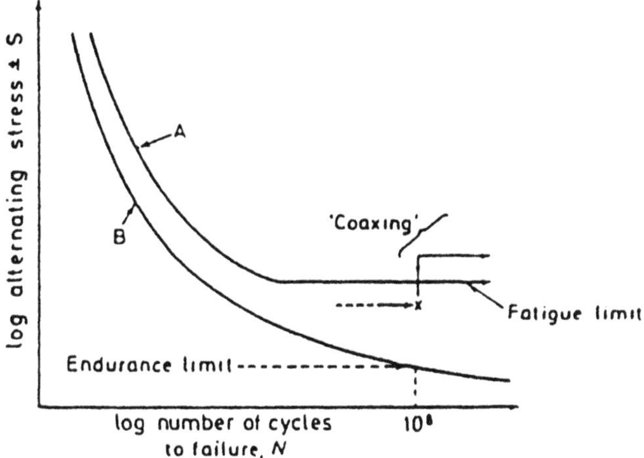

FIGURE 14.14 S–N fatigue curve. Curves of type A are typical of mild steel and alloys which strainage, and curves of type B are typical of nonaging alloys. (From Knott, 1973—reprinted with permission from Butterworth-Heinemann.)

A great deal of work has been carried out on the use of S–N curves, and a considerable amount of useful data has been accumulated on the effects of mean stress, environment, notches, and other factors. Such data have been used, and are still widely used, in the estimation of component lives in engineering structures. However, the S–N curve is empirical in nature, and it does not provide any fundamental understanding of the underlying fatigue processes in structures that may contain pre-existing flaws.

14.4.2 Strain-range Approach

Fatigue behavior in smooth specimens subjected to low-cycle fatigue is dependent on the plastic strain range (Coffin 1954; Manson, 1954). The amount of plastic strain imposed per cycle can be found from the hysteresis loop in the plot of stress versus strain over one cycle, as shown in Fig. 14.15. The effect of plastic range, $\Delta\varepsilon_p$, on the number of cycles to failure, N_f, is expressed by the Coffin–Manson relationship:

$$\Delta\varepsilon_p \cdot N_f^{\alpha_1} = C_1 \tag{14.1}$$

where $\alpha_1(\approx 0.5)$ and C_1 (≈ 1) are material constants. This relationshp was obtained empirically, and has been shown to hold for different materials (Fig. 14.16), under conditions of low-cycle fatigue. However, the Basquin

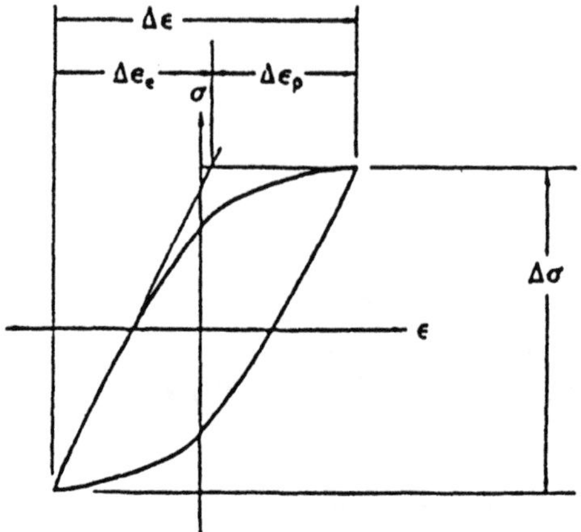

σ = Stress
ε = Strain
ε_e = Elastic strain
ε_p = Plastic strain
$\Delta\varepsilon_p = \Delta\varepsilon - (\Delta\sigma/E)$
E = Modulus of elasticity

Figure 14.15 Hysterisis loop of one fatigue cycle.

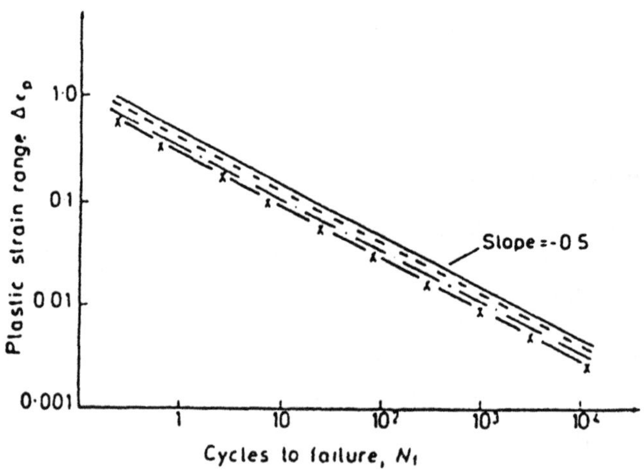

Figure 14.16 Coffin–Manson relationship: ——— C/Mn steel; - - - - - - Ni/Cr/Mo alloy steel; - - - · - - - Al–Cu alloy, - - - - x - - - - Al–Mg alloy. (From Knott, 1973—reprinted with permission from Butterworth-Heinemann.)

law (Basquin, 1910) is found to be more suitable for high-cycle fatigue. This relates the elastic strain range, $\Delta\varepsilon_e$ (see Fig. 14.15), to the number of cycles to failure, N_f, by the following expression:

$$\Delta\varepsilon_e \cdot N_f^{\alpha_2} = C_2 \tag{14.2}$$

where α_2 and C_2 are material constants.

It is also possible to obtain elastic and plastic strain range fatigue limits (Lukas et al. 1974), which have been shown to correspond to that fatigue limit obtained from stress-controlled tests (Kendall, 1986).

14.4.3 Effects of Mean Stress

Mean stress has been shown to have a marked effect on the endurance limit. Once the yield stress has been exceeded locally, and alternating plastic strain made possible, a mean tensile stress accelerates the fatigue fracture mechanisms. Therefore, since most *S–N* curves are obtained from tests conducted at zero mean stress, there is a need for an extra design criterion to account for the combined effects of mean and alternating stresses. Gerber (1874) and Goodman (1899) proposed relationships of the form (Figure 14.17):

$$\pm\sigma = \pm\sigma_0\left[1 - \left(\frac{\sigma_m}{\sigma_t}\right)^n\right] \tag{14.3}$$

where σ is the fatigue limit at a mean stress of σ_m, σ_0 is the fatigue limit for $\sigma_m = 0$, σ_t is the tensile strength of the material, and the exponent $n = 1$ in the Goodman expression, and $n = 2$ in Gerber's versionof Eq. (14.3). The resulting Goodman line and Gerber parabola are shown in Figs 14.17(a) and 14.17(b), respectively.

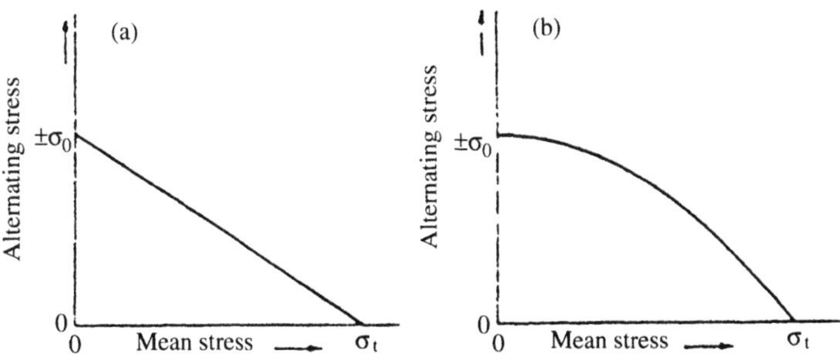

FIGURE 14.17 (a) Goodman line; (b) Gerber parabola.

14.4.4 Fatigue Behavior in Smooth Specimens

The fatigue behavior in smooth specimens is predominantly initiation controlled (Schijve, 1979). High cyclic stresses (> yield stress) are needed to cause alternating plastic deformation and hardening in the surface grains. These deformations are not fully reversible and they result in the formation of persistent slip bands (Ewing and Humfrey, 1903; Thompson et al., 1956), which develop into intrusions, and extrusions (Cottrell and Hull, 1957) that are usually associated with the nucleation of microcracks. The microcracks usually join up to form a single crack which propagates by Stage I crack growth (Forsyth, 1961) along an active slip band that is inclined at $\sim 45^\circ$ to the direction of the principal stress (Ham, 1966; Laird, 1967). Crack growth then continues by Stage II propagation (Forsyth, 1961), e.g., when the crack reaches a critical crack-tip opening (Frost et al., 1974), until the crack becomes sufficiently long for fast fracture or plastic collapse to take place.

The above processes may be divided into initiation and propagation stages, and the *S–N* curve can also be divided into initiation and propagation regions. The number of cycles for fatigue failure, N_f, is then regarded as the sum of the number of cycles for fatigue crack initiation, N_i, and the number of cycles for fatigue crack propagation, N_p. The *S–N* curves can, therefore, be regarded as the sum of two curves (S–N_i and S–N_p), as shown in Fig. 14.18. The extent to which either process contributes to the total number of cycles to failure depends on the stress level and the material. In ductile metals/alloys at low stress levels, N_f is governed by N_i, whereas at high stress levels, N_f is mainly determined by N_p.

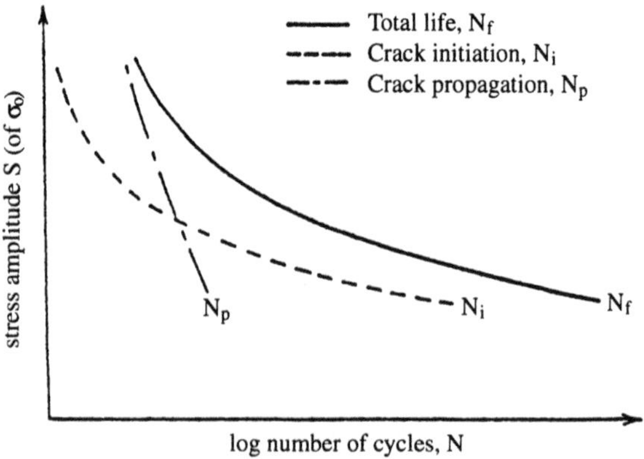

FIGURE 14.18 Initiation and propagation components of total fatigue life.

14.4.5 Limitations of Conventional Approach to Fatigue

The conventional approach to fatigue is based on the assumption that most engineering structures are flawless at the beginning of service—hence, the wide use of smooth specimens in conventional fatigue tests. This assumption is valid in the design of most machine components, which are more or less flawless. However, it is now universally accepted that most structures contain defects at the start of service. These defects must be accounted for in fatigue testing and design.

The relative importance of fatigue crack propagation compared to fatigue crack initiation has also been recognized for most practical cases. Although the *S–N* curves can accont for these two processes, they do not distinguish clearly between them. The results obtained from the conventional tests cannot, therefore, be used for prediction of the fatigue lives of most engineering structures with pre-existing flaws. Fatigue crack growth predictions in such structures require the use of fracture mechanics techniques, which are discussed in Sections 14.6, 14.8, and 14.11

14.5 DIFFERENTIAL APPROACH TO FATIGUE

Various workers have shown that the crack growth rate, da/dN, is a function of the applied stress range, $\Delta\sigma$, and the crack length (Head, 1956; Frost and Dugdale, 1958; McEvily and Illg, 1958; Liu, 1961). Head (1956) proposed that the crack growth rate is given by

$$\frac{da}{dN} = \frac{C_3 \Delta\sigma^3 a^{3/2}}{(\sigma_{ys} - \sigma)_{w_0^{1/2}}} \tag{14.4}$$

where C_3 is a material constant, w_0 is the plastic zone size, and a is half the crack length.

Similar expressions have also been obtained by Frost and Dugdale (1958) and Liu (1961), which can be written as

$$\frac{da}{dN} = C_4 \Delta\sigma^{\alpha_3} a^{\alpha_4} \tag{14.5}$$

where C_4 is a material constant, $\alpha_3 = 2$, and $\alpha_4 = 1$.

McEvily and Illg (1958) recognized the significance of the stress concentration at the crack tip, and proposed that the crack growth rate is a function of the maximum stress at the crack tip, σ_{max}, i.e.,

$$\frac{da}{dN} = f(K_t \sigma_{net}) \tag{14.6}$$

where K_t is the notch concentration factor, and σ_{net} is the net section stress.

However, although the success of the application of the differential method depends on the correlation of actual fatigue crack growth-rate data with predictions made using the above equations, the stress parameters used have not been shown to represent the local crack-tip driving force for crack extension. This is probably why their use has been superseded by the fracture mechanics parameters that are presented in the next section.

14.6 FATIGUE CRACK GROWTH IN DUCTILE SOLIDS

Fatigue crack growth in ductile solids can be categorized into the three regimes, as shown in Fig. 14.19. The first region (regime A) occurs at low ΔK and is called the near-threshold regime. This region corresponds to a

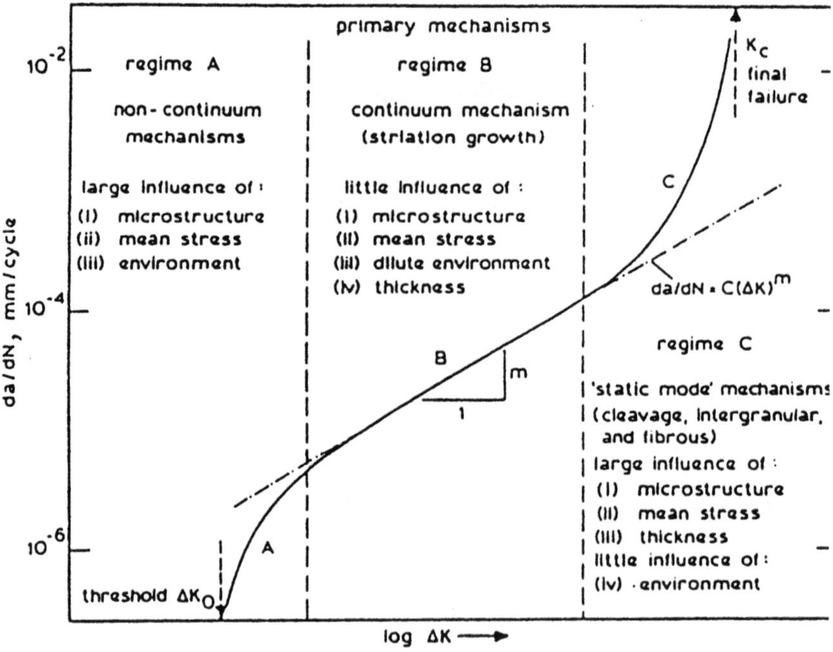

FIGURE 14.19 Schematic variation of da/dN with ΔK in steels showing primary mechanisms in the three distinct regimes of fatigue crack propagation. (From Ritchie, 1979—reprinted with permission of Academic Press.)

cleavage-like crack growth mechanism, where the crack follows preferred crystallographic directions. Below the fatigue threshold, ΔK_0, no crack propagation can be detected with existing instruments, and, in practice, the fatigue threshold is often defined as the ΔK that corresponds to a crack growth rate of 10^{-8} mm/cycle (approximately one lattice spacing per cycle). The middle regime (regime B) is a linear region of the plot which follows the Paris equation:

$$\frac{da}{dN} = C(\Delta K)^m \tag{14.7}$$

where da/dN is the fatigue crack growth rate, C is a material constant that is often called the Paris coefficient, ΔK is the stress intensity factor range, and m is the Paris exponent. Since the above power law expression applies largely to regime B, the mid-ΔK regime is often called the Paris regime (Paris and coworkers, 1961, 1963). Crack propagation usually proceeds by a mechanism of alternating slip and crack-tip blunting that often results in striation formation in this regime.

Environment has also been shown to have important effects on the fatigue crack growth and the formation of striations. It was first shown by Meyn (1968) that striation formation may be completey suppressed in vacuo in aluminum alloys which form well-defined striations in moist air. Pelloux (1969) suggested that the alternating shear process is reversible unless an oxide film is formed on the slip steps created at the crack tip. This oxide layer impedes slip on load reversal. A schematic illustration of the opening and closing of a crack during two fully reversed fatigue cycling in air and in vacuo is shown in Fig. 14.11.

The third regime (regime C) is called the high ΔK regime. An increase in crack growth rate is usually observed in this regime, and the material is generally close to fracture in this regime (Ritchie and Knott, 1973; Mercer et al., 1991a,b; Shademan, 2000). Accelerated crack growth occurs by a combination of fatigue and static fracture processes in this regime. Finally, fast fracture occurs when K_{max} is approximately equal to the fracture toughness, K_{Ic}, of the material.

Several studies have been carried out to investigate the factors that control the fatigue crack growth behavior in ductile solids. The interested reader is referred to the text by Suresh (1999). The major factors that affect fatigue crack growth in different regimes (A, B, and C) have also been identified in a review by Ritchie (1979).

In the near-threshold regime, fatigue crack growth is strongly affected by microstructure, mean stress, and environment. However, in regime B, these variables have a smaller effect compared (Figs 14.19 and 14.20) to those in regime A. In contrast, microstructure, means stress, and specimen

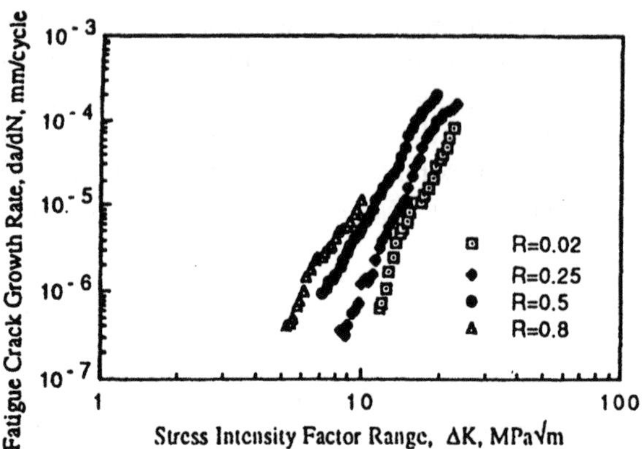

FIGURE 14.20 Effects of stress ratio on fatigue crack growth rate in mill-annealed Ti–6Al–4V. (Dubey et al., 1997—reprinted with permission of Elsevier Science.)

thickness have a strong effect on fatigue crack growth in regime C, where static fracture modes (cleavage), and interangular and ductile dimpled fracture modes are observed as K_{max} approaches the material fracture toughness K_{Ic}. In fact, the increase in the apparent slope in the da/dN–ΔK plot (in regime C) has been shown to be inversely related to the fracture toughness, K_{Ic} (Ritchie and Knott, 1973). This is shown in Fig. 14.21.

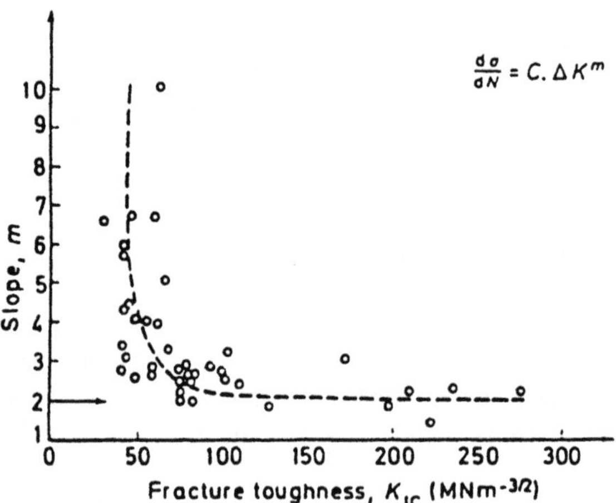

FIGURE 14.21 Variation of apparent slope, *m*, with monotonic fracture toughness. (From Ritchie and Knott, 1973—reprinted with permission of Elsevier Science.)

The mechanisms of fatigue crack growth in regimes, A, B, and C can be summarized on fatigue mechanism maps that show the domains of ΔK and K_{max} in which a given mechanism operates (Mercer et al., 199a,b; Shademan, 2000). Selected examples of fatigue maps are presented in Fig. 14.22 for single crystal and polycrystalline Inconel 718. These show plots of K_{max} (ordinate) against ΔK (abscissa). Constant-stress ratio domains correspond to straight lines in these plots. Hence, the transitions in fracture mechanism at a given stress ratio occur along these lines, as ΔK increases from regimes A, B, and C (Figs 14.19 and 14.22).

It is particularly important to note that the transitions in fracture modes correspond directly to the different regimes of crack growth. The changes in the slopes of the da/dN–ΔK plots are, therefore, associated with changes in the underying fatigue crack growth mechanisms. Furthermore, the transitions in the fatigue mechanisms occur gradually, along lines that radiate outwards. This point corresponds to the upper limit on the K_{max} axis, which also defines the upper limit for the triangle in which all fracture mode transitions can be described for positive stress ratios. The point corresponds clearly to the fracture toughness, K_{Ic} or K_c.

14.7 FATIGUE OF POLYMERS

A significant amount of work has been done on the fatigue behavior of plastics. Most of the important results have been summarized in monograph by Hertzberg and Manson (1980), and the interested reader is referred to their book for further details. Although the fatigue behavior of polymers exhibits several characteristics that are similar to those in metals, i.e., stable crack growth and S–N type behavior, there are several profound differences between the fatigue processes in polymers. These include hysteritic heating and molecular deformation processes that can give rise to the formation of shear bands and crazes in polymeric materials deformed under cyclic loading. Also, typical Paris exponents in polymers are between 4 and 20.

Fatigue crack growth rate data are presented in Fig. 14.23 for different polymeric materials. Note that the fatigue thresholds for polymers are generally very low. Furthermore, most polymers exhibit stable crack growth over only limited ranges of ΔK compared to those in metals. Polymers also exhibit significant sensitivity to frequency, with the crack growth rates being much faster at lower frequency than at higher frequency. The frequency sensitivity has been rationalized by considering the possible time-dependent interactions between the material and the test environment (Hertzberg and Manson, 1980).

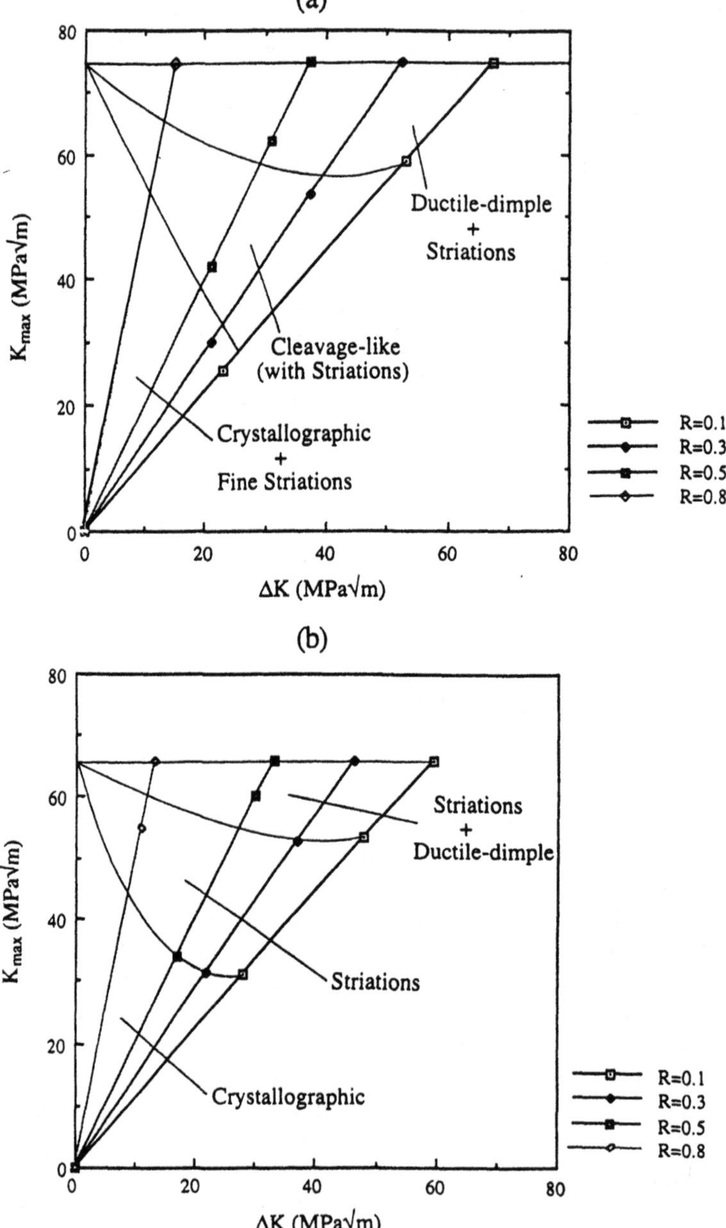

FIGURE 14.22 Fatigue fracture mechanism maps showing the transitions between fatigue fracture modes as a function of ΔK and K_{max} for (a) single crystal IN 718 and (b) polycrystalline IN 718.

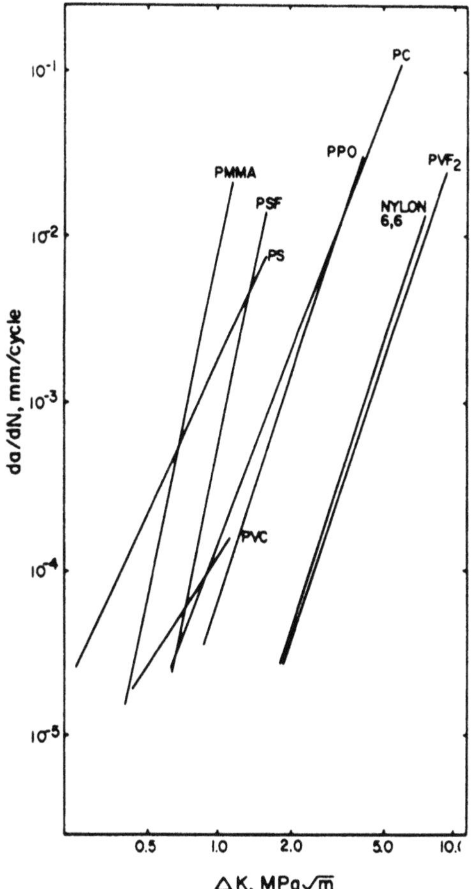

FIGURE 14.23 Fatigue crack growth rate data for selected polymers. (From Hertzberg and Manson, 1980—reprinted with permission from Academic Press.)

Fatigue crack growth in polymers occurs by a range of mechanisms. These include striation mechanisms that are somewhat analogous to those observed in metals and their alloys. In such cases, a one-to-one correpsondence has been shown to exist between the total number of striations and the number of fatigue cycles. However, polymers exhibit striated fatigue crack growth in the high-ΔK regime, while metals exhibit striated fatigue crack growth in the mid-ΔK regime where the crack growth rates are slower. Fatigue crack growth in polymers has also been shown to occur by crazing and the formation of discontinuous shear bands. In the case of the latter, the discontinuous shear bands are formed once every hundred of cycles. There is, therefore, not a one-to-one correspondence between the number of shear bands and the number of fatigue cycles. Further details on the mechanisms of crack growth in polymers can be found in texts by Hertzberg and Manson (1980) and Suresh (1999).

14.8 FATIGUE OF BRITTLE SOLIDS

14.8.1 Initiation of Cracks

For highly brittle solids with strong covalent or ionic bonding, and very little mobility of point defects and dislocations, defects such as pores, inclusions, or gas-bubble entrapments serve as potential sites for the nucleation of a dominant crack (Suresh, 1999). In most brittle solids, residual stress generated at grain boundary facets and interfaces gives rise to microcracking during cooling from the processing temperature. This occurs as a result of thermal contraction mismatch between adjacent grains or phases. These microcracks may nucleate as major cracks under extreme conditions. However, in general, a range of microcrack sizes will be nucleated, and the larger cracks will tend to dominate the behavior of the solid.

For semibrittle solid like MgO (Majumdar et al., 1987), microcracks may also form as a result of dislocation/microstructure interactions. In such solids, slip may initiate when the resolved shear stress exceeds a certain critical value on favorably oriented low-index planes. Dislocation sources (of the Frank–Read type) are activated, and subsequently, the glide of the dislocation loops, moving outwardly from the souces, is impeded by obstacles such as grain boundaries and/or inclusions. This results ultimately in dislocation pile-ups and microcrack nucleation, when critical conditions are reached (Cotterell, 1958).

14.8.2 Growth of Cracks

For most brittle solids, fatigue crack growth is very difficult to monitor, especially under tensile loading at room temperature. The Paris exponents, m, for brittle materials are generally very high, i.e., $\sim$ 20–200 (Dauskardt et al., 1990; Ritchie et al., 2000). However, stable crack growth, attributable solely to cyclic variations in applied loads, can occur at room temperature (even in the absence of an embrittling environment) in single-phase ceramics, transformation-toughened ceramics, and ceramic composites. This was demonstrated for cyclic compression loading of notched plates by Ewart and Suresh (1987).

Unlike ductile solids, where macroscopic fatigue can arise from cyclic slip, the driving forces for the crack growth at room temperature may involve the degradation of bridging zones behind the crack tip (Fig. 14.24), microcracking, martensitic transformations, and interfacial sliding (Sures, 1999). However, at elevated temperature, some semibrittle solids can exhibit characteristics of fatigue damage that are apparently similar to those found in ductile metals at elevanted temperature (Argon and Goodrich, 1969).

It is important to note here that several ceramics exhibit stable crack growth phenomena under static or quasistatic loads (Evans and Fuller, 1974; Widerhorn et al., 1980). Under such conditions, it is common to express the time rate of crack growth, da/dt, as a function of the linear elastic stress intensity factor, K:

$$\frac{da}{dt} = A(K)^p \tag{14.8}$$

where A and p are material constants that are obtained from crack growth experiments under static loading. It is also important to note here that stable crack growth in ceramics is often associated with toughening mechanisms such as crack bridging or transformation toughness by stress-induced mar-

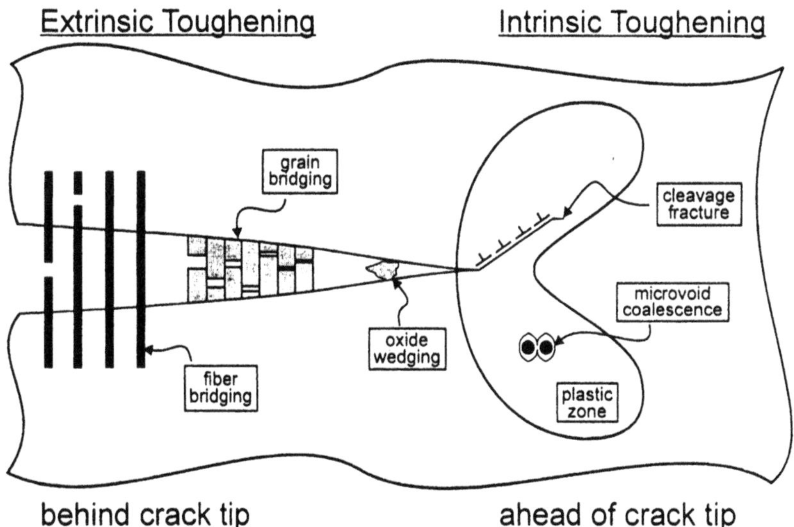

Figure 14.24 Schematic illustration of intrinsic and extrinsic fatigue damage mechanics. (From Ritchie et al., 2000—reprinted with permission from Elsevier Science.)

tensitic phase transformations. The effective stress intensity factors, K_{eff}, due to such mechanisms is given by

$$K_{eff} = K - K_s \tag{14.9}$$

where K is the applied stress intensity factor and K_S is the shielding SIF. In cases where crack-tip shielding levels are significant, K_{eff} should be used instead of K in Eq. (14.8). Further details on crack-tip shielding mechanisms have been presented in Chap. 13. It is simply sufficient to note here that stable crack growth in ceramics can be largely explained by the inelasticity associated with crack-tip shielding mechanisms under cyclic loading. This may involve cyclic breakdown of bridging ligaments due to wear phenomena behind the crack tip (Evans, 1980; Grathwohl, 1998; Dauskardt, 1990; Horibe and Hirihara, 1991; Lathabai et al., 1991; Kishimoto et al., 1994; Ramamurthy et al., 1994).

At elevated temperatures, glassy films that are added to most ceramics during processing often becomes viscous. Since these usually reside at grain boundaries, they can lead to the in-situ formation of viscous glassy films at grain bridges. Since such grain bridges undergo viscous flow under cyclic loading, they result in strong sensitivity to cyclic frequency and mean stress (Han and Suresh, 1989; Ewart and Suresh, 1992; Lin et al., 1992; Dey et al., 1995; Ritchie et al., 1997, 2000). Such viscoelastic grain bridging can contribute significantly to the occurrence of inelasticity during each fatigue cycles, and thus result in stable crack growth in ceramics under cyclic loading at elevated temperature.

In the case of intermetallics (compounds between metals and metals), the fatigue damage mechanisms are somewhat intermediate between those of metals and ceramics (Soboyejo et al., 1990; Rao et al., 1992; Davidson and Campbell, 1993; Mercer et al., 1997, 1999a; Ritchie et al., 1997, 2000, Stoloff, 1996. In such systems, fatigue crack growth occurs as a result of intrinsic crack-tip processes (partially reversible dislocation motion, deformation-induced twinning, and crack tip/environmental interactions) and crack wake processes (degradation of bridging zones), as shown schematically in Fig. 14.24. Evidence of crack-tip deformation is shown in Fig. 14.25, which is taken from a paper by Mercer and Soboyejo (1997). This shows clear evidence of crack-tip deformation-induced twinning in a gamma-based titanium aluminide intermetallic alloy. Stable fatigue crack growth may also occur in intermetallics due to the stress-induced martensitic transformations that occur in intermetallic composites reinforced with partially stabilized zirconia particles (Soboyejo et al., 1994; Ye, 1997; Ramasundaram et al., 1998).

However, the slowest fatigue crack growth rates are typically observed in intermetallics in which crack-tip dislocation processes are predominant. One example of such systems is the B2 Nb–15Al–40Ti intermetallic, which

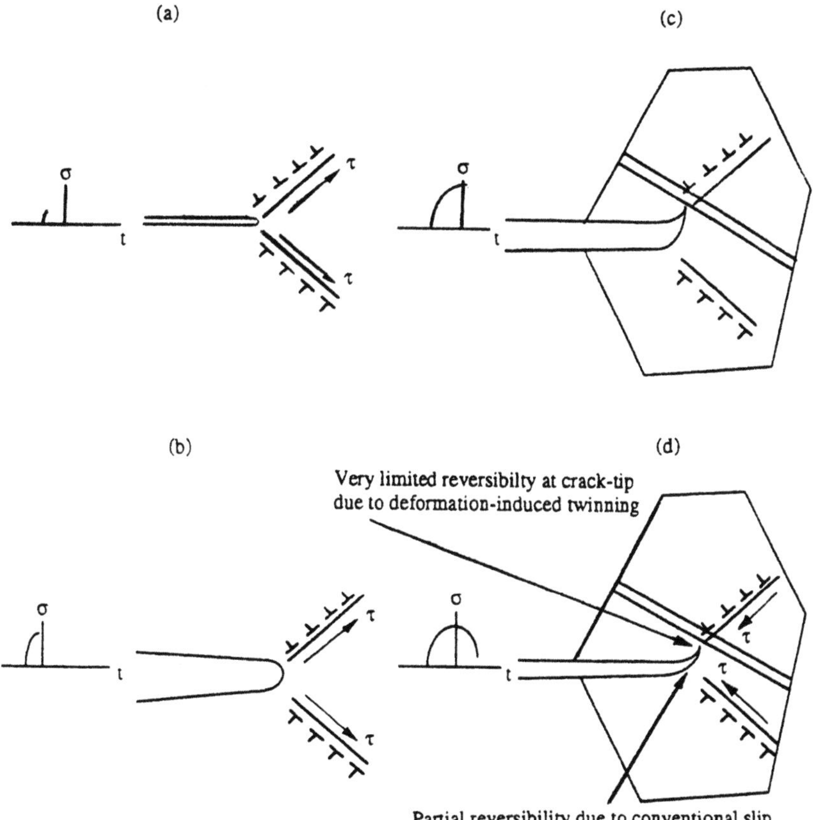

FIGURE 14.25 Crack-tip deformation by slip and deformation-induced twinning in gamma titanium aluminide intermetallics. (From Mercer and Soboyejo, 1997—reprinted with permission of Elsevier Science).

exhibits Paris constants of ~ 2–4, and comparable fatigue crack growth resistance to conventional structural alloys such as Ti–6Al–4V and Inconel 718 (Ye et al., 1998). However, most intermetallics have Paris exponents between 6 and 50 (Soboyejo et al., 1996; Mercer and Soboyejo, 1997, 2000; Ritchie et al., 1998, 1999). They also tend to exhibit fast fatigue crack growth rates when compared with those of structural metals and their alloys (Fig. 14.26).

Finally in this section, it is important to note that fatigue crack growth in brittle and ductile solids is affected by both ΔK and K_{max}. The relative contributions from each of these variables may be estimated from the modified two-parameter Paris equation (Jacobs and Chen, 1995; Vasudevan and

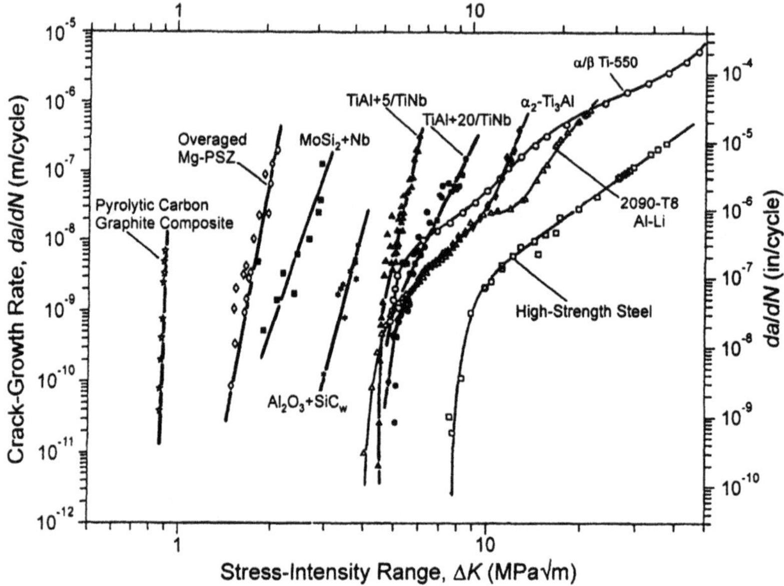

Figure 14.26 Fatigue crack growth rate data for selected intermetallics, metals, and ceramics. (From Ritchie et al., 2000—reprinted with permission of Elsevier Science.)

Sadananda, 1995; Soboyejo et al., 1998; Ritchie et al., 2000), which is given by

$$\frac{da}{dN} = C(\Delta K^m)(K_{max})^n \tag{14.10}$$

where m is the exponent of K_{max} and the other variables have their usual meaning. In the case of ductile metals and their alloys, the exponent m is usually much greater than n. However, brittle ceramics exhibit higher values of n than m, while intermetallics often have comparable values of m and n. The relative values of m and n, therefore, provide a useful indication of the contributions of ΔK and K_{max} to the fatigue crack growth process.

The above arguments are also consistent with fractographic data that show clear evidence of ΔK-controlled ductile (striations) fatigue fracture modes in metals, ΔK_{max}-controlled static fracture modes (cleavage-like fracture) in ceramics, and ΔK- and K_{max}-controlled mixed ductile plus static (mostly cleavage-like plus striations in some ductile intermetallics) fracture modes in intermetallics (Soboyejo et al., 1996; Stoloff, 1996; Mercer and Soboyejo, 1997; Ye et al., 1998; Ritchie et al., 2000). Typical fatigue fracture modes for intermetallic materials are shown in Figs 14.28–14.29.

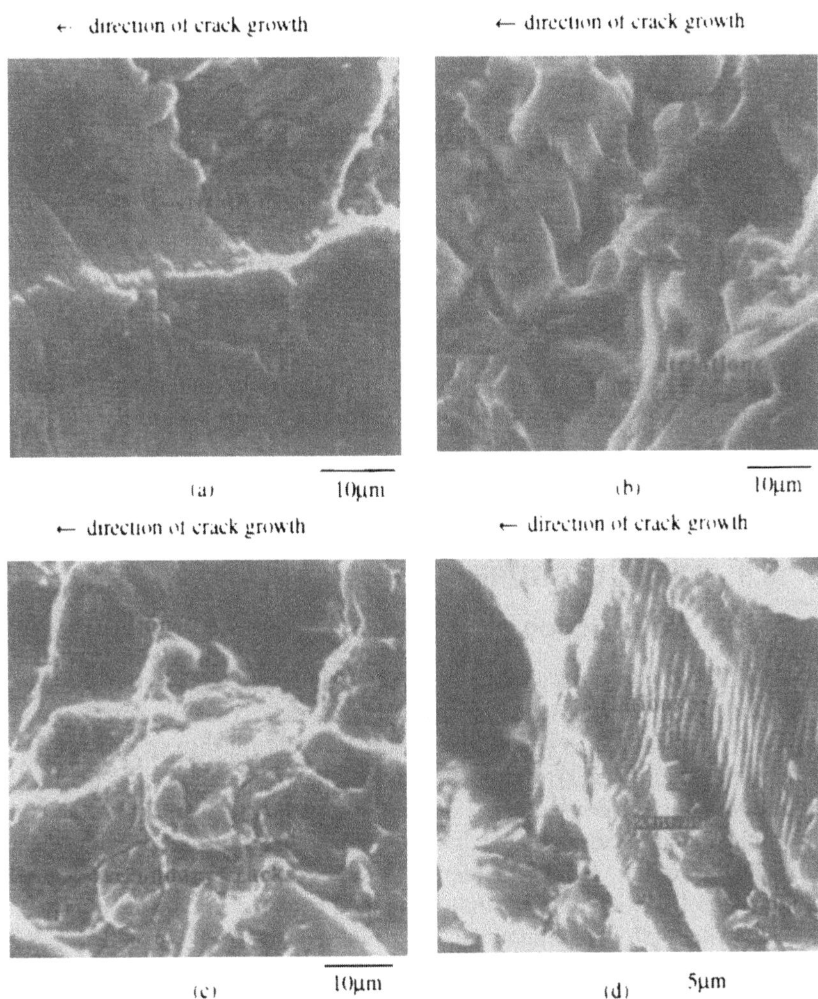

Figure 14.27 Fatigue fracture modes in mill-annealed Ti–6Al–4V. Paris regimes of the fracture surface showing ductile transgranular fracture mode: (a) $R = 0.02$; (b) $R = 0.25$; (c) $R = 0.5$; (d) $R = 0.8$. Evidence of fatigue striations and cleavage-like fracture surface can be clearly seen. Secondary cracks can also be seen in (c). (From Dubey et al., 1997—reprinted with permission of Elsevier Science.)

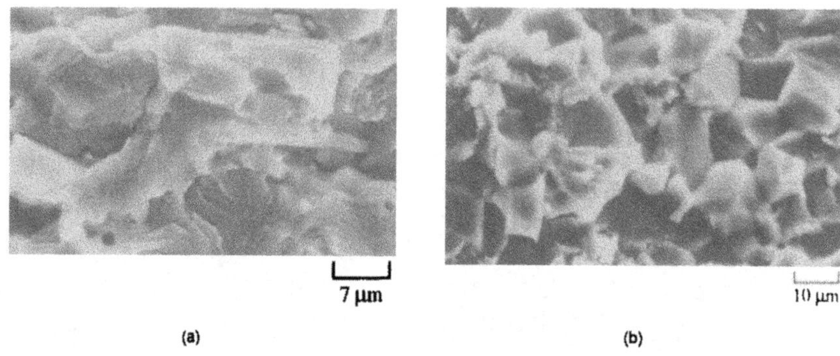

Figure 14.28 Fatigue fracture modes in TiAl-based alloys: (a) 25°C; (b) 700°C. (From Mercer and Soboyejo, 1997—reprinted with permission of Elsevier Science.)

14.9 CRACK CLOSURE

14.9.1 Introduction

The LEFM approach assumes that the fatigue crack growth rate, da/dN, is proportional to the stress intensity range, ΔK. The relationship between da/dN and ΔK is given by the power law relationship in Eq. (14.7). This assumes that a crack in an ideal elastic solid opens and closes at zero load, and is a direct consequence of the superposition principal that is often employed in linear elastic stress analysis (Timoshenko and Goodier, 1970). However, there is evidence to suggest that cracks close above zero load due to extra material that is wedged between the crack flanks.

The first mechanistic justification for this was provided by Elber (1970). Closure was considered to arise from the compressive residual plastic wade associated with fatigue growth,. This causes premature contact of the crack faces (plasticity-induced closure) before the minimum load, K_{min}, is attained in the reverse cycle, Fig. 14.30(a). Compressive stresses than exist in the lower part of the fatigue cycle, and the crack is "closed" during this period. This "closed" portion of the fatigue cycle cannot contribute to the fatigue cycle, and the effective stress intensity range, ΔK_{eff}, at the crack-tip is given by

$$\Delta K_{eff} = K_{max} - K_{op} \tag{14.11}$$

where K_{op} is the stress intensity at which the crack becomes fully open in the fatigue cycle. Elber (1970) postulated that the fatigue crack growth rate is then governed by ΔK_{eff}, such that

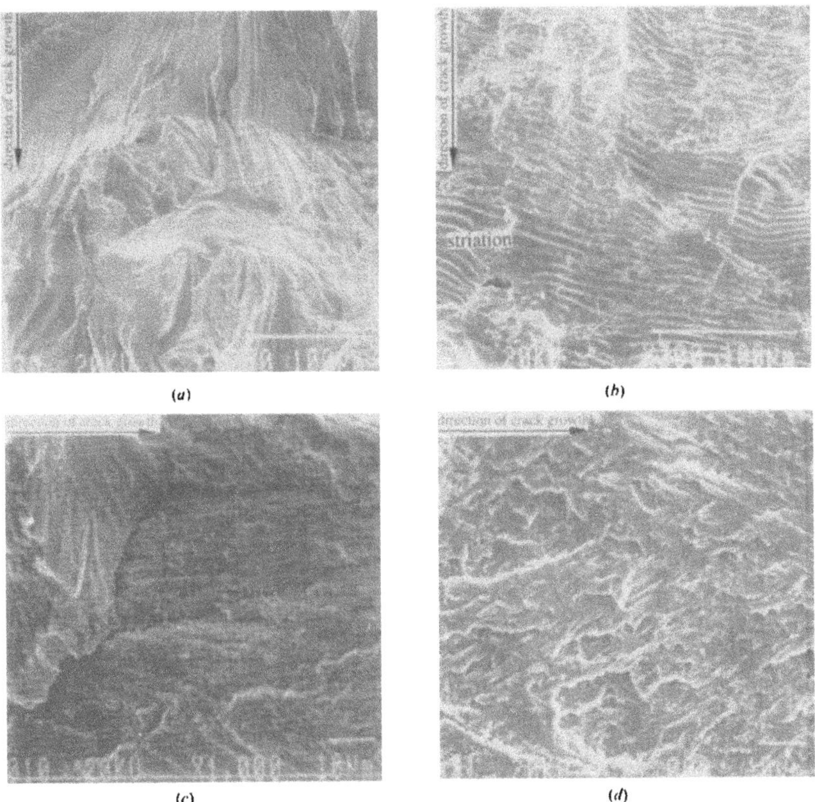

FIGURE 14.29 Fatigue fracture modes in mill-annealed Nb–Al–Ti: (a) low ΔK; (b) high ΔK; (c) secondary cracking in 40-Ti (DA); (d) relatively flat fracture surface of 40-Ti (STA). (From Ye et al., 1998—reprinted with permission of TMS, Warrendale, PA.)

$$\frac{da}{dN} = A(\Delta K_{eff})^m \tag{14.12}$$

Initially, plasticity-induced closure was thought to occur only under plane stress conditions (Lindley and Richards, 1974). However, there is now considerable evidence to show that plasticity-induced closure can also occur under plane strain conditions (Fleck and Smith, 1982). Suresh and Ritchie (1984a) have also shown that plasticity-induced closure is not always the principal mechanism of crack closure in the near-threshold regime. Similar conclusions have been reached by Newman (1982) using finite element studies which have confirmed that plasticity-induced closure, modeled for plane strain conditions, is insufficient to explain the marked effect of load ratio, R,

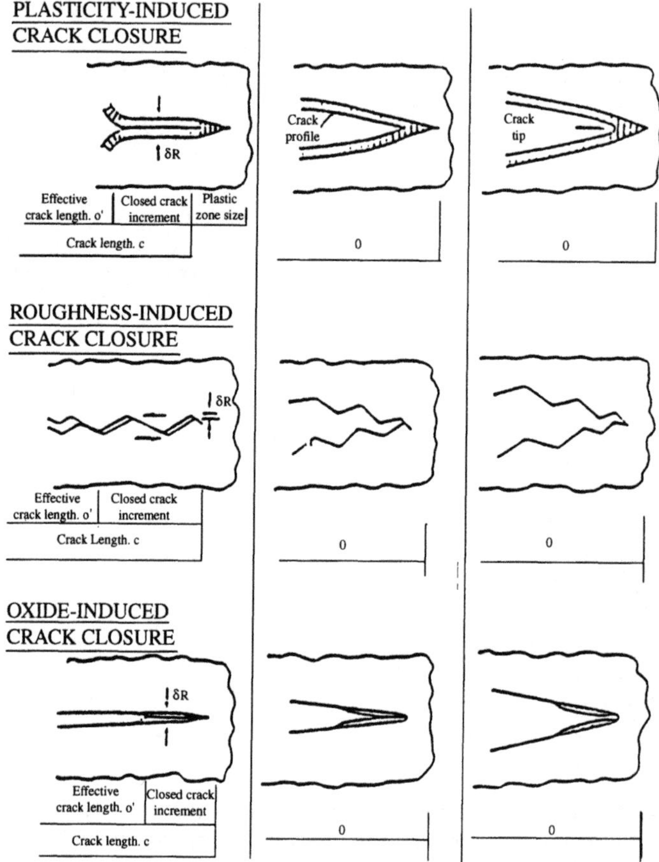

FIGURE 14.30 The three main mechanisms of crack closure: (a) plasticity-induced closure; (b) roughness-induced closure; (c) oxide-induced closure. The displacement, ∂_R, is the extra material wedged between the flanks. (From Fleck, 1983—reprinted with permission of Cambridge University.)

on the fatigue growth in the near-threshold regime. This was also the case even when the load-shedding sequence was acconted for in the finite element model (Newman, 1982).

Numerous independent studies have identified other mechanisms of closure. These include: (1) closure due to irregular fracture surface morphologies, i.e., roughness-induced closure [Fig. 14.30(b)], (2) closure due to corrosion deposits or the thickening of oxide layers, i.e., oxide-induced closure [Fig. 14.30(a)], (3) closure due to the penetration of viscous media into the crack (viscous fluid-induced closure), and (4) closure due to plasticity associated with phase transformation induced by the advancing crack tip (phase transformation-induced closure).

The major mechanisms of closure are reviewed in this section. The significance of the various components of closure is elucidated for near-threshold fatigue crack propagation, and the main methods of determining closure loads are also assessed.

14.9.2 Plasticity-induced Closure

Plasticity-induced closure (Elber, 1970, 1971) occurs as a result of residual plastic deformation in the wake of a propagating crack. Clamping stresses, due to the elastic constraint of surrounding material on the plastic wake, bring the crack faces into contact, and hence cause the closure of cracks even when the applied loads are tensile. This is illustrated in Fig. 14.30(a).

14.9.3 Roughness-induced Closure

Roughness-induced closure (Halliday and Beevers, 1979; Minakawa and McEvily, 1981) arises from contact between fracture surface asperities at discrete points behind the crack tip. These wedge open the crack at loads above the minimum load, as shown in Fig. 14.30(b). Also, the irreversible nature of crack-tip deformation and the possibility of slip step oxidation in most environments can lead to mismatch of fracture surface asperities and higher levels of roughness-induced closure (Suresh and Ritchie, 1982). In general, however, the level of roughness-induced closure depends on the height of the fracture surface asperities and the crack opening displacement. It will, therefore, be most significant in the near-threshold regime at low stress ratios, due to low crack opening displacements and rough fracture surface morphologies (Halliday and Beevers, 1979; King, 1981).

14.9.4 Oxide-induced Closure

Oxide-induced closure (Suresh et al., 1981; Soboyejo et al., 1990; Campbell et al., 1999) occurs in chemically active environments, e.g., water and moist laboratory air, where fretting debris or corrosion products may develop on the fracture surface behind the crack tip. Fretting is initiated by corrosion products and also by plasticity-induced and roughness-induced closure phenomena. This occurs by continual breaking and reforming of the oxide scale behind the crack tip due to repeated contact between the fracture surfaces caused by shear displacements (Mode II or Mode III) which are induced by contact between mating crack faces (Minakawa and McEvily, 1981). This results in the autocatalytic buildup (Romaniv et al., 1987) of hydrated oxide layers (Suresh et al., 1981). These thickened hydrated oxide layers, and the corrosion debris which can be formed in aggressive environments, wedge open the crack and enhance closure by causing premature contact between the fracture surfaces, as shown in Fig. 14.26c.

Oxide-induced closure is promoted by: (1) small crack-tip opening displacements (CTOD), such as in the near-threshold regime where the thickness of the excess debris is comparable with the CTOD; (2) highly oxidizing media such as water, where thermal oxidation is possible, in addition to, or in the absence of fretting oxidation (Suresh and Ritchie, 1983); (3) low load ratios where fretting is enhanced by repeated contact between fracture surfaces through low CTOD values (Suresh et al., 1981); (4) rough fracture surfaces, which at low load ratios, promote sliding and rubbing between mating fracture surfaces, thus enhancing fretting; and (5) lower strength materials where higher levels of plasticity-induced closure bring more asperities into contact to induce fretting associated with plasticity-induced and roughness-induced closure phenomena.

Evidence of oxide-induced closure in steels is often provided in the form of dark bands across the fracture surface. These usually occur in the near-threshold regime, as shown in Fig. 14.31. Measurements of the oxide thickness have also been made by Suresh and Ritchie (1983) using scanning Auger spectroscopy. Their peaks confirmed that oxide-induced closure is predominant when the oxide thickness is comparable with the CTOD.

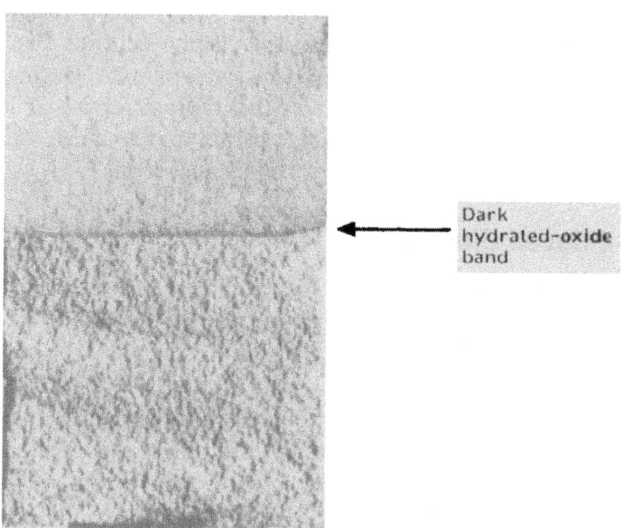

Figure 14.31 Dark hydrated oxide band (oxide thickening) in the near-threshold regime in Q1N (HY 80) pressure vessel steel.

14.9.5 Other Closure Mechanisms

It is also possible for crack closure to occur as a result of hydrodynamic wedging in the presence of viscous fluids (viscous fluid-induced closure) or from transformations induced by deformation ahead of the crack-tip (phase transformation-induced closure)—see Suresh and Ritchie (1984a) or Suresh (1999).

14.9.6 Closure Measurement Techniques

Conflicting results have been obtained by the different methods of measuring crack closure. In many cases, this is because the recorded closure load is a function of the sensitivity and type of closure instrumentation employed (Fleck, 1983). The three most popular methods of measuring crack closure are:

1. complianace techniques (Elber, 1970);
2. d.c. potential difference method (Irving et al., 1973);
3. ultrasonic techniques (Buck et al., 1973).

Crack opening and closing stress intensities, K_{op} and K_{cl}, can be determined from load-closure transducer traces obtained from any of these methods, as shown in Figs 14.32 and 14.33. It is important to note that precise locations of the points corresponding to the crack open-

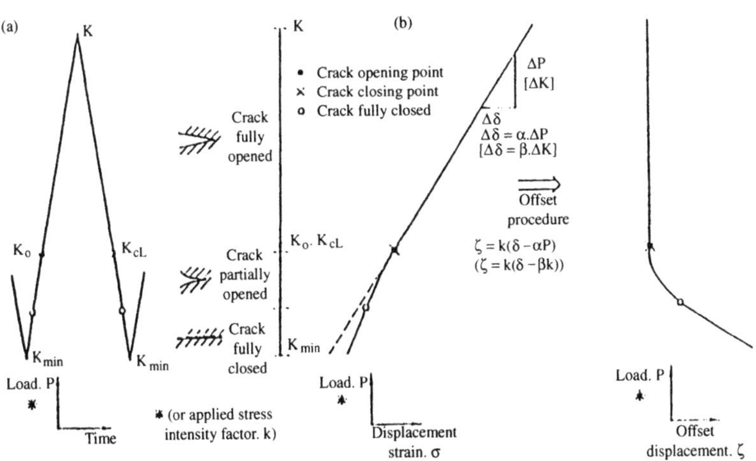

FIGURE 14.32 Determination of crack opening and closure loads by the compliance technique. (From Fleck, 1983—reprinted with permission of University of Cambridge.)

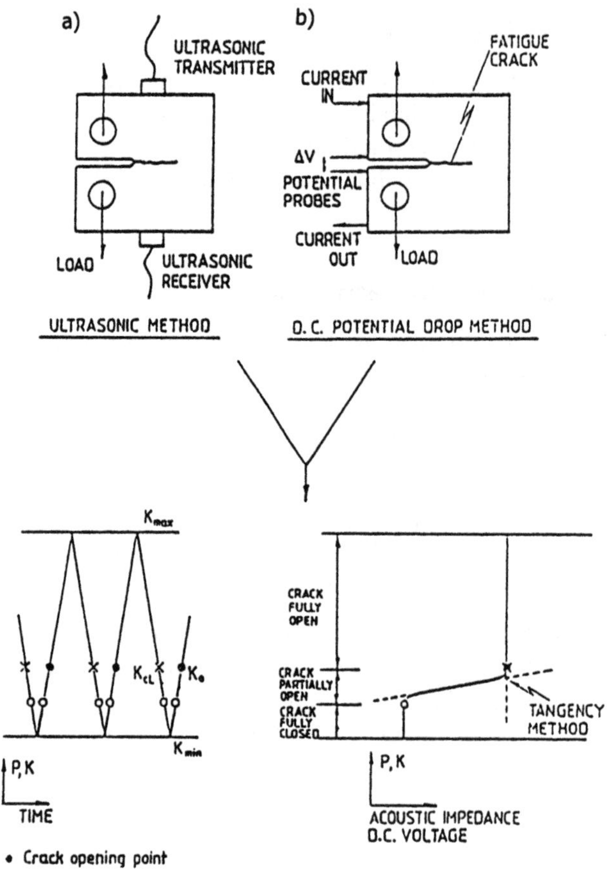

FIGURE 14.33 Determination of crack opening and closure loads: (a) ultrasonics; (b) d.c. potential drop method. (From Fleck, 1983—reprinted with permission of University of Cambridge.)

ing and closing loads on the load–transducer traces can be very difficult due to the tendency of cracks to open and close gradually. One was of overcoming this problem is to use the intersecting tangent method illustrated in Fig. 14.33(b). One tangent corresponds to the fully open crack, while the other corresponds to a fully closed crack. However, although this method may yield self-consistent closure data, it is questionable whether the resulting closure loads have any physical significance.

A more widely accepted method is illustrated in Fig. 14.32. The crack and closing loads, K_{op} and K_{cl}, are determined by identifying points corresponding to the deviation away from linearity in the load-closure traces (Dubey et al., 1997) during the forward and reverse cycles (Clerivet and Bathias, 1979; Gan and Weertmen, 1981; Fleck, 1983). These points are more difficult to obtain since cracks tend to open and close gradually. However, they do have some physical significance in that they correspond, respectively, within the limits of experimental error, to the point where the crack becomes fully open in the forward cycle and to the point of first contact between asperities in the reverse cycle. Improvements in the accuracy of the determination of the closure loads can also be obtained by incorporating an offset elastic displacement circuit into the compliance closure measurement system (Fleck, 1982). Typical plots of load versus offset displacement are shown in Fig. 14.32.

14.10 SHORT CRACK PROBLEM

Since the original discovery by Pearson (1975), the results of several experimental studies show "anomalous" short crack growth when compared with those obtained by fatigue growth in standard long-crack fracture mechanics specimens. Reviews are provided by Miller (1987), Smith (1983), and Suresh and Ritchie (1984b). Short cracks have been generally observed to grow anomalously, using along favored crystallographic planes, at stress intensities well below the long crack threshold. Typical sketches of da/dN versus ΔK are presented in Fig. 14.34.

In most cases, short crack growth rates exceed long crack growth rates at low ΔK values, although there is some evidence in steels of a mild reverse effect (Lankford, 1977; Romaniv et al., 1982). For very small cracks (10–100 μm), da/dN does not increase monotonically with ΔK. Instead, it usually decreases with increasing ΔK, until it reaches a minimum, which is often associated with a grain boundary (Lankford, 1982, 1985; Taylor and Knott, 1981). Some cracks become nonpropagating at the grain boundary if the crystallographic orientation in the next grain does not favor subsequent propagation. Other cracks exhibit the opposite trend, and accelerate to merge with long crack data in the early part of the Paris regime.

Anomalous short crack growth has been shown to be affected by stress ratio (Gerdes et al., 1984), grain size (Gerdes et al., 1984; Zurek et al., 1983), and environment (Gangloff and Wei, 1986). It has also been shown to be a stochastic process that can also be influenced by crack shape (Pineau, 1986; Wagner et al., 1986), microstructure

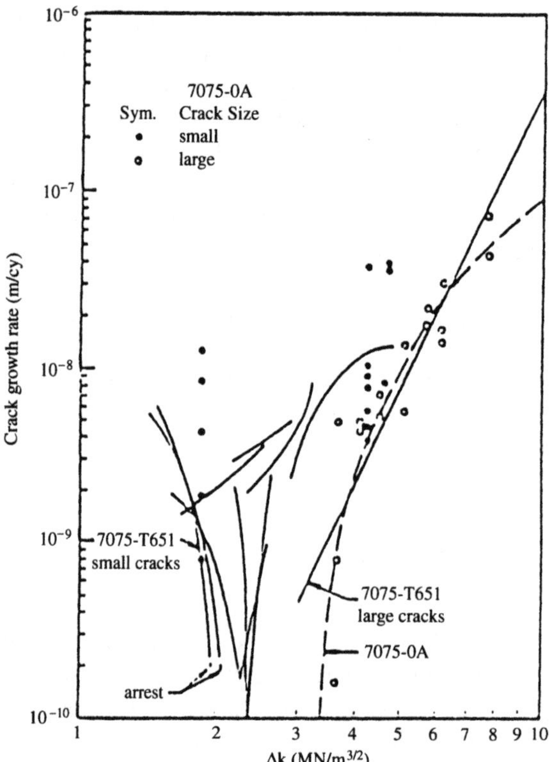

FIGURE 14.34 Typical anomalous short-crack growth rate data.

(Ravichandran, 1991), crack closure (Breat et al., 1983; Minakawa et al., 1983), enhanced crack-tip strains (Lankford and Davidson, 1986), and crack deflection (Suresh, 1983).

Ritchie and Lankford (1986) have classified the different types of short cracks: mechanically small cracks have lengths which are comparable with their plastic zones; microstructurally short cracks have lengths which are smaller than the controlling microstructural unit size; and physically small cracks are simply cracks whose lengths are less than 1 mm. Each of these cracks is associated with particular features that distinguishes it from long cracks, as shown in Table 14.1.

Table 14.1 also includes suggestions made by Ritchie and Lankford (1986) for the assessment of the different types of short cracks. These include the use of ΔJ, strain energy density, and ΔCTOD for mechanically short cracks; a probabilistic approach for microstructurally short cracks;

TABLE 14.1 Classes of Short Cracks

Type of small crack	Dimension	Responsible mechanism	Potential solution
Mechanically small	$a \lesssim r_y^a$	Excessive (active) plasticity	Use of ΔJ, ΔS, or crack-tip-opening displacement
Microstructurally small	$a \lesssim d_{lg}^b$	Crack tip shielding enhanced $\Delta\varepsilon_p$	Probabilistic approach
	$2c \lesssim (5\text{–}10)d_g$	Crack shape	
Physically small	$a \lesssim 1$ mm	Crack tip shielding (crack closure)	Use of ΔK_{eff}

[a] r_y is the plastic zone size or plastic field of notch.
[b] d_g is the critical microstructural dimension, e.g., grain size, a is the crack depth, and $2c$ the surface length.

and the use of effective SIFs, Δk_{eff}, for physically short cracks. Various parameters based on shear strains (Miller, 1987) and fatigue limits (Kitagawa and Takahashi, 1976) have also been proposed for the assessment of structures containing short cracks.

Appropriate analysis must, therefore, be used to characterize short crack growth. This is perhaps best illustrated using the "Kitagawa plot" of applied stress range, $\Delta\sigma_0$, versus crack length (Kitagawa and Takahashi, 1976), $2a$, shown in Fig. 14.35. Region III shows clearly S–N and LEFM long crack behavior. Region II represents the anomalous short-crack growth regime where LEFM and S–N curve data are generaly not applicable, and region I shows the conditions for nonpropagation cracks below the fatigue limit.

Also, failure to correlate short and long crack behavior does not necessarily indicate anomalous short crack behavior. In many cases, this may be due to the way in which the theory, usually LEFM, has been implemented. There are numerous examples of so-called anomalies which are simply the result of wrong or inappropriate applications of fracture mechanics criteria, e.g., due to the breakdown of similitude conditions (Ritchie and Suresh, 1983; Lies et al., 1986) or the use of load-controlled stress fields to describe the growth of short cracks from notches. The correct procedure in this case would be to superimpose the effect of the displacement-controlled stress field due to the notch on the crack-tip stress field (El Haddad et al., 1979; Lies et al., 1986).

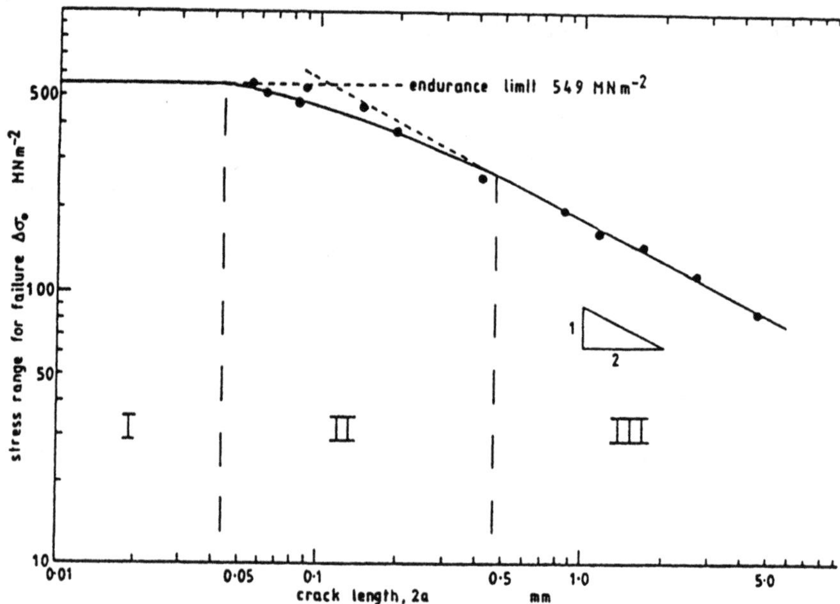

Figure 14.35 The "Kitagawa" plot. (From Kitagawa and Takahashi, 1976—Proc. ICM2.)

14.11 FATIGUE GROWTH LAWS AND FATIGUE LIFE PREDICTION

In addition to the Paris law [Eq. (14.7)], a number of other empirical fatigue crack growth laws have been proposed to relate fatigue crack growth rates to mechanical variables. These include crack growth laws by Walker (1970) and logarithmic and hyperbolic sine functions that have been used to fit fatigue crack growth rate data in the three regimes of fatigue crack growth. Most recently, Soboyejo et al. (1998, 2000b) have proposed a multiparameter fatigue crack growth law that can account for the effects of multiple variables, such as the SIF, K_{max}, temperature, T, and cyclic frequency. This gives the fatigue crack growth rate, da/dN, as

$$\frac{da}{dN} = \alpha_0\,(\Delta K)^{\alpha_1}(R)^{\alpha_2}(K_{max})^{\alpha_3}(T)^{\alpha_4}(f)^{\alpha_5} \tag{14.13}$$

where α_0, α_1, α_2, α_3, α_4, and α_5, are constants that can be determined by multiple linear regression (Soboyejo et al., 1998, 2000b). It is important to note here that the above constants need not be independent. In fact, Pearson correlation coefficients between each of the variables may be used to ascertain the degree to which each of the variables are linearly correlated. Furthermore, not all the variables need to be considered in Eq. 14.13. Hence, variables with exponents close to zero may be neglected, since any variable raised to the power zero is equal to 1.

The multiparameter law has been shown to apply to steels (Soboyejo et al., 1998), titanium alloys (Dubey et al., 1997; Soboyejo et al., 2000b), and nickel base alloys (Mercer et al., 1991a,b). It appears, therefore, to be generally applicable to the characterization of the effects of multiple variables on fatigue crack growth rate data.

In general, the multiparameter law may be expressed as

$$C = \left(\propto_0 \prod_{i=2}^{k} X_i^{\propto_i} \right) = \propto_0 X_2^{\propto_2} X_3^{\propto_3} \ldots, X_k^k \tag{14.14a}$$

and

$$\frac{da}{dN} = C(\Delta K)^m = \left(\propto_0 \prod_{i=1}^{k} X_i^{\propto_i} \right) (\Delta K)^m \tag{14.14b}$$

Note that $X_1 = \Delta K$ and $\propto_1 = m$ in the above formulation. The multiparameter law, therefore, provides a simple extension of the Paris law that reduces the problem of fatigue life prediction to the integration of a simple equation, Eq. (14.14b). This may be recognized by separating the variables in this equation and integrating between the appropriate limits:

$$\int_{a_0}^{a} da = \int_{0}^{N} \propto_o \prod_{i=1}^{k} X_i^{\propto_i} dN \tag{14.15}$$

where a_0 is the initial crack length (determined often via nondestructive inspection), a is the current crack length after N cycles of cycling load, and other parameters have their usual meaning. Since the X_i variables are often functions of crack length, a, numerical integration schemes are generally needed to solve Eq. (14.15). This may be done using available mathematical software packages.

For the simple single parameter case, in which ΔK is the dominant variable, Eq. (14.14b) may be expressed as

$$\int_{a_0}^{a} da = \int_{0}^{N} C(\Delta K)^m dN \tag{14.16}$$

where all the variables have their usual meaning. Since ΔK is also generally a function of a, numerical integration schemes are needed to solve Eq.

(14.15) or (14.16) in most cases. However, Eq. (14.16) may be solved analytically for only the simplest cases in which the geometric function, $F(a/W)$, in the equation for K, is not dependent on a.

One example is the case of a small edge crack of length a, subjected to Mode I loading through an applied stress range, $\Delta\sigma$. The SIF range, ΔK, is now given by (Tada et al., 1999):

$$\Delta K = 1.12(\Delta K)(\sqrt{\pi a}) \tag{14.17}$$

Substituting Eq. (14.17) into Eq. (14.16) gives

$$\int_{a_0}^{a} \mathrm{d}a = \int_{0}^{N} C(1.12(\Delta\sigma)(\sqrt{\pi a}))^m \mathrm{d}N \tag{14.18a}$$

or

$$\int_{a_0}^{a} \frac{\mathrm{d}a}{a^{m/2}} = \int_{0}^{N} C(1.12(\Delta\sigma)(\sqrt{\pi}))^m \mathrm{d}N \tag{14.18b}$$

Equation (14.18b) can be solved analytically to obtain:

$$N = \frac{2\left[a^{-\frac{m}{2}+1} - a_0^{-\frac{m}{2}+1}\right]}{C(2-m)[1.12(\Delta\sigma)(\sqrt{\pi})]^m} \tag{14.19}$$

Equation (14.19) gives the number of fatigue cycles required to reach a crack length, a, under remote loading at a stress of $\Delta\sigma$. Hence, it may be used to estimate the fatigue life if a corresponds to the critical crack length, a_c, at the onset of catastrophic failure, from the condition at which K_{max} (during the fatigue cycle) is equal to the material fracture toughness, K_{Ic}:

$$K_{Ic} = F\left(\frac{a}{W}\right)(\Delta\sigma)(\sqrt{\pi a_c}) \tag{14.20a}$$

or

$$a_c = \frac{1}{\pi}\left(\frac{K_{Ic}}{F\left(\frac{a}{W}\right)\Delta\sigma}\right)^2 \tag{14.20b}$$

Hence, the number of cycles to failure, N_f, may be estimated by using a_c as the upper integration limit in Eqs (14.15) and (14.16). This provides the basis for the estimation of fatigue life within a fracture mechanics framework.

14.12 FATIGUE OF COMPOSITES

Before closing, it is of interest to discuss the micromechanisms of fatigue in composite materials. As the reader can imagine, no unified theory of fatigue can be applied to the rationalization of fatigue in composite materials. However, it is possible to summarize some themes that have emerged from research over the past few decades. A significant fraction of prior work on composites has been done on titanium matrix composites (Harmon and Saff, 1989; Johnson et al., 1990; Sensemeier and Wright, 1990; Jeng et al., 1991; Kantzos et al., 1991; Walls et al., 1991; Majumdar and Newaz, 1992; Soboyejo and coworkers, 1994, 1997).

Most of the above studies have shown that the fibers tend to trap the cracks, giving rise ultimately to the formation of bridged cracks, as shown in Figs. 14.36 for a titanium matrix composite reinforced with TiB whiskers and $TiSi_2$ particles. The observed crack bridging mechanisms give rise to slower composite fatigue crack growth rates compared to those in the titanium matrix alloy (Fig. 14.37). Similar crack trapping and crack bridging phenomena have also been observed in other titanium matrix composites (Sensemeier and Wright, 1990; Jin and Soboyejo, 1997; Soboyejo et al., 1997).

To illustrate the multitude of processes that can occur during the cyclic deformation of composites, photomicrographs of the cyclically deformed gauge sections of a Ti–15Al–3Cr–3Al–3Sn/SiC composite are presented in Fig. 14.38, which is taken from a paper by Jin and Soboyejo (1997). In this composite, which is deformed at a stress range corresponding to ~ 0.5 of its ultimate tensile strength (UTS), interfacial debonding and interfacial radial cracking occur within 10% of the fatigue life, Fig. 14.38(a). This is followed by stress-induced α-phase precipitation, and cracking of the molybdenum cross weave that is used to hold the composite together, at $0.2N_f$, Fig. 14.38(b). The carbon coating on the SiC then cracks [Fig. 14.38(c)] and further stress-induced precipitation is observed at $0.3N_f$, Fig. 14.38(d). Fiber cracking is then observed at $\sim 0.5N_f$ [Fig. 14.38(e)] along with stress-induced microvoid formation at the interface between the matrix and the TiC interphase, Fig. 14.38(f). Damage in composite materials may, therefore, involve interfacial cracking, interfacial debonding, matrix and/or fracture, fiber pull-out. The actual sequence of fatigue damage will, therefore, vary widely, depending on the properties of the matrix, fiber, interface, and interphases.

In cases where crack bridging is observed in composites under cyclic loading, significant efforts have been made to develop bridging laws for the estimation of the shielding components due to fiber bridging. The pioneering work in this area was done by Marshall et al. (1985), who developed a

(a)

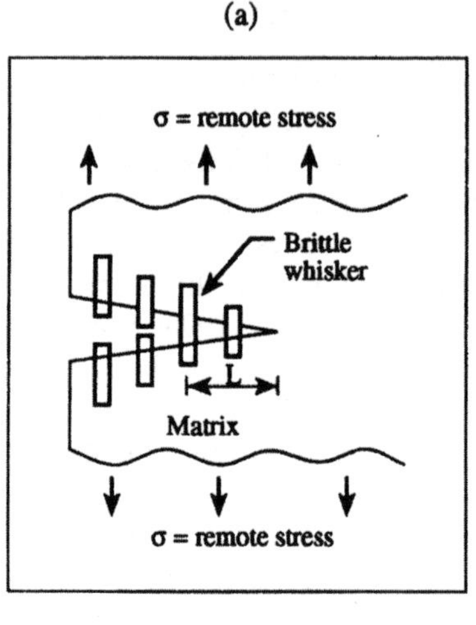

(b)

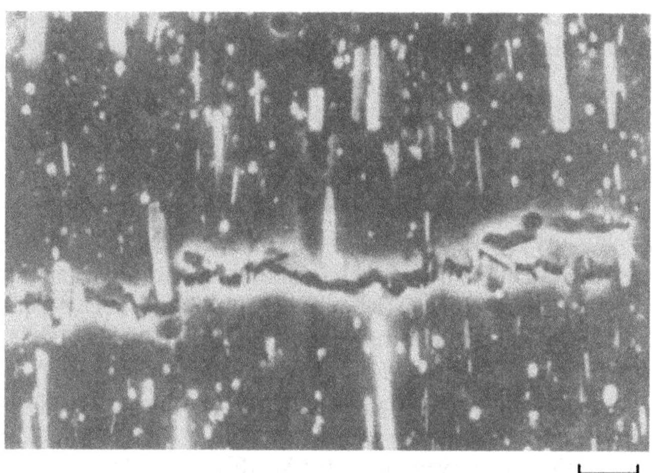

FIGURE 14.36 Elastic crack bridging in Ti–8.5Al–1B–1Si: (a) schematic illustration; (b) actual crack-tip region.

modeling framework for analyzing the shielding associated with fiber bridging in cases where frictional bonding exists between the matrix and the fibers. Following the analysis of Marshall et al. (1985), the bridging traction, $p(x)$, on penny cracks of size, a, that are greater than the fiber spacing is given by

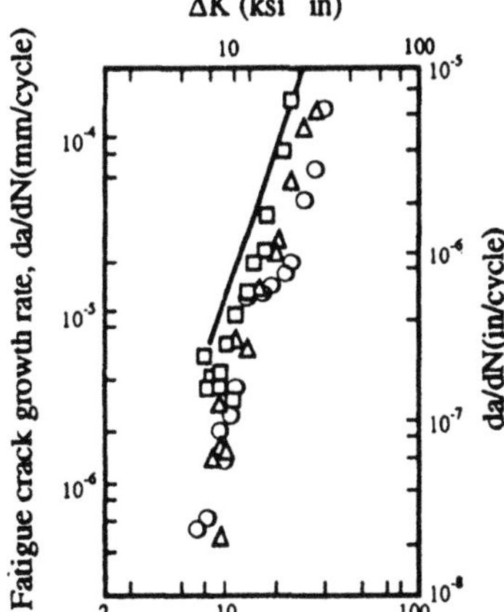

Figure 14.37 Comparison of fatigue crack growth rates in titanium matrix alloy and titanium matrix composites. (From Soboyejo et al., 1994—reprinted with permission of Elsevier Science.)

$$p(x) = V_f \sigma_f(x) \tag{14.21}$$

where V_f is the volume fraction of fibers, and $\sigma_f(x)$ is the stress in the fibers expressed as a function of distance, x, from the crack-tip (Fig. 14.39). The stress intensity factor, K, at the crack tip is thus reduced by the closure pressure via:

$$K = 2\sqrt{\frac{a}{n}} \int_0^1 \frac{[\sigma^\infty - p(x)]X\mathrm{d}X}{\sqrt{1 - x^2}} \tag{14.22}$$

where $X = x/a$, and the closure pressure is related to the crack opening displacement via:

$$p(x) = \propto \sqrt{u(x)} \qquad (\text{for } x > a_0) \tag{14.23a}$$

or

$$p(x) = 0 \qquad (\text{for } x < a_0) \tag{14.23b}$$

where

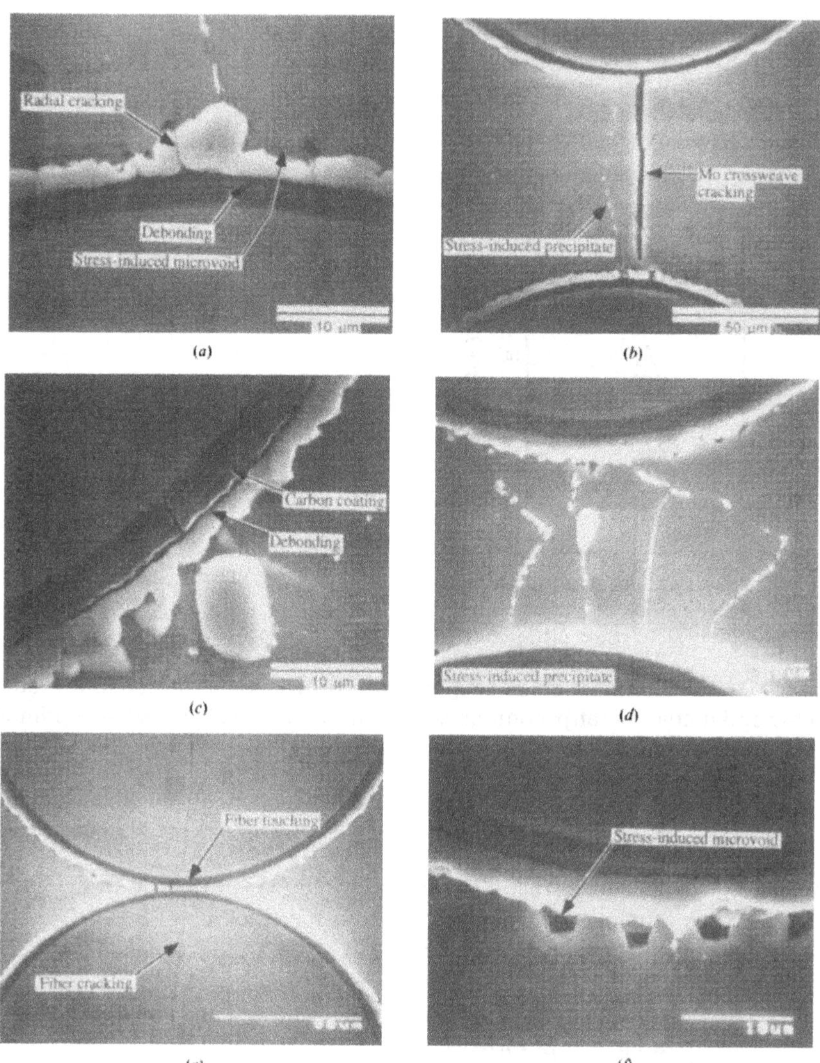

FIGURE 14.38 Damage mechanisms in four-ply [75] Ti–15V–3Cr–3Al–3Sn composite reinforced with carbon-coated SiC fibers and cyclically deformed at $\Delta\sigma = 0.5\ \sigma_{UTS}$: (a) debonding and reaction layer cracking after 0.1 N_f; (b) cracking of Mo wire and some stress-induced precipitate after 0.2 N_j; (c) crack propagation along the reaction layer and the interface between coating and SiC fiber after 0.3 N_f; (d) stress-induced precipitation (silicon-riched precipitates) along the beta grain boundaries after 0.3 N_f; (e) fiber cracking after 0.5 N_f, (f) stress-induced microvoid formation around the fiber after 0.6 N_f. (Reprinted with permission of TMS, Warrendale, PA.)

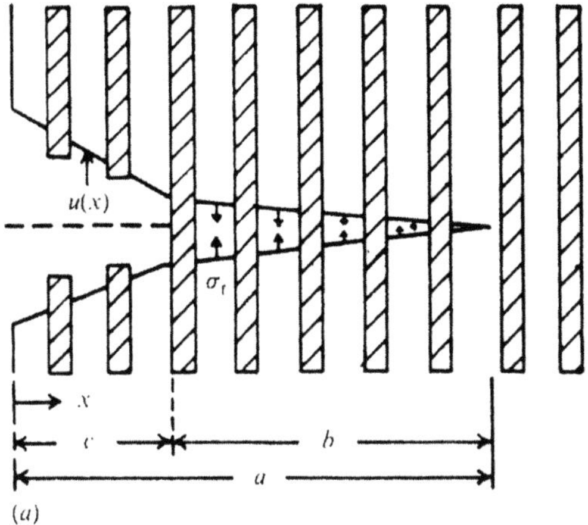

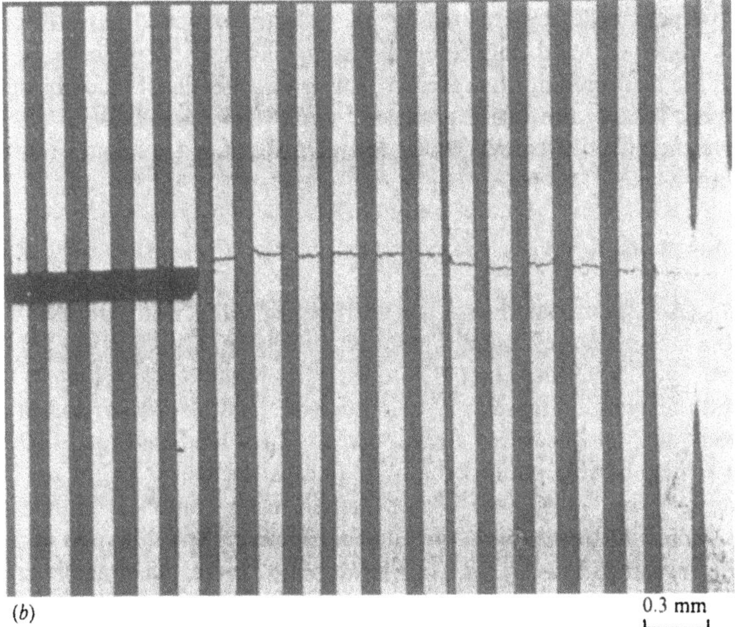

Figure 14.39 Schematic illustration of fiber bridging. (From Sensemeier and Wright, 1990—reprinted with permission of TMS, Warrendale, PA.)

$$\propto = \frac{4\tau F^2 E_f E_c}{R E_m (1 - V_f)} \tag{14.24}$$

where R is the factor radius, τ is the frictional stress, E_f is the fiber Young's modulus, E_c is the composite Young's modulus, E_m is the matrix Young's modulus, and V_f is the fiber volume fraction. The crack opening displacement, $u(x)$, behind the crack tip is given by

$$u(X) = \frac{4(1-\nu^2)a}{\pi E_c} \int_x^1 \frac{1}{\sqrt{s^2 - x^2}} \left(\int_0^s \frac{[\sigma^\infty - p(\xi)]\xi}{\sqrt{s^2 - \xi^2}} \right) ds \tag{14.25}$$

where s and ξ are normalized position co-ordinates, and ν is Poisson's ratio for the composite. The displacement function above may be evaluated readily using numerical integration schemes. The effective SIF may thus be evaluated by numerical integration of Eqs (14.25) and (14.28).

However, under cyclic loading, the effective crack driving force is different under forward and revered loading. This is partly because the tractions are recovered during revered loading. The details of crack bridging under cyclic loading have been modeled by McMeeking and Evans (1990). They give the effective crack-tip SIF range, ΔK_{tip}, for a zero mean stress/stress ratio as

$$\Delta K_{tip} = 2K_{tip}\left(\frac{\Delta\sigma}{2}\right) \tag{14.26}$$

where $K_{tip}(\Delta\sigma/2)$ corresponds to the near-tip SIF for a bridged crack subjected to a stress of $(\Delta\sigma/2)$ under monotonic loading. Equation (14.26) has been used by Soboyejo et al. (1997) to estimate the effective crack-tip SIF range and the overall fatigue to a wide range of stress. A comparison of the predictions and the measured fatigue lives is presented in Fig. 14.40 for possible values of frictional strength, τ, and initial crack length a_0. The predictions in Fig. 14.40 have been obtained using numerical integration schemes similar to those alluded to earlier. However, ΔK_{tip} is used instead of ΔK in the simulations, in an effort to account for the shielding effects of crack bridging. The values of τ have also been varied to stimulate the variations due to fatigue degradation of the interfacial strength (Eldridge, 1991). The bounds in the predicted stress-life behavior are generally consistent with the measured stress-life behavior.

14.13 SUMMARY

An overview of the fatigue of materials is presented in this chapter. Following a brief historical perspective and a review of empirical

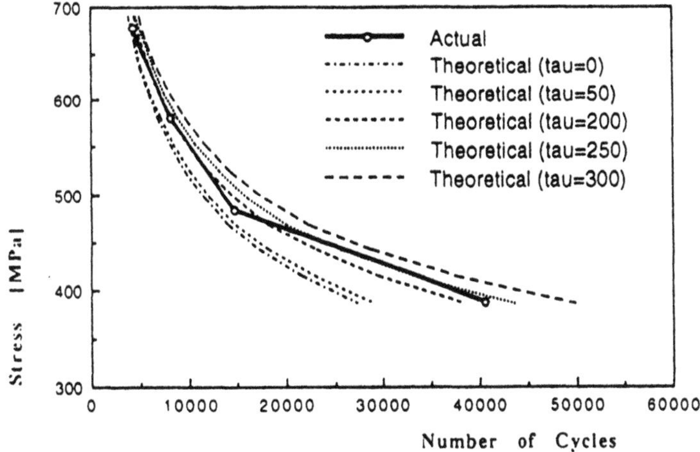

FIGURE 14.40 Comparison of measured and predicted stress-life behavior obtained for Ti–15V–3Cr–3Al–3Sn/SiC composite. (From Soboyejo et al., 1996—reprinted with permission of TMS, Warrendale, PA.)

approaches to fatigue, fracture mechanics approaches are presented along with descriptions of the underlying mechanisms of fatigue in different classes of materials. Mechanisms of crack closure are also elucidated before discussing the so-called short-crack problem. Fatigue crack growth laws are then presented within the context of fatigue life prediction. Finally, more complex mechanisms of fatigue and crack-tip shielding are examined in selected composite materials.

BIBLIOGRAPHY

Albert, W.A. (1838) J Arch Miner. vol. 10, p 215.

Argon, A.S. and Goodrich, J.A. (1969) In: P.L. Pratt, ed. Fracture. Chapman Hall, London, pp 576–586.

Basquin, O.H. (1910) ASTM. vol. 10, p 625.

British Standards Institution BS 3518 (1963).

Buck, O., Ho, C.L., and Marcus, H.L. (1973) Eng Fract Mech. vol. 5, p 23.

Clerivet, A and Bathias, C. (1979) Eng Fract Mech. vol. 12, p 599.

Coffin, L.F., (1954) Trans ASME. vol. 76, p 923.

Cottis, R.A. (1986) In: R.O. Ritchie and D.L. Davidson, eds. Small Fatigue Cracks. p 265.

Cottrell (1958) Trans Metall Soc. pp 192–203.

Cottrell, A.H. and Hull, D. (1957) Proc Roy Soc. vol. 242A, p 211.

Dauskardt, R.H. (1993) Acta Metall Mater. vol. 41, pp 2165–2181.

Dauskardt, R.H. (1990) J Am Ceram Soc. vol. 73, pp 893–903.

Davidson, D.L. and Campbell, J.B. (1993) Metall Trans. vol. 24A, p 1555.

Davidson, D.L. and Suresh, S., eds (1884) Fatigue Crack Growth Threshold Concepts. The Metallurgical Society of The American Institute of Mining, Metallurgical and Petroleum Engineers, Warrendale, PA.

Dowling, N.E. and Begley, J.A. (1976) Mechanics of Crack Growth, ASTM STP 590. American Society for Testing and Materials, Conshohocken, PA, pp 82–103.

Dubey, S., Soboyejo, A.B.O., and Soboyejo, W.O. (1997) Acta Metall Mater.

Elber, W. (1970) Eng Fract Mech. vol. 2, p 37.

Elderidge, J. (1991) Desktop Fiber Push-Out Apparatus, NASA Technical Report Memorandum 105341. NASA Glenn Research Center, Cleveland, OH.

El Haddad, M.H., Topper, T.H., and Smith, T.N. (1979) Eng Fract Mech. vol. 11, p 572.

Evans, A.G. (1980) Int J Fract. vol. 16, pp 485–498.

Evans, A.G. and Fuller, E. (1974) Metall Trans. vol. SA, pp 27–33.

Ewing, J.A. and Humfrey, J.C.W. (1903) Proc Roy Soc. vol. 200A, p 241.

Fine, M.E. and Ritchie, R.O. (1979) Proceedings of the ASM Materials Science Seminar, St. Louis, MO, p 245.

Fleck, N.A. (1983) PhD thesis, University of Cambridge, Cambridge, UK.

Fleck, N.A. and Smith, R.A. (1982) Int J Fatigue. July, p 157.

Forsyth, P.J.E. (1961) Proceedings of the Crack Propagation Symposium, vol. 1. Cranfield College of Aeronautics, Cranfield, UK, p 76.

Forsyth, P.J.E. and Ryder, D.A. (1961) Metallurgia. vol. 63, p 117.

Frost, N.E. and Dugdale, D.S. (1958) J Mech Phys Solids. vol. 6, p 92.

Frost, N.E., Marsh, K.J., and Pook, L.P. (1974) Metal Fatigue. Clarenden Press, Oxford, UK.

Gan, D. and Weertman, J. (1981) Eng Fract Mech. vol. 15, p 87.

Gangloff, R.P. and Wei, R.P. 1986) In: R.O. Ritchie and D.L. Davidson, eds. Small Fatigue Cracks.

Gerber, W. (1874) Z Bayer Archit. Ing Ver. vol. 6, p 101.

Gerdes, C., Gysler, A., and Lutjering, G., eds (1984) Fatigue Crack Growth Threshold Concepts.

Goodman, J. (1899) Mechanics Applied to Engineering. Longman, Green.

Grathwohl, G. (1998) Materialwissenschaft Werkstoffech vold. vol. 19, pp 113–124.

Hall, E.O. (1951) Proc Phys Soc. vol. 64B, p 747.

Halliday, M.D. and Beevers, C.J. (1979) Int J Fract. vol. 15, p R27.

Ham, R.K. (1966) Can Metall Q. vol. 5, p 161.

Harmon, D. and Saff, C.R. (1989) In: W.S. Johnson, ed. Metal Matrix Composites Testing, Analysis and Failure Modes, ASTM STP 1032. ASTM, Conshohocken, PA, p 194.

Head, A.K. (1956). J. Appl. Mech., 23, 78, p. 407.

Hertzberg, R. and Manson, J.A. (1980) Fatigue of Engineering Plastics. Academic Press, New York.

Horibe, S. and Hirihara, R. (1991) Fatigue Fract Eng Mater Struct. vol. 14, p 863.

Illston, J.M., Dinwoodie, J.M., and Smith, A.A. (1979) Concrete, Timber and Metals. Van Nostrand Reinhold, Crystal City, VA.

Irving, P.E., Robinson, J.L., and Beevers, C.J. (1973) Int J Fract. vol. 9, p 105.

Irwin, G.R. (1957) J Appl Mech. vol. 24, p 361.

Jeng, S.M., Yang, J.M., and Yang, C.J. (1991) Mater Sci Eng A. vol. A138, p 169.

Jin, O., Li, Y. and Soboyejo, W.O. (1997) Metall Mater Trans A. vol. 28A, p 2583.

Johnson, W.S., Lubowinski, S.J., and Highsmith, A.L. (1990) In: J.M. Kennedy, H.H. Mueller, and W.S. Johnson, eds. Thermal and Mechanical Behavior of Metal Matrix and Ceramic Matrix Composites. ASTM STP 1080, ASTM, Conshohocken, PA, p 193.

Kantzos, P., Telesman, J., and Ghosn, L. (1991) In: T.K. O'Brien, ed. Composite Materials; Fatigue and Fracture. ASTM STP 1110, ASTM, Conshohocken, PA, p 711.

Kendall, J.M. (1986) PhD thesis, University of Cambridge.

King, J.E. (1981) Fatigue Eng Mater Struct. vol. 4, p 311.

Kishimoto, K., Ueno, A., Okawara, S., and Kawamoto, H. (1994) J Am Ceram Soc. vol. 77, pp 1324–1328.

Kitagawa, H. and Takahashi, S. (1976) Proc ICM2. Boston, MA, p 627.

Knott, J.F. (1973) Fundamentals of Fracture Mechanics. Butterworth-Heinemann, Oxford, UK.

Kuo, A.S. and Liu, H.W. (1976) Scripta Met. vol. 10, p 723.

Laird, C. (1967) ASTM STP 415, p 139.

Laird, C. and Smith, G.C. (1962) Phil Mag. vol. 7, p 847.

Laird, C. and Smith, G.C. (1963) Phil Mag. vol. 8, p 1945.

Lankford, J. (1977) Eng Fract Mech. vol. 9, p 617.

Lankford, J. (1982) Fatigue Eng Mater Struct. vol. 5, p 233.

Lankford, J. (1985) Fatigue Eng Mater Struct. vol. 8, p 161.

Lankford, J. and Davidson, D.L. (1986). In: R.O. Ritchie and J.Lankford, eds. Small Fatigue Cracks. TMS, Warrendale, PA.

Lathabai, S., Rodel, J., and Lawn, B.R. (1991) J Am Ceram Soc. vol. 74, pp 1340–1348.

Lewandowski, J.J., Hipplsey, C.A., Ellis, M.B.D., and Knott, J.F. (1987) Acta Met. vol. 35, p 593.

Lies, B.N., Topper, A.T., Ahmed, J., Broek, D., and Kannine, M.F. (1986) Eng Fract Mech. vol. 5, p 883.

Lindley, T.C. and Richards, C.E. (1974) Mater Sci Eng., vol. 14, p 281.

Liu, H.W. (1961) Trans ASME. vol. 83, p 23.

Lukas, P., Klesnil, M., and Polak, J. (1974) Mater Sci Eng., vol. 15, p 239.

Majumdar, B.S. and Burns, S.J. (1987) J Mater Sci. vol. 22, pp 1157–1162.

Majumdar, B.S. and Newaz, G.M. (1992) Phil Mag A. vol. 66, p 187.

Manson, S.S. (1954) US Nat Advis Cttee Aeronaut, Tech. Note 2933.

Marshall, D.B., Cox, B.N., and Evans, A.G. (1985) Acta Metall. vol. 33, pp 2013–2021.

McEvily, A.J., Jr. and Illg, W. (1958) The Rate of Propagation in Two Aluminum Alloys. NASA TN 4394, September.

McMeeking, R.M. and Evans, A.G. (1990) Mech Mater. vol. 9, pp 217–232.
Mercer, C. and Soboyejo, W.O. (1995) Scripta Mater. vol. 33, p 1169.
Mercer, E. and Soboyejo, W.O. (1997) Acta Metall Mater. vol. 45, pp 961–971.
Mercer, C., Soboyejo, A.B.O., and Soboyejo, W.O. (1999a) Acta Mater. vol. 47, p 2727.
Mercer, C., Soboyejo, A.B.O., and Soboyejo, W.O. (1999b) Mater Sci Eng A. vol. A270, p 308.
Meyn, D.A. (1968) Trans ASM. vol. 61, pp 42–51.
Miller, K.J. (1987) Fatigue Fract Eng Mater Struct. vol. 10, p 75.
Minakawa, K. and McEvily, A.J. (1981) Scripta Met. vol. 15, p 633.
Mulvihill, P.M. and Beevers, C.J. (1986) In: K.J. Miller and E.R. de las Rios, eds. The Behaviour of Short Fatigue Cracks. Institute of Mechanical Engineers, London, UK, p 203.
Neumann, P. (1969) Acta Met. vol. 17, p 1219.
Neumann, P. (1974) Acta Met., vol. 22, p 1155.
Newman, J.C. (1982) Proceedings of AGARD 55th Specialists' Meeting on Behaviour of Short Cracks in Airframe Components. Toronto, AGARD-CP-328.
Orowan, E. (1949) Rep Phys. vol. 12, p 185.
Paris, P.C. and Erdogan, F. (1963) Trans ASME, J Basic Eng. vol. 85, p 528.
Paris, P.C., Gomez, M. and Anderson, W.E. (1961) A Rational Analytic Theory of Fatigue, The Trend in Engineering, vol. 13. University of Washington, Seattle, WA.
Pearson, S. (1975) Eng Fract Mech. vol. 7, p 235.
Pelloux, R.M.N. (1969) Trans ASM. vol. 62, p 281.
Pelloux, R.M.N. (1970) Eng Fract Mech. vol. 1, p 697.
Petch, N.J. (1953) J Iron Steel Inst. vol. 173, p 25.
Pineau, A. (1986) In: R.O. Ritchie and D.L. Davidson, eds. Small Fatigue Cracks. AIME, Warrendale, PA, p 191.
Ramasundaram, P., Bowman, R., and Soboyejo, W.O. (1998) Metall Mater Trans. vol. 29A, p 493.
Rankine, W.J. (1843) Proc Inst Civ Engrs. vol. 2, p 105.
Rao, V., Soboyejo, W.O., and Ritchie, R.O. (1992) Metall Mater Trans. vol. 23A, pp 2249–2257.
Ravichandran, K.S. (1991) Acta Mater. vol. 39, p 420.
Ritchie, R.O. (1979) Int. Met Rev. vol. 20, p 205.
Ritchie, R.O. and Knott, J.F. (1973) Acta Met. vol. 21, p 639.
Ritchie, R.O. and Lankford, J. (1986) Mater Sci Eng. vol. 84, p 11.
Ritchie, R.O. and Suresh, S. (1983) Mater Sci Eng. vol. 57, p 27.
Ritchie, R.O., Gilbert, C.J., and McNaney, J.M. (2000) Int J Solids Struct. vol. 37, pp 311–329.
Romaniv, O.N., Tkach, A.N., and Lenetz, Yu. N. (1987) Fatigue Eng Mater Struct. vol. 10, p 203.
Schijve, J. (1979) Eng Fract Mech. vol. 11, p 167.
Sensemeier, M.D. and Wright, P.K. (1990) In: D.K. Liaw and M Gungor, eds. Fundamental Relationships between Microstructure and Mechanical Properties of Metal Matrix Composites. TMS, Warrendale, PA, p 441.

Shademan, S. (2000) PhD thesis, The Ohio State University, Columbus, OH. Smith, R.A. (1983) ASTM STP 811, p 264.

Smith, R.A. (1984) In: R.A. Smith, ed. Fatigue Crack Growth—Thirty Years of Progress. Pergamon Press, Oxford, UK, p 1.

Soboyejo, W.O., Aswath, P.B., and Deffeyes, J. (1990) Mater Sci Eng. vol. A138, p 95.

Soboyejo, W.O., Lederich, R.J., and Sastry, S.M.L. (1994) Acta Metall Mater. vol. 30, p 1515.

Soboyejo, W.O., Chen, L.-C., Bahtishi, N., Ye, F., Schwartz, D., and Lederich, R.J. (1996) Acta Metall Mater. vol. 44, p 2027.

Soboyejo, W.O., Rabeeh, B., Li, Y.-L., Chu, W., Lavrentenyev, A., and Rokhlin, S. (1997) Metall Mater Trans. vol. 28A, p 1667.

Soboyejo, W.O., Ni, Y., Li, Y., Soboyejo, A.B.O., and Knott, J.F. (1998) Fatigue Eng Mater Struct. vol. 21, p 541.

Soboyejo, W.O., Mercer, C., Schymanski, J., and Van der Laan, S. (2001) An investigation of thermal shock in a high temperature refractory ceramic. J Am Ceram Soc. vol. 84, pp 1309–1314.

Soboyejo, A.B.O., Foster, M.A., Mercer, C., Papritan, J.C., and Soboyejo, W.O. (2000b) In: P.C. Paris and K.L. Ferina, eds. Fatigue and Fracture Mechanics. ASTM STP 1360, ASTM, Conshohocken, PA, p 327.

Stoloff, N.S. (1996) Fatigue and fracture of high-temperature intermetallics. In: N.S. Stoloff and R.H. Jones, eds. Processing and Design Issues in High-Temperature Materials. TMS, Warrendale, PA, p. 195.

Suresh, S. (1983) Met Trans. vol. 14A, p 2375.

Suresh, S. (1999) Fatigue of Materials. 2nd ed. Cambridge University Press, Cambridge, UK.

Suresh, S. and Ritchie, R.O. (1982) Met Trans. vol. 13A, p 1627.

Suresh, S. and Ritchie, R.O. (1983) Eng Fract Mech. vol. 18, p 785.

Suresh, S. and Ritchie, R.O. (1984a) In: D.L. Davidson and S. Suresh, eds. Fatigue Crack Growth Threshold Concepts. The Metallurgical Society of the American Institute of Mining, Metallurgical and Petroleum Engineering, Warrendale, PA.

Suresh, S. and Ritchie, R.O. (1984b) Int Met Rev. vol. 29, p 445.

Suresh, S., Zamiski, G.F., and Ritchie, R.O. (1981) Metall Trans. vol. 12A, pp 1435–1443.

Tada, H., Paris, P.C., and Irwin, G. (1999) The Stress Analysis of Cracks Handbook. ASME, New York.

Taylor, D. and Knott, J.F. (1981) Fatigue Eng Mater Struct. vol. 4, p 147.

Timoshenko, S.P. and Goodier, J.N. (1970) Theory of Elasticity. 3rd ed. McGraw-Hill, New York, NY.

Thompson, N., Wardsworth, N.J., and Louat, N. (1956) Phil. Mag. vol. 1, p 113.

Tomkins, B. (1968) Phil Mag. vol. 18, p 1041.

Tomkins, B. and Biggs, W.D. (1969) J Mater Sci. vol. 4, p 544.

Wagner, L., Gregory, J.K., Gysler, A., and Lutjering, G. (1986) In: R.O. Ritchie and J. Lankford, eds. Small Fatigue Cracks. TMS, Warrendale, PA, p 117.

Walker, K. (1970) Effects of Environment and Complex Load History for Fatigue Life. ASTM STP 462, American Society for Testing and Materials, Conshohocken, PA, pp 1–15.

Walls, D., Bao, G., and Zok, F. (1991) Scripta Metall. vol. 25, p 911.
Westergaard, H.M. (1939) Trans ASME. vol. 61, pp A-49–A53.
Wiederhorn, S. (1980) Metal Sci. Aug–Sept, p 450.
Williams, M.L. (1957) J. Appl Mech. vol. 24, p 109.
Wohler, A. (1858) Z Bauw. vol. 8, p 642.
Wohler, A. (1860) Z Bauw. vol. 10, p 583.
Wohler, A. (1863) Z Bauw. vol. 13, p 233.
Wohler, A. (1866) Z Bauw. vol. 16, p 67.
Wohler, A. (1870) Z Bauw. vol. 20, p 74.
Wohler, A. (1871) Engineering vol. 11, p 199.
Wood, W.A. (1958) Phil Mag. vol. 3, p 692.
Ye, F., Mercer, C., and Soboyejo, W.O. (1998) Metall Mater Trans. vol. 30A, p 2361.
Ye, F., Farkas, D. and Soboyejo , W.O. (1999a) Mater Sci Eng A vol. A264, p 81.
Ye, F., Li, M. and Soboyejo, W.O. (1999b) J Am Ceram Soc. vol 82, p 2460.
Zappfe, C.A. and Worden, C.O. (1951) Trans ASM. vol. 43,1 p 958.
Zurek, A.K., James, M.R., and Morris, M.L. (1983) Met Trans. vol. 14A, p 1697.

15

Introduction to Viscoelasticity, Creep, and Creep Crack Growth

15.1 INTRODUCTION

So far, our discussion on the mechanical behavior has considered only time-independent deformation. However, the mechanical behavior of materials may also be time dependent. This can give rise to time-dependent strains or crack growth that can result, ultimately, in component failure or damage. For several materials deformed at temperatures above about 0.3–0.5 of their melting temperatures, T_m (in K), time-dependent deformation can occur by creep or stress relaxation. This may result ultimately in a range of failure mechanisms that are illustrated schematically in Fig. 15.1. In crystalline metals and their alloys, creep damage can occur by stress-assisted diffusion and/or dislocation motion. Microvoids may also form and coalesce by the same mechanisms during the final stages of creep deformation. Furthermore, creep damage mechanisms may occur at crack tips, giving rise ultimately to creep crack growth phenomena.

Time-dependent creep deformation has also been observed in polymeric materials by viscous flow processes. These can result in time-dependent elastic (viscoelastic) or time-dependent plastic (viscoplastic) processes. Such time-dependent flow can happen at temperatures above the so-called glass transition temperature, T_g. Time-dependent deformation may also occur in crystalline materials. Depending on the crystal structure and tem-

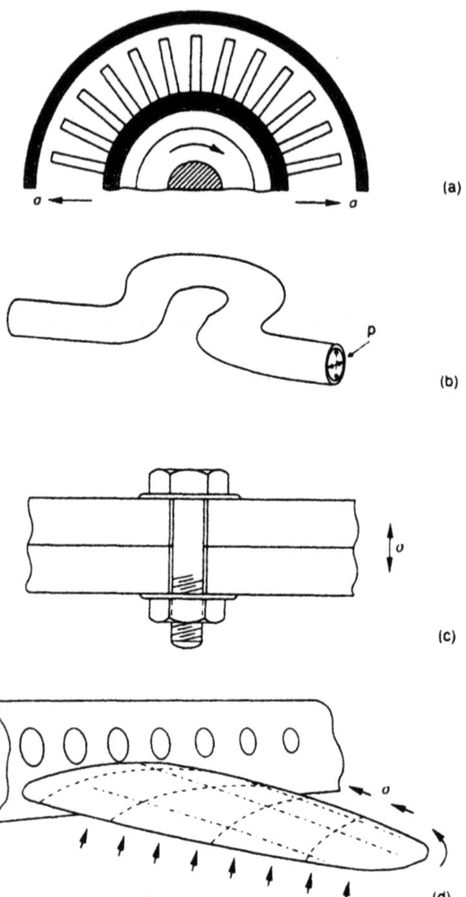

Figure 15.1 Creep is important in four classes of design: (a) displacement limited; (b) failure-limited; (c) relaxation limited; (d) buckling limited. [From Ashby and Jones (1996) with permission from Butterworth-Heinemann.]

perature, these can give rise to stress-assisted movement of interstitials and vacancies, and an elastic deformation.

This chapter presents an introduction to time-dependent deformation in crystalline and amorphous materials. Time-dependent deformation/creep of polymers is described along with the temperature dependence of deformation in polymers. Phenomenological approaches are then described for the characterization of the different stages of creep deformation. These are followed by an overview of the creep deformation mechanisms. The creep mechanisms are summarized in deformation maps before discussing some engineering approaches for creep design and the prediction of the creep lives of engineering structures and components. Finally, a brief introduction to superplasticity is then presented before concluding with an introduction to time-dependent fracture mechanics and the mechanisms of creep crack growth.

15.2 CREEP AND VISCOELASTICITY IN POLYMERS

15.2.1 Introduction

In general, time-dependent deformation occurs in materials at temperatures between 0.3 and 0.5 of T_m, the melting point (in K). In the case of polymeric materials, which have relatively low melting points, considerable time-dependent deformation has been observed, even at room temperature. The resulting deformation in polymers exhibits much stronger dependence on temperature and time, when compared to that in metallic and ceramic materials. This is due largely to Van der Waals forces that exist between polymers chains (Fig. 1.8). Since the Van der Waals forces are relatively weak, significant time-dependent deformation can occur by chain-sliding mechanisms. (Chap. 1).

15.2.2 Maxwell and Voigt Models

In general, the time-dependent deformation of polymers can be described in terms of creep and stress relaxation, Fig. 15.1(a) and (c). Creep is the time-dependent deformation that occurs under constant stress conditions, Fig. 15.1(a), while stress relaxation is a measure of the stress response under constant strain conditions, Fig. 15.1(c). The underlying mechanics of the time-dependent response of polymers will be described in this section.

Time-independent deformation and relaxation in polymers can be modeled using various combinations of springs and dashpots arranged in series and/or parallel. Time-independent elastic deformation can be modeled solely by springs that respond instantaneously to applied stress, according to Hooke's law, Fig. 15.2(a). This gives the initial elastic stress, σ_o, as the product of Young's modules, E, and the instantaneous elastic strain, ε_1, i.e., $\sigma_o = E\varepsilon_o$, where ε_o is the instantaneous/initial strain. Similarly, purely time dependent strain–time response can be described by the viscous response of a dashpot. This gives the dashpot time-dependent stress, σ_d, as the product of the viscosity, η, and the strain rate, $d\varepsilon/dt$ [Fig. 15.2(b), i.e., $\sigma_d = \eta \, d\varepsilon/dt$.

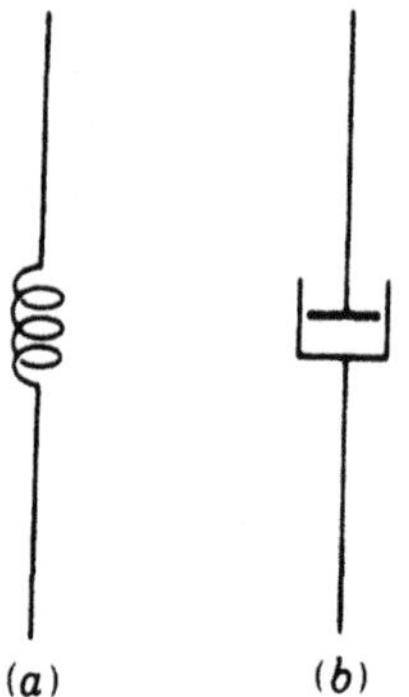

Figure 15.2 Schematic illustration of (a) spring model and (b) dashpot model. [From Hertzberg (1996) with permission from John Wiley.]

The simplest of the spring–dashpot models are the so-called Maxwell and Voigt models, which are illustrated schematically in Fig. 15.3(a) and (b), respectively.

15.2.2.1 Maxwell Model

The *Maxwell model* involves the arrangement of a spring and a dashpot series, Fig. 15.3(a). Under these conditions, the total strain in the system, ε, is the sum of the strains in the spring, ε_1, and the strain in the dashpot, ε_2. This gives

$$\varepsilon = \varepsilon_1 + \varepsilon_2 \tag{15.1}$$

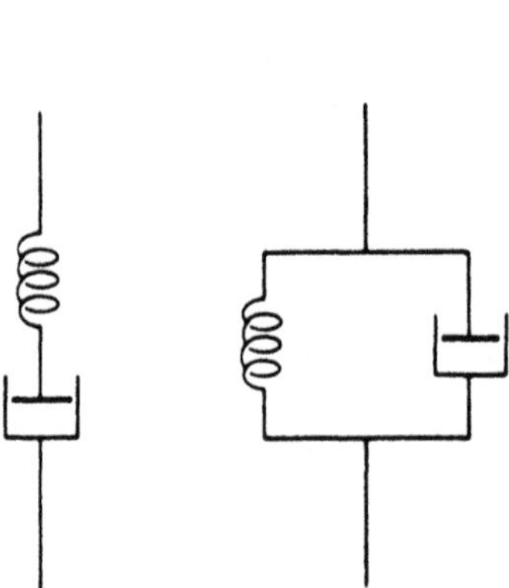

Figure 15.3 Schematic illustration of (a) Maxwell model and (b) Voigt model. [From Hertzberg (1996) with permission from John Wiley.]

Since the spring and the dashpot are in series, the stresses are equal in the Maxwell model. Hence, $\sigma_1 = \sigma_2 = \sigma$. Taking the first derivative of strain with respect to time, we can show from Eq. (15.1) that

$$\frac{d\varepsilon}{dt} = \frac{d\varepsilon_1}{dt} + \frac{d\varepsilon_2}{dt} = \frac{1}{E}\frac{d\sigma}{dt} + \frac{\sigma}{\eta} \tag{15.2}$$

where $\varepsilon_1 = \sigma/E$ and $d\varepsilon_2/dt = \sigma/\eta$. In general, the Maxwell model predicts an initial instantaneous elastic deformation, ε_1, followed by a linear time-dependent plastic deformation stage, ε_2, under constant stress conditions (Fig. 15.4).

However, it is important to note that the strain–time response in most materials is not linear under constant stress $\sigma = \sigma_o$, i.e., creep conditions. Instead, most polymeric materials exhibit a strain rate response that increases with time. Nevertheless, the Maxwell model does provide a good model of stress relaxation, which occurs under conditions of constant strain, $\varepsilon = \varepsilon_o$, and strain rate, $d\varepsilon/dt = 0$. Applying these conditions to Eq. (15.2) gives

$$0 = \frac{1}{E}\frac{d\sigma}{dt} + \frac{\sigma}{\eta} \tag{15.3}$$

Separating variables, rearranging Eq. (15.3), and integrating between the appropriate limits gives

$$\int_{\sigma_o}^{\sigma} \frac{d\sigma}{\sigma} = \int_0^{\sigma_1} -\frac{E}{\eta}\,dt \tag{15.4a}$$

or

$$\sigma = \sigma_0 \exp\left(-\frac{Et}{\eta}\right) \tag{15.4b}$$

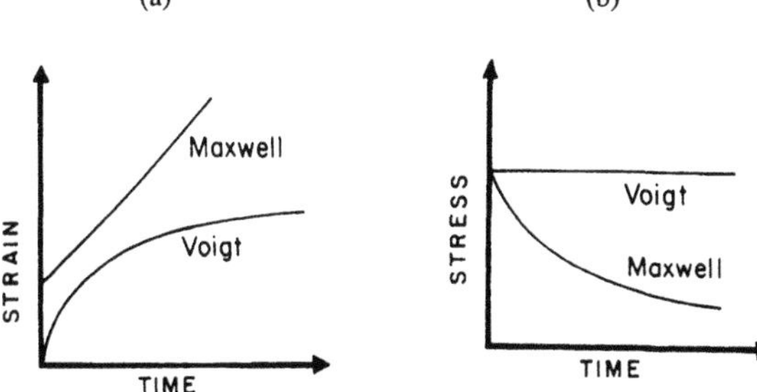

FIGURE 15.4 (a) Strain–time and (b) stress–time predictions for Maxwell and Voigt models. [From Meyers and Chawla (1998) with permission from Prentice Hall.]

Equation (15.4b) shows that the initial stress, σ_o, decays exponentially with time (Fig. 15.5). The time required for the stress to relax to a stress of magnitude σ/e (e is exp 1 or 2.718) is known as the relaxation time, τ. This is given by the ratio, η/E. Equation (15.4) may, therefore, be expressed as

$$\sigma = \sigma_0 \exp\left(-\frac{t}{\tau}\right) \tag{15.5}$$

Equation (15.5) suggests that stress relaxation occurs indefinitely by an exponential decay process. However, stress relaxation does not go on indefinitely in real materials. In addition to polymers, stress relaxation can occur in ceramics and glasses at elevated temperature (Soboyejo et al., 2001) and in metallic materials (Baker et al., 2002). However, the mechanisms of stress relaxation in metals are different from those in polymers. Stress relaxation in metals involves the movement of defects such as vacancies and dislocations.

15.2.2.2 Voigt Model

The second model illustrated in Fig. 15.3 is the *Voigt Model*. This involves the arrangement of a spring and a dashpot in parallel, as shown schematically in Fig. 15.3(b). For the spring and the dashpot in parallel, the strain in the spring and strain in the dashpot are the same, i.e., $\varepsilon = \varepsilon_1 = \varepsilon_2$. However,

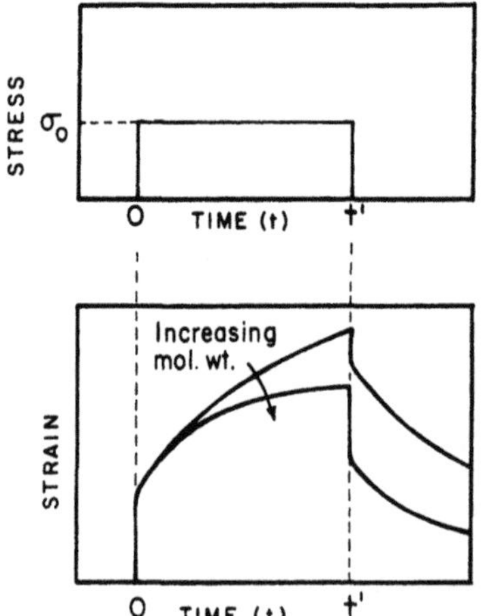

Figure 15.5 Effects of increasing molecular weight on the time-dependence of strain in a viscoelastic polymer. [From Meyers and Chawla (1999) with permission from Prentice Hall.]

the total stress is the sum of the stress in the spring and the dashpot. This gives

$$\sigma = \sigma_1 + \sigma_2 \tag{15.6}$$

where σ_1 is the stress in the spring and σ_2 is the stress in the dashpot. Substituting the relationships for σ_1 and σ_2 into Eq. (15.6) now gives

$$\sigma = E\varepsilon_1 + \eta \frac{d\varepsilon_2}{dt} \tag{15.7a}$$

Furthermore, since $\varepsilon = \varepsilon_1 + \varepsilon_2$, we can write:

$$\sigma = E\varepsilon + \eta \frac{d\varepsilon}{dt} \tag{15.7b}$$

Under constant stress (creep) conditions, $d\sigma/dt = 0$. Hence, differentiating Eq. (15.7b) now gives

$$\frac{d\sigma}{dt} = 0 = E\frac{d\varepsilon}{dt} + \eta \frac{d^2\varepsilon}{dt} \tag{15.8}$$

Equation (15.8) can be solved by setting $v = d\varepsilon/dt$. This gives

$$0 = Ev + \eta \frac{dv}{dt} \tag{15.9a}$$

or

$$\frac{dv}{v} = -\frac{E}{\eta}\,dt \tag{15.9b}$$

Integrating Eq. (15.9) between the appropriate limits gives

$$\int_{v_o}^{v} \frac{dv}{v} = \int_0^t -\frac{E}{\eta}\,dt \tag{15.10a}$$

or

$$\ln\frac{v}{v_0} = -\frac{Et}{\eta} \tag{15.10b}$$

Taking exponentials of both sides of Eq. (15.10) gives

$$v = v_0 \exp\left(-\frac{Et}{\eta}\right) \tag{15.11}$$

However, substituting $v = d\varepsilon/dt$, $v_o = d\varepsilon_o/dt = \sigma/\eta$, and $\tau = \eta/E$ into Eq. (15.11) results in

$$\dot{\varepsilon} = \dot{\varepsilon}_0 \exp\left(-\frac{t}{\tau}\right) \tag{15.12a}$$

or

$$\frac{d\varepsilon}{dt} = \frac{d\varepsilon_0}{dt} \exp\left(\frac{-t}{\tau}\right) = \frac{\sigma}{\eta} \exp\left(\frac{-t}{\tau}\right) \tag{15.12b}$$

Separating variables and integrating Equation (15.12b) between limits gives

$$\int_0^\varepsilon d\,\varepsilon = \int_0^t \frac{\sigma}{\eta} \exp\left(\frac{-t}{\tau}\right) dt \tag{15.13a}$$

or

$$\varepsilon = -\frac{\sigma}{\eta} \tau \left[\exp\left(\frac{-t}{\tau}\right) - 1\right] \tag{15.13b}$$

Now since $\tau = \eta/E$, we can simplify Eq. (15.13b) to give

$$\varepsilon = \frac{\sigma}{E}\left[1 - \exp\left(\frac{-t}{\tau}\right)\right] \tag{15.13c}$$

The strain–time dependence associated with Eq. (15.13c) is illustrated schematically in Fig. 15.4. In general, the predictions from the Voigt model are consistent with experimental results. Also, as $t = \infty$ the $\varepsilon \rightarrow \sigma_o/E$. Furthermore, for the stress relaxation case, $\varepsilon/\varepsilon_o = \text{constant}$. Hence, $d\varepsilon/dt = 0$. Therefore, from Eq. (15.7) we have

$$\sigma = E\varepsilon_0 \tag{15.14}$$

Equation (15.14) gives the constant stress response shown schematically in Fig. 15.4(b). It is important to note that the molecular weight and structure of a polymer can strongly affect its time-dependent response. Hence, increasing the molecular weight (Fig. 15.5) or the degree of cross-linking in the chain structure tends to increase the creep resistance. This is because increased molecular weight and cross-linking tend to increase the volume density of secondary bonds, and thus improve the creep resistance.

Similarly, including side groups that provide structural hindrance (steric hindrance) to the sliding chains will also increase the creep resistance. However, in such polymeric systems, the two-component (Voigt or Maxwell) models do not provide adequate descriptions of the stress–time or strain–time responses. Instead, multicomponent spring and dashpot models are used to characterize the deformation response of such polymeric systems. The challenge of the polymer scientist/engineer is to determine the appropriate combination of springs and dashpots that are needed to characterize the deformation response of complex polymeric structures.

Example 15.1

An example of a more complex spring–dashpot model is the four-element one shown in Fig. 15.6. This consists of a Maxwell model in series with a Voigt model. The overall strain experienced by this model is given by the sum of the Maxwell and Voigt strain components. This may, therefore, be expressed as

$$\varepsilon = \frac{\sigma}{E_1} + \frac{\sigma}{\eta_1} t + \frac{\sigma}{E_2}\left[1 - \exp\left(\frac{-t}{\tau_2}\right)\right] \tag{15.15}$$

The resulting strain–time response associated with the combined model is presented in Fig. 15.7. Upon loading, this shows the initial elastic response, ε_1, associated with the Maxwell spring element at time $t = 0$. This

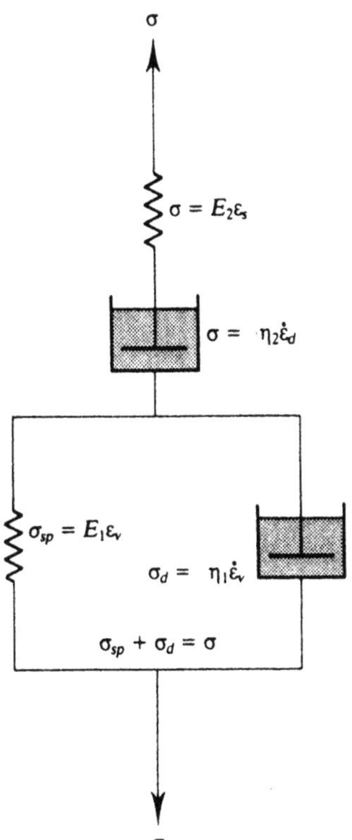

FIGURE 15.6 Four-element model consisting of a Maxwell model in series with a Voigt model. [From Courtney (1990) with permission from McGraw-Hill.]

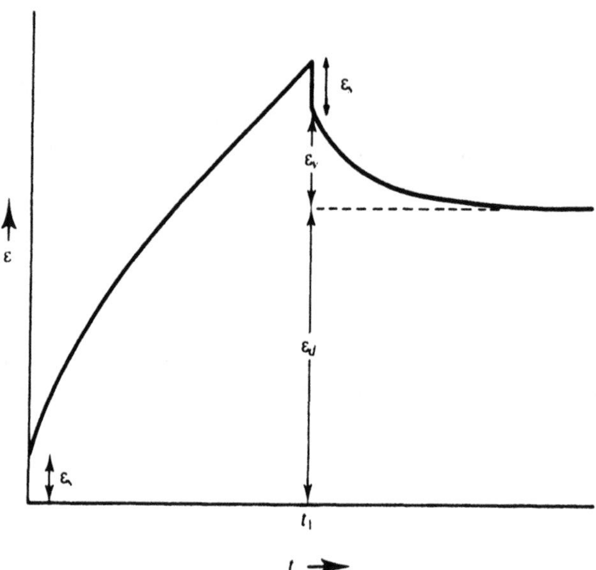

Figure 15.7 Strain–time response of combined Maxwell and Voigt model (four-element model). [From Courtney (1990) with permission from McGraw-Hill.]

is followed by the combined viscoelastic and viscoplastic deformation of the Maxwell and Voigt spring and dashpot elements. It is important to note here that the strain at long times is $\sim \sigma/\eta_1$. This is so because the exponential term in Eq. (15.15) tends towards zero at long times. Hence, the strain–time function exhibits an almost linear relationship at long times.

Upon unloading (Fig. 15.7), the elastic strain in the Maxwell spring is recovered instantaneously. This is followed by the recovery of the viscoelastic Voigt dashpot strain components. However, the strains in the Maxwell dashpot are not recovered. Since the strains in the Maxwell dashpot are permanent in nature, they are characterized as viscoplastic strains.

15.3 MECHANICAL DAMPING

Under cyclic loading conditions, the strain can lag behind the stress. This gives rise to mechanical hysterisis. The resulting time-dependent elastic behavior is associated with mechanical damping of vibrations. Such damping phenomena may cause induced vibrations, due to applied stress pulses, to die out quickly. The hysterisis associated with the cyclic deformation of a viscoelastic solid can be modeled by eliminating the Maxwell dashpot from the four-element model (Figs. 15.6 and 15.7).

Upon loading, there are two limiting values of the modulus. The first corresponds to the so-called *unrelaxed modulus*, E_u, at $t = 0$. This is greater than the *relaxed modulus* that is reached after some viscoelastic flow has occurred over a period of time (Fig. 15.8). Upon unloading, the elastic strain, ε_1, is recovered instantaneously. This is followed by the gradual recovery of the Voigt viscoelastic strain.

For a material subjected to cyclically varying stress and strain [Fig. 15.9(a)], the time available for the above flow processes is controlled by the cyclic frequency, f. This time period, T, is given by the inverse of the cyclic frequency, $T = 1/f$. Hence, if $T = 1/f$ is much greater than the time required for viscoelastic response to occur, then the stress and the strain will be in phase, and the stress–strain profile will correspond to that of a linear elastic solid with a modulus equal to the unrelaxed modulus, E_u, Fig. 15.9(b). In contrast, when the cyclic frequency is slow, there is enough time for viscoelastic deformation to occur. The stress and strain will also be in phase under these conditions. However, the modulus will be equal to the relaxed modulus E_r (Fig. 15.9).

At intermediate cyclic frequencies, where the cycle period is comparable to the inverse of the viscoelastic time constant, the strain lags the stress,

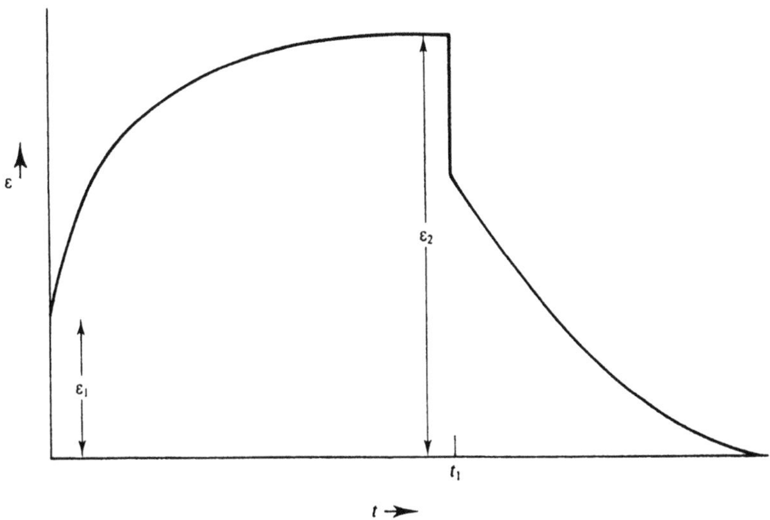

Figure 15.8 Strain–time response of a three-element viscoelastic model. [From Courtney (1990) with permission from McGraw-Hill.]

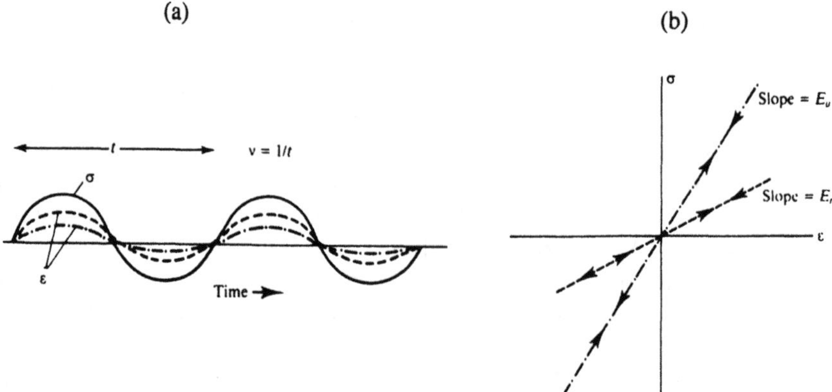

FIGURE 15.9 Schematic illustration of (a) the time dependence of applied stress and applied strain under cyclic loading and (b) relaxed and unrelaxed moduli observed respectively at slow and high frequencies. [From Courtney (1990) with permission from McGraw-Hill.]

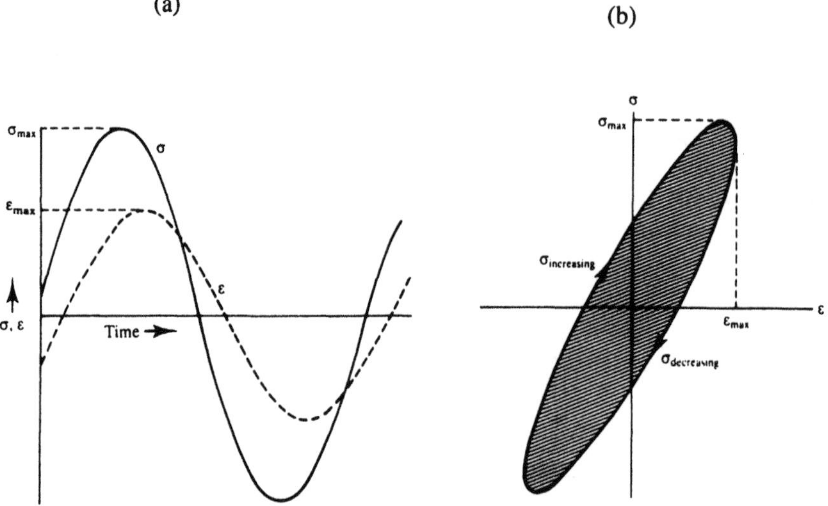

FIGURE 15.10 Schematic of (a) strain lagging stress at intermediate cyclic frequencies and (b) resulting hysteresis cross-plot of stress and strain. [From Courtney (1990) with permission from McGraw-Hill.]

and hysterisis occurs. Under these conditions, the stress–strain paths are rather different during loading and unloading, and the average modulus is in between the relaxed and the unrelaxed moduli, i.e., between E_r and E_u (Fig. 15.10). The shaded area in Fig. 15.10(b) represents the hysteritic/irreversible energy loss per cycle. This can lead to considerable hysteretic heating during the cyclic deformation of polymeric materials.

15.4 TEMPERATURE DEPENDENCE OF TIME-DEPENDENT FLOW IN POLYMERS

Since the time-dependent flow of polymer chains is a thermally activated process, it can be described by the Arrhenius equation. Time-dependent flow of polymer chains occurs readily above the glass transition temperature, T_g (Fig. 1.9). Above T_g, the network structure can breakdown and reform locally. However, below T_g, there is not enough thermal energy for this to occur. Hence, the material cannot flow easily below T_g.

As discussed earlier, the deformation response can be studied under constant stress (creep) or constant strain (stress relaxation) conditions. If we define the creep compliance, C, as the ratio of the strain (at any given time) and the stress, then we can describe the deformation–time response in terms of compliance versus time, as illustrated in Fig. 15.11(a). Note that this shows different creep curves at different temperatures, above or below the T_g. However, in general, it is much easier to design with a single master curve.

A single master curve can be achieved by shifting the curves along the log time axis to form a single curve, Figs. 15.11(a) and (b). This is of great practical advantage since the individual creep curves may require years to obtain. In any case, Williams et al. (1955) have proposed the so-called *WLF equation* to describe the *time-shift factor*, a_T. This is given by

$$\log a_T = \frac{-C_1(T - T_s)}{(C_2 + T - T_s)} \tag{15.16}$$

where C_1 and C_2 are material constants, T is the temperature, and T_s is a reference temperature. The amount of the shift can be calculated by setting the reference temperature, and using techniques that are illustrated schematically in Fig. 15.11(a) and (b). Alternatively, we may also define a stress relaxation modulus (the inverse of the compliance) as a function of time. This is shown in Fig. 15.11(c), in which the relaxation modulus is plotted against temperature. This may also be reduced to a single master curve [Fig. 15.11(d)] using similar techniques to those described earlier.

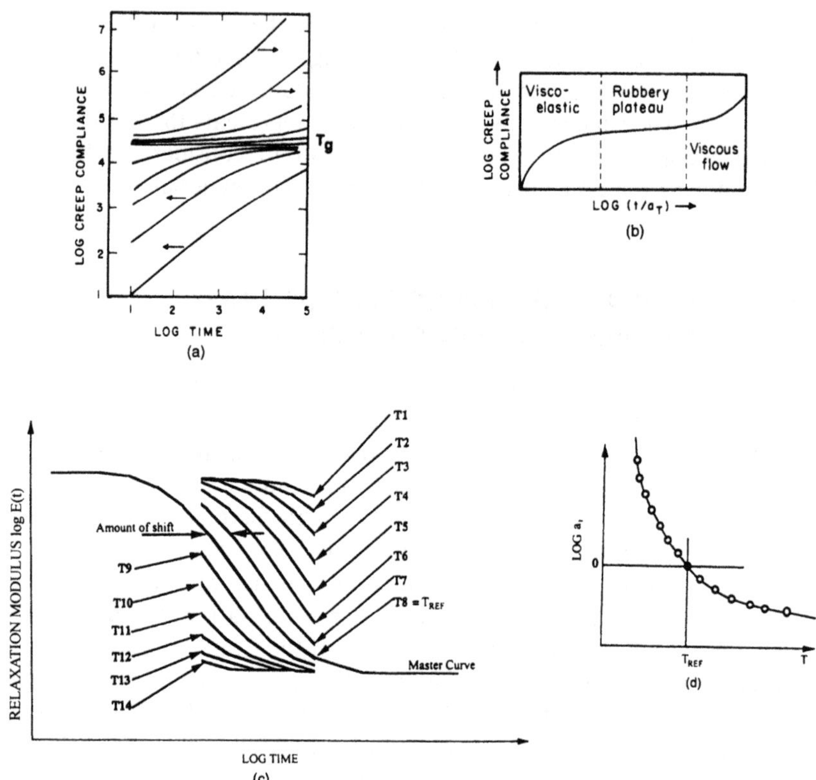

FIGURE 15.11 (a) Dependence of creep compliance on time and temperature; (b) superposition of different curves by horizontal shifting along the time axis by an amount log a_t (c) amount of shift to produce master curve; (d) experimentally determined shift factor. [From Meyers and Chawla (1999) with permission from Prentice Hall.]

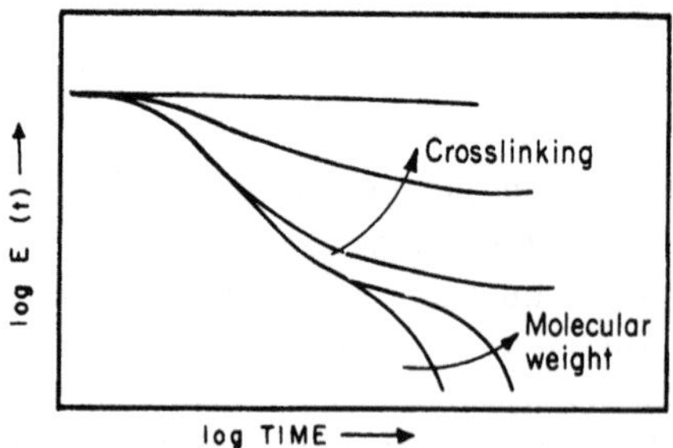

FIGURE 15.12 Dependence of relaxation modulus on time and polymer structure. [From Meyers and Chawla (1999) with permission from Prentice Hall.]

Finally, it is important to note that the temperature dependence of the relaxation modulus is highly dependent on molecular weight and the extent of cross-linking of the polymer chains. This is illustrated in Fig. 15.12, which shows the time dependence of the relaxation modulus. This can be exploited in the management of residual stresses in engineering structures and components that are fabricated from polymeric materials.

15.5 INTRODUCTION TO CREEP IN METALLIC AND CERAMIC MATERIALS

The above discussion has focused largely on creep and viscoelasticity in polymeric materials. We will now turn our attention to the mechanisms and phenomenology of creep in metallic and ceramic materials with crystalline and noncrystalline structures. Creep in such materials is often studied by applying a constant load to a specimen that is heated in a furnace. When constant stresses are required, instead of constant load, then the weight can be immersed in a fluid that decreases the effective load with increasing length in a way that maintains a constant applied true stress (Andrade, 1911). This is achieved by a simple use of the Archimedes principle. Similarly, a constant stress may be applied to the speicmen using a variable lever arm that applies a force that is dependent on specimen length (Fig. 15.13). In any case, the strain–time results obtained from constant stress or constant load tests are qualitatively similar. However, the precision of constant stress tests may be desirable in carefully controlled experiments.

Upon the application of a load, the instantaneous deformation is elastic in nature at time $t = 0$. This is followed by a three-stage deformation process. The three stages (I, II, and III) are characterized as the primary (stage I), secondary (stage II), and tertiary (stage III) creep regimes. These are shown schematically in Fig. 15.14(a). It is also important to note that the magnitude of the creep strains increases significantly with increasing temperature and stress, Fig. 15.14(a). Furthermore, the relative fractions of the differrent stages of creep deformation change significantly with increasing temperature and stress, Fig. 15.14(b).

In many practical engineering problems, e.g., the design of turbine blades in aeroengines and the design of land-based engine components, the applied stresses and temperatures are usually sufficient to induce creep at temperatures above ~ 0.3–0.5 of the melting temperature, T_m (in K). Under such conditions, creep deformation may lead to the loss of tolerance and component/structural failure after many years of service. In such cases, the time to creep rupture (the creep rupture life) is an important design

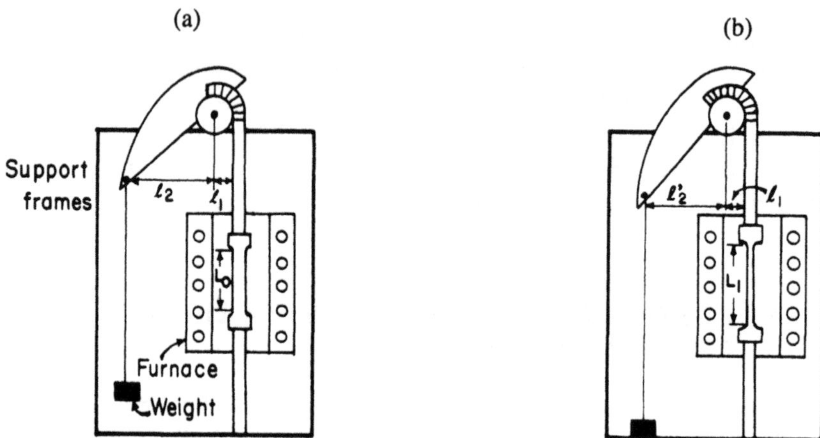

Figure 15.13 Schematic of creep testing under constant stress conditions. Creep machine with variable lever arms to ensure constant stress on specimen; note that l_2 decreases as the length of specimen increases. (a) Initial position; (b) length of specimen has increased from L_0 to L_1. [From Meyers and Chawla (1999) with permission from Prentice Hall.]

parameter. In general, this decreases with increasing stress and temperature, Fig. 15.14(a).

Alternatively, we may plot the strain rate [the slope of Fig. 15.14(a)] against time, as shown in Fig. 15.14(b). This shows that the strain rate decreases continuously with time in stage I, the primary or transient regime. This is followed by the stage II creep regime (secondary creep regime) in

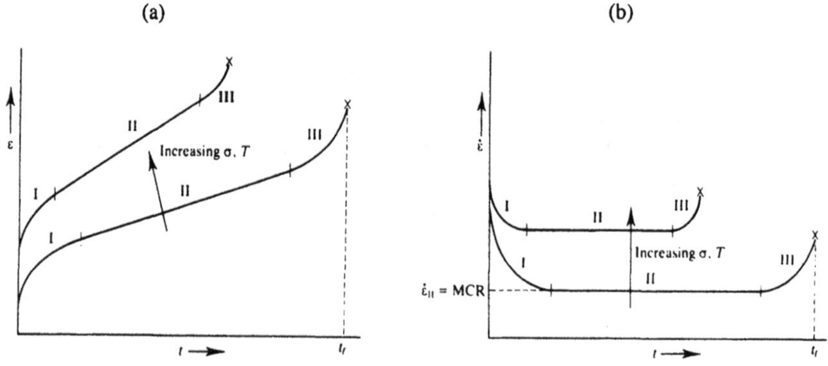

Figure 15.14 Schematics showing the time dependence of (a) strain and (b) strain rate. [(b) Taken from Courtney (1990) with permission from McGraw-Hill.]

which the strain rate is constant.* Finally, in stage III (the tertiary creep regime), the strain rate increases continuously with time, until creep rupture occurs at t_f. Figure 15.14(b) also shows that the overall creep strain rates increase with increasing stress and temperature.

The occurrence of the different stages of creep may be explained by considering the relative contributions of hardening and recovery phenomena to the overall creep deformation. In stage I, the creep strain rate decreases continuously with increasing strain. This is due to work hardening phenomena that are associated with increased dislocation density and the possible formation of dislocation subgrain structures. In stage II, the steady-state creep regime is associated with the formation of an invariant microstructure/dislocation substructure in which the hardening due to dislocation substructure evolution is balanced by recovery or softening mechanisms. Finally, in stage III (the tertiary creep regime), the creep strain rate increases with increasing time. This is due largely to the onset of creep damage by microvoid nucleation/growth mechanisms. Stage III creep may also occur as a result of dynamic recrystallization (Allameh et al., 2001a). Mechanisms such as these lead to the accelerating creep rates that result ultimately in creep fracture at $t = t_f$.

In many engineering scenarios, the time to creep rupture provides some useful guidelines for component design, e.g., in the hot sections of gas turbines in aeroengines and land-based engines that operate at temperatures of up to $\sim$ 1000°–1150°C. However, it is more common to design these structures to limit creep to the primary or secondary creep regimes. Furthermore, the overall damage in the material may be exacerbated in the presence of aggressive chemical environments, e.g., in chemical or petrochemical structures or the components of aeroengines or land-based gas turbines, where chemical degradation can occur by chlorination or sulfidation reactions and high-temperature oxidation mechanisms (Li et al., 1998).

Before closing, it is important to note that there is increasing evidence that a number of structural materials do not exhibit a true steady-state creep regime (Poirier, 1985). Instead, these materials are thought to exhibit a minimum creep rate, $\dot{\varepsilon}_{\min}$, which is followed by a gradual increase in creep rates until rupture. Apparently, such materials do not develop the steady-state microstructures/substructures that are needed for true steady-state creep. In any case, it is common in such materials to consider the conditions for $\dot{\varepsilon}_{\min}$ instead of the secondary creep rate, $\dot{\varepsilon}_{ss}$, in regime II.

---

*Note that a minimum strain rate may occur in some materials in which a true steady-state creep regime is not achieved.

15.6 FUNCTIONAL FORMS IN THE DIFFERENT CREEP REGIMES

As discussed in Sec. 15.5, the functional forms of the strain–time relationships are different in the primary, secondary, and tertiary regimes. These are generally characterized by empirical or mechanism-based expressions. This section will focus on the mechanism-based mathematical relationships.

In the primary creep regime, the dependence of creep strain, ε, on time, t, has been shown by Andrade (1911) to be given by

$$\varepsilon = \alpha t^{1/3} \tag{15.17a}$$

where α is a constant. Equation (15.17a) has been shown to apply to a wide range of materials since the pioneering work by Andrade (1911) almost a century ago. The extent of primary creep deformation is generally of interest in the design of the hot sections of aeroengines and land-based gas turbines. However, the portion of the creep curve that is generally of greatest engineering interest is the secondary creep regime.

In the secondary creep regime, the overall creep strain, ε, may be expressed mathematically as

$$\varepsilon = \varepsilon^{o} + \varepsilon[1 - \exp(-mt)] + \dot{\varepsilon}_{ss} t \tag{15.17b}$$

where the ε^{o} is the instantaneous elastic strain, $\varepsilon[1 - \exp(-mt)]$ is the primary creep term, $\dot{\varepsilon}_{ss}t$ represents the secondary creep strain component, m is the exponential parameter that characterizes the decay in strain rate in the primary creep regime, ε is the peak strain in the primary creep regime, and t corresponds to time. Similarly, we may express the secondary/steady state creep strain rate, $\dot{\varepsilon}_{ss}$, as

$$\dot{\varepsilon}_{ss} = A\sigma^{n} \exp\left(\frac{-Q}{RT}\right) \tag{15.18a}$$

where A is a constant, σ is the applied mean stress, n is the creep exponent, Q is the activation energy, R is the universal gas constant (8.317 J/mol K), and T is the absolute temperature in kelvins. Taking logarithms of Eq. (15.18) gives

$$\log \dot{\varepsilon}_{ss} = \log A + n \log \sigma - \frac{Q}{RT} \tag{15.18b}$$

Equation (15.18) can be used to extract some important creep parameters. First, if we conduct constant stress tests at the same temperature, then all of the terms in Eq. (15.18) remain constant, except for σ. We may now plot the measured secondary strain rates, $\dot{\varepsilon}_{ss}$, as a function of the applied mean stress, σ. A typical plot is presented in Fig. 15.15. This

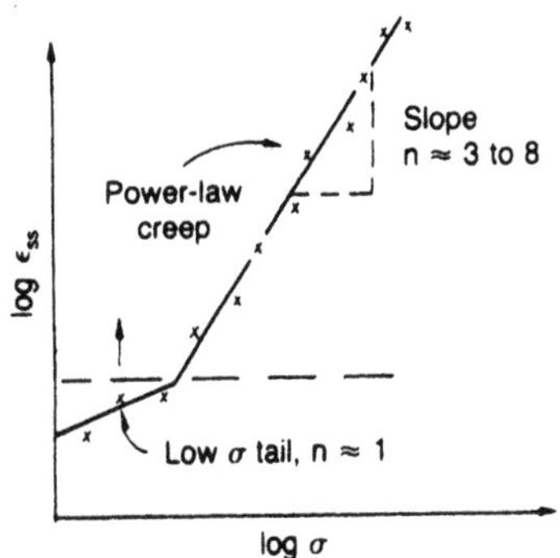

Figure 15.15 Plot of secondary creep strain rate versus mean stress to determine the creep exponent. [From Ashby and Jones (1996) with permission from Butterworth-Heinemann.]

shows a linear plot of $\log \dot{\varepsilon}_{ss}$ versus $\log \sigma$, which is expected from Eq. (15.15). The slope of this line is the secondary creep exponent, n.*

The parameter n is important because it provides some important clues into the underlying mechanisms of secondary creep deformation. When $n = 1$, then creep is thought to occur by diffusion-controlled creep. Similarly, for m between 3 and 8, secondary creep is dislocation controlled. Values of m greater than 7 have also been reported in the literature, especially for some intermetallics (Allameh et al., 2001a). These are generally associated with dynamic recrystallization phenomena (unstable microstructures) and constant structure creep phenomena that result in $\dot{\varepsilon}_{ss}$ values that scale with $(\sigma^2)^n$ (Gregory and Nix, 1987). Further details on creep deformation mechanisms will be presented in Sec. 15.8.

Another important secondary creep parameter is the activation energy. This can also be used to provide some useful insights into the underlying processes responsible for thermally activated and stress-activated creep processes. From Eq. (15.18), and for all Arrhenius processes, it is clear that the activation energy, Q, can be determined by plotting the secondary or the minimum creep rate, $\dot{\varepsilon}_{ss}$, as a function of the inverse of the absolute temperature $1/T$ (in K). A typical plot is shown in Fig. 15.16. From Eq. (15.18), the negative slope of the line corresponds to $-Q/R$. Hence, the activation energy, Q, can be determined by multiplying the slope by the universal gas constant, R.

*Note that the exponent, n, may vary depending on the magnitude of the applied stress, with $n \approx 1$ at low stresses and $n \approx 3$–8 in the higher stress regime.

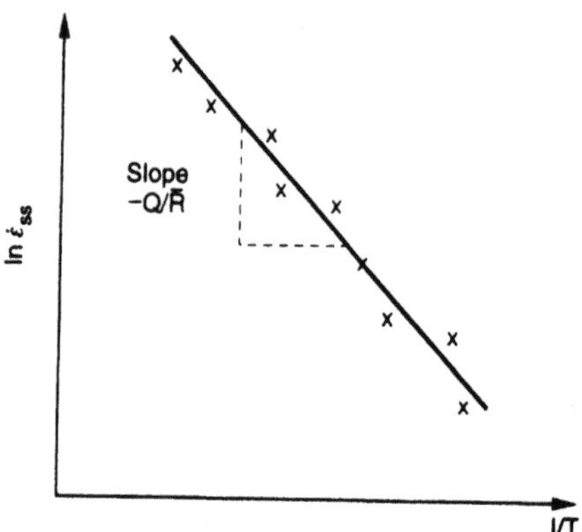

Figure 15.16 Determination of the activation energy from a plot of secondary creep strain rate versus the inverse of the absolute temperature. [From Ashby and Jones (1996) with permission from Butterworth-Heinemann.]

In general, the constants A, n, and Q can vary with temperature and stress. This is because the underlying mechanisms of creep deformation can vary significantly with stress and temperature. In most crystalline materials, diffusion-controlled creep is more common at low stresses and moderate/high temperatures, while dislocation-controlled creep is more common at moderate/high temperatures and higher stress levels.

Furthermore, the transition from the primary to the secondary creep regime appears as an inflection point when the creep data are collected via constant-load experiments. The strain rate corresponding to this inflection point is often characterized as minimum strain rate, $\dot{\varepsilon}_{\text{min}}$. However, Orowan (1947) has noted that the minimum strain rate obtained from the inflection point, at the end of the primary creep regime (under constant-load conditions) may not have much to do with the subsequent steady state under constant-stress conditions. Nevertheless, the practice is very widespread, and may explain some of the higher creep exponents that have been reported in the literature.

In general, the rate-controlling process in the secondary creep regime involves volume diffusion at high homologous temperatures (T/T_{m}) or dislocation core diffusion at lower homologous temperatures. Furthermore, although the magnitudes of the activation energies, Q, differ from material to material, it has been shown that $Q_{\text{CD}} = 0.6Q_{\text{v}}$ (where subscript CD corresponds to core diffusion and subscript v corresponds to vacancy diffusion (Burton and Greenwood, 1970a, 1970b).

At higher stress levels, the power law expression breaks down. In this regime, Garafalo (1963) has found that the creep rate can be represented by

$$\dot{\varepsilon} = B\sin(c\sigma)^m \exp\left(-\frac{Q}{RT}\right) \tag{15.18c}$$

where B and c are constants, m is the creep exponent, and the other terms have their usual meaning. In this regime, more of the creep strain rate is derived from thermally activated glide, and the creep exponents show an increasing stress dependence from typical values between 3 and 8.

Finally, the overall time dependence of the creep curves can be modeled using the so-called θ approach (Evans and Wiltshire, 1985). This gives the time dependence of the creep strain as

$$\varepsilon = \theta_1(1 - e^{-\theta_2 t}) + \theta_3(e^{\theta_4 t} - 1) \tag{15.18d}$$

where θ_1 and θ_3 scale the strains, and θ_2 and θ_4 are the curvatures in the primary and tertiary creep regimes. The θ parameters are also dependent on temperature (Evans et al., 1993; Wiltshire, 1997), as shown in Fig. 15.17. In any case, the applicability of the so-called θ approach has been demonstrated for a wide range of materials. The θ approach can also be used to determine the activation energies associated with the primary and tertiary creep rate constants using techniques developed by Evans and Wiltshire (1985).

15.7 SECONDARY CREEP DEFORMATION AND DIFFUSION

The steady-state creep rate, $\dot{\varepsilon}_{ss}$, may be characterized by the following expression of Mukherjee et al. (1969):

$$\dot{\varepsilon}_{ss} = \frac{\alpha G b}{kT} D_0 \exp\left(\frac{-Q}{RT}\right)\left(\frac{b}{d}\right)^p\left(\frac{\sigma}{G}\right) \tag{15.19}$$

where α is a dimensionless constant, G is the shear modulus, b is the Burgers vectors, k is the Boltzmann constant, T is the absolute temperature, D_o is the diffusion coefficient, Q is the activation energy, R is the universal gas constant, d is the grain size, and σ is the applied mean stress. Equation (15.19) suggests that the diffusion coefficient, D, is an important parameter. This is given by

$$D = D_0 \exp\left(\frac{-Q_D}{RT}\right) = zb^2\nu_D \exp\left(\frac{-Q_D}{RT}\right) \tag{15.20}$$

where D_o is a constant, Q_D is the activation energy for diffusion, R is the gas constant, T is the absolute temperature, z is the coordination number of atoms, b is the Burgers vector, and ν_D is the atomic jump frequency. The creep activation energy has been shown to correlate well with the diffusion activation energy, Q_D (Sherby and Burke, 1968). This is shown in Fig. 15.18,

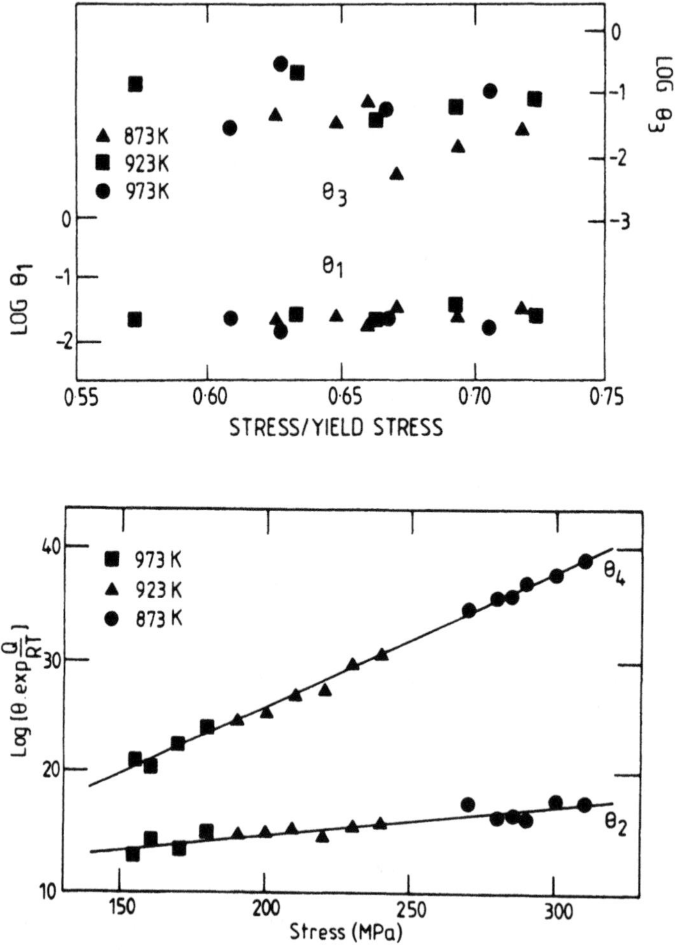

FIGURE 15.17 Variations of θ parameters (in DT2203Y05) with stress and temperature. [From Wiltshire (1997) with permission from TMS.]

in which Q and Q_D are shown to be highly correlated for a large number of metallic and nonmetallic materials.

However, in some materials, the activation energy for creep can be up to half of that required for bulk/lattice diffusion. Such low creep activation energies are often associated with grain boundary diffusion processes ($Q_{GB} \sim 0.5$–$0.6Q_D$). In any case, it is clear that diffusion processes play an important role in creep processes, as discussed by Burton and Greenwood (1970a,b). Also, since diffusion involves vacancy motion, an

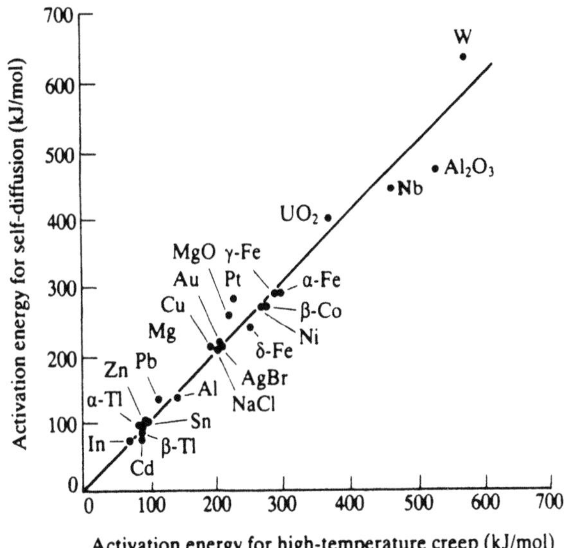

FIGURE 15.18 Correlation of the creep activation energy with the self-diffusion activation energy. [From Sherby and Burke, 1968.]

understanding of this subject is important in developing a basic understanding of microscopic creep processes. These will be described in some detail in the next section.

15.8 MECHANISMS OF CREEP DEFORMATION

15.8.1 Introduction

As discussed earlier, creep deformation may occur by dislocation-controlled or diffusion-controlled mechanisms. In cases where creep is dislocation controlled, the underlying mechanisms may involve the unlocking of dislocations that are pinned by precipitates or solute atoms/interstitials. In such cases, the unlocking may occur by the exchange of atoms and vacancies, which gives rise to dislocation climb, Fig. 15.19(a). Such climb processes tend to occur at temperatures above $\sim 0.3T_m$ (T_m is the melting temperature in kelvins). Climb may also occur by core diffusion, as illustrated schematically in Fig. 15.19(b).

Alternatively, dislocation creep may occur by a sequence of glide and climb mechanisms, as illustrated in Fig. 15.20. These involve the detachment of dislocations from local obstacles by climb processes. This is followed by

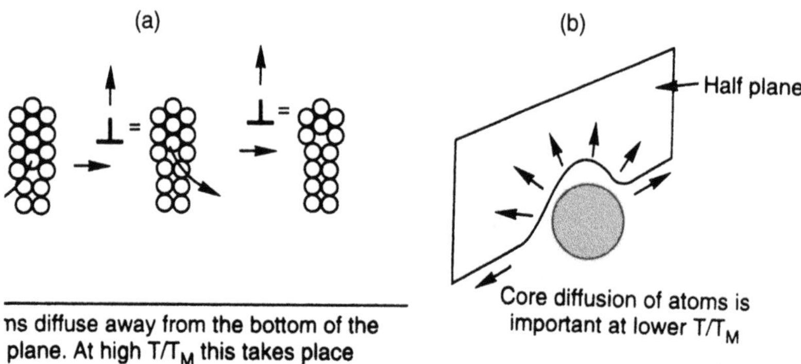

Figure 15.19 How diffusion leads to dislocation climb: (a) atoms diffuse from the bottom of the half-plane; (b) core diffusion of atoms. [From Ashby and Jones (1996) with permission from Butterworth-Heinemann.]

dislocation glide. The process then repeats itself when the gliding dislocation encounters another obstacle. In this way, significant creep deformation can occur by combinations of dislocation glide and climb processes.

At lower stresses and higher temperatures (~ 0.5–$0.99T_m$), creep may occur by bulk or grain boundary diffusion mechanisms (Figs 5.21 and 15.22). Bulk diffusion mechanisms involve a flux of vacancies that produce a net increase in length in the direction of the applied tensile stress (Fig. 15.21) or a net decrease in length in the direction of applied compressive

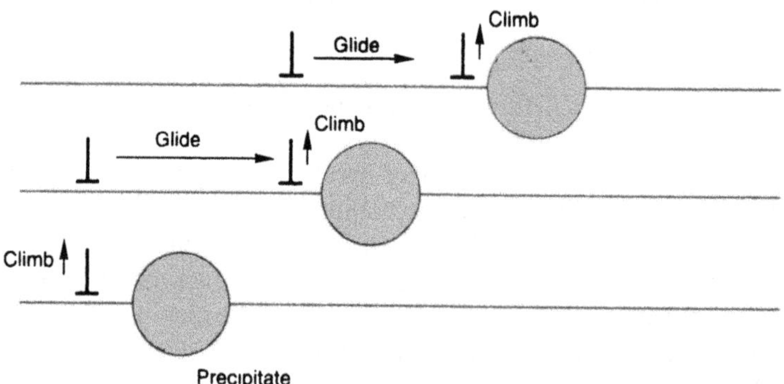

Figure 15.20 Schematic illustration of dislocation glide and climb processes. [From Ashby and Jones (1996) with permission from Butterworth-Heinemann.]

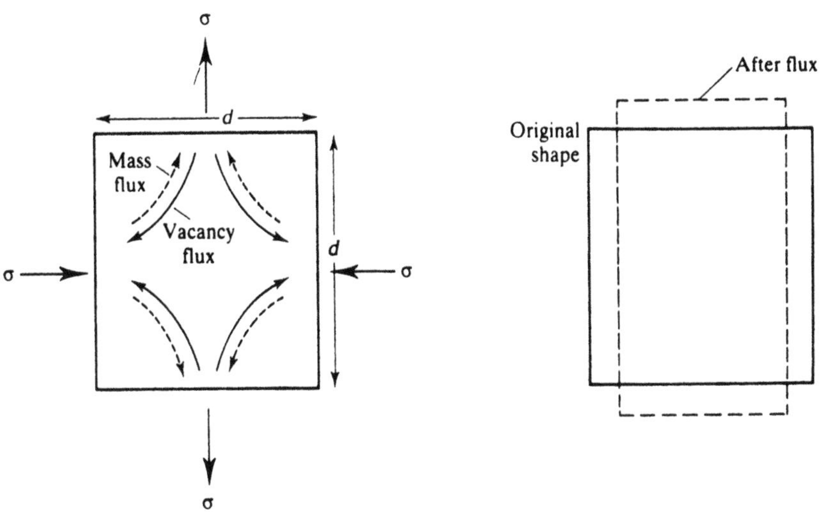

Figure 15.21 Schematic illustration of bulk diffusion. [From Courtney (1990) with permission from McGraw-Hill.]

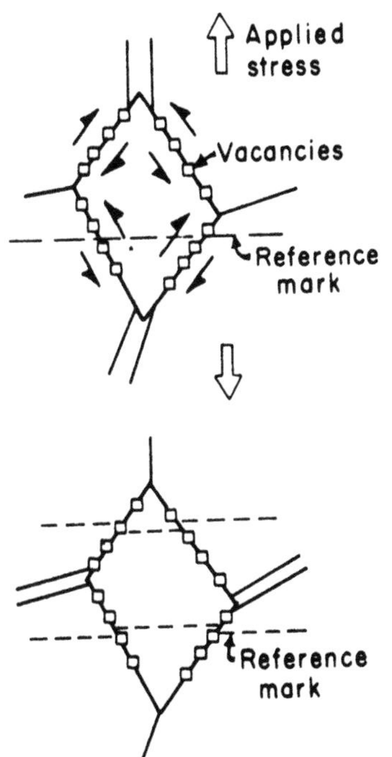

Figure 15.22 Schematic of grain boundary diffusion (Coble) creep. [From Meyers and Chawla (1998) with permission from Prentice Hall.]

stresses. Creep by bulk diffusion is often characterized as *Nabarro–Herring creep*, named after Nabarro (1948) and Herring (1950).

Alternatively, in materials with smaller grain sizes, creep deformation may occur by fast diffusion along grain boundary channels, Fig. 2.12(b). This gives rise to a creep strain-rate dependence that is very sensitive to grain size. Furthermore, since unchecked diffusion along grain boundaries is likely to lead to microvoid nucleation, grain boundary sliding mechanisms are often needed to accommodate the diffusion or deformation at grain boundaries. This section reviews the basic mechanisms of creep deformation in crystalline and noncrystalline metals and ceramics.

15.8.2 Dislocation Creep

As discussed earlier, dislocation creep may occur by glide and/or climb processes (Fig. 15.20). In the case of pure glide-controlled creep, Orowan (1947) has proposed that creep involves a balance between work hardening and recovery during deformation at high temperature. This gives

$$d\sigma = \left(\frac{\partial\sigma}{\partial\varepsilon}\right)_{t,\sigma} d\varepsilon + \left(\frac{\partial\sigma}{\partial t}\right)_{\sigma,\varepsilon} dt \tag{15.21}$$

where $\left(\frac{\partial\sigma}{\partial\varepsilon}\right)_{t,\sigma}$ is the rate of hardening, $\left(\frac{\partial\sigma}{\partial t}\right)_{\sigma,\varepsilon}$ is the rate of recovery, and the other variables have their usual meaning. Theories have been proposed for the modeling of dislocation-controlled creep (Weertman, 1955, 1957, Mukherjee et al., 1969). The models by Weertman (1955, 1957) consider edge dislocation climb away from dislocation barriers. In the first model (Weertman, 1955), Lomer–Cottrell locks are considered as barriers. The dislocations overcome these barriers by climb processes aided by vacancy or interstitial generation. The second model by Weertman (1957) was developed for hexagonal closed packed (h.c.p) metals. Subsequent work by Bird et al. (1969) showed that the secondary creep rate due to the climb of edge dislocations may be expressed as

$$\dot{\varepsilon}_{ss} = \frac{AGb}{kT}\left(\frac{\sigma}{G}\right)^n \tag{15.22}$$

where A is a constant, and the creep exponent, n, is equal to 5. In general, however, creep exponents between 3 and 8 have been associated with dislocation glide–climb mechanisms. Furthermore, creep exponents greater than 8 have been reported in cases where constant structure creep occurs. These give rise to strain rates that scale with σ^2 (and not just σ) in Eq. (15.22) (Gregory and Nix, 1987). Hence, the apparent creep exponents associated with constant structure creep may be between ~ 6 and 16 (Soboyejo et al., 1993).

Before closing, it is important to discuss the so-called Harper–Dorn creep mechanism. This was first proposed by Harper and Dorn (1957) for creep in aluminium with large grain size ($d > 400\,\mu m$). They concluded that creep in such material at low stresses and high temperatures can occur by dislocation climb. The resulting expression for the strain rate due to Harper–Dorn creep is

$$\dot{\varepsilon}_{HD} = A_{HD} \frac{D_L G b}{kT} \left(\frac{\sigma}{G}\right) \tag{15.23}$$

where A_{HD} is a constant, D_L is the lattice diffusion coefficient, G is the shear modulus, and all other constants have their usual meaning. Harper–Dorn creep has been reported in a number of systems, including some ceramics, where other mechanisms of creep can occur. However, a number of subsequent investigators have found it difficult to replicate the original work of Harper and Dorn.

15.8.3 Diffusion Creep and Grain Boundary Sliding

Creep may occur by lattice or grain boundary at lower stresses ($\sigma/G \sim 10^{-4}$) and moderate/high temperatures. The mechanisms proposed for bulk/lattice diffusion-controlled creep are illustrated in Fig. 15.21. Under the application of stress, vacancies move from boundary sources to boundary sinks. A corresponding flux of atoms also occurs in the opposite direction. This was first studied by Nabarro (1948) and Herring (1950). The strain rate due to Nabarro–Herring creep can be expressed as

$$\dot{\varepsilon}_{ss} = A_{NH} D_L \frac{Gb}{kT} \left(\frac{b}{d}\right)^2 \left(\frac{\sigma}{G}\right) \tag{15.24}$$

where A_{NH} is a Nabarro–Herring creep constant, D_L is the lattice diffusion coefficient, and the other constants have their usual meaning.

Alternatively, diffusion-controlled creep may also occur by grain boundary diffusion. This is illustrated in Fig. 15.22. This was first recognized by Coble (1963). The grain boundary diffusion is driven by the same type of vacancy concentration gradient that causes Nabarro–Herring creep. However, in the case of Coble creep, mass transport occurs by diffusion along grain boundaries in a polycrystalline structure, or diffusion along the surfaces of a single crystal. The expression for the strain rate due to Coble creep is given by

$$\dot{\varepsilon}_c = A_c D_{GB} \frac{Gb}{kT} \left(\frac{\delta}{b}\right) \left(\frac{b}{d}\right)^3 \left(\frac{\sigma}{G}\right) \qquad (15.25)$$

where A_c is the Coble creep constant, D_{GB} is the grain boundary diffusion coefficient, δ is the effective thickness of the grain boundary, and the other terms have their usual meaning. It is important to note here that the Coble creep rate is proportional to $1/d^3$ (where d is the grain size). In contrast, the Nabarro–Herring creep rate is proportional to $1/d^2$. Coble creep is, therefore, much more sensitive to grain size than is Nabarro–Herring creep. Furthermore, Coble creep is more likely to occur in materials with finer grain sizes.

The strong sensitivity of Coble creep and Nabarro–Herring creep to grain size suggests that larger grain sizes are needed to improve the resistance to diffusional creep. This has prompted the development of large grain superalloys for thermostructural applications in aeroengines and land-based engines. The most extreme examples of such large grained structures are single-crystal alloys or directionally solidified alloys with all their grain boundaries aligned parallel to the applied loads. Issues related to the design of such creep resistant microstructures will be described later on in this chapter.

Before concluding this section on diffusion-controlled creep, it is important to discuss the importance of grain boundary sliding phenomena. These are needed to prevent microvoid or microcrack formation due to the mass transfer associated with grain boundary or bulk diffusion (Figs 15.21 and 15.22). Hence, the diffusion creep rates must be balanced exactly by grain boundary sliding rates to avoid the opening up cracks or voids. This is illustrated in Figs 15.23(a) and (b), which are taken from a review by Evans and Langdon (1976). Note that the grain boundary sliding heals the crack/voids [Fig. 15.23(c)] that would otherwise open up due to grain boundary diffusion, Fig. 15.23(b). Conversely, we may also consider the accommodation of grain boundary sliding by diffusional flow processes. This can be visualized by considering an idealized interface with a sinusoidal profile (Fig. 15.24). Note that the sliding of the grain boundaries (due to an applied shear stress) must be coupled with diffusional accommodation to avoid opening up cracks or microvoids. Diffusional creep and grain boundary sliding are, therefore, sequential processes. As with most sequential creep processes, the slower of the two processes will control the creep rate.

Grain boundary sliding is particularly important in superplasticity. In fact, superplasticity is generally thought to occur largely by grain boundary sliding. However, large amounts of such sliding may lead, ultimately, to microvoid nucleation and creep rupture in the tertiary creep regime.

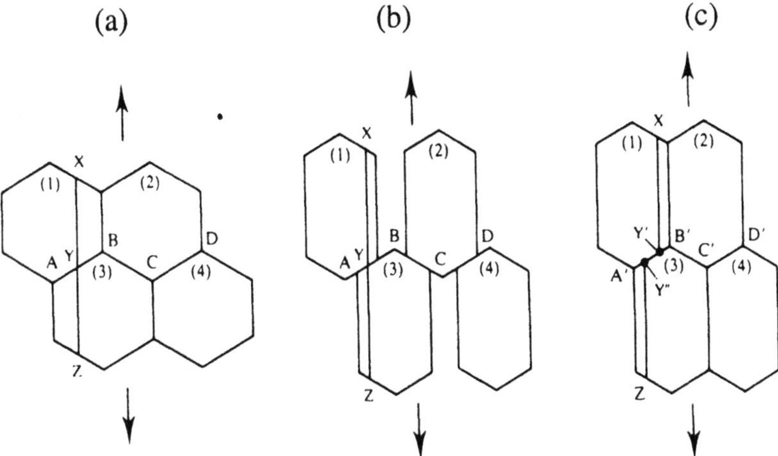

FIGURE 15.23 (a) Four grains in a hexagonal array before creep deformation; (b) after deforming by diffusional creep, one dimension of the grain is increased and the other is decreased, and "voids" are formed between the grains; (c) the voids are removed by grain boundary sliding. The extent of sliding displacement is quantified by the distance Y′Y″, which is the offset along the boundary between grains 1 and 3 of the original vertical scribe line XYZ. [From Evans and Langdon, 1976.]

15.8.4 Deformation Mechanism Maps

The above discussion has shown that creep deformation may occur by a range of dislocation-controlled and diffusion-controlled mechanisms. However, the mechanisms that prevail depend strongly on the applied stress, the environmental temperature, the strain rate and the grain size. The prevailing mechanisms can be summarized conveniently on so-called *deformation mechanism maps* (Fig. 15.25). These show that parametric ranges of normalized parameters associated with different deformation mechanisms.

Examples of deformation mechanism maps are presented in Figs 15.25a and 15.25b. Figure 15.25(a) shows the domains of elastic and plastic deformation on a plot of normalized stress (σ/G) versus the homologous temperature (T/T_m). Note that elastic deformation is the dominant mechanism at lower stresses and temperatures. Also, conventional plastic flow is the dominant mechanism at higher stresses and lower temperatures. However, at intermediate and higher temperatures, dislocation creep is the most likely mechanism. In contrast, at lower stresses and higher temperatures, grain boundary diffusion (Coble creep) is the dominant mechanism. The Coble creep domain also increases with decreasing grain size. At the highest tem-

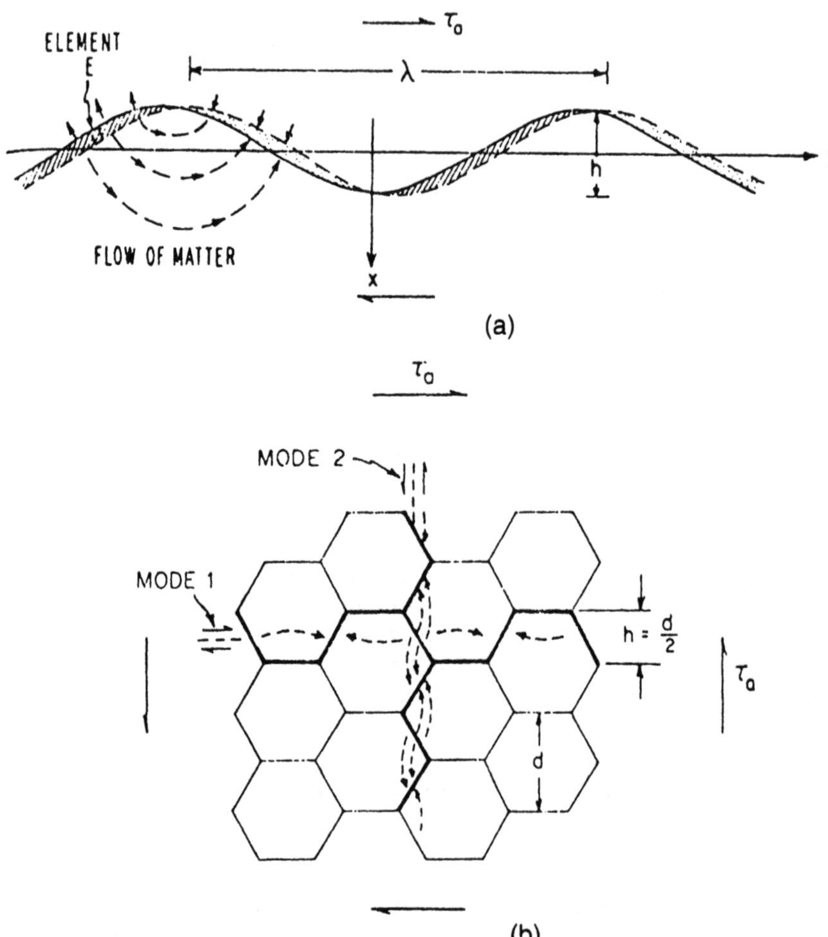

Figure 15.24 (a) Steady-state grain boundary sliding with diffusional accommodations; (b) same process as in (a), in an idealized polycrystal—the dashed lines show the flow of vacancies. [Reprinted with permission from Raj and Ashby, 1971.]

peratures and lower stresses, pure viscous flow (creep exponent of 1) occurs by Nabarro–Herring creep (bulk diffusion).

For comparison, a different type of deformation mechanism map is presented in Fig. 15.25(b). This shows a plot of strain rate against σ/G, in which the domains for the different deformation mechanisms are clearly identified. Quite clearly, other types of plots may be used to show the ranges of nondimensional parameters corresponding to different grain sizes, strain

(a)

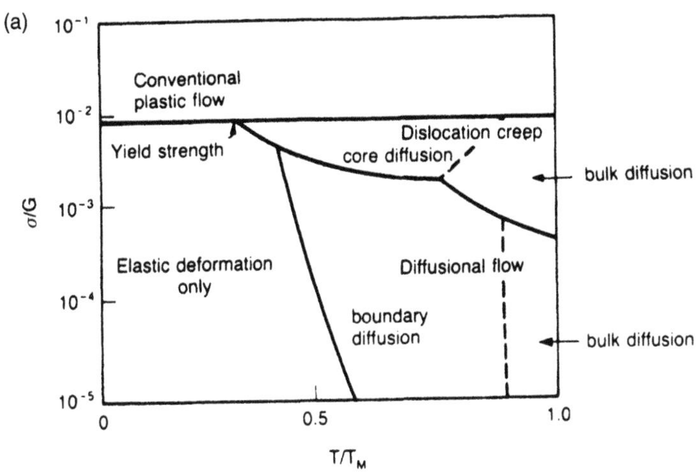

(b)

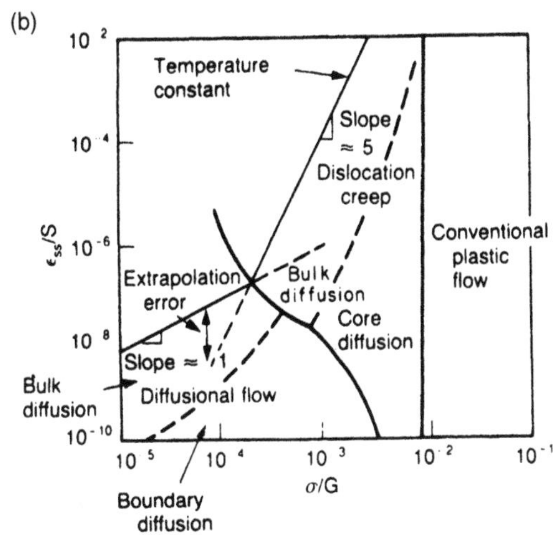

FIGURE 15.25 Deformation mechanism maps: (a) σ/G versus T/T_m; (b) ε_s versus σ/G. [From Ashby and Jones (1996) with permission from Butterworth-Heinemann.]

rates, temperatures, and applied stresses. The boundaries in the different maps and the underlying creep deformation mechanisms will also depend strongly on the structure of the different materials.

A book containing deformation mechanism maps for selected materials has been published by Frost and Ashby (1982). Such maps are extremely useful in the selection of materials for high-temperature structural applications. However, the users of such maps must be careful not to attribute too much precision to these maps. This is due to the fact that the equations and experimental data from which the maps are constructed are only approximate in nature. Hence, the maps are no better or worse than the equations/data from which they were constructed (Frost and Ashby, 1982).

15.9 CREEP LIFE PREDICTION

A large number of engineering structures are designed to operate without creep failure over time scales that can extend over decades. Since the creep tests that are used to predict the lives of such structures are typically carried out over shorter time scales (usually months), it is important to develop methods for the extrapolation of the test data to the intended service durations. Such approaches assume that the underlying creep mechanisms are the same in the short-term creep tests and the long-term service conditions. However, this cannot be guaranteed without testing for extended periods of time, which is often impractical.

Nevertheless, it is common to use measured creep data to estimate the service creep lives of several engineering components and structures. One approach involves the use of experimental results of creep lives obtained at different stresses and temperatures. An example of such results is presented in Fig. 15.26. This shows different creep curves obtained over a range of

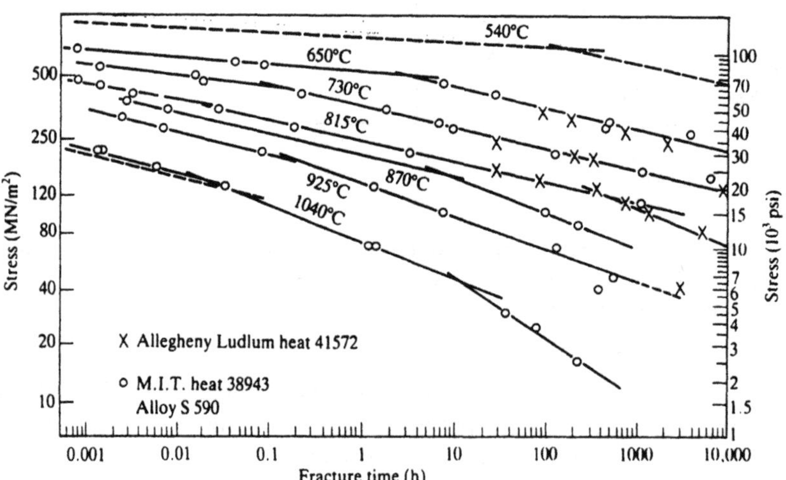

Figure 15.26 Plots of stress versus fracture time obtained from iron-based alloy (S590) at various temperatures. [Original data from Grant and Bucklin (1950); adapted from Courtney (1990) with permission from McGraw-Hill.]

testing temperatures. Such data can be difficult to use in practice due to the inherent difficulties associated with interpolation and extrapolation between measured data obtained at different temperatures and stresses. It is, therefore, common to use creep master curves that reduce all the measured creep data to a single creep curve.

The essential idea behind the master curve is that the time to failure, t_f, is dominated by the secondary creep regime. If this is the case, then we may take the natural logarithms of Eq. (15.18a) to obtain Eq. (15.18b). If we now express $\log A + n \log \sigma$ as $f(\sigma)$, we may rewrite Eq. (15.18) as

$$\log \dot{\varepsilon}_{ss} = -\frac{Q}{RT} + f(\sigma) \tag{15.26}$$

where log refers to natural logarithms (ln), $\dot{\varepsilon}_{ss}$ is the steady-state strain rate, Q is the activation energy, R is the universal gas constant, and T is the absolute temperature. Furthermore, if we now assume that the time to failure, t_f, is inversely proportional to the secondary strain rate, $\dot{\varepsilon}_{ss}$, then $t_f = k\dot{\varepsilon}_{ss}$ or $\dot{\varepsilon}_{ss} t_f = k'$, where k and k' are proportionality constants (Monkman and Grant, 1956). Substituting this into Eq. (15.26) and rearranging gives

$$T[\log t_f - \log k + f(\sigma)] = \frac{Q}{R} \tag{15.27}$$

Since Q/R is a constant, when the same mechanism prevails in the short- and long-term tests, then the left-hand side of Eq. (15.27) must be equal to a constant for a given creep mechanism. This constant is known as the Larsen–Miller parameter (*LMP*). It is a measure of the creep resistance of a material, and is often expressed as

$$LMP = T(\log t_f + C) \tag{15.28}$$

where $\log t_f$ is the log to the base 10 of t_f in hours, and C is a constant determined by the analysis of experimental data. The Larsen–Miller plot for the experimental data presented in Fig. 15.26 is shown in Fig. 15.27. This shows that the family of curves in Fig. 15.26 reduces to a single master curve that can be used to estimate the service creep life over a wide range of conditions. Since the Larsen–Miller parameter varies with stress, the implicit assumption is that different temperature and time combinations will have the same Larsen–Miller parameter at the same stress. This will only be a good assumption if the underlying creep mechanisms are the same under the experimental and projected service conditions.

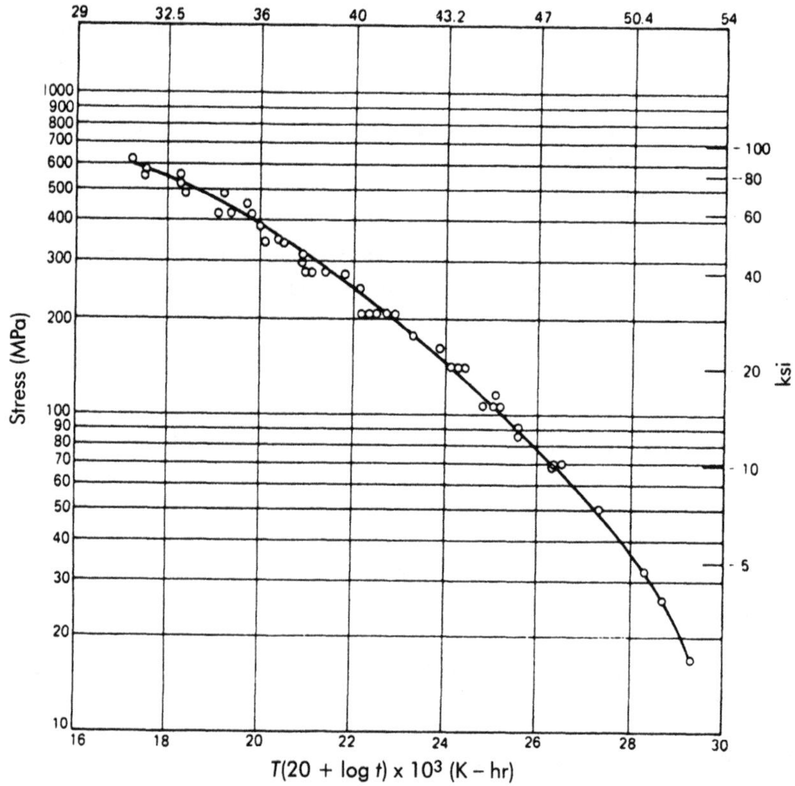

Figure 15.27 Larsen–Miller plot of the data presented in Figure 15.26 for S-590 alloy. [From Hertzberg (1996) with permission from John Wiley.]

15.10 CREEP DESIGN APPROACHES

Since creep generally occurs at temperatures above ~ (0.3–0.5) of the melting temperature, T_m (in K), it is common to choose higher melting-point metals and ceramics in the initial stages of creep design (see Fig. 15.28). However, designing against creep involves much more than simply choosing a higher melting-point solid. Hence, beyond the initial choice of such a solid, the internal structure of the material can be designed to provide significant resistance to creep deformation. In the case of designing against power law creep in metals and ceramics, there are two primary considerations.

First, the materials of choice are those that resist dislocation motion. Hence, materials that contain obstacles to dislocation motion are generally of interest. These include precipitation strengthened and solid solution

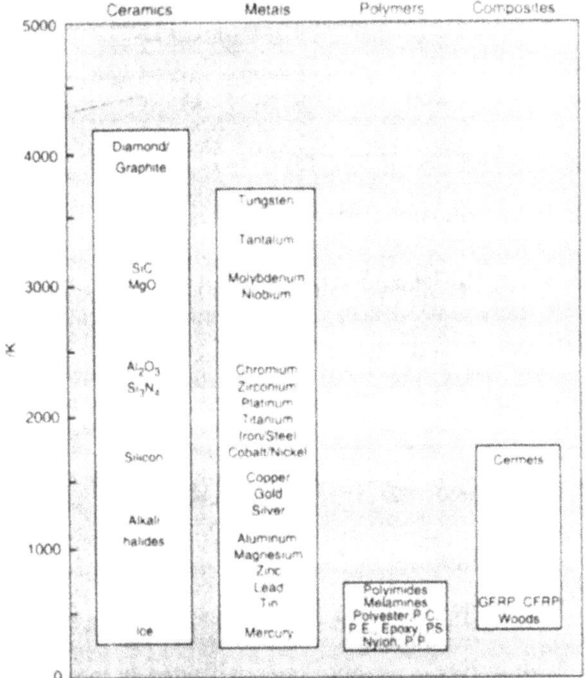

Figure 15.28 Melting or softening temperatures for different solids. [From Ashby and Jones (1996) with permission from Butterworth-Heinemann.]

strengthened materials. However, the precipitates or second phases must be stable at elevated temperature to be effective creep strengtheners.

The second approach to creep strengthening of power law creeping materials involves the selection of materials with high lattice resistance to dislocation motion. Such materials are generally covalently bonded solids. They include oxides, carbides, and nitrides. Unfortunately, however, these materials are brittle in nature. They therefore present a different set of problems to the designer.

Most designs are done for power law creeping solids. However, in materials with relatively small grain sizes, diffusional creep may become life limiting. This is particularly true in materials subjected to low stresses and elevated temperature. Material design against creep in such materials may be accomplished by: (1) heat treatments that increase the grain size, (2) the use of grain boundary precipitates to resist grain boundary sliding, and (3) the choice of materials with lower diffusion coefficients. Diffusional creep considerations are particularly important in the design of structures fabricated from structural ceramics in which power law creep is suppressed by their high lattice resistance, and diffusion creep is promoted by their small grain sizes.

15.11 THRESHOLD STRESS EFFECTS

In the case of oxide-dispersion-strengthened (ODS) alloys, a number of researchers (Williams and Wiltshire, 1973; Parker and Wiltshire, 1975; Srolovitz et al., 1983, 1984; Rösler and Arzt, 1990; Arzt, 1991) have suggested that a threshold stress is needed to detach grazing dislocations that are pinned to the interfaces of oxide dispersions and other types of dispersoids/precipitates. In such cases, the creep strain rate is controlled by a reduced stress $(\sigma - \sigma_o)$, such that

$$\dot{\varepsilon} = A'(\sigma - \sigma_0)^p \exp\left(\frac{-Q_C}{RT}\right) \tag{15.29}$$

where A' is a constant, $A' \neq A$, $p \sim 4$, σ_o is a back stress or the so-called threshold stress, Q is the activation energy corresponding to the minimum creep rate at a constant $(\sigma - \sigma_o)$, and the other constants have their usual meaning. Unfortunately, however, σ_o cannot be independently measured or predicted. Hence, further research is needed to understand better the creep behavior of ODS alloys.

It is important to note that a wide range of dislocation substructures have been observed in ODS alloys (Mishra et al., 1993; DeMestral et al., 1996). Mishra (1992) has also developed dislocation mechanism maps that show the rate-controlling dislocation processes that are likely to be associated with different microscale features. As with a number of creep researchers that have examined the creep behavior of ODS alloys, Mishra et al. (1993) have used the concept of a threshold stress in the analysis of creep data obtained for a number of materials. Their analysis also separates out the effects of interparticle spacing from the threshold stress effects.

Finally in this section, it is important to note that the threshold stress in several aluminum-base particle-hardened systems has also been shown to exhibit a temperature dependence (Sherby et al., 1997). This temperature dependence has been explained by dislocation/solute interactions akin to those observed in dilute solid solutions. In particular, Fe and Mg have been suggested as the two elements that contribute to the observed threshold stress effects in aluminum-base particle-hardened systems.

15.12 CREEP IN COMPOSITE MATERIALS

Composite structures can be designed against creep deformation. This may be accomplished by the use of stiff elastic reinforcements that resist plastic flow. The resistance provided by the reinforcements depends strongly on their shape and elastic/plastic properties. Hence, rod-like reinforcements provide a different amount of creep strengthening compared to disk-shaped reinforcements. Also, the geometry of rod-like or disk-shaped reinforcements can significantly affect the overall creep strength of a composite.

The strengthening associated with composites reinforced with rod-like and disk-shaped reinforcements has been modeled by Rösler et al. (1991) (see Fig. 15.29). They describe the overall creep strengthening in terms of a strengthening parameter, $\lambda = \sigma/\sigma_o$, that represents the ratio of the

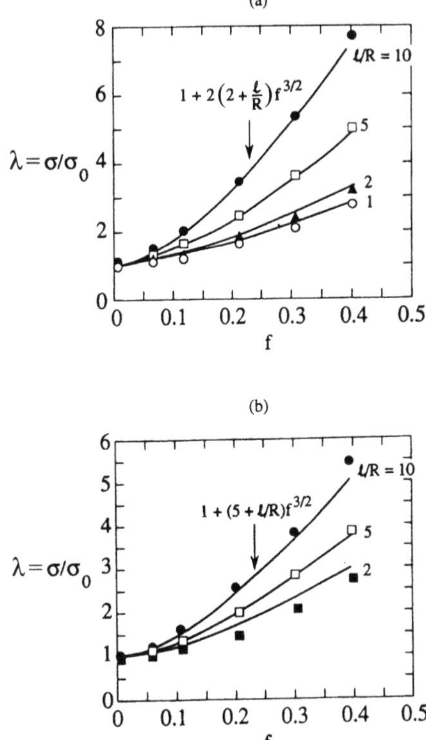

FIGURE 15.29 Predictions of creep strengthening in composites reinforced with (a) rod-like reinforcements and (b) disk-shaped reinforcements. [From Rösler et al. (1991) with permission from Elsevier.]

composite strength to the matrix strength. In the case of rod-like reinforcements, i.e., whiskers or fibers, λ is given by

$$\lambda = 1 + 2\left(2 + \frac{l}{R}\right) f^{3/2} \tag{15.30a}$$

where l is the length of the fibers, R is the fiber radius, and f is the volume fraction of fibers. The parameter, λ, is embedded in the power law creep expression for a composite $\dot{\varepsilon}_c = \dot{\varepsilon}_0(\sigma/\lambda\sigma_0)^n$, where σ is the stress, n is the creep exponent, σ_o is the reference stress, $\dot{\varepsilon}_0$ is the creep coefficient, and $\dot{\varepsilon}_c$ is the composite creep rate. Similarly, λ is given by

$$\lambda = 1 + \left(5 + \frac{l}{R}\right) f^{3/2} \tag{15.30b}$$

where l and R correspond to the length and radius of the plate, and f is the plate volume fraction. Plots of λ versus fiber volume fraction are presented in Fig. 15.29(a) and (b) for composites with different reinforcement volume fraction, f. These plots show the strong dependence of the creep strengthening parameter, λ, on f and reinforcement shape. The predictions of creep strengthening have also been shown to provide reasonable estimates of creep strengthening in in-situ titanium matrix composites reinforced with TiB whiskers (Soboyejo et al., 1994). However, the TiB whiskers also appear to resist grain boundary sliding in a way that is not considered in the existing continuum models.

Before closing, it is important to note that the deformation restraint provided by the stiff elastic whiskers may be relaxed by diffusional relaxation mechanisms (Rösler et al., 1991). Such diffusional relaxation processes are promoted by the stress gradients that exist between the top and middle sections of fibers/whiskers. These drive a flux that relaxes the deformation restraint provided by the stiff fibers/whiskers. The process is illustrated schematically in Fig. 15.30, which is taken from a paper by Rösler et al. (1991).

15.13 THERMOSTRUCTURAL MATERIALS

Although composite materials are of considerable scientific interest, most of the existing elevated-temperature structural materials (thermostructural mateirals) are actually monolithic, metallic, or ceramic materials with relatively high melting points. The common high-temperature metallic systems include: superalloys and refractory alloys; while the ceramics include silicon nitride, silicon carbide, and some oxides.

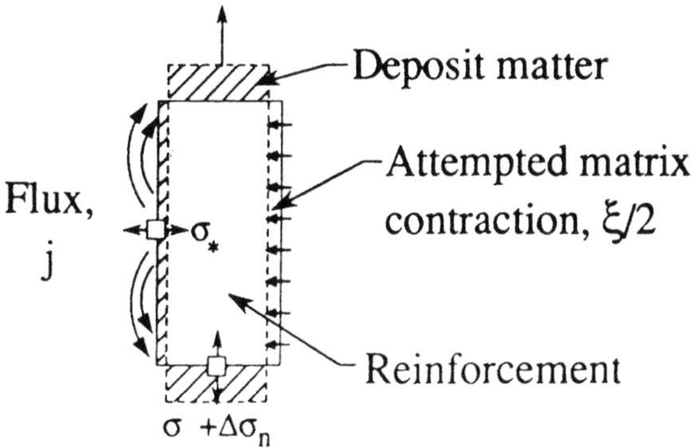

Figure 15.30 Schematic illustration of diffusional relaxation. [From Rösler et al. (1991) with permission from Elsevier.]

In recent years, however, high-temperature structural intermetallics have also been developed for potential structural applications in the intermediate-temperature regime. These include gamma-based titanium aluminides (based on TiAl), nickel aluminides (based on NiAl and Ni_3Al), and niobium aluminides (based on Nb_3Al). The structure and properties of these systems will be discussed briefly in this section on heat-resistant materials.

The superalloys are alloys of Fe, Co, or Ni that have excellent combinations of creep strength, hot corrosion resistance, and thermal fatigue resistance (Bradley, 1979; Gell et al., 1984; Loria 1989, 1997). These are typically alloyed with Nb, Mo, W, and Ta to obtain alloys with a remarkable balance of mechanical properties and resistance to environmental degradation, which are used extensively in aeroengines, land-based engines, and nuclear reactors.

Superalloys have been produced by a range of processing techniques. These include: casting, powder processing, mechanical alloying, forging, rolling, extrusion, and directional solidification of columnar and single-crystal structures. In the case of nickel-base superalloys, Mo, W, and Ti are effective solid solution strengtheners; Cr and Co also result in some solid solution strengthening. However, the main purpose of Co is to stabilize the Ni_3Al γ' phase in the γ nickel solid solution (Fig. 15.31). Furthermore, Nb or Ta may also substitute for Al in the ordered Ni_3Al γ' structure, which exhibits a three- to six-fold increase in strength with increasing temperature between room temperature and $\sim 700°C$ (Stoloff 1971; Jensen and Tien 1981a, 1981b). Most recently, single-crystal IN 718 alloys containing Ni_3Nb γ'' precipitates have been produced in a γ Ni solid solution matrix (Mercer et al., 1999). These have remarkable combinations of creep and fatigue resistance.

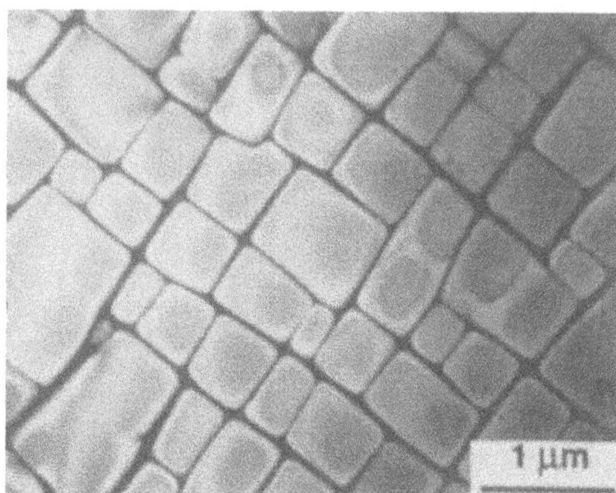

FIGURE 15.31 Transmission electron microscope superlattice dark-field images of γ precipitates in MXON. [From Kahn et al. (1984) with permission from TMS.]

The use of single-crystal nickel-base superalloys has increased largely as a reuslt of their superior creep resistance with respect to coarse-grained versions of these alloys. Furthermore, the absence of grain boundary elements (which reduce the incipient melting temperature) and the appropriate use of heat treatments can be used to achieve strength levels that are close to maximum possible values. In most cases, the crystals are oriented such that their normals correspond with the [001] orientation, which is generally the direction of the centrifugal force vector in blade and vane applications. However, some deviations in crystal orientation (typically controlled less than 8°) are observed in cast single crystals, giving rise to some variability in the measured creep date.

A number of researchers have studied the mechanisms of creep deformation in single-crystal nickel base superalloys. These include Pollock and Argon (1992), Moss et al. (1996), and Nabarro (1996). In general, the γ' phase is difficult to shear. The initial stages of deformation, therefore, occur by the filling of the γ channels with dislocations. In some systems, the γ' phase may also undergo creep deformation. This gives rise to a rafted morphology that is shown in Fig. 15.32. Evidence of such rafting mechanisms has been reported (Pearson et al., 1980; Pollock and Argon, 1992; Nabarro, 1996). The rafted morphology is generally associated with a degradation in creep resistance.

Due to their improved creep resistance, single-crystal nickel-base superalloys are used extensively in blade and vane applications in the gas turbines of aeroengines. They are used mostly in the hottest sections of the

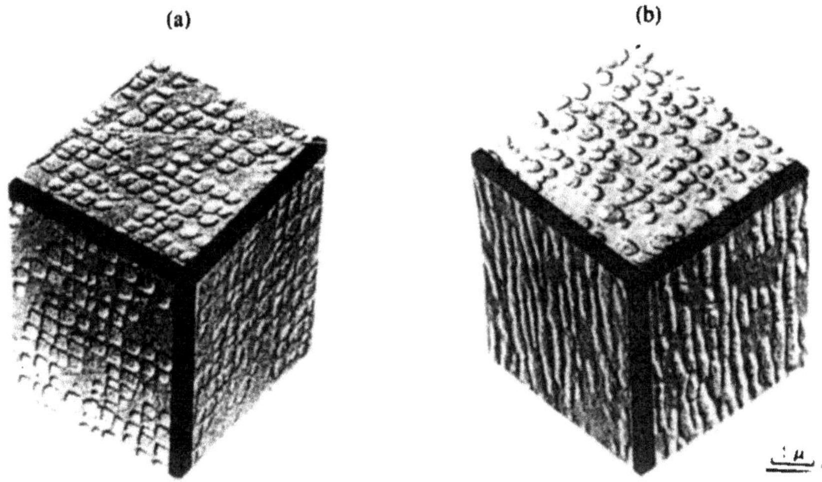

FIGURE 15.32 Electron micrographs of Ni_xAl γ' precipitates in a γ solid solution matrix: (a) cubic γ' phase in a γ matrix before creep deformation and (b) rafted structure after creep deformation. [Reprinted with permission from Fredholm and Strudel.]

turbine at temperatures below ~ 110°C. This temperature limit is due partly to possible creep failure that can occur during thermal exposures above 1100°C (Fig. 15.33). However, since this upper limit defines the limits in the efficiency and thrust that can be derived from gas turbines, engine designers have gone to great lengths to introduce internal cooling passages (Fig. 15.34) and thermal barrier coatings (TBCs) (Fig. 15.35) that can be used to increase the actual surface temperatures of blades and vanes that are fabricated from single-crystal nickel-base alloys.

In the schematic shown in Fig. 15.35, the thermal barrier is an yttria-stabilized zirconia coating (typically $ZrO_2 + 7.5\,\text{mol}\,\%\,Y_2O_3$) with a cubic crystal structure. This can be used to promote a temperature drop of ~ 200°C from the surface of the blade to the surface of the nickel-base superalloy. The TBC is bonded to the nickel-base superalloy with a PtAl or FeCrAlY bond coat. As a result of interdiffusion mechanisms, a thermally grown oxide (TGO) develops between the bond coat and the zirconia coating. The thickness of this bond coat increases parabolically (Walter et al., 2000; Chang et al., 2001) until failure initiates by film buckling or cracking (Choi et al., 2000) from morphological imperfections at the interface between the bond coat and the TGO (Mumm and Evans, 2001) or ratcheting due to incremental deformation in the TBC system (Karlsson and Evans, 2001).

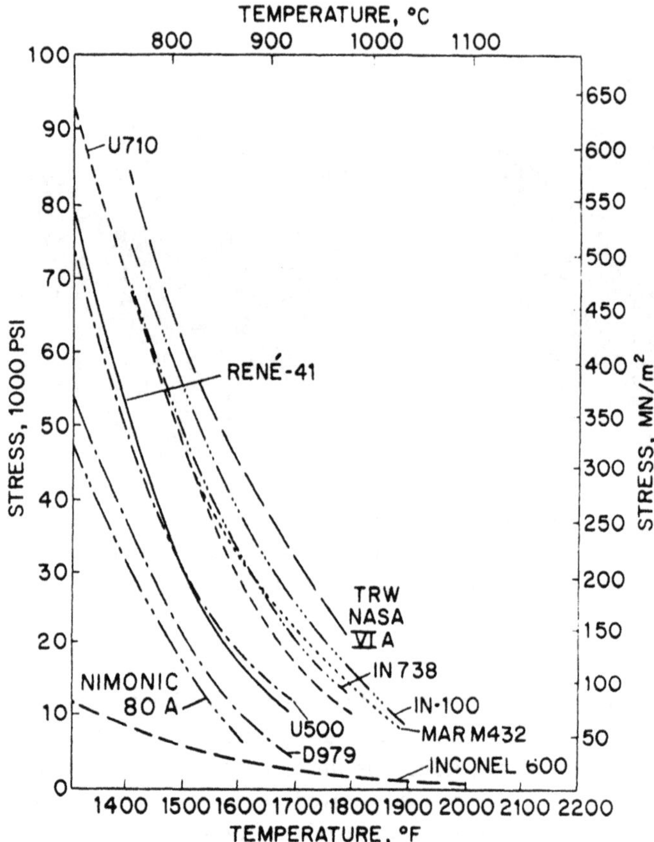

Figure 15.33 Stress versus temperature curves for rupture in 1000 h for selected nickel-based superalloys. [From Sims and Hagel (1972) with permission from John Wiley.]

In addition to their applications in blades and vanes, polycrystalline ingot or powder metallurgy nickel-base superalloys are used extensively in disk applications within the turbine sections of aeroengines and land-based engines. Since the temperatures in the disk sections are much lower than those in the blade/vane sections, the polycrystalline nickel-base superalloys are generally suitable for applications at temperatures below ~ 650°C. In this temperature regime, IN 718, IN 625, and IN 600 are some of the "work horses" of the aeroengine industry (Loria, 1989, 1997). These alloys have good combinations of creep and thermal fatigue resistance. However, there have been efforts to replace them with powder metallurgy nickel-base superalloys in recent years.

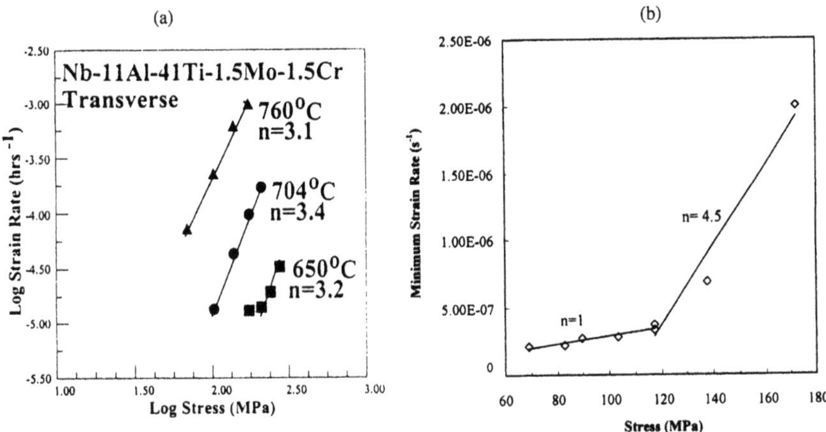

FIGURE 15.34 Secondary creep behavior in (a) Nb–11Al–41Ti–1.5 Mo–1.5 Cr alloy in atom % [from Hayes and Soboyejo, 2001) and (b) 44Nb–35Ti–6Al–5Cr–8V–1W–0.5Mo–0.3Hf–0.5O–0.3C alloy [from Allameh et al., 2001c.]

It should be clear that the upper temperature limits to the applications of nickel are set by the intrinsic limits (melting point) of nickel. Hence, efforts to produce engines with greater efficiency have focused largely on the exploration of materials with inherently greater melting points than nickel. The systems that have been explored include: ceramics (Si_3N_4, SiC, and oxides), ceramic/ceramic composites (Si/SiC and oxide/oxide), and

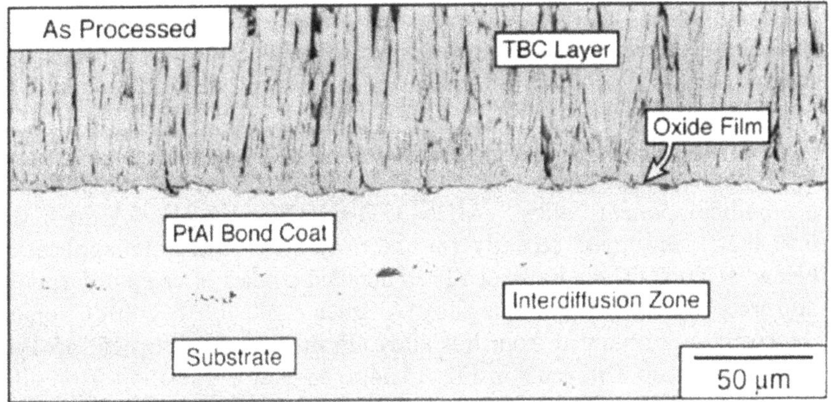

FIGURE 15.35 Photograph of thermal barrier coating. [Courtesy of Dr. Daniel R. Mumm.]

intermetallics (niobium silicides, NiAl, and Mo–Si–B). Unfortunately, however, none of these materials have the combination of damage tolerance and creep resistance required for near-term applications in engines. There is, therefore, a need for further materials research and development efforts to produce the next generation of gas-turbine materials/systems.

In the intermediate-temperature regime between 500° and 700°C, gamma-based titanium aluminides, Ni_3Al-base alloys, and niobium-base alloys are being developed for potential applications. The gamma-based titanium aluminides have attractive combinations of creep resistance (Martin et al., 1983; Hayes and London, 1991; Maruyama et al., 1992; Oikawa, 1992; Wheeler et al., 1992; Bartels et al., 1993; Es-Souni et al., 1993; Soboyejo and Lederich, 1993; Bartholomeusz and Wert, 1994; Jin and Bieler, 1995; Lu and Hemker, 1997; Skrotzki, 2000; Allameh et al., 2001a) at temperatures up to ~ 760°C. However, gamma-based titanium aluminides are limited by oxidation phenomena above this temperature regime and by brittleness at room temperature (Kim and Dimmiduk, 1991; Chan, 1992; Davidson and Campbell, 1993; Campbell et al., 2000; Lou and Soboyejo, 2001). Similar problems with brittleness have been encountered with the NiAl intermetallic system (Noebe et al., 1991; Ramasundaram et al., 1998).

However, in the case of Nb_3Al- and Ni_3Al-based intermetallics, alloys with attractive combinations of creep and oxidation resistance have been designed for intermediate-temperature applications below ~ 700°C. This has been achieved largely by alloying with boron in Ni_3Al-based systems (Aoki and Izumi, 1979; Liu, 1993; Sikka, 1997). Similarly, in the case of Nb_3Al-based intermetallics alloyed with 40 atom % Ti (Nb–15Al–40Ti), attractive combinations of fatigue and fracture behavior (Ye et al., 1998; Soboyejo et al., 1999) and creep resistance [Fig. 15.34(a)] have been designed. The creep exponents in the Nb–15Al–40Ti alloy also suggest diffusion-controlled creep at lower stresses ($n \sim 1$) and dislocation-controlled creep at higher stresses ($n \sim 5$).

Adequate combinations of fatigue and fracture resistance (Loria, 1998, 1999) and creep resistance (Allameh et al., 2001b, c) have been reported for a multicomponent alloy (44Nb–35Ti–6Al–5Cr–8V–1W–0.5Mo–0.3Hf–0.5O–0.3C) developed recently for intermediate-temperature applications below ~ 700°C. There have also been detailed studies of creep deformation and creep strengthening in this alloy (Allameh et al., 2001b, 2001c). Some of the creep data obtained from this alloy are presented in Figs 15.34(b) and 15.35. The creep exponents in Fig. 15.34(a) suggest a transition from diffusion-controlled creep to dislocation-controlled creep at higher stresses. Also, as with other b.c.c. metals and their alloys (Wadsworth et al., 1992), an inverted primary is observed along with truncated secondary and extended tertiary creep regime in the multicomponent alloy 44Nb–35Ti–6Al–5Cr–8V–1W–0.5Mo–0.3Hf–0.5O–0.3C.

Tranmission electron microscopy studies of the above-mentioned alloy have revealed Orowan-type dislocation/particle interactions [Fig. 15.36(a)], dislocation grazing of particles (which can also be considered in terms of the detachment of the dislocations from particles) [Fig. 15.36(b)], the glide of individual dislocations [Fig. 15.36(c)], and what appears to be dislocation pairs [Fig. 15.36(d)]. The possible strengthening contributions from different

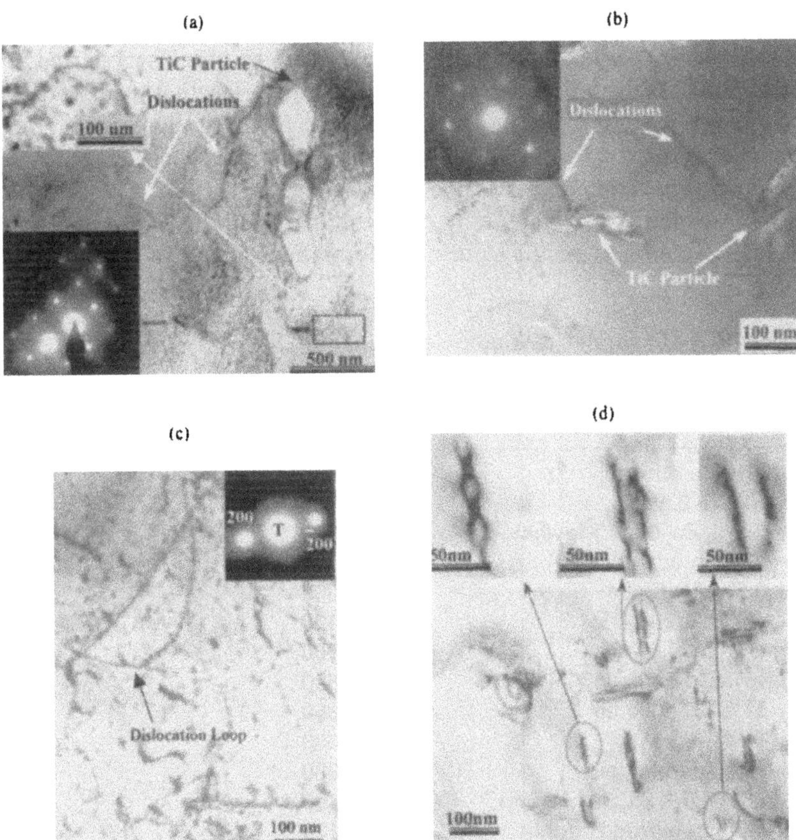

Figure 15.36 Transmission electron micrographs of dislocation substructure in a 44Nb–35Ti–6Al–5Cr–8V–1W–0.5Mo–0.3Hf–0.5O–0.3C alloy: (a) Orowan-type dislocations; (b) dislocation grazing of TiC particles; (c) glide of individual dislocations; (d) possible evidence of dislocation pairs. [From Allameh et al., 2001a, 2001b.]

types of dislocation interactions have been modeled in a paper by Allameh et al. (2001b).

15.14 INTRODUCTION TO SUPERPLASTICITY

Some fine-grained metals and ceramics exhibit plastic strains between a few hundred and a thousand per cent due to a phenomenon that is known as superplasticity. This has been observed to occur in certain ranges of temperature and strain rate in which resistance to necking is significant. Evidence of superplasticity has been observed in titanium alloys such as Ti–6Al–4V, aluminium alloys, and some ceramics (Table 15.1). Furthermore, the occurrence of superplasticity has been recognized as an opportunity for forming complex parts from materials that would otherwise be difficult to shape. This has been true especially in the aerospace industry, where superplastic forming techniques are being used increasingly to form complex aeroengine and airframe parts.

The ability of a material to deform superplastically is strongly related to its resistance to necking during deformation. This has been correlated

Table 15.1 Examples of Materials Exhibiting Superplastic Behavior

Material	Maximum strain (%)
Al–33% Cu eutectic	1500
Al–6% Cu–0.5% Zr	1200
Al–10.7% Zn–0.9% Mg–0.4% Zr	1500
Bi–44% Sn eutectic	1950
Cu–9.5% Al–4% Fe	800
Mg–33% Al eutectic	2100
Mg–6% Zn–0.6% Zr	1700
Pb–18% Cd eutectic	1500
Pb–62% Sn eutectic	4850
Ti–6% Al–4% V	1000
Zn–22% Al eutectoid	2900
Al(6061)–20% SiC (whiskers)	1400
Partially stabilized zirconia	120
Lithium aluminosilicate	400
Cu–10% Al	5500
Zirconia	350
Zirconia + SiO_2	1000

Source: Taplin et al. (1979).

with the strain rate sensitivity, m, for various metals and their alloys (Fig. 15.37). This is defined in Eq. (5.12). Possible values of m are between 0 and 1. However, in the case of superplastic materials, the values of m are closer to 1. This is due largely to their resistance to necking under certain conditions of strain rate and temperature.

The dependence of stress, σ, and strain rate sensitivity, m, on strain rate is shown in Fig. 15.38. This shows that stress increases more rapidly than strain rate sensitivity in response to changes in strain rate. Furthermore, the peak in the strain rate sensitivity, m, occurs in a regime, in which the stress exhibits the strongest rise with increasing strain rate. This is the regime in which superplasticity is most likely to occur.

Superplasticity is not observed in regimes I and III in which the strain rate sensitivities are low. Furthermore, in the regimes where superplasticity is observed, the amount of superplastic deformation tends to increase with increasing temperature and smaller grain sizes. This is shown in Fig. 13.39 using data obtained for a Zr–22Al alloy. The possible effects of decreasing grain size and increasing temperature on stress and strain rate sensitivity are also illustrated in Fig. 15.40.

Most of the mechanistic efforts to explain superplasticity involve the accommodation of grain boundary sliding by other plastic flow processes. As with creep deformation, grain boundary sliding is accommodated either by diffusional flow and/or dislocation-based mechanisms. The accommoda-

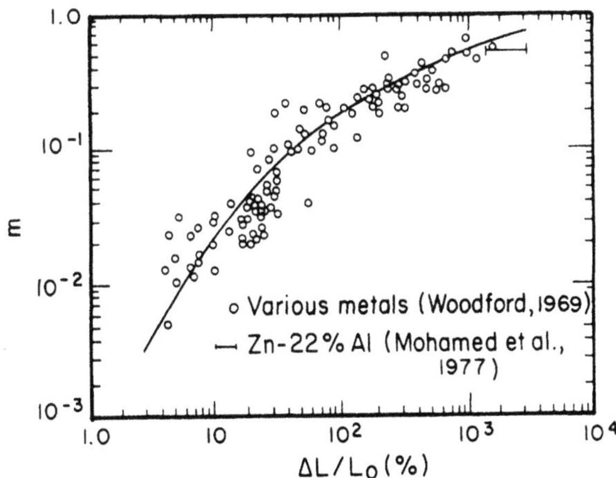

FIGURE 15.37 Effects of strain rate sensitivity, m, on the maximum tensile strain to failure for different alloys (Fe, Mg, Pu, Pb-Sr, Ti, Zn, and Zr-based alloys). [From Taplin et al., 1979.]

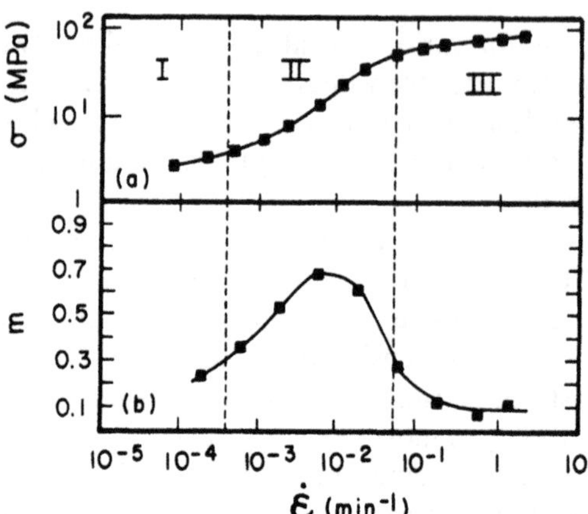

Figure 15.38 Strain rate dependence of (a) stress and (b) strain rate sensitivity in a Mg–Al eutectic alloy (grain size 10 μm) tested at 350°C. [From Lee, 1969.]

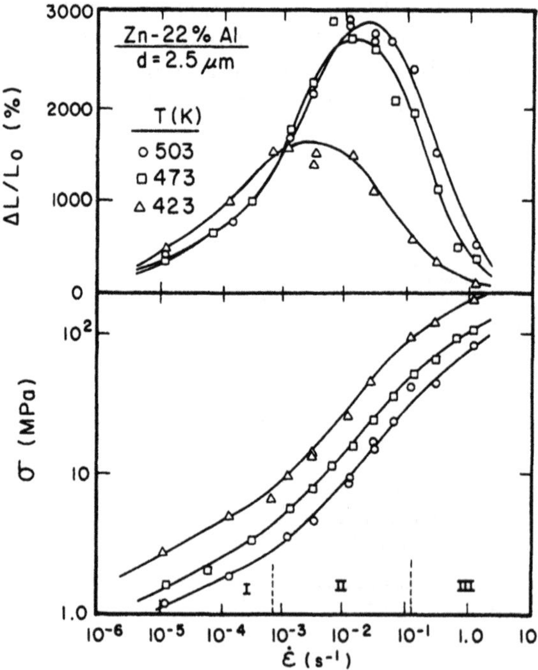

Figure 15.39 Dependence of tensile fracture strain and stress on strain rate. [From Mohamed et al., 1977.]

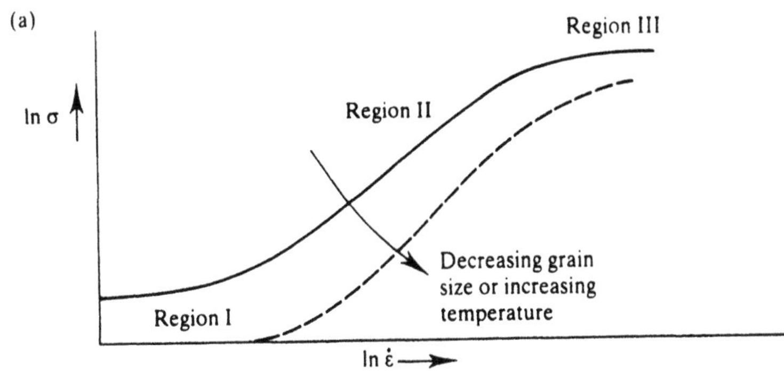

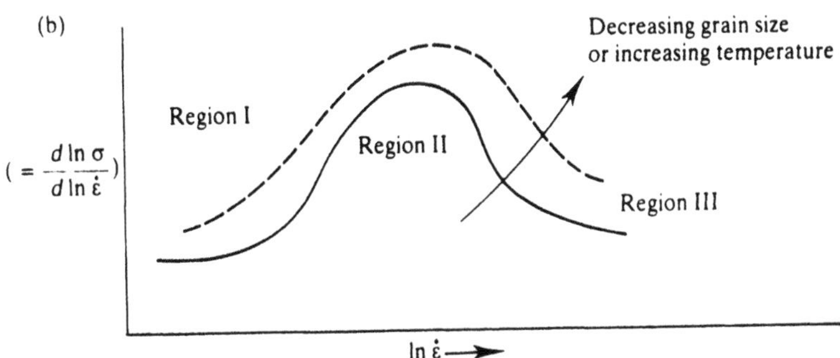

Figure 15.40 (a) Low-stress–strain rate behavior of a material manifesting superplasticity. In regions I and III, the strain rate sensitivity (b) is fairly small, whereas it is high in Region II where superplasticity is observed. As indicated in (a), increases in temperature or decreases in grain size shift the σ–$\dot{\varepsilon}$ curve downward and to the right. The same changes produce a somewhat higher value on m as shown in (b).

tion by diffusional flow was originally proposed by Ashby and Verall (1973). This uses a grain switching mechanism to explain how grain shape is preserved during superplastic deformation. This is illustrated in Fig. 15.41.

In the grain switching model, the grains in the initial state [Fig. 15.41(a)] undergo an increase in grain boundary area in the intermediate state, Fig. 15.41(b). This is followed by diffusional accommodation of the shape change in the intermediate stage by bulk or grain boundary diffusion, Fig. 15.42. Provided that the applied stress exceeds the threshold stress required for grain switching, the strain rate for grain switching is considerably greater than that required for conventional creep.

Furthermore, the shape accommodation may also occur by dislocation motion. However, since concentrated dislocation activity is generally not

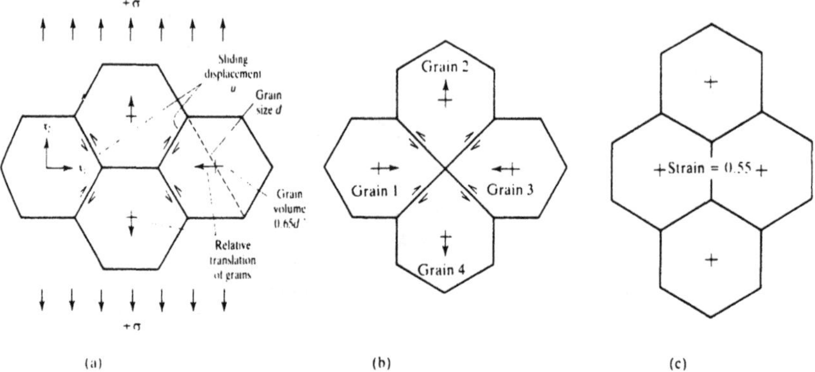

FIGURE 15.41 Grain switching mechanism of Ashby and Verrall: (a) initial state; (b) intermediate state; (c) final state. [From Ashby and Verall, 1973.]

observed during superplastic deformation, it may occur as a transitional mechanism. Also, dislocation glide–climb processes may occur in addition to bulk or grain boundary diffusion processes that preserve shape during superplastic deformation.

One of the problems that can arise during superplastic deformation is the problem of cavitation. This can result from incompatible deformation of adjacent grains that leads ultimately to microvoid formation. The problem of cavitation can be overcome, to some extent, by the application of hydrostatic stresses during superplastic forming. Otherwise, cavitation can lead to premature failure and defective parts.

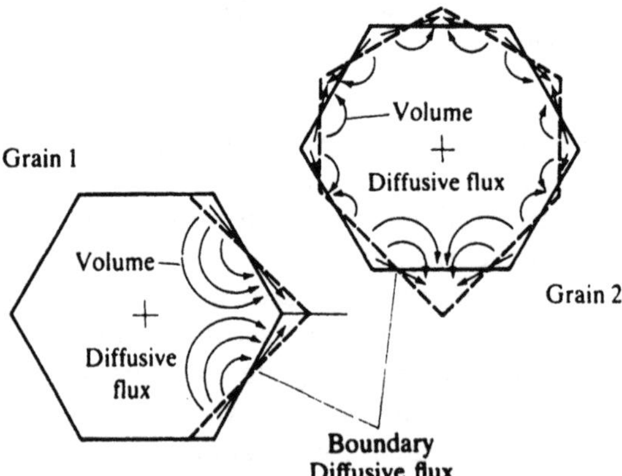

FIGURE 15.42 Accommodation of intermediate stage of grain switching by bulk and grain boundary diffusion. [From Ashby and Verall, 1973.]

Nevertheless, superplastic forming has gained increasing acceptance in the aerospace and automotive industries, where it is being used increasingly in the fabrication of components with complex shapes. However, the relatively slow strain rates that are used in superplastic forming result in relatively low production rates and higher production costs. This may be a serious issue in cost-driven automotive components, but less of an issue in performance-driven aerospace components.

Before closing, it is important to note that grain refinement has been proposed as a method for achieving high strain rates during superplastic forming. This can be achieved by mechanical alloying (MA) or powder metallurgy (PM) processing to obtain alloys with refined grain sizes. In the case of aluminum alloys produced by PM and MA techniques, a number of researchers (Bieler and Mukherjee, 1990; Higashi et al., 1991; Matsuki et al., 1991; Kim et al., 1995) have shown that high strain rate superplastic forming can be achieved by microstructural refinement to grain sizes of $\sim 1\ \mu m$ or less.

Kim et al. (1995) have also shown that the high strain rate superplasticity can be explained by a model that is analogous to a "core mantle" grain boundary sliding model proposed originally by Gifkins (1976). In this model, the mantle corresponds to the outer shell of subgrains where grain boundary sliding accommodation occurs. However, no deformation occurs in the inner core subgrains, which remain equiaxed and invariant with strain.

Kim et al. (1995) have also modified an original theory by Ball and Hutchison (1969) to obtain the following constructive equation for the strain rate:

$$\dot{\varepsilon} = K\left(\frac{D_L}{\bar{L}\bar{\lambda}}\right)\left(\frac{\sigma - \sigma_0}{E}\right)^2 \tag{15.31}$$

where K is a proportionality constant, D_L is the lattice diffusion coefficient, $\bar{L}$ is the average grain size, $\bar{\lambda}$ is the average climb distance (which is related to the average subgrain size), σ is the applied mean stress, E is the unrelaxed modulus, and σ_o is a threshold stress that must be exceeded before a grain boundary sliding can occur at low stresses. It is associated with the pinning of grain boundaries by fine particles (Kim et al., 1995).

The occurrence of high strain rate superplasticity in fine-grained alloys has significant implications for the manufacturing of superplastically formed parts at high production rates. It is also likely that the enhancement in production rates will be increased further as improved techniques are developed for the processing of materials with sub–micrometer and nanoscale grain sizes (Langdon, 2001).

15.15 INTRODUCTION TO CREEP DAMAGE AND TIME-DEPENDENT FRACTURE MECHANICS

15.15.1 Creep Damage

The discussion above has shown that stress- and temperature-induced deformation can give rise to creep. Hence, upon the application of a load to a material or structure, the initial instantaneous elastic deformation is followed by the stages of primary, secondary, and tertiary creep. Such time-dependent deformation can occur at relatively low strains ($\sim$ 5–6%) due to formation growth of voids at grain boundaries.

The voids appear predominantly on boundaries that are perpendicular to the direction of the applied tensile stress. The voids are formed by a combination of grain boundary and bulk diffusional creep (Fig. 15.43). However, pre-existing voids may also act as sources of atoms, giving rise to void growth until the remaining sections can no longer support the applied loads. When this occurs, the voids grow at an increasing rate until catastrophic failure occurs.

Grain boundary microvoids or cracks may also form as a result of incomplete accommodation of shape changes between adjacent grains undergoing grain boundary sliding phenomena. When these occur, the voids and cracks can also grow by diffusional creep and dislocation-induced plasticity until catastrophic failure occurs. Such cavity growth has been modeled by Ghosh et al. (1999). Cavity nucleation and growth can give rise to the rapid increase in the creep strain rates that are typically observed in stage III, the tertiary creep regime (Fig. 15.14). Microscopic creep damage processes may also give rise to stable crack growth processes

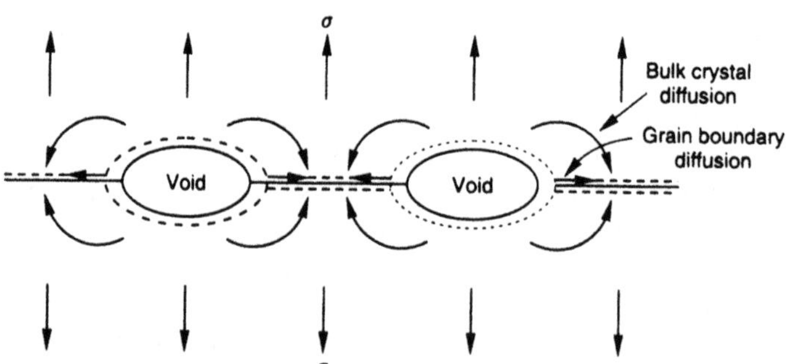

FIGURE 15.43 Formation of grain boundary microvoids due to a combination of bulk and grain boundary diffusion. [From Ashby and Jones, 1976.]

associated with crack extension from notches or pre-existing cracks (Argon et al., 1991).

Continuum creep deformation and diffusional flow have been modeled by Needleman and Rice (1980) for cases in which deformations occur in regions close to the boundaries. However, experimental studies have revealed some surprising modes of crack growth that are not predicted by the continuum models. For example, in their studies conducted on 304 stainless steel, Ozmat et al. (1991) have shown that creep crack growth from prior fatigue precracks (with prior fatigue history) does not result in in-plane crack growth. Instead, the cracks bifurcate along paths that are inclined at $\sim 45^\circ$ to the initial crack plane. In contrast, in cases where the cracks initiate from blunt notches, they meander up and down, and grow essentially as extensions of the plane of the original notch. Such phenomena have been modeled using mechanism-based continuum models within a finite element framework (Argon et al., 1991).

The regimes and modes of creep damage have been studied by a number of investigators. The experiments have identified two extreme types of behavior. At one extreme, alloys with inhomogeneous distributions of grain boundary particles have been shown to exhibit inhomogeneous cavity or microcrack nucleation. This has been observed in solid solution alloys, such as 304 stainless steel. The growth of cavities and cracks in such materials has been shown to give rise to relatively isolated grain boundary facets microcracks that grow, coalesce, and result ultimately in final fracture (Chen and Argon, 1981; Don and Majumdar, 1986). The other limiting type of behavior has been observed in creep-resistant alloys with inherent distributions of heterogeneous phases, e.g., Nimonic 80A (Dyson and McLean, 1977) and Astroloy (Capano et al., 1989). This leads to homogeneous intergranular cavitation on most of the grain boundaries, resulting in final fracture, without the intermediate formation of a significant density of grain boundary cracks (Capano et al., 1989).

The above processes have been modeled by Hayhurst et al. (1984a, b) using phenomenological approaches. Tvergaard (1984, 1985a, 1985b, 1986) and Mohan and Brust (1998) have also developed mechanistic models for the prediction of creep damage. These are beyond the scope of the current text. However, the interested reader is referred to the above references, which are listed at the end of this chapter. A review of prior theoretical work in this area can also be found in a paper by Hsia et al. (1990), while Ashby and Dyson (1984) provide a general catalog of the microscopic phenomena that can contribute to creep crack growth. We will now turn our attention to the subject of time-dependent fracture mechanics.

15.15.2 Time-Dependent Fracture Mechanics

In cases where dominant cracks are present, a range of microscopic crack-tip damage processes can occur in a small zone close to the crack tip (Fig. 15.44). Since the material at the crack tip is undergoing local failure, the tip of a growing crack is most likely to be in the tertiary creep regime. However, depending on the local stress/strain distributions, the regions remote from the crack tip may be undergoing elastic deformation, or primary and secondary stages of creep deformation (Fig. 15.44).

Due to the wide range of possible stress states, it is common in most analytical treatments to consider a few limiting cases. In cases where the deformation is predominantly elastic in the specimen/component, the crack tip field can be characterized by linear elastic fracture mechanics, i.e., the stress intensity factor provides a measure of the crack driving force. However, for materials undergoing global deformation by steady-state creep, the crack driving force can be defined by the C^* integral (Fig. 15.45). This is defined by replacing displacements with displacement rates, and strains with strain rates in the definition of the J integral. This gives

$$C^* = \int_{\Gamma} \left(\dot{w}\, \mathrm{d}y - \sigma_{ij} n_j \frac{\partial \dot{u}_i}{\partial x} \mathrm{d}s \right) \tag{15.32}$$

where $\dot{w}$ is the time derivative of the strain energy density (the stress work rate or power density); $\sigma_{ij}n_j$ corresponds to the traction stresses, T_i, acting on the contour boundaries; u_i are the displacement vector components; ds

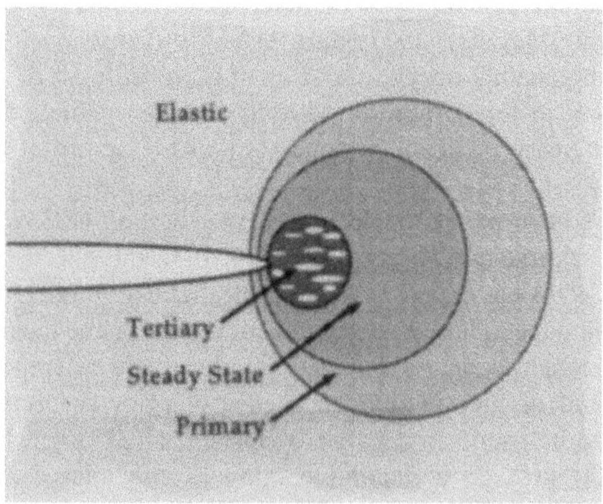

Figure 15.44 Schematic illustration of crack-tip damage zones at the tip of a crack. [From Anderson, 1995.]

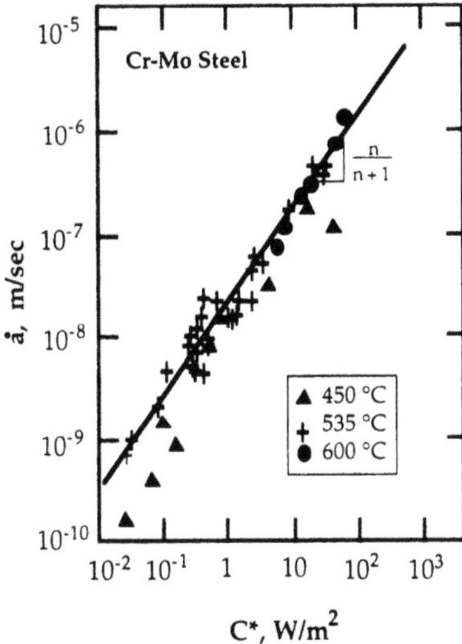

FIGURE 15.45 Typical creep crack growth rate data. [From Riedel, 1989.

corresponds to increments in length on the arbitrary contour, Γ; σ_{ij} are the components of the stress tensor; and n_j are the components of the normal vector to Γ. The stress work rate, $\dot{w}$, is given by

$$\dot{w} = \int_0^{\varepsilon_{kl}} \sigma_{ij}\, \mathrm{d}\dot{\varepsilon}_{ij} \tag{15.33}$$

For steady-state creep, $\dot{\varepsilon}_{ij} = A\sigma_{ij}^n$, and the HRR-type local crack-tip stresses and strain rates are given by

$$\sigma_{ij} = \left(\frac{C^*}{AI_n r}\right)^{\frac{1}{n+1}} \tilde{\sigma}_{ij}(n,\theta) \tag{15.34a}$$

and

$$\varepsilon_{ij} = \left(\frac{C^*}{AI_n r}\right)^{\frac{n}{n+1}} \tilde{\varepsilon}_{ij}(n,\theta) \tag{15.34b}$$

where C^* is the amplitude of the crack-tip fields, A is the coefficient in the power law creep expression ($\dot{\varepsilon}_{ij} = A\sigma_{ij}^n$), n is the creep exponent (and not the hardening exponent), and the other terms (I_n, r, $\tilde{\sigma}_{ij}$, θ) have the same meaning as in the HRR field expressions (see Eqs. 11.85a–c). The C^* parameter represents the crack driving force under steady-state creep conditions (Landes and Begley, 1976; Nikbin et al., 1976; Ohji et al., 1976). It has been shown to correlate well with creep crack growth data obtained from a range of structural engineering materials.

As with long fatigue crack growth in the Paris regime, the relationship between the creep crack growth rate da/dt can be related to the C^* integral (Fig. 15.45). This gives the following power law expression:

$$\frac{da}{dt} = \beta(C^*)^{\gamma} \tag{15.35}$$

where β and γ are constants. Equation (15.35) is only applicable when global deformation occurs by steady-state creep. However, long-term deformation is required before steady-state creep conditions can be achieved. Conversely, short-term deformation occurs by elastic deformation, in which the crack driving force is characterized by the stress intensity factor, K. In between the initial elastic and steady-state conditions, the crack driving force is characterized by a transition parameter $C(t)$. For small-scale deformation, $C(t)$ is given by (Riedel and Rice, 1980):

$$C(t) = \frac{K_I^2(1-\nu^2)}{(n+1)Et} \tag{15.36}$$

where K_I is the Mode I stress factor, ν is Poisson's ratio, n is the creep exponent, E is Young's modulus, and t corresponds to time. The transition time, t_1, from short- to long-time behavior is given by

$$t_1 = \frac{K_I^2(1-\nu^2)}{(n+1)EC^*} = \frac{J}{(n+1)C^*} \tag{15.37}$$

where the above terms have their usual meaning. Between the short- and the long-time regimes, the parameter $C(t)$ can be estimated from

$$C(t) = C^*\left(\frac{t_1}{t} + 1\right) \tag{15.38}$$

The parameter $C(t)$ has been found to provide good correlations with measured creep crack growth rate data in the transitional regime (Saxena, 1989). However, unlike K and C^*, it is difficult to measure $C(t)$ experimentally. This has led Saxena (1986) to propose a C_t parameter that can be measured by separating the overall/global displacement, Δ, into elastic components, Δ_e, and time-dependent creep components, Δ_t. This gives

$$\Delta = \Delta_e + \Delta_t \tag{15.39}$$

For fixed loads, $\dot{\Delta}_t = \dot{\Delta}$, and C_t is given by

$$C_t = -\frac{1}{b}\left[\frac{\partial}{\partial a}\int_0^{\dot{\Delta}_t} P\,\mathrm{d}\dot{\Delta}_t\right]_{\dot{\Delta}_t} \tag{15.40}$$

where C_t is the creep component of the power release rate, b is the specimen breath, a is the crack length, P is the applied load, and $\dot{\Delta}_t$ is the displacement rate due to creep phenomena. Alternative expressions have been derived for small-scale and transitional creep behavior in work by Saxena (1986). Furthermore, Bassani et al. (1995) have shown that C_t characterizes creep crack growth rates better than $C(t)$, C^*, and K, using experimental data over a wide range of C^*/C_t ratios. Chun-Pok and McDowell (1990) have also included the effects of primary creep in the C_t parameter.

15.16 SUMMARY

This chapter presents an introduction to time-dependent deformation and damage in polymers, metals, intermetallics, and ceramics. The chapter started with an introduction to creep and viscoelasticity in polymers. Idealized spring and dashpot models were described for the estimation of time-dependent and time-independent components of deformation polymeric materials. Mechanical damping was also elucidated before discussing the time-dependent flow of polymer chains associated with creep and stress relaxation phenomena. This was followed by a brief review of single master curves represented by the Williams, Landel, and Ferry (WLF) equation. Finally, the section on polymer deformation ended with some comments on the effects of temperature on the relaxation modulus. Following the initial focus on polymer deformation, the rest of the chapter explored the physics of time-dependent deformation and damage in metals, intermetallics, and ceramics. Phenomenological approaches to creep deformation were presented along with a discussion on creep deformation mechanisms, i.e., diffusional and dislocation-controlled creep mechanisms, and grain boundary sliding phenomena. The mechanisms were summarized in deformation mechanism maps before describing engineering approaches to the prediction of creep life. Creep design approaches were then discussed before reviewing the creep deformation in oxide-dispersion strengthened and composite materials. This was followed by a review of thermostructural materials and coatings that are actually used in a number of high-temperature applications. Finally, a brief introduction to superplasticity was presented before concluding with a final section on creep damage and time-dependent fracture mechanics.

BIBLIOGRAPHY

Allameh, S. M., Li, M. et al. (2001a) J Mater Sci 36: 3539–3547.

Allameh, S. M., Hayes, R. P., Srolovitz, D., and Soboyejo, W. O. (2001b) Creep behavior in a Nb–Ti base alloy. Mater Sci Eng A.

Allameh, S. M., Hayes, R. P., Loria, E., and Soboyejo, W. O. (2001c) Microstructure and mechanical properties of a Nb–Ti-based alloy. Mater Sci Eng.

Anderson, T. L. (1995) Fracture Mechanics. Boca Raton, FL: CRC Press.

Andrade, E. N. da. L. (1911) Proc Roy Soc London A84: 1.

Aoki, K. and Izumi, O. (1979) Nippon Kinzoku Gakkaishi 43: 1190.

Argon, A. S., Hsia, K. J., et al. (1991) In: Argon, A. S. ed. Topics in Fracture and Fatigue. Springer-Verlag, pp. 235–270.

Ashby, M. F. and Dyson, B. F. (1984) Advances in fracture research. Proceedings of ICF6, Pergamon Press.

Ashby, M. F. and Jones, D. R. H. (1996) An Introduction to their Properties and Applications. Woburn, MA: Butterworth-Heinemann.

Ashby, M. F. and Verall, R. A. (1973) Acta Metall 21: 149.

Baker, S. P., Vinci, R. P. and Arias, T. (2002) Mater Res Bull 27: 26–28.

Ball, A. and Hutchison, M. M. (1969) Metal Sci 3: 1.

Bartholomeusz, M. F. and Wert, J. A. (1994) Metall Trans 25A: 2161.

Bassani, J. L., Hawk, D. E., et al. (1995) Evaluation of the C_t Parameter for Characterizing Creep Crack Growth Rate in the Transient Regime. West Conshohocken, PA: American Society for Testing and Materials, pp. 112–130.

Bieler, T. R. and Mukherjee, A. K. (1990) Mater Sci Eng A 128: 171.

Bird, J. E., Mukherjee, A. K., et al. (1969) Quantitative Relation Between Properties and Microstructure. Haifa, Israel: Israel University Press, p. 255.

Burton, B. and Greenwood, G. W. (1970a) Metal Sci 4: 215.

Burton, B. and Greenwood, G. W. (1970b) Acta Metall 19: 1237.

Campbell, J. B., Venkateswara Rao, K. T., et al. (2000) Metall Mater Trans 30A: 563–577.

Capano, M., Argon, A. S., et al. (1989) Acta Metall 37: 3195.

Chan, K. S. (1992) Metall Mater Trans 24A: 569–583.

Chang, T., Walter, M., Mercer, C., and Soboyejo, W. O. (2001) In: D. R. Mumm, Walter, M., Popoola, O., and Soboyejo, W. O. eds. Durable Surfaces. Zurich: Trans-Tech, pp. 185–198.

Chen, I.-W. and Argon, A. S. (1981) Acta Metall 29: 1321.

Choi, S. R., Hutchison, J. W., et al. (2000) Acta Mater 48: 1815–1827.

Chun-Pok, L. and McDowell, D. L. (1990) Int J Fract 11: 81–104.

Coble, R. L. (1963) J Appl Phys 34: 1679.

Courtney, T. H. (1990) Mechanical Behavior of Materials. New York: McGraw-Hill.

Davidson, D. L. and Campbell, J. B. (1993) Metall Mater Trans 24A: 1555.

DeMestral, B., Eggeler, G., and Klam, H-J. (1996) Metall Mater Trans 27A: 879–890.

Don, J. and Majumdar, S. (1986) Acta Metall 34: 961.

Dyson, B. F. and McLean, D. (1977) Metal Sci 11: 37.

Es-Souni, M., Barterls, A., et al. (1993) In: Darolia, R. Lewandoski, J. J. and Liu, C. T., et al., eds. Structural Intermetallics. The Minerals, Metals, and Materials Society: 335.
Evans, A. G. and Langdon, T. G. (1976) Progr Mater Sci 21: 171–411.
Evans, R. W., Preston, J. A., Wiltshire, B., and Little, E. A. (1993) Mater Eng A167: 65.
Evans, R. W. and Wilshire, B. (1985) Creep of Metals and Alloys. London: Institute of Metals.
Frost, H. J. and Ashby, M. F. (1982) Deformation Mechanism Maps. Oxford, UK: Pergamon Press.
Garafalo, F. (1963) Trans Metal Soc, AIME 227: 35.
Gell, M., Kortovich, C. S., Bricknell, R. H., Kent, W. B., and Radovich, J. F. (1984) Superalloys 84. Warrendale, PA: TMS.
Ghosh, A. K., Bae, D. H., et al. (1999) Materials Science Forum. Zurich: Trans Tech.
Gifkins, R. C. (1976) Metall Trans 7A: 1225.
Gregory, J. and Nix, W. D. (1987) Metall Trans.
Harper, J. and Dorn, J. E. (1957) Acta Metall 5: 654.
Hayes, R. P. and London, B. (1991) Acta Metall Mater 40: 2167.
Hayes, R. P. and Soboyejo, W. O. (2001) An investigation of the creep behavior of an ordered Nb–11Al-31Ti–1.5Cr intermetallic alloy. Mater Sci Eng, in press.
Hayhurst, D. R., Dimmer, P. R., et al. (1978) Phil Trans Roy Soc London A311: 103.
Hayhurst, D. R., Brown, P. R., and Morrison, C. J. (1984a) Phil Trans Roy Soc London A311: 131.
Herring, C. (1950) J Appl Phys 21: 437.
Herzberg, R. W. (1996) Deformation and Fracture Mechanics of Engineering Materials. New York: John Wiley.
Higashi, K., Okada T., et al. (1991) In: Hori, S. Tokizane, M., and Furushiro, N., eds. Superplasticity in Advanced Materials. The Japanese Society for Research on Superplasticity, p. 569.
Jensen, R. R. and Tien, J. K. (1981a) In: Tien J. K. and Elliot, J. F., eds. Metallurgica Treatises. Warrendale, PA: TMS: 526.
Jensen, R. R. and Tien, J. K. (1981b) Source Book in Materials for Elevated Temperature. Materials Park, OH: ASM International.
Jin, Z. and Bieler, T. R. (1995) Phil Mag A 71: 925.
Kahn, T., Caron, P., and Duret, C. (1984) In: Gell, M. Kortovich, C. S., Bricknell, R. H., Kent, W. B., and Radovich, J. F., eds. Superalloys 1984. Warrendale, PA: TMS-AIME, p. 145.
Karlsson, A. and Evans, A. G. (2001) Acta Mater 49: 1793–1804.
Kim, W.-J., Taleff, E., et al. (1995) Scripta Metall Mater 32: 1625–1630.
Kim, Y. W. and Dimiduk, D. M. (1991) J Metals 43: 40–47.
Landes, J. D. and Begley, J. A. (1976) A Fracture Mechanics Approach to Creep Crack Growth. ASTM STP 590. West Conshohocken, PA: ASTM, pp. 128–148.
Lee, D. (1969) Acta Metall 17: 17.

Li, Y., Soboyejo, W. O., et al. (1998) Metall Mater Trans B 30B: 495–504.
Liu, C. T. (1993) Ni_3Al aluminide alloys. In: Darolia, R. Lewandoski, J. J., Liu, C. T., et al., eds. Structural Intermetallics. Warrendale, PA: TMS, pp. 367–377.
Loria, E. (1989) Superalloy 718. Warrendale, PA: TMS.
Loria, E. (1997) Superalloy 718, 625 and 706 and Various Derivatives. Warrendale, PA: TMS.
Loria, E. (1998) Mater Sci Eng A254: 63–68.
Loria, E. (1999) Mater Sci Eng A271: 430–438.
Lou, J. and Soboyejo, W. O. (2001) Metall Mater Trans 32A: 325–337.
Lu, M. and Hemker, K. (1997) Acta Mater 45: 3573.
Martin, P. L., Mendiratta, M. G., and Lipsitt, H. (1983) Metall Trans A14: 2170.
Matsuki, K., Matsumoto, H., et al. (1991) In: Hori, S. Tokizane, M., and Furushiro, N., eds. Superplasticity in Advanced Materials. The Japanese Society for Research on Superplasticity: 551.
Mercer, C., Soboyejo, A. B. O., and Soboyejo, W. O. (1999) Acta Materialia 47: 2727–2740.
Meyers, M. A. and Chawla, K. K. (1999) Mechanical Behavior of Materials, Uppersaddle River, NJ: Prentice Hall.
Mishra, R. S. (1992) Scripta Metall Mater 26: 309.
Mishra, R. S., Baradkar, A. G., et al. (1993) Acta Metall 41: 1993–2243.
Mohan, R. and Brust, F. W. (1998) Fatigue Fract Eng Mater Struct 21: 569–581.
Monkman, F. C. and Grant, N. J. (1956) ASTM STP 56: 593.
Moss, S. J., Webster, G. A., and Fleury, E. (1996) Metall Mater Trans 27A: 829–837.
Mukherjee, A. K., Bird, J. E., et al. (1969) Trans ASM 62: 155.
Mumm, D. R. and Evans, A. G. (2000) Acta Mater 48: 1815–1827.
Nabarro, F. R. N. (1948) Deformation of crystals by motions of single ions. Report of a Concurrence of Strength of Solids. London: Physical Society.
Nabarro, F. R. N. (1996) Metall Mater Trans 27A: 513–530.
Needleman, A. and Rice, J. R. (1980) Acta Metall 28: 1315.
Nikbin, K. M., Webster, G. A., et al. (1976) ASTM STP 601. West Conshohocken, PA: ASTM, pp. 47–62.
Noebe, R., Misra, A., et al. (1991) Iron Steel Inst Japan 31: 1172–1185.
Ohji, K., Ogura, K., et al. (1976) Trans Jap Soc Mech Eng 42: 350–358.
Oikawa, H. (1992) Mater Sci Eng A153: 427.
Orowan, E. (1947) The Creep of Metals. West of Scotland Iron and Steel Institute, p. 45.
Ozmat, B., Argon, A. S., et al. (1991) Mech Mater 11: 1.
Pearson, D. D., Lemkey, F. D., et al. (1980) Superalloys 1980. Materials Park, OH: ASM International.
Poirier, J. P. (1985) Creep of crystals: high temperature deformation processes in metals. In: Ceramics and Minerals. Cambridge, UK: Cambridge University Press.
Pollock, T. and Argon, A. S. (1992) Acta Metall Mater 40: 1.
Raj, R. and Ashby, M. F. (1971) Metall Trans 2: 1113.
Ramasundaram, R., Ye, F., et al. (1998) Metall Mater Trans 29A: 493–505.

Riedel, H. and Rice, J. R. (1980) Tensile Cracks in Creeping Solids. ASTM STP 1020. West Conshohocken, PA: ASTM, pp. 185–201.
Rösler, J., Bao, G., et al. (1991) Acta Metall Mater 39: 2733.
Saxena, A. (1986) Creep Crack Growth Under Non-Steady-State Conditions. ASTM STP 905. West Conshohocken, PA: ASTM, pp. 185–201.
Saxena, A. (1989) Recent advances in elevated temperature crack growth and models for life prediction. Adv Fract Res 2: 1675–1688.
Sherby, O. D. and Burke, P. M. (1968) Progr Mater Sci 13: 325.
Sherby, O. G., Gonzalez-Doncel, G., et al. (1997) Threshold stresses in particle-hardened materials. In: Earthman J. C. and Mohamed, F., eds. Creep and Fracture of Engineering Materials and Structures. Warrendale, PA: TMS, 9–18.
Sikka, V. (1997) Mater Sci Eng A239: 564–569.
Sims, C. T. and Hagel, W. C. (1972) The Superalloys. New York: John Wiley.
Skrotzki, B. (2000) Proceedings of the 8th Symposium on the Creep and Fracture of Materials, Tsukuba, Japan. Trans Tech Publications, Uetikon-Zurich, Switzerland.
Soboyejo, W. O. and Lederich, R. J. (1993) Novel approaches to the assessment of creep deformation in gamma-based titanium aluminides. First International Symposium on Structural Intermetallics. Warrendale, PA: TMS.
Soboyejo, W. O., Lederich, R. J., et al. (1994) Acta Metall Mater 42: 2579–2591.
Soboyejo, W. O., DiPasquale, J., et al. (1999) Metall and Materials Transactions 30A: 1025–1045.
Soboyejo, W. O., Schymanski, J., et al. (2001) J Am Ceram Soc 84: 1309–1314.
Stoloff, N. S. (1971) In: A. Kelloy and Nicholson, R. B., eds. Strengthening Methods in Crystals. New York: John Wiley.
Taplin, D. M. R., Dunlap, G. L., et al. (1979) Annu Rev Mater Sci 9: 151–180.
Tien, J. K. and Gamble, R. P. (1972) Metall Mater Trans 3: 2157–2162.
Tvergaard, V. (1984) Acta Metall 32: 1977.
Tvergaard, V. (1985a) Mech Mater 4: 181.
Tvergaard, V. (1985b) Int J Solids Struct 21: 279.
Tvegaard, V. (1986) Int J Fract 31: 183.
Wadsworth, J., Doughtery, S. E., et al. (1992) Scripta Metall Mater 27: 71–76.
Walter, M., Onipede, B., Mercer, C., and Soboyejo, W. O. (2000) J Eng Mater Technol 122: 333–337.
Weertman, J. (1955) J Appl Phys 26: 1213.
Weertman, J. (1957) J Appl Phys 28: 362.
Wheeler, D. A., London, B., et al. (1992) Scripta Metall Mater 26: 939.
Williams, M. L., Landel, R. F., and Ferry (1955) J Am Ceram Soc 77: 3071.
Wiltshire, B. (1997) Creep mechanisms in oxide dispersion strengthened alloys. In: J. C. Earthman and Mohammed, F. A., eds. Creep and Fracture of Engineering Materials and Structures. Warrendale, PA: TMS, pp. 19–28.
Ye, F., Mercer, C., et al. (1998) Metall Mater Trans 29A: 2361–2374.

Index

Milton Keynes UK
Ingram Content Group UK Ltd.
UKHW021931071024
449327UK00022B/1762

9 780367 446932